Quantum Theory and Its Stochastic Limit

Springer
Berlin
Heidelberg
New York
Barcelona
Hong Kong
London
Milan
Paris
Tokyo

Luigi Accardi Yun Gang Lu
Igor Volovich

Quantum Theory and Its Stochastic Limit

Springer

Professor Luigi Accardi
Centro Vito Volterra
Facolta di Economia
Università di Roma
Via di Torvergata
00133 Roma, Italy

Professor Yun Gang Lu
Dipartimento di Matematica
Università degli Studi di Bari
Via Orabona, 4
70125 Bari, Italy

Professor Igor Volovich
Steklov Mathematical Institute
Gubkina 8
117966 Moscow, GSP-1, Russia

Library of Congress Cataloging-in-Publication Data
Accardi, L. (Luigi), 1947–
Quantum theory and its stochastic limit/Luigi Accardi, Yun Gang Lu, Igor Volovich. p.cm.
Includes bibliographical references.
ISBN 3540419284 (alk. paper)
1. Quantum theory. 2. Stochastic processes. I. Lu, Yun Gang, 1955– II. Volovich, I.V. III. Title.
QC174.17.S76 A33 2001 530.12–dc21 2001049456

ISBN 3-540-41928-4 Springer-Verlag Berlin Heidelberg New York

Springer-Verlag Berlin Heidelberg New York
a member of BertelsmannSpringer Science+Business Media GmbH

http://www.springer.de

Typesetting by the authors
Data conversion by LE-TEX, Leipzig
Cover design: *design & production* GmbH, Heidelberg

Printed on acid-free paper SPIN 10834825 57/3141/tr 5 4 3 2 1 0

Preface

Nowadays it is becoming clearer and clearer that, in the description of natural phenomena, the triadic scheme – *microscopic, mesoscopic, macroscopic* – is only a rough approximation and that there are many levels of description, probably an infinite hierarchy, in which the specific properties of a given level express some kind of *cumulative* or *collective* behaviour of properties or systems corresponding to the lower levels. One of the most interesting challenges for contemporary natural sciences is the comprehension of the connections among these different levels of description of reality and the deduction of the laws of higher levels in this hierarchy from basic laws corresponding to lower levels.

Since these cumulative or collective phenomena are, typically, nonlinear effects, the transition from this general program to concrete scientific achievements requires the developement of techniques which allow physical information to be extracted from nonlinear quantum systems. Explicitly integrable examples of such systems are rare, and the most interesting physical phenomena are not captured by them. Even in the case of linear systems the fact that an explicit solution is formally available is often useless, since it is impossible to interpret interesting physical phenomena from it.

In the absence of generally applicable methods one introduces approximations whose role is to capture those phenomena which become dominant at different orders of magnitude of some physical parameters (e.g., long times, low densities, weak couplings, low or high energies, low or high temperatures, large masses). The usual asymptotic methods of quantum theory (perturbation theory, scattering, semi-classical limit, etc.) provide useful tools to approximate the values of individual quantities of physical interest; however, in the past ten years a new idea has begun to emerge independently in several fields of physics and mathematics: *instead of approximating individual quantities, let us look for an approximating theory* from which the individual approximations can be obtained by standard procedures. Such a theory should work simultaneously as a *magnifying glass* and as a *filter*, in the sense that all phenomena pertaining to the scales of magnitudes we are interested in should be magnified while those pertaining to all the remaining scales should be filtered away. One way to achieve this goal is through a new technique, *the stochastic limit*: the purpose of the present book is to explain this technique in a simple and self-contained way and to describe how the new ideas and structures, which emerge from it, apply to concrete physical problems.

The main result of the stochastic limit is that, combining the basic ideas of scattering (long times) and perturbation theory (small parameter), it *automatically selects from the dynamics the dominating terms (in this regime) and shows that they can be resummed, giving rise to a new evolution operator which is again unitary* and such that, in many cases, the corresponding evolution equations are explicitly integrable. This unitary evolution is an approximation of the original one; however, although simpler, it preserves much nontrivial information on the original complex system. A first indication of this complexity is the fact that *these equations are singular*: the limit Hamiltonian is a functional of some white noise whose explicit form is uniquely determined by the original system. So, in order to deal with them, one needs white noise calculus. This explains the name "stochastic limit". We develop a new technique to bring such singular equations to normal order (which differs from the usual normal order because it involves not the usual but the *causal commutator*); once this is done, most quantities of physical interest can be calculated simply by solving a linear equation. A remarkable feature of this procedure is that *a normally ordered white noise Hamiltonian equation is a stochastic Schrödinger equation*, i.e. a stochastic differential equation. In the past ten years this type of equation has been widely used to build a multiplicity of phenomenological models in quantum optics, solid-state physics, quantum field theory, quantum measurement theory, etc. Thus the stochastic limit allows these phenomenological models to be deduced from the basic laws of physics. In fact it should be emphasized that, in the stochastic limit, randomness is not postulated a priori but it is derived from the microscopic quantum equations. In this sense we can say that *the stochastic limit describes the microscopic structure of quantum noise by identifying it with the fast degrees of freedom of the original system.* The intuitions that the fast degrees of freedom of a system can act as driving random forces on the slow ones (a generalization of Haken's "slaving principle"), that chaos can be an infinite reservoir creating ordered structures and forms by means of a *stochastic resonance principle*, etc., not only become exact in the stochastic limit, but also find in it a precise quantititative formulation in the sense that *the distinction between fast and slow degrees of freedom is not determined from the outside but realized by the dynamics itself*; the fast degrees of freedom are those which in the limit become quantum white noises (also called "master fields"). Another advantage of the transition from phenomenological models to models deduced from the basic laws is that the wealth, beauty and variety of the new mathematical structures, hidden in these basic laws and made explicit by the stochastic limit, by far exceed those used in the phenomenological models. The stochastic limit also provides a general approach to the phenomenon of quantum decoherence which is important, in particular for quantum computers.

The stochastic limit takes inspiration from the pioneering studies of quantum dynamical systems by Fermi, Bogoliubov, van Hove and Prigogine, and its main goal is a detailed qualitative study of quantum dynamics, in analogy to Poincaré's qualitative study of classical dynamics.

Summing up: the basic philosophy of the stochastic limit can be formulated in a single sentence: *if we look at the fast degrees of freedom of a nonlinear system with a clock, adapted to the slow ones, then the former look like an independent increment process, typically a white noise* (the highest degree in the chaos hierarchy!). The universality of this technique is proved by the fact that it can be applied to a great variety of Hamiltonians. Since in the simplest situations a quantum white noise is given by a pair of noncommuting classical white noises, it follows that the stochastic limit also provides a link between classical probability and quantum theory *in real time*, without the need for imaginary times or analytical continuations. In particular the quantum process obtained in the stochastic limit, when restricted to some special sets of compatible observables (abelian algebras) gives rise to interesting classical stochastic processes (e.g. birth and death processes describing the population dynamics of atomic levels), thus explaining the efficiency of classical probabilistic techniques in the description of several quantum phenomena such as stimulated emission in lasers, cosmic ray cascades, branching in neutron diffusions, etc. Another example of this phenomenon is Glauber's dynamics of the open Ising model which emerges here as the restriction of a more interesting quantum Hamiltonian flow. The symmetry of the resulting quantum Markov semigroup (which takes place if the initial state of the field is an equilibium state, but not in the Fock case) is related to the symmetry expressed by the Onsager relations.

Once the stochastic Schrödinger equation has been obtained, it is relatively easy to obtain the Langevin equation (stochastic limit of the Heisenberg evolution). *All the known types of master equations (and several new ones) are obtained just by taking the partial expectation of Langevin equations with respect to the reference state of the master field.* This procedure corresponds to the adiabatic elimination of the fastly relaxing variables, a technique also called "coarse graining", and establishes a connection between the stochastic limit and the traditional "projection techniques" used to provide a microscopic Hamiltonian foundation to these equations and to derive the Kubo–Mori theory. This connection evidences how strongly irreversible and dissipative behaviours can be perfectly described by Hamiltonian dynamics, thus providing a unified approach for reversible (Hamiltonian) and irreversible (master equation) description of quantum systems in the spirit of Prigogine's paradigm about the fundamental (as opposed to phenomenic) nature of irreversibility: if we think of decay phenomena as basic observable features of irreversibility, we see that they are perfectly compatible with a reversible dynamical evolution.

It should be underlined, however, that the stochastic limit goes far beyond the master equation because it does not eliminates the fast degrees of freedom. This allows one to estimate the probabilities of some collective states, or more generally the behaviour of a complex (nonlinear) system with many degrees of freedom, in terms of relatively few functions of the microscopic characteristics of the quickly relaxing degrees of freedom (according to the interpretation these functions are called "order parameters", "kinetic" or "susceptibility" or "transport coefficients", etc.).

The method is of very simple applicability in the sense that, for a large class of physically meaningful models, in addition to the energy shifts, broadening and lifetimes, which can also be obtained with other methods, it allows one to guess the limit equations directly by inspection of the initial Hamiltonian system and, from them, to deduce easily much information about multiparticle transitions, correlations, particle statistics, etc. We have tried to condense the main results of the stochastic limit into the so-called *stochastic golden rules*, which are a generalization of the Fermi golden rule and which allow one to solve, just by inspection of the interaction Hamiltonian, the following problem: *given a quantum Hamiltonian system, write down immediately the associated stochastic Schrödinger equation* (this, as explained above, automatically gives also the Langevin and the master equation).

This rule is formulated in Chap. 4, and the reader already familiar with the basic formalism of quantum field theory can begin reading this book directly from this chapter. The examples covered there and in the following chapter are variations of the spin–boson Hamiltonian. They are sufficient to illustrate two of the basic principles (i.e. general statements independent of the specific model) which emerge from the stochastic limit, namely:

- the stochastic resonance principle (see Sect. 6.2);
- the time consecutive principle (see Sects. 8.3 and 8.4).

In Chap. 11, which concludes Part I, four other basic principles of the stochastic limit are formulated without proofs (these will be given in Part III):

- the stochastic universality class principle;
- the block principle;
- the orthogonalization principle;
- the stochastic bosonization principle.

These principles are used to extend the stochastic golden rule to polynomial interactions and to fermions. The block principle is particularly interesting because it states that if the interaction is a monomial of degree n then some *special configurations in the n-particle space of the original field coalesce and behave like a single pseudo-particle whose second quantization gives the master field.* Stochastic bosonization (in dimensions strictly higher than two) is a particular case of the block principle for fermions. Super-symmetric structures arise when one starts from a fermi Hamiltonian including polynomials of both even and odd degree.

The fact that the stochastic golden rule allows one to obtain quite nontrivial results with practically no mathematical effort supports our expectation that the stochastic limit could become an everyday tool for a multiplicity of physicists. In view of this, one could even forgive the fact that the results are also mathematically rigorous. In any case, to separate the new physical ideas and effects deduced from this method from their mathematical proof, which in some cases can be heavy, we have relegated the proofs of the main estimates and principles of the stochastic limit to Part III. In Parts I and II we emphasize the main ideas which allow one to obtain the correct answer quickly, at a physical level of rigor. Since the Planck constant is one of the parameters which can be involved in the time rescaling, the present method also provides a natural second step after semiclassical approximation (for this reason it is sometimes called *semiquantum approximation*) in the sense that, just as in the semiclassical limit one obtains a deterministic classical system (analogy with the law of large numbers), one could say that the stochastic limit captures a new typically quantum leading term in a different regime: analogy with the central limit theorem. In the usual semiclassical approximation one obtains, in the limit, classical trajectories. In the semiquantum approximation one obtains Brownian motion trajectories in the simplest examples and new types of classical or quantum white noises in the more sophisticated ones. Another difference with the semiclassical approximation (as well as with relativity theory) is that, while in these cases the old theory is recovered from the new for a limiting value of a parameter (Planck constant, velocity of light), in the stochastic limit it is the new theory which is recovered from the old for a limiting value of some parameters (coupling constant, time, density, energy, etc.).

In Part II we deal with strongly nonlinear interactions and illustrate the qualitatively new phenomena which arise in connection with them. The main new feature here is *the breaking of the standard commutation relations and consequently of the usual statistics.* This is a consequence of the domination of the contribution of the *noncrossing diagrams* to the dynamics of these more complex systems (the crossing diagrams are shown to tend to zero in the stochastic limit). This leads to the emergence of a multiplicity of nonstandard (i.e. neither classical nor boson nor fermion) master fields (white noises) as approximations of usual Hamiltonian systems: for example, in several cases the master field, even if coming from a usual Boson field, exhibits a kind of *superlocalization* phenomenon and becomes a bounded random variable (such as a fermion). Among the prototype models in which these phenomena appear we mention quantum electrodynamics without dipole approximation, the polaron model (Chap. 12), the Anderson model (Chap. 13), and many other models could be added. In fact, in Chap. 14 we prove that this type of behaviour is universal among the field–field interactions with conservation of momentum.

The notion of *quantum entanglement* acquires with the stochastic limit a meaning which goes beyond the familiar notion of superposition and leads to the conclusion, supported by a large number of examples, that under appropriate physical situations *nonlinearly interacting quantum systems cannot be separated even at a kinematical level and behave as a single new quantum object satisfying new types of commutation relations and therefore new statistics.* (Historically the first example of this phenomenon appeared in nonrelativistic quantum electrodynamics without dipole approximation, where the atomic degrees of freedom commute with the field operators before the limit, but after they develop nontrivial commutation relations which account for the nonlinearity of the interaction.) This qualitative statement has a mathematical counterpart in the two notions of the "Hilbert module" and the *interacting Fock space* which have emerged, from the stochastic limit, as natural candidates for the description of the state space of interacting systems.

From the mathematical point of view, the stochastic limit has brought a unification as well as a deep innovation in the theories of classical, quantum stochastic and white noise calculus as developed respectively by Ito, Hudson–Parthasaraty and Hida, and also in the theory of generalized functions (the developement of the theory of distributions on the standard simplex). More precisely, it has motivated the developement of a *white noise approach to stochastic calculus* which is completely new even at a classical level. The introduction of the Ito formula in a white noise, i.e. generalized functions, context leads to nontrivial generalizations of both classical and quantum stochastic calculus which turn out to correspond to first-order (for the Wiener process) or normally ordered second-order (for jump processes) powers of the Fock white noise. Higher powers of white noise cannot be dealt with usual (classical or quantum) probabilistic techniques and require renormalization. Thus we can say that the present approach also reveals some unsuspected probabilistic aspects of renormalization theory.

The stochastic limit is a generally applicable approximating theory which captures the dominant phenomena in the weak coupling–long time regime. Recently a general method for studying quantum dynamics has been developed and corrections to the stochastic limit have been computed. Moreover an exact general expression for matrix elements of the evolution operator ("ABC-formula") was obtained [ArVo00].

The present book is addressed to researchers and students, physicists, mathematicians, experts in quantum communication and information engineering and all those interested in nonlinear problems of quantum theory. Much of the material presented here has been published in several articles in the past ten years, but the general presentation and most of the proofs have been simplified and appear for the first time in book form. Many of the results obtained as applications of the stochastic limit technique have not been included in this book: the most interesting applications of stochastic bosonization in higher dimensions and of quantum interacting particle systems (barely mentioned in Sect. 5.19), many concrete applications of Belavkin's theory of quantum filtering, the whole body of results pertaining to the nonequilibrium and transport phenomena, in particular the temperature dependence of the conductivity tensor, the fractional quantum Hall effect, the low-density limit and the deduction of the kinetic equations (Boltzmann, Vlasov, etc.) from the stochastic limit. The basic idea of the low-density limit is illustrated, in the simplest (i.e. Fock) case, in Chap. 10. The really interesting case (finite temperature) is much richer in structure and more difficult.

Some of the heaviest mathematical parts have been omitted. In particular the proofs of the possibility of exchanging the summation of the iterated series and the stochastic limit and taking the term-by-term limit of the series (in those models where this is possible). The details of such proofs are strongly model dependent and do not give much insight. We have preferred to give, in Chaps. 15 and 16, detailed proofs of the general, model-independent estimates from which the estimates needed in each single model can be deduced by rather standard arguments. The same can be said for the estimate of the error, which, beyond its mathematical interest, is crucial for defining the range of applicability of the whole theory.

Since a discussion of all these topics would have exceeded the limitations of this book, we have decided to publish these results in a separate book. The present book synthesizes several results obtained by the authors in collaboration with several other coauthors. In particular we mention Irina Arefeva (Chap. 14), Vieri Mastropietro (Chap. 13) and Sergei Kozyrev (Chaps. 4, 5 and 12).

Finally it is our pleasure to thank several people who read parts of the numerous versions of the present book and encouraged us with comments and suggestions. They are Irina Arefeva, Fabio Bagarello, Luigi Galgani, Kentaro Imafuku, Sergei Kozyrev, Vieri Mastropietro, K.R. Parthasarathy, Alexander Pechen, Guido Rizzi, Michael Skeide, Tadashi Toyoda and Wilhelm von Waldenfels. Special thanks are due to Maria Minotti, who carefully typed the manuscript.

Rome, March 2002 *Luigi Accardi, Yun Gang Lu, Igor Volovich*

Contents

Part I. Statement of the Problem and Simplest Models

1. Notations and Statement of the Problem 3
1.1 The Schrödinger Equation 3
1.2 White Noise Approximation for a Free Particle: The Basic Formula 5
1.3 The Interaction Representation: Propagators 9
1.4 The Heisenberg Equation 10
1.5 Dynamical Systems and Their Perturbations 11
1.6 Asymptotic Behaviour of Dynamical Systems: The Stochastic Limit 12
1.7 Slow and Fast Degrees of Freedom 13
1.8 The Quantum Transport Coefficient: Why Just the t/λ^2 Scaling? 15
1.9 Emergence of White Noise Hamiltonian Equations from the Stochastic Limit 18
1.10 From Interaction to Heisenberg Evolutions: Conditions on the State 21
1.11 From Interaction to Heisenberg Evolutions: Conditions on the Observable 22
1.12 Forward and Backward Langevin Equations 23
1.13 The Open System Approach to Dissipation and Irreversibility: Master Equation 24
1.14 Classical Processes Driving Quantum Phenomena 27
1.15 The Basic Steps of the Stochastic Limit 27
1.16 Connection with the Central Limit Theorems 30
1.17 Classical Systems 32
1.18 The Backward Transport Coefficient and the Arrow of Time 33
1.19 The Master and Fokker–Planck Equations: Projection Techniques 34
1.20 The Master Equation in Open Systems: an Heuristic Derivation 36

1.21 The Semiclassical Approximation for the Master Equation 38
1.22 Beyond the Master Equation 40
1.23 On the Meaning of the Decomposition $H = H_0 + H_{\mathrm{I}}$: Discrete Spectrum 42
1.24 Other Rescalings when the Time Correlations Are not Integrable 44
1.25 Connections with the Classical Homogeneization Problem 45
1.26 Algebraic Formulation of the Stochastic Limit 46
1.27 Notes 49

2. Quantum Fields 57
2.1 Creation and Annihilation Operators 57
2.2 Gaussianity 59
2.3 Types of Gaussian States: Gauge-Invariant, Squeezed, Fock and Anti-Fock 60
2.4 Free Evolutions of Quantum Fields 61
2.5 States Invariant Under Free Evolutions 63
2.6 Existence of Squeezed Stationary Fields 65
2.7 Positivity of the Covariance 67
2.8 Dynamical Systems in Equilibrium: the KMS Condition 68
2.9 Equilibrium States: the KMS Condition 69
2.10 q-Gaussian Equilibrium States 70
2.11 Boson Gaussianity 71
2.12 Boson Fock Fields 72
2.13 Free Hamiltonians for Boson Fock Fields 75
2.14 White Noises 76
2.15 Boson Fock White Noises and Classical White Noises 76
2.16 Boson Fock White Noises and Classical Wiener Processes 77
2.17 Boson Thermal Statistics and Thermal White Noises 77
2.18 Canonical Representation of the Boson Thermal States 79
2.19 Spectral Representation of Quantum White Noise 81
2.20 Locality of Quantum Fields and Ultralocality of Quantum White Noises 83

3. Those Kinds of Fields We Call Noises 85
3.1 Convergence of Fields in the Sense of Correlators 85
3.2 Generalized White Noises as the Stochastic Limit of Gaussian Fields 86
3.3 Existence of Fock, Temperature and Squeezed White Noises 90
3.4 Convergence of the Field Operator to a Classical White Noise 92
3.5 Beyond the Master Equation: The Master Field 93
3.6 Discrete Spectrum Embedded in the Continuum 94
3.7 The Stochastic Limit of a Classical Gaussian Random Field 97

3.8 Semiclassical Versus Semiquantum Approximation 98
3.9 An Historical Example: The Damped Harmonic Oscillator.. 100
3.10 Emergence of the White Noise: A Traditional Derivation ... 102
3.11 Heuristic Origins of Quantum White Noise 103
3.12 Relativistic Quantum White Noises 104
3.13 Space–Time Rescalings: Multidimensional White Noises.... 106
3.14 The Chronological Stochastic Limit 108
3.15 Notes.. 111

4. Open Systems 113
4.1 The Nonrelativistic QED Hamiltonian 113
4.2 The Dipole Approximation 116
4.3 The Rotating-Wave Approximation 117
4.4 Composite Systems 118
4.5 Assumptions on the Environment (Field, Gas, Reservoir, etc.) 119
4.6 Assumptions on the System Hamiltonian.................... 120
4.7 The Free Hamiltonian 121
4.8 Multiplicative (Dipole-Type) Interactions:Canonical Form .. 121
4.9 Approximations of the Multiplicative Hamiltonian 122
4.9.1 Rotating-Wave Approximation Hamiltonians 123
4.9.2 No Rotating-Wave Approximation Hamiltonians with Cutoff 123
4.9.3 Neither Dipole nor Rotating-Wave Approximation Hamiltonians Without Cutoff.................... 124
4.10 The Generalized Rotating-Wave Approximation........... 125
4.11 The Stochastic Limit of the Multiplicative Interaction 126
4.12 The Normal Form of the White Noise Hamiltonian Equation 127
4.13 Invariance of the Ito Correction Term Under Free System Evolution.......................... 128
4.14 The Stochastic Golden Rule: Langevin and Master Equations 129
4.15 Classical Stochastic Processes Underlying Quantum Processes 132
4.16 The Fluctuation–Dissipation Relation 133
4.17 Vacuum Transition Amplitudes 134
4.18 Mass Gap of D^+D and Speed of Decay 136
4.19 How to Avoid Decoherence 137
4.20 The Energy Shell Scalar Product: Linewidths 138
4.21 Dispersion Relations and the Ito Correction Term 140
4.22 The Case: $H_1 = |k^2|$ 141
4.23 Multiplicative Coupling with the Rotating Wave Approximation: Arbitrary Gaussian State 142
4.24 Multiplicative Coupling with RWA: Gauge Invariant State . 143
4.25 Red Shifts and Blue Shifts 144
4.26 The Free Evolution of the Master Field 145

4.27 Algebras Invariant Under the Flow 147
4.28 Notes 148

5. Spin–Boson Systems 153
5.1 Dropping the Rotating-Wave Approximation 154
5.2 The Master Field 154
5.3 The White Noise Hamiltonian Equation 157
5.4 The Operator Transport Coefficient: no Rotating-Wave Approximation, Arbitrary Gaussian Reference State 158
5.5 Different Roles of the Positive and Negative Bohr Frequencies 159
5.6 No Rotating-Wave Approximation with Cutoff: Gauge Invariant States 161
5.7 No Rotating-Wave Approximation with Cutoff: Squeezing States 161
5.8 The Stochastic Golden Rule for Dipole Type Interactions and Gauge-Invariant States 162
5.9 The Stochastic Golden Rule 164
5.10 The Langevin Equation 170
5.11 Subalgebras Invariant Under the Generator 172
5.12 The Langevin Equation: Generic Systems 173
5.13 The Stochastic Golden Rule Versus Standard Perturbation Theory 176
5.14 Spin–Boson Hamiltonian 178
5.15 The Damping and Oscillating Regimes: Fock Case 181
5.16 The Damping and the Oscillating Regimes: Nonzero Temperature 183
5.17 No Rotating-Wave Approximation Without Cutoff 184
5.18 The Drift Term for Gauge-Invariant States 185
5.19 The Free Evolution of the Master Field 187
5.20 The Stochastic Limit of the Generalized Spin–Boson Hamiltonian 187
5.21 The Langevin Equation 190
5.22 Convergence to Equilibrium: Connections with Quantum Measurement 191
5.23 Control of Coherence 194
5.24 Dynamics of Spin Systems 196
5.25 Nonstationary White Noises 201
5.26 Notes 202

6. Measurements and Filtering Theory 205
6.1 Input–Output Channels 205
6.2 The Filtering Problem in Classical Probability 206
6.3 Field Measurements 207
6.4 Properties of the Input and Output Processes 210

6.5 The Filtering Problem in Quantum Theory 213
6.6 Filtering of a Quantum System Over a Classical Process ... 214
6.7 Nondemolition Processes 215
6.8 Standard Scheme to Construct Examples of Nondemolition Measurements 217
6.9 Discrete Time Nondemolition Processes 217
6.10 Notes 218

7. Idea of the Proof and Causal Normal Order 219
7.1 Term-by-Term Convergence of the Series 219
7.2 Vacuum Transition Amplitude: The Fourth-Order Term ... 220
7.3 Vacuum Transition Amplitude: Non-Time-Consecutive Diagrams 223
7.4 The Causal δ-Function and the Time-Consecutive Principle 225
7.5 Theory of Distributions on the Standard Simplex 227
7.6 The Second-Order Term of the Limit Vacuum Amplitude .. 230
7.7 The Fourth-Order Term of the Limit Vacuum Amplitude .. 230
7.8 Higher-Order Terms of the Vacuum–Vacuum Amplitude ... 231
7.9 Proof of the Normal Form of the White Noise Hamiltonian Equation 231
7.10 The Unitarity Condition for the Limit Equation 233
7.11 Normal Form of the Thermal White Noise Equation: Boson Case 234
7.12 From White Noise Calculus to Stochastic Calculus 235

8. Chronological Product Approach to the Stochastic Limit . 237
8.1 Chronological Products 237
8.2 Chronological Product Approach to the Stochastic Limit ... 238
8.3 The Limit of the nth Term, Time-Ordered Product Approach: Vacuum Expectation 242
8.4 The Stochastic Limit, Time-Ordered Product Approach: General Case 244

9. Functional Integral Approach to the Stochastic Limit 247
9.1 Statement of the Problem 247
9.2 The Stochastic Limit of the Free Massive Scalar Field 248
9.3 The Stochastic Limit of the Free Massless Scalar Field 250
9.4 Polynomial Interactions 251
9.5 The Stochastic Limit of the Electromagnetic Field 253

10. Low-Density Limit: The Basic Idea 255
10.1 The Low-Density Limit: Fock Case, No System 255
10.2 The Low-Density Limit: Fock Case, Arbitrary System Operator 257

10.3 Comparison of the Distribution and the Stochastic Approach ... 258
10.4 LDL General, Fock Case, No System Operator, $\omega = 0$ ... 259

11. Six Basic Principles of the Stochastic Limit ... 261
11.1 Polynomial Interactions with Cutoff ... 262
11.2 Assumptions on the Dynamics: Standard Models ... 263
11.3 Polynomial Interactions: Canonical Forms, Fock Case ... 266
11.4 Polynomial Interactions: Canonical Forms, Gauge-Invariant Case ... 269
11.5 The Stochastic Universality Class Principle ... 275
11.6 The Case of Many Independent Fields ... 277
11.7 The Block Principle: Fock Case ... 278
11.8 The Stochastic Resonance Principle ... 280
11.9 The Orthogonalization Principle ... 280
11.10 The Stochastic Bosonization Principle ... 280
11.11 The Time-Consecutive Principle ... 282

Part II. Strongly Nonlinear Regimes

12. Particles Interacting with a Boson Field ... 285
12.1 A Single Particle Interacting with a Boson Field ... 287
12.2 Dynamical q-Deformation: Emergence of the Entangled Commutation Relations ... 289
12.3 The Two-Point and Four-Point Correlators ... 291
12.4 The Stochastic Limit of the N-Point Correlator ... 293
12.5 The q-Deformed Module Wick Theorem ... 296
12.6 The Wick Theorem for the QED Module ... 301
12.7 The Limit White Noise Hamiltonian Equation ... 302
12.8 Free Independence of the Increments of the Master Field ... 304
12.9 Boltzmannian White Noise Hamiltonian Equations: Normal Form ... 306
12.10 Unitarity Conditions ... 309
12.11 Matrix Elements of the Solution ... 310
12.12 Normal Form of the QED Module Hamiltonian Equation ... 312
12.13 Unitarity of the Solution: Direct Proof ... 313
12.14 Matrix Elements of the Limit Evolution Operator ... 314
12.15 Nonexponential Decays ... 317
12.16 Equilibrium States ... 321
12.17 The Master Field ... 321
12.18 Proof of the Result for the Two- and Four-Point Correlators ... 322
12.19 The Vanishing of the Crossing Diagrams ... 324
12.20 The Hot Free Algebra ... 333

12.21 Interaction of the QEM Field with a Nonfree Particle 335
12.22 The Limit Two-Point Function 338
12.23 The Limit Four-Point Function 342
12.24 The Limit Hilbert Module 344
12.25 The Limit Stochastic Process 347
12.26 The Stochastic Differential Equation 348
12.27 Notes 349

13. The Anderson Model 351
13.1 Nonrelativistic Fermions in External Potential: The Anderson Model 352
13.2 The Limit of the Connected Correlators 355
13.3 The Four-Point Function 356
13.4 The Limit of the Connected Transition Amplitude 359
13.5 Proof of (13.4.3) 362
13.6 Solution of the Nonlinear Equation (13.4.2) 365
13.7 Notes 367

14. Field–Field Interactions 369
14.1 Interacting Commutation Relations 369
14.2 The Tri-linear Hamiltonian with Momentum Conservation . 373
14.3 Proof of Theorem 14.2.1 375
14.4 Example: Four Internal Lines 379
14.5 The Stochastic Limit for Green Functions 380
14.6 Second Quantized Representation of the Nonrelativistic QED Hamiltonian 381
14.7 Interacting Commutation Relations and QED Module Algebra 383
14.8 Decay and the Universality Class of the QED Hamiltonian . 384
14.9 Photon Splitting Cascades and New Statistics 386

Part III. Estimates and Proofs

15. Analytical Theory of Feynman Diagrams 393
15.1 The Connected Component Theorem 395
15.2 The Factorization Theorem 402
15.3 The Case of Many Independent Fields 411
15.4 The Fermi Block Theorem 411
15.5 Non-Time-Consecutive Terms: The First Vanishing Theorem 416
15.6 Non-Time Consecutive Terms: The Second Vanishing Theorem 419
15.7 The Type-I Term Theorem 423
15.8 The Double Integral Lemma 431

15.9 The Multiple-Simplex Theorem 435
15.10 The Multiple Integral Lemma 438
15.11 The Second Multiple-Simplex Theorem 439
15.12 Some Combinatorial Facts and the Block Normal Ordering Theorem 446

16. Term-by-Term Convergence 453
16.1 The Universality Class Principle and Effective Interaction Hamiltonians 455
16.2 Block and Orthogonalization Principles 458
16.3 The Stochastic Resonance Principle 459

References 461

Index 469

Part I

Statement of the Problem and Simplest Models

1. Notations and Statement of the Problem

In this chapter we first introduce some general notations that shall be used throughout the book and then explain the main ideas of the stochastic limit in a general framework (i.e. independent of any specific model). Additional notations shall be introduced when needed.

1.1 The Schrödinger Equation

The pure states of a quantum mechanical system are described by rays in a Hilbert space[(1)] $\mathcal{H}$, i.e. unit vectors modulo an arbitrary phase, identified as rank-one projections acting on this space. Convex combinations of such projections or limits thereof (in the *norm topology*) are called *density operators* (or *density matrices*). If ρ is a density operator and Tr denotes the trace on the algebra of all bounded operators on $\mathcal{H}$ (always denoted $\mathcal{B}(\mathcal{H})$ in the following), then the map

$$A \mapsto \mathrm{Tr}(\rho A) =: \langle A \rangle$$

is well defined for any $A \in \mathcal{B}(\mathcal{H})$ and is called an *expectation value* or a *state*, a *pure* state if ρ has rank 1 and a *mixture* otherwise.

Self-adjoint operators on $\mathcal{H}$ are also called *observables* and $\langle A \rangle$ is called the *expectation value of the observable* A *in the state* ρ. If ρ has rank 1, then there exists a unit vector ψ in $\mathcal{H}$ such that

$$\langle A \rangle = \langle \psi, A\psi \rangle,$$

where the right-hand side denotes the scalar product in $\mathcal{H}$. By a *Hamiltonian* we mean a positive (or bounded below) observable, i.e. a self-adjoint operator on the Hilbert space $\mathcal{H}$ for which

$$H \geq 0. \tag{1.1.1}$$

---

(1) In the following, reference marks like this in the text refer to the comments listed in the section "Notes" at the end of the chapter.

The associated *Schrödinger equation* is

$$i\hbar\,\partial_t\psi_t = H\psi_t, \tag{1.1.2}$$

and, unless stated otherwise, we set

$$\hbar = 1. \tag{1.1.3}$$

The solution of (1.1.2) with initial condition ψ_0 at time t_0 is formally given by

$$\psi_t = e^{-i(t-t_0)H}\psi_0 = V_{t-t_0}\psi_0, \tag{1.1.4}$$

where the *quantum evolution*

$$V_t := e^{-itH} \tag{1.1.5}$$

is a *one-parameter unitary group* on the Hilbert space $\mathcal{H}$ in the sense that it satisfies the equation

$$V_tV_s = V_{t+s}, \qquad V_t^* = V_{-t}, \tag{1.1.6}$$

whose *infinitesimal generator* is H, i.e. V_t satisfies the (*operator*)*Schrödinger equation*

$$\partial_t V_t = -iHV_t \quad , \quad V_0 = 1. \tag{1.1.7}$$

If $\mathcal{H}$ is the space $L^2(\mathbb{R}^d)$ of square integrable functions with respect to the Lebesgue measure, then the *position* $q = (q_1 \dots q_d)$ and *momentum* $p = (p_1 \dots p_d)$ operators are defined, for $f \in \mathcal{H} = L^2(\mathbb{R}^d)$ differentiable, $x = (x_1 \dots x_d) \in \mathbb{R}^d$, by

$$q_j f(x) := x_j f(x), \quad p_j f(x) = \frac{1}{i}\frac{\partial}{\partial x_j} f(x), \tag{1.1.8}$$

and satisfy the *Heisenberg* (or *canonical*) *commutation relations*:

$$[q_k, q_j] = [p_k, p_j] = 0$$

$$[q_k, p_j] = q_kp_j - p_jq_k = i\delta_{kj} = \begin{cases} i & \text{if} \quad k = j \\ 0 & \text{if} \quad k \neq j \end{cases} .$$

If the Hamiltonian H is a function of the momentum operator $p = (p_1 \dots p_d)$, then it is called a *free Hamiltonian*. The Fourier transform implements a representation of the canonical commutation relations in which p, and therefore any function of it, *acts as multiplication*.

1.2 White Noise Approximation for a Free Particle: The Basic Formula

The basic idea of the stochastic limit is the following: the fast degrees of freedom of a quantum system, depending on a (small) parameter λ, in time scales of order t/λ^2, behave like white noise. In this section we illustrate the simplest occurrence of such a situation, which takes place when the small parameter is $\hbar$ and the system is free. The following relation, and its generalizations given in Lemma 1.9.1, play an important role in the stochastic limit (see [Vl83] where similar relations are proved). In the following $\mathcal{S}(\mathbb{R}^d)$ will denote the space of infinitely differentiable functions whose derivatives of any order decrease to zero faster than any polynomial (Schwartz space).

Proposition 1.2.1. *The following limit holds in the sense of Schwartz distributions in* $\mathcal{S}'(\mathbb{R}^2)$*:*

$$\lim_{\varepsilon\to 0} \frac{1}{\varepsilon} e^{itx/\varepsilon} = 2\pi\delta(t)\delta(x). \tag{1.2.1}$$

Proof 1.2.1. Let us fix two test functions φ, $\psi \in \mathcal{S}(\mathbb{R}^1)$ and consider the integral

$$I = \frac{1}{\varepsilon} \int dt\psi(t) \int dx\varphi(x) e^{itx/\varepsilon}.$$

By the change of variables $t = \varepsilon\tau$ we obtain

$$I = \int d\tau\psi(\varepsilon\tau) \int dx\varphi(x) e^{i\tau x} = \sqrt{2\pi} \int d\tau\psi(\varepsilon\tau)\hat{\varphi}(\tau),$$

where $\hat{\varphi}$ denotes the Fourier transform of φ. Now using the fact that $\hat{\varphi}(\tau) \in \mathcal{S}(\mathbb{R}^1)$ and using the Lebesgue dominated convergence theorem, we obtain:

$$\lim_{\varepsilon\to 0} \sqrt{2\pi} \int d\tau\psi(\varepsilon\tau)\tilde{\varphi}(\tau) = \sqrt{2\pi}\psi(0) \int d\tau\hat{\varphi}(\tau) = 2\pi\psi(0)\varphi(0),$$

which proves the statement.

The following example shows how white noise correlations can arise from ordinary quantum mechanics as an application of Proposition 1.2.1 above. Consider a quantum system, with n degrees of freedom, whose Hamiltonian is a function of the momentum operator $p = (p_1 \dots p_n)$:

$$H = E_p.$$

The associated Schrödinger equation in the momentum representation is

$$\partial_t\psi(t,k) = -\frac{i}{\hbar} E_k\psi(t,k), \tag{1.2.2}$$

where $k = (k_1 \dots k_n)$ and k_j, t are real numbers. Therefore, if the initial condition is chosen to be

$$\psi_\hbar(0,k) = \hbar^{-1/2}\psi_0(k), \quad \|\psi_0\|^2 = \int |\psi_0(k)|^2 dk = 1, \tag{1.2.3}$$

the solution of (1.2.2) is

$$\psi_\hbar(t,k) = \frac{e^{-itE_k/\hbar}}{\hbar^{1/2}}\,\psi_0(k).$$

Consider now the time correlations

$$\langle \psi_\hbar(t',\cdot), \psi_\hbar(t,\cdot)\rangle = \int \frac{e^{-i(t-t')E_k/\hbar}}{\hbar}\,|\psi_0(k)|^2 dk.$$

Then, because of Proposition 1.2.1

$$\lim_{\hbar\to 0} \frac{e^{-i(t-t')E_k/\hbar}}{\hbar} = 2\pi\delta(t-t')\delta(E_k), \tag{1.2.4}$$

the limit being meant in the sense of distributions (in the variables t, t', k). It follows that

$$\lim_{\hbar\to 0} \langle \psi_\hbar(t',\cdot), \psi_\hbar(t,\cdot)\rangle = 2\pi\delta(t-t') \int dk\delta(E_k)|\psi_0(k)|^2, \tag{1.2.5a}$$

the limit being meant in the sense of distributions (in the variables t, t').

Notice that in this argument $\hbar$ enters only as a parameter, and the interpretation of the above approximation as a new kind of semiclassical approximation is quite natural (see Sect. 3.8 for further discussion). Another interpretation is the following: if the energy function E_k is replaced with λE_k ($\lambda > 0$), this is equivalent to replacing $\hbar$ with $\hbar/\lambda$ in the above discussion. In this sense one could say that the limit $\hbar \to 0$ is a *strong coupling limit*.

The normalization condition (1.2.3) simply amounts to a *time rescaling*. To convince oneself of this fact, it is sufficient to notice that the solution $\psi_\hbar(t,k)$ is equal to $\hbar^{-1/2}\psi_0(t,k)$, where $\psi_0(t,k)$ is the solution of the Schrödinger equation (1.2.2) with initial condition $\psi_0(k)$, and to use the identity

$$\langle \psi_\hbar(t',\cdot), \psi_\hbar(t,\cdot)\rangle = \hbar^{-1}\langle \psi_0, e^{-i(t-t')E_p/\hbar}\psi_0\rangle. \tag{1.2.5b}$$

Then, recalling that for convergence in the distribution sense we have to multiply by smooth test functions $\varphi(t)$, $\psi(t')$ and integrate, we obtain

$$\int dt \int dt' \varphi(t)\psi(t')\langle \psi_\hbar(t',\cdot), \psi_\hbar(t,\cdot)\rangle$$
$$= \frac{1}{\hbar}\int dt \int dt' \varphi(t)\psi(t')\langle \psi_0, e^{-i(t-t')E_p/\hbar}\psi_0\rangle,$$

and with the change of variables $(t-t')/\hbar = \tau$ and $t' = t'$, we find

$$\int d\tau \int dt' \varphi(\hbar\tau + t')\psi(t')\langle\psi_0, e^{-i\tau E_p}\psi_0\rangle. \tag{1.2.6}$$

So the limit $\hbar \to 0$ means that we consider time intervals for which $\hbar\tau(= t-t')$ is small. Under this assumption, and supposing in addition that the Hamiltonian H has integrable time correlations in the state ψ_0, i.e.

$$\int_{-\infty}^{+\infty} d\tau |\langle\psi_0, e^{-i\tau E_p}\psi_0\rangle| < +\infty,$$

and the test functions have compact support, we can replace $\varphi(\hbar\tau + t')$ with $\varphi(t')$ in (1.2.6) (dominated convergence). Then, performing the integration over $d\tau$, we find

$$\langle\psi_0, \delta(E_p)\psi_0\rangle \int_{-\infty}^{+\infty} dt' \varphi(t')\psi(t'), \tag{1.2.7}$$

which is equivalent to (1.2.5a) (the Dirac δ-function) of an operator is defined at the end of the present section). Note that, under our assumptions, the error in replacing (1.2.6) by (1.2.7) is of order $\hbar$. Therefore, using (1.2.5b) we see that (1.2.6), multiplied by $\hbar$, is the dominating contribution to the correlation $\langle\psi_0, e^{-i(t-t')E_p/\hbar}\psi_0\rangle$ and that the error is of order $(\hbar)^2$.

In summary, just as the semiclassical limit finds classical mechanics as a first-order approximation to quantum mechanics, the white noise approximation captures the second-order approximation. In this sense we can say that it describes *the fluctuations around the classical solution.*

The limit (1.2.5a) has two interesting implications:

(i) In time intervals which are small with respect to the dimensions of $\hbar$, the quantum mechanical time correlations tend to become *white noise* (i.e. δ) correlations (see Sect. 2.15 for the definition).

(ii) In the same regime, as far as momentum is concerned, the whole interaction is concentrated on the *resonance surface*:

$$\{k : E_k = 0\}. \tag{1.2.8}$$

The surface (1.2.8) is called a *resonance surface* because it illustrates in the simplest situation the *stochastic resonance* principle, which will dealt with in Chap. 5.

The above considerations suggest that, over appropriate time scales, one can obtain a good approximation of standard quantum mechanics by considering a quantum field in the variables (t, k), which is white noise (i.e. δ correlated) in the time variable and, with respect to the momentum variable, is concentrated on the energy surface (1.2.8).

We shall see that a suitable generalization of such a picture is not only correct, but also universal in the sense of being applicable to almost all quantum systems with a continuous spectrum and that it leads to the conclusion that the dominating contributions to the usual Hamiltonian theory are given by a singular Hamiltonian which is a functional of some white noise. The advantage of the new Hamiltonian theory is that the new dynamical equation defines the physical parameters, relevant in the time scale under consideration, in a simple and intuitive way that gives much more insight than the original Hamiltonian.

The associated white noise Hamiltonian calculus is a generalization of both classical and quantum stochastic calculus. Its most remarkable feature is that it shows how it is concretely possible to combine the unitarity of the quantum mechanical evolution with an elegant and easy-to-apply *general theory of quantum dissipative and irreversible phenomena.*

Let us conclude this section with a general definition of the δ-function of an operator, which has already appeared in formula (1.2.7) above and which will often be used in the following.

Definition 1.2.1. *For a self-adjoint operator H, we define*

$$\delta(H) = \frac{1}{2\pi}\int_{\mathbb{R}} dte^{itH} = \frac{1}{2\pi}\int_{\mathbb{R}} dt \int_{\mathbb{R}} E^H(d\lambda)e^{it\lambda} = \int_{\mathbb{R}} E^H(d\lambda)\delta(\lambda)$$

weakly in terms of the spectral theorem. If ξ, η are in the absolutely continuous spectrum of H, and the associated spectral measure of H has a smooth density (say, in the Schwartz space) denoted by $\mu_{\xi,\eta}$, then by definition

$$\langle \xi, \delta(H)\eta\rangle = \frac{1}{2\pi}\int\int e^{it\lambda}\langle \xi, E^H(d\lambda)\eta\rangle dt$$
$$= \int \mu_{\xi,\eta}(\lambda)\delta(\lambda)d\lambda = \mu_{\xi,\eta}(0) = \langle \delta, \mu_{\xi,\eta}\rangle.$$

From this formula it easily follows that, for any hermitean operator H, $\delta(H)$ is a positive, possibly degenerate, sesquilinear form. For example, let $\mathcal{H}_1 = L^2(\mathbb{R}^d)$ be interpreted, to fix our ideas, as a momentum representation and let

$$H = p = \int k \cdot E(dk) = \int \sum_{j=1}^{d} k_j E_j(dk)$$

be the momentum operator on $\mathcal{H}_1$. If $\omega(\cdot)$ is a Schwartz distribution, then for any f, $g \in \mathcal{S}(\mathbb{R}^d)$, the sesquilinear form

$$\langle f, \omega(H)g\rangle = \langle \omega, \overline{f}g\rangle$$

makes sense, since $\mathcal{S}(\mathbb{R}^d)$ is an algebra and we take it as the definition of

$$\omega(H) = \int \omega(k)E(dk).$$

1.3 The Interaction Representation: Propagators

If the Hamiltonian H has the form

$$H = H_0 + H_I \tag{1.3.1}$$

(see Sect. 1.23 for comments on this decomposition) the wave function $\psi_I(t)$ in the interaction representation is defined by

$$\psi_I(t) = e^{itH_0}\psi(t) = e^{itH_0}e^{-it(H_0+H_I)}\psi_0, \tag{1.3.2}$$

and it satisfies the *Schrödinger equation in the interaction representation*

$$\partial_t\psi_I(t) = -iH_I(t)\psi_I(t), \tag{1.3.3}$$

where $H_I(t)$ is defined by

$$H_I(t) := e^{itH_0}H_Ie^{-itH_0}. \tag{1.3.4}$$

The solution of (1.3.3) with initial condition $\psi_I(t_0)$ at time t_0 is given by

$$\psi_I(t) = U(t,t_0)\psi_I(t_0), \tag{1.3.5}$$

where the evolution operator $U(t,t_0)$,

$$U(t,t_0) = e^{itH_0}e^{-i(t-t_0)(H_0+H_I)}e^{-it_0H_0}, \tag{1.3.6}$$

also called *a propagator*, satisfies

$$U(t,t_0) = U(t,s)U(s,t_0),\ U(t,t_0)^* = U(t_0,t),\ U(t,t) = 1. \tag{1.3.7}$$

One has

$$U(t,t_0) = e^{it_0H_0}U(t-t_0,0)e^{-it_0H_0}. \tag{1.3.8}$$

The evolution operator $U(t,t_0)$ is the solution of

$$\partial_t\,U(t,t_0) = -iH_I(t)U(t,t_0)\ ,\quad U(t_0,t_0) = I, \tag{1.3.9}$$

with H_I given by (1.3.4). The S matrix is formally defined as the limit

$$S = \lim_{t\to+\infty} U(t,-t). \tag{1.3.10}$$

Remark 1.3.1. For the applications of the stochastic limit the converse of the above construction will be of crucial importance. Namely: let $U(t,t_0)$ be a 2-parameter family of unitary operators satisfying conditions (1.3.6) and (1.3.9) above and let $\tilde{H}_0$ be a Hamiltonian, then the one-parameter unitary family

$$e^{-it\tilde{H}_0}U(t,0) =: V_t$$

is a quantum evolution (i.e. $V_{t+s} = V_tV_s$ and $V_t^* = V_{-t}$) if, and only if, (1.3.9) holds with H_0 replaced by $\tilde{H}_0$. This is equivalent to the so-called *cocycle condition with respect to* $\exp(-it\tilde{H}_0)$:

$$U(t-t_0,0) = e^{-i(t_0-s)\tilde{H}_0}U(t-s,0)e^{i(t_0-s)\tilde{H}_0}U(s-t_0,0). \tag{1.3.11}$$

1.4 The Heisenberg Equation

The *Heisenberg evolution* of an observable (more generally, any non-necessarily self-adjoint operator) A, associated with the Schrödinger evolution (1.1.4) with $t_0 = 0$, is defined by

$$\langle \psi_t, A\psi_t \rangle = \langle \psi_0, V_t^* A V_t \psi_0 \rangle =: \langle \psi_0, A_t \psi_0 \rangle. \tag{1.4.1}$$

Therefore,

$$A_t := V_t^* A V_t = e^{itH} A e^{-itH} = u_t(A). \tag{1.4.2}$$

The dual evolution, for a density matrix ρ, is defined by the identity

$$\mathrm{Tr}\{\rho A_t\} = \mathrm{Tr}\{\rho_t A\}.$$

Therefore,

$$\rho_t := e^{-itH} \rho e^{itH} = V_t \rho V_t^* = u_t^*(\rho). \tag{1.4.3}$$

The map $A \mapsto A_t =: u_t(A)$ satisfies the identities

$$u_t u_s = u_{t+s}, \quad u_t(AB) = u_t(A) u_t(B), \quad u_t(A^*) = u_t(A)^*, \quad u_t(1) = 1. \tag{1.4.4}$$

Any family u_t, satisfying (1.4.4) is called a *Heisenberg evolution* or a *one-parameter automorphism group* of the *algebra of the observables*, and the one-parameter family A_t, defined by (1.4.2), satisfies the *generalized Heisenberg equation* (or *quantum Liouville equation*):

$$\partial_t A_t = L(A_t), \tag{1.4.5}$$

whose *generator* L (also called the *Liouvillian*) is a $*$-derivation, i.e.

$$L(AB) = L(A)B + AL(B), \; L(A^*) = L(A)^*.$$

When u_t has the form (1.4.2), its generator has the form $L = i[H, \,\cdot\,]$, corresponding to the *Heisenberg equation* of motion:

$$\partial_t A_t = i[H, A_t], \quad A_0 = A. \tag{1.4.6}$$

The dual equation, for the density matrix, is

$$\partial_t \rho_t = -i[H, \rho_t]. \tag{1.4.7}$$

Similarly, if $U(t, t_0)$ is the evolution operator in interaction representation [see (1.3.7)] the associated Heisenberg evolution is

$$A_t := j_{(t,t_0)}(A) = U(t, t_0)^* A U(t, t_0)$$

and satisfies the *flow equation*

$$\jmath_{(t,t_0)}(A) = \jmath_{(s,t_0)}\jmath_{(t,s)}(A), \tag{1.4.8}$$

whose differential form is the Heisenberg equation in the interaction representation

$$\partial_t A_t = -i[H_{\mathrm{I}}(t), A_t], \tag{1.4.9}$$

where $H_{\mathrm{I}}(t)$ given by (1.3.4).

1.5 Dynamical Systems and Their Perturbations

In this section we introduce the notion of a *dynamical system* that will be used in the present book. More generally, purely algebraic definitions are possible, but in most cases of physical interest one fixes a priori a representation, i.e. a concrete Hilbert space where the algebraic objects are realized, and one works in that context. By a *dynamical system* we mean a pair

$$\{\mathcal{H}, H\} \tag{1.5.1}$$

consisting of an Hilbert space $\mathcal{H}$ (state space) and an Hamiltonian H which defines the dynamical evolution through the operator Schrödinger equation (1.1.7). A dynamical system is said to be *integrable* if the associated dynamical evolution can be solved explicitly. We start from an integrable system

$$\{\mathcal{H}, H_0\},$$

called the *free system*. Then, given another self-adjoint operator H_{I}, called the *interaction Hamiltonian*, we can always define the time-evolved operator $H_{\mathrm{I}}(t)$ using (1.3.4), i.e.

$$H_{\mathrm{I}}(t) := e^{itH_0} H_{\mathrm{I}} e^{-itH_0}. \tag{1.5.2}$$

Therefore, the Schrödinger equation in the interaction representation

$$\partial_t U(t, t_0) = -iH_{\mathrm{I}}(t)U(t, t_0), \quad U(t_0, t_0) = 1, \tag{1.5.3}$$

is well defined, as described in Sect. 1.3, and its solution, when it exists, gives the propagator $U(t, t_0)$, which contains all the physically relevant information on the dynamics of the system. As explained in Sect. 1.3, knowledge of this propagator is formally equivalent to knowledge of the solution of the Schrödinger equation with a *total Hamiltonian*, $H_0 + H_{\mathrm{I}}$, but this formulation of the problem is sometimes more convenient. In almost all models of physical interest, one starts from a free system $\{\mathcal{H}, H_0\}$ and perturbs it by means of an interaction Hamiltonian H_{I}. Therefore we slightly generalize the notion of a dynamical system by defining it as a triple system

$$\{\mathcal{H}_0, H_0, U(t,s)\},$$

where $\{\mathcal{H}, H_0\}$ is an integrable dynamical system and $U(t,s)$ is a propagator satisfying the cocycle equations (1.3.6) and (1.3.9). If $U(t,s)$ is differentiable (in t), then the new definition is equivalent to the one given at the beginning of this section. However, the generalized definition also allows the possibility for $U(t,s)$ to satisfy a *stochastic differential* or *white noise Hamiltonian equation*, and in the following we shall show how these equations arise in physics.

Usually, the total Hamiltonian $H_0 + H_\mathrm{I}$, even when it is well defined, is not integrable, and one looks for some reasonably good approximation to the physically measurable quantities which typically are expressions of the form

$$|\langle\phi_{\mathrm{out}}, U(t,t_0)\phi_{\mathrm{in}}\rangle|^2 \tag{1.5.4}$$

or of the slightly more general form

$$\langle\phi_{\mathrm{in}}, U^*(t,t_0)\ X\ U(t,t_0)\phi_{\mathrm{in}}\rangle, \tag{1.5.5}$$

where $\phi_{\mathrm{out}}, \phi_{\mathrm{in}}$ are state vectors and X is an operator.

Remark 1.5.1. We have not introduced explicitly any privileged state of the dynamical system. However, for infinite systems the introduction of an Hilbert space $\mathcal{H}$, where the Hamiltonian H_0 acts, requires the choice of a specific representation, and in all concrete cases this is done by fixing a state. Thus, at least for infinite systems, the choice of a privileged state is implicit in our definition.

1.6 Asymptotic Behaviour of Dynamical Systems: The Stochastic Limit

Two basic asymptotic techniques of quantum physics are: (i) *scattering theory*, which is concerned with the *long-term behaviour* of physical systems, for which

$$t \to \infty \text{ (scattering theory)},$$

and (ii) *perturbation theory*, which is concerned with *weak effects* (for example, weak coupling or low density), for which

$$\lambda \to 0 \text{ (perturbation theory)},$$

where λ is a parameter upon which the interaction Hamiltonian (and in some cases, e.g. low density, even the free system) depends.

The stochastic limit puts together the scattering and perturbation theories by studying *the long-term cumulative effects* of weak actions. The *long term* $(t/\lambda^2 \to +\infty)$ compensates for the small parameter $(\lambda \to 0)$, and new rich structures with nontrivial interactions emerge. This picture will be confirmed in the following by a detailed study of several models.[(1)] The *stochastic limit of quantum theory* starts from the usual quantum Hamiltonian dynamics depending on a parameter λ, rescales some parameters in the associated evolution (typically *time*)[(2)], and obtains in the limit $\lambda \to 0$ a new dynamical system driven by a singular Hamiltonian which represents the decay and shift contributions to the dynamics. In the usual approximations used in quantum theory (perturbation theory, semiclassical approximation, etc.), one approximates the expectation values (1.5.4) and (1.5.5) for different observables. In the stochastic limit approach the fundamental laws themselves are approximated in the following sense: one starts with a family of quantum dynamical systems depending on a parameter λ,

$$\left\{\mathcal{H}_\lambda, H_0^{(\lambda)}, U_t^{(\lambda)}\right\} \tag{1.6.1}$$

(in many examples considered in this book λ is just the coupling constant and neither the state space $\mathcal{H}_\lambda$ nor the free Hamiltonian $H_0^{(\lambda)}$ depend on λ). With these notations the basic content of *the stochastic limit* is described by the following symbolic relation:

$$\lim_{\lambda\to 0}\left\{\mathcal{H}_\lambda, H_0^{(\lambda)}, U_{t/\lambda^2}^{(\lambda)}\right\} = \left\{\mathcal{H}, \tilde{H}_0, U_t\right\}, \tag{1.6.2}$$

where $\{\mathcal{H}, \tilde{H}_0, U_t\}$ is a new quantum dynamical system in the sense of the preceeding section. In Sect. 1.15 and Sect. 3.1 we describe in what sense the limit (1.6.2) has to be understood. The basic physical idea of the stochastic limit is expressed by the *approximate equality*[(3)]

$$U_{t/\lambda^2}^{(\lambda)} \approx U_t, \qquad \text{equivalently} \qquad U_t^{(\lambda)} \approx U_{\lambda^2 t}, \tag{1.6.3}$$

which is valid for large times t and for small, but finite, values of the parameter λ. The precise meaning of the approximation (1.6.3) is that a large class of measurable and physically interesting quantities of the λ-dependent system are approximated by the corresponding quantities of the limit system with an error which vanishes as $\lambda \to 0$ and which can be explicitly estimated (see Sects. 1.10 and 1.11 for examples of such quantities and Chaps. 15 and 16 for the estimates).

We emphasize that the limit system is a quantum dynamical system in the sense of Sect 1.1. In particular the unitarity of the evolution operator is preserved under the stochastic limit.[(4)]

1.7 Slow and Fast Degrees of Freedom

One of the basic physical ideas behind the stochastic limit program is *the separation of the natural time scales of a complex system.* In order to explain this idea let us consider an Hamiltonian of the form

$$H = H_0 + H_{\mathrm{I}}^{(\lambda)}. \tag{1.7.1}$$

For every fixed reference state (e.g. a vacuum) one can associate with this Hamiltonian a natural characteristic time scale,[(1)] namely:

(i) The *fast relaxation time* t_{R}. This is a characteristic time such that the time correlation

$$\langle H_{\mathrm{I}}(0)H_{\mathrm{I}}(t)\rangle \tag{1.7.2}$$

practically vanishes after a few multiples of t_{R}. Here $H_{\mathrm{I}}(t)$ is the Heisenberg evolution of H_{I} under the H_0 evolution and $\langle \,\cdot\, \rangle$ denotes the expectation with respect to this state.
One can define the *fast degrees of freedom* by those observables X whose correlations

$$\langle X(0)X(t)\rangle - \langle X(0)\rangle\langle X(t)\rangle \tag{1.7.3}$$

have a decay rate of the same order of t_{R} or faster. Here too $X(t)$ is the Heisenberg evolution of an observable quantity under the H_0 evolution.

(ii) The *slow degrees of freedom* are defined by those observables whose correlations have a decay rate t_{S} (*slow relaxation time*) of an order much larger than t_{R}.

(iii) There exist other natural characteristic time scales, for example, t_{int}, the *interaction time* between the slow and the fast degrees of freedom. This is a characteristic time for appreciable effects of the interaction, in the sense that it describes the decay rate of correlations (1.7.3) but where $X(t)$ is the Heisenberg evolution of X under H evolution and not H_0 evolution.

(iv) Finally there is t_{Eq}, the equilibrium time.

Typically one has

$$t_{\mathrm{R}} \ll t_{\mathrm{int}} \ll t_{\mathrm{S}} \ll t_{\mathrm{Eq}}. \tag{1.7.4}$$

For example, for relaxation of spin systems (slow degrees of freedom) in a dilute atomic vapor (fast degrees of freedom), t_{R} may be taken to be the mean time between collisions and is typically of the order of 10^{-8} s at ordinary temperatures, whereas t_{S} may be as long as a few seconds.

Moreover, since $t_{\mathrm{S}} \gg t_{\mathrm{R}}$, the fast degrees of freedom, looked at with respect to the time scale of the slow degrees of freedom, will appear to be almost completely uncorrelated, i.e. *totally uncorrelated noise.*

It is plausible that different interactions give rise to different kinds of noise. Thus a natural question to be answered is how to describe these kinds of noises.

One may think of at least two physical situations in which the condition $t_{\mathrm{S}} \gg t_{\mathrm{R}}$ is satisfied:

(i) *Weak coupling*, characterized by

$$H_{\mathrm{I}}^{(\lambda)} = \lambda H_{\mathrm{I}}.$$

This means that the interaction is feeble, so that each individual collision has a very small effect. On the other hand, there are *very many* collisions (infinitely many in the long-term limit) and this situation is the physical counterpart of the Lindberg condition in the central limit theorem of classical probability, where one has a sum of weakly dependent terms, each of which becomes infinitesimal. In our case the weak dependence corresponds to the fast decrease of the correlations of the fast degrees of freedom.
This analogy suggests that the situation should be well approximated by some kind of central limit effect, leading to some diffusion process.

(ii) *Low density.* Collisions only happen at *infrequent time instants.* However, in contrast to the weak coupling limit situation, each individual collision does not have a very small effect (i.e. the interaction is not supposed to be weak).
When speaking of rare events it is natural to think of a form of Poisson process, so in this case we expect Poisson-like behaviour.

In both cases an appreciable influence of the fast degrees of freedom on the slow degrees of freedom can only be observed as a long-term cumulative effect of many collisions; this means that we will have to consider the time evolution of observables up to very large microscopic times in order to see an appreciable macroscopic effect.

1.8 The Quantum Transport Coefficient: Why Just the t/λ^2 Scaling?

It remains to be explained why one chooses the rescaling $t \to t/\lambda^2$ as the new time scale and what the physical meaning of this rescaling is. The reason why (in many important cases) the appropriate time rescaling is t/λ^2 (and not, say, $t/\varphi(\lambda)$ for some other function φ of λ) is best understood by considering second-order perturbation theory. Let us consider the equation

$$\partial_t U_t^{(\lambda)} = -i\lambda H_{\mathrm{I}}(t) U_t^{(\lambda)}, \quad U_0^{(\lambda)} = 1. \tag{1.8.1}$$

Lemma 1.8.1. *Denote by $\langle\cdot\rangle$ the expectation value with respect to a fixed vector Φ. Suppose that the Hamiltonian has mean zero, is time-translation invariant and such that the function $s \mapsto \langle H_{\mathrm{I}}(0)H_{\mathrm{I}}(s)\rangle$ is integrable (integrable time correlations), i.e.*

$$\langle H_{\rm I}(t)\rangle = 0 \tag{1.8.2}$$

$$\langle H_{\rm I}(t_1+s)\dots H_{\rm I}(t_n+s)\rangle = \langle H_{\rm I}(t_1)\dots H_{\rm I}(t_n)\rangle \tag{1.8.3}$$

$$\int_{-\infty}^{+\infty} |\langle H_{\rm I}(0)H_{\rm I}(t)\rangle| dt < +\infty. \tag{1.8.4}$$

Then the expectation value of the second-order term of the iterated series for $U_t^{(\lambda)}$*, i.e.*

$$-\lambda^2 \int_0^t dt_1 \int_0^{t_1} dt_2 \langle H_{\rm I}(t_1)H_{\rm I}(t_2)\rangle, \tag{1.8.5}$$

has a finite nonzero limit, as $\lambda \to 0$ *and* $t \to +\infty$*, if and only if*

$$\lim_{\lambda\to 0, t\to+\infty} \lambda^2 t = \tau = const. \neq 0 \quad (< +\infty). \tag{1.8.6}$$

In this case the limit of (1.8.5), as $\lambda \to 0$*, is equal to*

$$-\tau \int_{-\infty}^0 ds \langle H_{\rm I}(0)H_{\rm I}(s)\rangle. \tag{1.8.7}$$

Proof 1.8.1. By time-translation invariance, the second-order term (1.8.5) has the form

$$-\lambda^2 \int_0^t dt_1 \int_0^{t_1} dt_2 \langle H_{\rm I}(0)H_{\rm I}(t_2-t_1)\rangle.$$

By the change of variable $t_2 - t_1 =: s_2$, this becomes

$$-\lambda^2 \int_0^t dt_1 \int_{-t_1}^0 ds_2 \langle H_{\rm I}(0)H_{\rm I}(s_2)\rangle,$$

and with the change of variable $\lambda^2 t_1 =: s_1$, we find

$$-\int_0^{\lambda^2 t} ds_1 \int_{-s_1/\lambda^2}^0 ds_2 \langle H_{\rm I}(0)H_{\rm I}(s_2)\rangle. \tag{1.8.8}$$

Since for all $s_1 > 0$, as $\lambda \to 0$,

$$-\int_{-s_1/\lambda^2}^0 ds_2 \langle H_{\rm I}(0)H_{\rm I}(s_2)\rangle \to -\int_{-\infty}^0 ds \langle H_{\rm I}(0)H_{\rm I}(s)\rangle = \text{const.},$$

it follows that, as $\lambda \to 0$, (1.8.8) tends to zero by dominated convergence if $\lambda^2 t \to 0$, while if (1.8.6) holds, it tends to (1.8.7).

Remark 1.8.1. The integral in (1.8.7) can be interpreted as a *quantum transport coefficient*. It plays an important role in the stochastic limit (see Sect. 4.14) and it provides a natural bridge between this theory and both the Green–Kubo–Mori theory of nonequilibrium phenomena, which is a first-order effect, and the Fermi golden rule, which is a second-order effect. More precisely, we shall see (see Sect. 5.8) that in several cases (e.g. in the weak coupling but not in the low-density case) the transport coefficient allows us to compute the lifetimes and energy shifts in agreement with the Fermi golden rule. In this sense we may say that the stochastic golden rule is a rigorous formulation of the Fermi golden rule. However, even in these simple cases, the amount of information contained in the stochastic limit is much greater than that obtained by these standard perturbation techniques.

Note, moreover, that the assumptions of Lemma 1.8.1 do not require that the expectation value $\langle\cdot\rangle$ is scalar valued, i.e. it can also be a partial expectation involving only some of the degrees of freedom of the system. In this case the transport coefficient is an operator and one speaks of the *operator valued transport coefficient.* These coefficients shall play an important role in the following.

Remark 1.8.2. Lemma 1.8.1 shows that, under (1.8.2), (1.8.3) and (1.8.4), (1.8.6) is essentially the only one which guarantees the existence and nontriviality of the limit of the vacuum expectation of the second-order of the series term. The limit (1.8.6) means that in times of order t/λ^2 the interaction produces effects of order τ. Thus λ provides a natural time scale for the observable effects of the interaction. But $\lambda^2 t \to \tau \Leftrightarrow t \sim \tau/\lambda^2$; therefore, (1.8.6) is equivalent to

$$\lambda \to 0 \,, \quad t \mapsto t/\lambda^2 . \tag{1.8.9}$$

Note that the rescaling (1.8.9) has been obtained

(i) taking expectations for an expectation value satisfying the time stationarity and the analytical condition (1.8.6); and
(ii) only for the second-order term.

It is therefore natural to ask oneself the following questions:

(i) Is (1.8.9) (or more generally the rescaling suggested by second-order perturbation theory) sufficient to guarantee the convergence, at least term by term and in a topology to be specified, of the whole time-rescaled perturbative series?[(1)]
(ii) If the answer to (i) is affirmative, can we conclude that the term-by-term limit of the time-rescaled perturbative series can be re-summed into a new evolution operator, U_t? Is U_t unitary?
(iii) Which evolution equation is satisfied by U_t?
(iv) What physical information *about the original system* can one extract from U_t?

The stochastic limit of quantum theory has been developed to answer these questions.

Answer to (i): Yes, the limit exists for a great multiplicity of models and the convergence is the quantum analogue of convergence in probability theory law. In many models the convergence is not only term by term, but it holds for the whole perturbation series.

Answer to (ii): In all cases in which the limit was proved to exist it turned out to be unitary.

Answer to (iii): In all cases in which the limit exists, the evolution equation has been found and proved to be a singular Hamiltonian equation driven by some form of quantum white noise or, equivalently, in the linear case, by a quantum stochastic differential equation.

Answer to (iv): The advantage of the new equation versus the original one is that we obtain a simplification of the model in the sense that many physical effects are more easily read in this new type of quantum stochastic equation, which is a natural generalization of the usual Schrödinger equation. For example, in most models the vacuum expectation value of the evolution operator (vacuum transition amplitude) satisfies a simple ordinary differential equation which can be solved explicitly – a thing that one cannot do for the original evolution operator.

1.9 Emergence of White Noise Hamiltonian Equations from the Stochastic Limit

In this section we describe a very simple heuristic procedure which allows us to quickly guess the form of the white noise equation satisfied by the limit evolution operator associated, a given interaction Hamiltonian. The equation satisfied by the dynamical evolution $U_t^{(\lambda)}$ is

$$\partial_t U_t^{(\lambda)} = -iH_{\mathrm{I}}^{(\lambda)}(t)U_t^{(\lambda)}, \tag{1.9.1}$$

with the initial condition

$$U_0^{(\lambda)} = 1, \tag{1.9.2}$$

which can be interpreted either as the usual Schrödinger equation with time-dependent Hamiltonian $H_{\mathrm{I}}^{(\lambda)}(t)$ or as a *Schrödinger equation in the interaction picture*. In this book we shall be mainly concerned with the latter interpretation. One starts from a model with a time-independent Hamiltonian

$$H^{(\lambda)} = H_0^{(\lambda)} + H_{\mathrm{I}}^{(\lambda)}, \tag{1.9.3}$$

where $H_0^{(\lambda)}$ describes the dynamics of a free system and $H_{\mathrm{I}}^{(\lambda)}$ an interaction. Then one defines $U_t^{(\lambda)}$ and $H_{\mathrm{I}}^{(\lambda)}(t)$, respectively, by

$$U_t^{(\lambda)} = e^{itH_0^{(\lambda)}}e^{-itH^{(\lambda)}}, \tag{1.9.4}$$

$$H_{\mathrm{I}}^{(\lambda)}(t) = e^{itH_0^{(\lambda)}}H_{\mathrm{I}}^{(\lambda)}e^{-itH_0^{(\lambda)}}. \tag{1.9.5}$$

Note, for future use, that $U_t^{(\lambda)}$ is the adjoint of the *backward wave operator* at time t, i.e.

$$U_t^{(\lambda)} =: \Omega_-^{(\lambda)*}(t) \to \Omega_-^{(\lambda)*}, \tag{1.9.6}$$

where $\Omega_-^{(\lambda)}$ is the backward wave operator. In these notations $U_t^{(\lambda)}$ satisfies (1.9.1) and (1.9.2) with $H_{\mathrm{I}}^{(\lambda)}(t)$ given by (1.9.5). With the change of variables $t_1 \mapsto \lambda^2 t_1$ (1.9.1) takes the form

$$\partial_t U_{t/\lambda^2}^{(\lambda)} = -i\frac{1}{\lambda^2} H_{\mathrm{I}}^{(\lambda)}(t/\lambda^2) U_{t/\lambda^2}^{(\lambda)}, \tag{1.9.7}$$

which is similar to (1.9.1) but with the interaction Hamiltonian $H_{\mathrm{I}}(t)$ replaced by the *rescaled interaction Hamiltonian,* $\frac{1}{\lambda^2} H_{\mathrm{I}}^{(\lambda)}(t/\lambda^2)$. The first step of the stochastic limit approach is to prove that the rescaled interaction Hamiltonian converges, in an appropriate sense, to a new Hamiltonian H_t, i.e.

$$\lim_{\lambda \to 0} \frac{1}{\lambda^2} H_{\mathrm{I}}^{(\lambda)} \left(\frac{t}{\lambda^2} \right) = H_t. \tag{1.9.8}$$

If the convergence is strong enough then the relationship (1.9.8) should imply that

$$U_{t/\lambda^2}^{(\lambda)} \to U_t, \tag{1.9.9}$$

where the convergence is meant in the sense explained in Sect. 1.6 and U_t is the solution of

$$\partial_t U_t = -iH_t U_t. \tag{1.9.10}$$

The limit relation (1.9.8) suggests that the limit Hamiltonian H_t is a rather singular object. For example, if $H_{\mathrm{I}}^{(\lambda)}(t)$ were independent of λ and equal to an integrable function of t, then (1.9.8) would simply be a multiple of the δ-function in t characterizing the white noise correlations.(1) This is due to the following lemma that we shall constantly use.

Lemma 1.9.1. *For any integrable function F on $\mathbb{R}$, one has*

$$\lim_{\lambda \to 0} \frac{1}{\lambda^2} F\left(\frac{\tau - t}{\lambda^2} \right) = \delta(\tau - t) \int_{-\infty}^{+\infty} F(\sigma) d\sigma, \tag{1.9.11}$$

in the sense that, for any bounded continuous function φ on $\mathbb{R}$, one has

$$\lim_{\lambda \to 0} \int_{-\infty}^{+\infty} \frac{1}{\lambda^2} F\left(\frac{\tau - t}{\lambda^2} \right) \varphi(\tau) d\tau = \varphi(t) \int_{-\infty}^{+\infty} F(\sigma) d\sigma. \tag{1.9.12}$$

Proof 1.9.1. With the change of variable $(\tau - t)/\lambda^2 = \sigma$, the left-hand side of (1.9.12) becomes equal to

$$\lim_{\lambda \to 0} \int_{-\infty}^{+\infty} F(\sigma)\varphi(\lambda^2\sigma + t)d\sigma = \varphi(t) \int_{-\infty}^{+\infty} F(\sigma)d\sigma,$$

and the last equality follows from dominated convergence.

We shall see that the intuition described before Lemma 1.9.1 is essentially correct. More precisely, the rescaled Hamiltonian $(1/\lambda^2)H_{\mathrm{I}}^{(\lambda)}(t/\lambda^2)$ will be expressed as a functional of new operators,

$$\frac{1}{\lambda} a\left(\frac{t}{\lambda^2}, k\right), \tag{1.9.13}$$

called the *rescaled fields*, and the limit Hamiltonian (1.9.8) will be a functional of some quantum white noise whose specific form (covariance and commutation relations) depends on the form of the initial Hamiltonian $H_{\mathrm{I}}^{(\lambda)}(t)$. This gives the basic idea of the emergence of white noise in the stochastic limit. A precise formulation in the simplest case is given in Theorem 3.7.1, and further discussion of the rescaled fields is given in Sect. 3.5.

Complementary Material

1.10 From Interaction to Heisenberg Evolutions: Conditions on the State

The evolution operator U_t is the limit of evolution operators $U^{(\lambda)}_{t/\lambda^2}$ *in the interaction representation*, but the physical predictions of the dynamics of the original Hamiltonian model are expressed in terms of *the Heisenberg evolution*, which is given by (1.4.2). More precisely, we consider an observable X and its rescaled time evolution

$$X^{(\lambda)}(t/\lambda^2) = e^{i(t/\lambda^2)H^{(\lambda)}} X e^{-i(t/\lambda^2)H^{(\lambda)}}, \tag{1.10.1}$$

and we are interested in expectation values of the form

$$\langle \xi, X^{(\lambda)}(t/\lambda^2)\xi\rangle = \left\langle \xi, e^{i(t/\lambda^2)H^{(\lambda)}} X e^{-i(t/\lambda^2)H^{(\lambda)}} \xi \right\rangle . \tag{1.10.2}$$

But the right-hand side of (1.10.2) is equal to

$$\begin{aligned}&\left\langle \xi, e^{i(t/\lambda^2)H_0^{(\lambda)}} e^{-i(t/\lambda^2)H_0^{(\lambda)}} e^{i(t/\lambda^2)H^{(\lambda)}} X e^{-i(t/\lambda^2)H^{(\lambda)}} e^{i(t/\lambda^2)H_0^{(\lambda)}} e^{-i(t/\lambda^2)H_0^{(\lambda)}} \xi \right\rangle \\ &\qquad = \left\langle \xi, e^{i(t/\lambda^2)H_0^{(\lambda)}} U^{(\lambda)}_{-t/\lambda^2} X U^{(\lambda)*}_{-t/\lambda^2} e^{-i(t/\lambda^2)H_0^{(\lambda)}} \xi \right\rangle ,\end{aligned} \tag{1.10.3}$$

and the problem is that, while one can usually control the limit of the interaction term $U^{(\lambda)}_{-t/\lambda^2}$, the rescaled free evolution $\exp\left[-i(t/\lambda^2)H_0^{(\lambda)}\right]$ in general will have no limit as $\lambda \to 0$. If one restricts one's attention to expectation values of the form (1.10.2), where ξ *is an eigenvector of the free Hamiltonian*, then there is no problem because the right-hand side of (1.10.3) becomes equal to

$$\left\langle \xi, U^{(\lambda)}_{-t/\lambda^2} X U^{(\lambda)*}_{-t/\lambda^2} \xi \right\rangle . \tag{1.10.4}$$

So, apart from a change in the sign of time (see (1.4.2)), we are reduced to an expectation value in the interaction picture.

There are any many important examples where the limit (1.10.4) exists. The typical situation is when one is in the open *system* framework, described in Chap. 4, and

$$\xi = \xi_0 \otimes \Phi, \tag{1.10.5}$$

where ξ_0 is an eigenvector of the system Hamiltonian and Φ is a reservoir vector state invariant by the reservoir Hamiltonian (e.g. the Fock vacuum, an equilibrium state, etc.). This is the typical case of a discrete eigenvalue embedded in the continuous spectrum.

For several problems (for example, all problems in which only information on the master equation is required), vectors such as (1.10.5) give sufficient information, but in general the condition that ξ in (1.10.5) is an eigenvector of the free Hamiltonian is too restrictive and one has to face the following problem: *for which observables X does the limit* (1.10.1), *of their rescaled Heisenberg evolutions exist?* In the following section this question is answered for a large class of observables of physical interest.

1.11 From Interaction to Heisenberg Evolutions: Conditions on the Observable

Another case in which the Heisenberg evolution can be reduced to the interaction picture evolution is when the observable X in (1.10.1) commutes with the free Hamiltonian $H_0^{(\lambda)}$. In this case in fact, in the notation of (1.10.1),

$$X^{(\lambda)}(t) = e^{itH^{(\lambda)}} X e^{-itH^{(\lambda)}} = e^{itH^{(\lambda)}} e^{-itH_0^{(\lambda)}} X e^{itH_0^{(\lambda)}} e^{-itH^{(\lambda)}} = U_t^{(\lambda)*} X U_t^{(\lambda)}, \tag{1.11.1}$$

and we see that this case is, in some sense, the *dual* of the preceeding one. If X does not commute with the free Hamiltonian, but satisfies the condition

$$e^{-tH_0^{(\lambda)}} X e^{itH_0^{(\lambda)}} = e^{it\omega} X \tag{1.11.2}$$

for all times t and for some real number ω, then one can use the stochastic limit to estimate the behaviour of $U_{t/\lambda^2}^{(\lambda)*} X U_{t/\lambda^2}^{(\lambda)}$, and from the identity

$$e^{-it\omega/\lambda^2} U_{t/\lambda^2}^{(\lambda)*} X U_{t/\lambda^2}^{(\lambda)} = X^{(\lambda)}(t/\lambda^2)$$

and knowledge of the asymptotic behaviour of the left-hand side, practically all the quantities of physical interest can be estimated.

In fact we shall see in Sect. 1.22 that, in order to obtain the largest class of observables for which the limit

$$\lim_{\lambda\to 0} U_{t/\lambda^2}^{(\lambda)*} X U_{t/\lambda^2}^{(\lambda)} = \lim_{\lambda\to 0} X^{(\lambda)}(t/\lambda^2)$$

exists, we have to take into consideration operators $X = X^{(\lambda)}$ which depend on λ in a very special way.

1.12 Forward and Backward Langevin Equations

From the previous two sections we have seen that there are two important cases in which information on the interacting evolution can be translated immediately into information on the Heisenberg evolution:

(i) when one has information on invariant states of the free evolution (Sect. 1.10)
(ii) when one has information on *eigen operators* of the free evolution [see (1.11.2)].

It is clear that these pieces of information are dual to each other. More precisely, the considerations in Sect. 1.11 lead one to consider the Heisenberg evolution equation

$$X_{-t}^{(-)} := U_{-t}^{(\lambda)} X U_{-t}^{(\lambda)*}, \tag{1.12.1a}$$

which is the *time-reflected* form of the equation considered in Sect. 1.2, i.e.

$$X_t^{(+)} := U_t^{(\lambda)*} X U_t^{(\lambda)}. \tag{1.12.1b}$$

The map $t \mapsto X_t^{(+)}$ is called the *forward evolution* of the operator X; the map $t \mapsto X_t^{(-)}$ is the *backward evolution.* Note that the time-reflected evolution (1.12.1b) is $X_{-t}^{(-)}$. In what follows we shall analyze the asymptotic behaviour under the rescaling $t \mapsto t/\lambda^2$ of both types of evolutions.

The existence of the limit (1.9.9) and the fact that it satisfies equation (1.9.10) suggests that, for appropriately chosen families of observables $X^{(\lambda)}$, also the evolution (1.12.1b) is such that, for some observable X in the limit space and in the sense of convergence of the matrix elements (see Sect. 1.15),

$$\lim_{\lambda\to 0} U_{t/\lambda^2}^{(\lambda)*} X U_{t/\lambda^2}^{(\lambda)} = U_t^* X U_t =: j_t(X), \tag{1.12.2}$$

where U_t is the unitary operator defined by the limit (1.9.9). Moreover the family j_t, defined by (1.12.2), is a flow (in the sense of Sect. 1.26), and $j_t(X)$ satisfies the *forward* or *inner* white noise Langevin equation

$$\partial_t j_t(X) = j_t([iH_t, X]). \tag{1.12.3}$$

The flow dual to (1.12.1a,b) describes the evolution of states. So, if we represent a state by a density matrix $\rho^{(\lambda)}$, its evolution will be

$$\rho^{(\lambda)}(t) := U_t^{(\lambda)} \rho^{(\lambda)} U_t^{(\lambda*)}. \tag{1.12.4}$$

(The reason why it is essential to consider λ-dependent initial states is explained in Sect. 1.15.)

For many choices of $\rho^{(\lambda)}$ one can show that as $\lambda \to 0$ the limit $\rho(t)$ of (1.12.4) exists and satisfies the *backward* or *outer* white noise Langevin equation

$$\partial_t \rho(t) = i[H_t, \rho(t)]. \tag{1.12.5}$$

Remark 1.12.1. Apparently there is a perfect symmetry between the inner equation (1.12.3) (because the commutator is *inside* the flow) and the outer equation (1.12.5). However from the analytical point of view there can be a big difference. For example, in the system–reservoir scheme of Sect. 1.13, if the system operators do not depend on time and X is a system observable, then the commutator in (1.12.3) does not depend on time and is a system operator. But if we consider the dual (backward or outer) equation (1.12.5), this is never the case.

As a consequence, to prove the convergence (1.12.2) one can use the iterated series expansion of (1.12.3), but to prove the convergence of (1.12.4), as $\lambda \to 0$, one has to expand separately in series $U_t^{(\lambda)}$ and $U_t^{(\lambda)^*}$, and this multiple expansion causes the emergence of multiple-simplex integrals and the necessity to develop a technique to control them (the *multiple-simplex theorem* of Sect. 15.9).

1.13 The Open System Approach to Dissipation and Irreversibility: Master Equation

A standard scheme to describe dissipation passes through the so-called *open system* (or *system-reservoir*) approach, the basic physical idea of which is that dissipation and irreversible behaviour arise when two systems, traditionally denoted S (*system*) and R (*reservoir*), interact and one of them exhibits some *macroscopic* or *chaotic* features.

Typically, in the open system approach, S is a spatially confined system located in a fixed region of space (in particular, its free Hamiltonian has a discrete spectrum). One usually takes R to be infinitely extended (in particular, its time evolution has a continuous spectrum with rapidly decaying, e.g. integrable, correlations) in order to eliminate Poincaré recurrences and to obtain a well-defined asymptotic behaviour in the limit as $t \to \infty$. In other words, the reservoir is physically distinguished from the system not only because it is infinitely extended (in contrast to the spatial confinement of S), but also because *it is more chaotic* in the sense described above; one expects that, by the effect of the interaction, and in some limiting situations, energy will flow from the system to the reservoir and never come back.

The system R can model several different physical systems and, according to the interpretation, is called *reservoir*, *gas*, *heat bath*, *field*, *background radiation*, *environment*, *measurement apparatus*, *noise*, *etc.*, and is usually modelled either by a classical stochastic process or by a second quantized quantum system. Typical examples of system–reservoir pairs are the following: atom–EM field, electron–phonon field, signal–receiver, particle–gas and Fermi field (in a finite volume)–Bose field.

In some sense the open system scheme can be considered not as a particular case of the general scheme of the quantum dynamical system described in Sect. 1.6, but as a *generalization of it* which represents the specific form of the interaction. In fact it is sufficient to take the system S to be trivial (i.e. its state space is the complex numbers) to recover the generic quantum system.

The system–reservoir models can be described as random systems in a random environment. At the moment there are two main approaches to model a random environment:

(i) Try and guess a plausible phenomenological description of the environment, usually in terms of a classical or quantum stochastic process.
(ii) Write an exact Hamiltonian model, based on the fundamental laws of quantum mechanics and deduce the equations for the random environment as a suitable limit of the basic Hamiltonian equations.

The models of quantum measurement recently discussed by several authors belong to category (i). The stochastic limit approach belongs to category (ii). The advantage of the latter approach over the former lies in the fact that the phenomenological parameters, describing the environment after the limit, are explicitly expressed in terms of the original Hamiltonian model and therefore they can be measured or controlled.

For system–reservoir models, between the Hamiltonian and the stochastic limit description there is an intermediate level called *the master equation* which can be described as follows: At time 0 the two systems are uncoupled; hence their initial state is a product state

$$\varphi_S \otimes \varphi_R = \mathrm{Tr}_{S,R}(W_S \otimes W_R \,\cdot) = \langle \,\cdot\, \rangle_S \otimes \langle \,\cdot\, \rangle_R \tag{1.13.1}$$

where $\varphi_S = \langle \,\cdot\, \rangle_S$ is an arbitrary state of the system$^{(1)}$ and $\varphi_R = \langle \,\cdot\, \rangle_R$ is a fixed reference state of the reservoir. In all the examples considered up to now, φ_R is a Gaussian (quasi-free) boson or fermion state.

The composite system evolves according to the usual Hamiltonian dynamics in the interaction representation, and due to the interaction, if X is an observable of the system S, then its time evolved in the Heisenberg picture, $X^{(\lambda)}(t)$, will depend also on the degrees of freedom of the reservoir according to

$$\begin{aligned} X^{(\lambda)}(t) &= e^{iH^{(\lambda)}t/\lambda^2} \cdot e^{-iH_0t/\lambda^2}(X \otimes 1)\, e^{iH_0t/\lambda^2} \cdot e^{-iH^{(\lambda)}t/\lambda^2} \\ &= U^{(\lambda)+}_{t/\lambda^2}(X \otimes 1) U^{(\lambda)}_{t/\lambda^2}, \end{aligned} \tag{1.13.2}$$

where

$$U^{(\lambda)}_{t/\lambda^2} = e^{iH_0t/\lambda^2} \cdot e^{-iH^{(\lambda)}t/\lambda^2} \tag{1.13.3}$$

is the *wave operator at time* t. If the reservoir degrees of freedom are averaged out, one obtains a new observable of the system S defined by

$$\overline{\varphi}_R \left(X^{(\lambda)}(r) \right) =: \overline{X}^{(\lambda)}(r) \tag{1.13.4}$$

where $\overline{\varphi}_R$ denotes the *partial expectation* with respect to the reservoir degrees of freedom, i.e. it is the map from the observables of the composite (S, R) system to the observables of the S system, uniquely defined by continuous linear extension of the map

$$\overline{\varphi}_R(X_S \otimes X_R) := X_S \varphi_R(X_R) = X_S \langle X_R \rangle_R \tag{1.13.5}$$

(X_S: an S-observable, X_R: an R-observable).

The map $X \to \overline{X}^{(\lambda)}(t)$ is called the *reduced dynamics*. The observables of the form $\overline{X}^{(\lambda)}(t)$ no longer obey a reversible evolution, and in fact their evolution is described by complicated (integro-differential) equations which give little insight, because the details of the evolution of the reservoir are hardly controllable.

One is therefore led to study the evolution $\overline{X}^{(\lambda)}(t)$ in some limiting situation, in the hope that the limiting procedure will sweep away the inessential details and produce more controllable equations in which the information on the reservoir is reduced to a few macroscopic parameters (such as temperature, relaxation times, leading frequencies and coupling constants between system and reservoir).

Many such limiting procedures have been considered in the physical literature, especially in quantum optics and in the theory of damped quantum systems (weak coupling, singular coupling, low density, hydrodynamical limit, etc.), and they describe different physical situations. In this section we shall discuss the weak coupling limit approach to the master equation (in the low-density case, also the reservoir depends on λ).

More concretely one considers the limit

$$\lim_{\lambda \to 0} (\varphi_S \otimes \varphi_R) \left(X^{(\lambda)}(t) \right) = \lim_{\lambda \to 0} \left\langle X^{(\lambda)}(t) \right\rangle_R \tag{1.13.6}$$

for arbitrary states φ_S, while the reservoir state φ_R is kept fixed;[(2)] then one shows that, under suitable assumptions on the free and interaction Hamiltonians, there exists a *quantum Markovian* (or *dynamical*) semigroup such that as $\lambda \to 0$ the limit (1.13.6) exists and is equal to

$$\varphi_S \left(P^t(X) \right) . \tag{1.13.7}$$

Moreover, the infinitesimal generator G, of P^t, is given by

$$G(X) = K^+ X + XK + \sum_{j=1}^{n} D_j^+ X D_j , \tag{1.13.8}$$

with D_j, K being system operators satisfying

$$K^+ + K + \sum_{j=1}^{n} D_j^+ D_j = 0; \tag{1.13.9}$$

therefore the reduced evolution of the observable

$$\bar{X}(t) := P^t(X)$$

is described by

$$\frac{d}{dt}\bar{X}(t) = G\left(\bar{X}(t)\right) = K^+\bar{X}(t) + \bar{X}(t)K + \sum_{j=1}^{n} D_j^+\bar{X}(t)D_j \tag{1.13.10}$$

The differential equation (1.13.10), satisfied by the reduced evolution, is called *a master equation* (see Sect. 4.14).

1.14 Classical Processes Driving Quantum Phenomena

Several quantum phenomena, from neutron fusion, to multiphoton production in cascades originated by cosmic rays, to photon statistics of the electro–magnetic field, have been historically described first by classical stochastic processes, and only much later a fully quantum treatment has been developed. That these classical stochastic descriptions of quantum phenomena are good approximations is shown by the fact that these classical (or semiclassical) approximations are still now in use in several applications of quantum theory to technology.

One of the basic advantages of the stochastic limit approach with respect to the previous derivations of master equations (Markovian approximation) is that it provides a dynamical explanation of the mechanism through which these classical processes can drive quantum phenomena. This too is a universal phenomenon (see Sects. 4.27 and 5.18).

1.15 The Basic Steps of the Stochastic Limit

In this section we summarize the basic qualitative steps of the stochastic limit, the nature of which is common to almost all the models considered up to now and which can be described by the following scheme, where the arrow $\to$ means convergence:

Field	→ Master field (quantum noise)
Effective free field evolution	→ Time shift in the noise space
Field Hilbert space	→ Noise (master) space
Evolution operator $U_t^{(\lambda)}$	→ Limit evolution operator U_t
Schrödinger eq. in interaction representation	→ White noise Schrödinger eq.
$\partial_t V_t^\lambda = -iH_{\mathrm{I}}(t)V_t^\lambda$	→ $\partial_t U_t = -iH(t)U_t$
or equivalently	or Quantum stochastic diff. eq.
$\partial_t V_t^\lambda = -iH_{\mathrm{I}}(t)V_t^\lambda$	→ $dU_t = F_\alpha(t)dM_t^\alpha$
Heisenberg evolution	
$U_t^{(\lambda)} \cdot u_t^0(a) \cdot U_t^{(\lambda)^*} =: u_t^\lambda(a)$	→ Stochastic flow: $u_t(a) = j_t(\tau_t(a))$
Heisenberg equation	→ Quantum Langevin equation
Reduced Heisenberg equation	→ Master equation

After having determined the structure of the master field and its algebra (for strongly nonlinear interactions new commutation relations appear) the main steps to be done are the following:

(i) To establish the convergence of the rescaled evolution operator as $\lambda \to 0$,

$$U_{t/\lambda^2}^{(\lambda)} \to U_t. \tag{1.15.1}$$

The limit (1.15.1) does not exist in any of the usual operator topologies; alternatively if it exists, it is trivial. In the stochastic limit one considers the limit of matrix elements of U_{t/λ^2} with respect to vectors (collective vectors, see Chap. 16) obtained by applying to the cyclic vector Φ_λ some functionals (typically polynomials or exponentials) of the rescaled fields (1.15.1). Denoting Φ'_λ, Φ''_λ two such vectors, one proves that as $\lambda \to 0$ the matrix elements of the time-rescaled evolution operator

$$\left\langle \Phi'_\lambda, U_{t/\lambda^2}^{(\lambda)} \Phi''_\lambda \right\rangle \tag{1.15.2}$$

converge to the corresponding matrix elements

$$\langle \Psi', U_t \Psi'' \rangle, \tag{1.15.3}$$

where U_t is a unitary operator in the master Hilbert space and the two vectors Ψ', Ψ'' are uniquely determined by the vectors $\Phi'_\lambda, \Phi''_\lambda$. The *convergence* of a family of operators X_λ, in the sense specified above, will be called *convergence in the sense of matrix elements.*

(ii) The last step is to establish the limit

$$U_{t/\lambda^2}^{(\lambda)*} X_\lambda U_{t/\lambda^2}^{(\lambda)} \to U_t^* X U_t, \tag{1.15.4}$$

i.e. one shows that if a family of observables X_λ converges in the sense of matrix elements in the collective vectors as $\lambda \to 0$ to an observable X on the master Hilbert space then the limit

$$\left\langle \Phi'_\lambda, U^{(\lambda)*}_{t/\lambda^2} X_\lambda U^{(\lambda)}_{t/\lambda^2} \Phi''_\lambda \right\rangle \tag{1.15.5}$$

exists and is equal to

$$\langle \Psi', U_t^* X U_t \Psi'' \rangle \tag{1.15.6}$$

where $\Phi'_\lambda, \Phi''_\lambda, \Psi', \Psi''$ are as in (ii). The limit (1.15.6) is more fundamental than the limit (1.15.3), and it can be shown to exist in cases in which (1.15.3) does not: a typical example is the *stochastic Ising model* in infinite volume (see Sect. 5.19).

(iii) The basic goal of the theory is to prove that as $\lambda \to 0$ the one parameter families of operators

$$U_t, \quad X(t) := U_t^* X U_t, \tag{1.15.7}$$

satisfy white noise (or equivalently quantum stochastic differential) equations whose explicit form is uniquely determined in terms of the initial model.

The advantage of this procedure is that the main physical properties are more easily readable from these equations rather than from the original ones. The equation satisfied by U_t is the stochastic generalization of the Schrödinger equation in interaction representation (1.10.1) and the equation satisfied by $X(t) = U_t^* X U_t$ is the generalization of the Heisenberg equation in interaction representation (1.4.9) and is called *the quantum Langevin equation.*

(iv) The connection with the original van Hove approach leading to the quantum master equation is easily obtained. It corresponds to (1.15.5) in the case of a discrete system interacting with a continuous field (or reservoir), and X is an observable of the system alone. Moreover the limit is taken with respect to a single vector of the field,

$$\Phi'_\lambda = \Phi''_\lambda = \Phi, \tag{1.15.8}$$

namely the vacuum vector Φ. In this case the limit expectation value (1.15.6) is not a scalar, because one is averaging only on the field degrees of freedom and not on the system ones; instead, it is an operator of the system

$$\langle \Psi, U_t^* X U_t \Psi \rangle = P^t(X) = \bar{X}(t), \tag{1.15.9}$$

where Ψ is the vacuum vector of the master field. The quantum master equation is simply the equation satisfied by $\bar{X}(t)$, i.e.

$$\frac{d}{dt}\bar{X}(t) = L\left(\bar{X}(t)\right), \tag{1.15.10}$$

where L is the generator of the semigroup P^t. The fact that P^t is a semigroup and the explicit form of the master equation follow from the Langevin equation by a standard technique, the quantum Feynman–Kac formula of [Ac78].

(v) With few exceptions (cf. the functional integral approach to the stochastic limit, discussed in Chap. 9) the limit, as $\lambda \to 0$, of $U^{(\lambda)}_{t/\lambda^2}$ is investigated term by term, and one gives a definite meaning to these terms by introducing some cut–offs (form factors).

(vi) The main analytical difficulties to be overcome are:
- to determine the explicit form of the limit of the single terms;
- to prove a uniform estimate in order to exchange the limit $\lim_{\lambda\to 0}$ and the sum $\sum_{n=0}^{\infty}$ of the iterated series (such an estimate is not always available, and in these cases our results should be interpreted as term-by-term results);
- to re-sum all the contributions of the limit series, and to prove that this sum satisfies the desired stochastic equation; and
- to remove the cutoff.

These analytical questions are dealt with in Part III.

In the following sections the main results are formulated, and the main ideas of their proofs as well as of some applications of the final formulae are outlined.

1.16 Connection with the Central Limit Theorems

The mechanism through which the quantum white noise arises in the stochastic limit is a quantum generalization of the *central limit theorem* of classical probability theory (see, for example, [PrRo67]). According to this theorem, if (A_j) is a sequence of classical, independent, identically distributed real-valued random variables, with finite variance, then the sequence

$$B_N := \frac{1}{\sqrt{N}} \sum_{j=1}^{N} A_j \tag{1.16.1}$$

converges in the sense of correlators (in fact, in a stronger sense, in the classical case) to a Gaussian random variable B with the same variance. The *functional central limit theorem* (sometimes also called invariance principle) generalizes the central limit theorem by considering the sums, depending on the real parameter t,

$$B_N(t) := \frac{1}{\sqrt{N}} \sum_{j=1}^{[Nt]} A_j, \tag{1.16.2}$$

where $[Nt]$ denotes the integer part of Nt. For each fixed N, $B_N(t)$ is a stochastic process with index $t \in \mathbb{R}$, and one studies the convergence of this sequence of processes in the sense of correlators (or in stronger topologies). The result is that the limiting family (B_t) is a classical Brownian motion.

The assumptions of independence and of stationarity can also be weakened. The same result remains true if the sums in (1.16.2) are replaced by integrals according to the rule

$$\frac{1}{\sqrt{N}} \sum_{j=1}^{[Nt]} A_j \to \lambda \int_0^{t/\lambda^2} A_s ds, \tag{1.16.3a}$$

$$\frac{1}{\sqrt{N}} =: \lambda, \quad \frac{1}{\lambda^2} = N. \tag{1.16.3b}$$

In addition, in the continuous case, even if A_t are not independent, the functional central limit theorem still holds (both in the classical and the quantum case) if they are Gaussian with rapidly decaying (e.g. integrable) correlations. In fact the central limit theorem for Gaussian random variables is very simple, because the only thing one has to prove is the convergence of the two-point function.

Now consider the iterated series for the evolution operator $U^{(\lambda)}_{t/\lambda^2}$ of the rescaled Schrödinger equation (1.9.7), i.e.

$$U^{(\lambda)}_{t/\lambda^2} = 1 + \sum_{n=1}^{\infty} (-i)^n \int_0^{t/\lambda^2} dt_1 \dots \int_0^{t_{n-1}} dt_n H^{(\lambda)}_{\mathrm{I}}(t_1) \dots H^{(\lambda)}_{\mathrm{I}}(t_n), \tag{1.16.4}$$

in the case (weak coupling) in which $H^{(\lambda)}_{\mathrm{I}}(t)$ has the form

$$H^{(\lambda)}_{\mathrm{I}}(t) = \lambda H_{\mathrm{I}}(t),$$

then the iterated series (1.16.4) becomes

$$U^{(\lambda)}_{t/\lambda^2} = 1 + \sum_{n=1}^{\infty} (-i)^n \lambda^n \int_0^{t/\lambda^2} dt_1 \dots \int_0^{t_{n-1}} dt_n H_{\mathrm{I}}(t_1) \dots H_{\mathrm{I}}(t_n); \tag{1.16.5}$$

its first-order term is precisely the integral term in (1.16.3a) (multiplied by $-i$) after the replacement $A_t = H_{\mathrm{I}}(t)$. Note, moreover, that if $H_{\mathrm{I}}(t)$ were a classical process, i.e. if the variables $H_{\mathrm{I}}(t)$ would commute for different t (this is unrealistic in quantum theory, but it includes the whole of classical probability), then the integrand in (1.16.5) would be a symmetric function of $t_1 \dots t_n$, and therefore (1.16.5) would become

$$\begin{aligned} U^{(\lambda)}_{t/\lambda^2} &= 1 + \sum_{n=1}^{\infty} (-i)^n \lambda^n \frac{1}{n!} \left(\int_0^{t/\lambda^2} dt_1 H_{\mathrm{I}}(t_1) \right)^n \\ &= \exp\left(-i\lambda \int_0^{t/\lambda^2} dt_1 H_{\mathrm{I}}(t_1) \right), \end{aligned} \tag{1.16.6}$$

its expectation is nothing but the characteristic function of the process

$$W_\lambda(t) := \lambda \int_0^{t/\lambda^2} dt_1 H_{\rm I}(t_1). \tag{1.16.7}$$

This process also has a quantum mechanical meaning (i.e. even when $H_{\rm I}(t)$ do not commute for different t); in fact in this case the identity (1.16.6) still holds, provided the exponential is interpreted as a time-ordered exponential. In this case the quantity (1.16.7) it is just the first-order (in λ) term of the series (1.16.5).

Note, moreover, that in classical probability the convergence as $\lambda \to 0$ of the characteristic function is a much stronger property than the convergence of correlators, since it uniquely determines the probability distribution of the process.

Comparing (1.16.7) with (1.16.3a) we conclude that: *the problem with finding the limit as* $\lambda \to 0$ *of the rescaled evolution operator* (1.16.4), *is the quantum generalization of the classical functional central limit theorem.* Since, even in the simplest quantum models, the random variables $H_{\rm I}(t)$ do not commute, the methods of classical probability are not sufficient to solve this problem, and one has to develop new tools.

1.17 Classical Systems

The use of the Heisenberg equation (1.4.6) is convenient because it allows one to deal, within a unified formalism, with classical and quantum Hamiltonian systems.

The observables of a classical Hamiltonian system are identified to (measurable) functions on its phase space. If the phase space is the usual Euclidean space $\mathbb{R}^{2n}$ (but for a general smooth manifold all what follows remains true if formulated in a slightly more elaborated language), and if h denotes the Hamiltonian of the system, the flow

$$T_t : x = (q,p) \in \mathbb{R}^{2n} \mapsto T_t x = (q_t, p_t) \in \mathbb{R}^{2n},$$

which, to the generic point (q,p) in the phase space $\mathbb{R}^{2n}$, associates the solution at time t of the Hamilton equations of motion

$$\dot{q} = \partial_p h, \quad \dot{p} = -\partial_q h$$

with initial condition equal to (q,p), preserves the Lebesgue measure and therefore induces a unitary operator V_t on $L^2(\mathbb{R}^{2n})$, defined by

$$V_t f(x) := f(T_t x), \quad x \in \mathbb{R}^{2n}. \tag{1.17.1}$$

The identity

$$\partial_t f(q_t, p_t) = \partial_q f \cdot \dot{q} + \partial_p f \cdot \dot{p} = \partial_q f \partial_p h - \partial_p f \partial_q h \qquad (1.17.2)$$

shows that the generator L of this unitary evolution, the *Liouvillian*, is given by the Poisson brackets with the Hamiltonian:

$$\partial_p h \partial_q f - \partial_q h \partial_p f =: L(f). \qquad (1.17.3)$$

On the other hand the unitarity of V_t, defined by (1.17.1), implies that L has the form $L = iH$ for some self-adjoint operator H. Therefore

$$L = iH = i\left(\partial_p h \frac{1}{i}\partial_q - \partial_q h \frac{1}{i}\partial_p\right). \qquad (1.17.4)$$

For example, for free Hamiltonians of the form $h_0 = p^2/2m$, one has $\partial_p h_0 = p/m$, $\partial_q h_0 = 0$, and therefore

$$H = \frac{p}{m}\frac{1}{i}\partial_q.$$

While for Hamiltonians of the form $h_0 + V$, where $V = V(q)$ is a function of position, one has

$$H = \frac{p}{m}\frac{1}{i}\partial_q + (\partial_q V)\frac{1}{i}\partial_p.$$

Using the above remark, all the techniques developed for quantum theory can be automatically applied to classical Hamiltonian systems.[(1)]

1.18 The Backward Transport Coefficient and the Arrow of Time

Given an operator U_t which satisfies (1.3.10), its inverse operator, U_t^*, satisfies

$$\partial_s U_s^{(\lambda)*} = i\lambda U_s^{(\lambda)*} H_{\mathrm{I}}(s), \quad U_0^{(\lambda)*} = 1, \qquad (1.18.1)$$

and the solution is considered to be in the interval $[t, 0]$ (meaning that t is negative and that the initial condition is at time 0). After the rescaling $t \to t/\lambda^2$, the equation satisfied by $U_{t/\lambda^2}^{(\lambda)*}$ becomes, in integral form,

$$U_{t/\lambda^2}^{(\lambda)*} = 1 - i\int_t^0 dt_1 U_{t_1/\lambda^2}^{(\lambda)*}\frac{1}{\lambda}H_{\mathrm{I}}(t_1/\lambda^2).$$

Therefore the second-order term of the series is

$$-\frac{1}{\lambda^2}\int_t^0 dt_1 \int_{t_1}^0 dt_2 H_{\mathrm{I}}(t_2/\lambda^2) H_{\mathrm{I}}(t_1/\lambda^2).$$

Taking the expectation value with respect to a time-stationary state, we now find

$$-\frac{1}{\lambda^2}\int_t^0 dt_1 \int_{t_1}^0 dt_2 \left\langle H_{\rm I}\left(\frac{t_2-t_1}{\lambda^2}\right) H_{\rm I}(0)\right\rangle = -\int_t^0 dt_1 \int_0^{-t_1/\lambda^2} d\tau \langle H_{\rm I}(\tau) H_{\rm I}(0)\rangle,$$

which converges (making $t = 1$) to

$$\int_0^{-\infty} d\tau \langle H_{\rm I}(\tau) H_{\rm I}(0)\rangle = -\int_{-\infty}^0 \langle H_{\rm I}(\tau) H_{\rm I}(0)\rangle d\tau. \tag{1.18.2}$$

Comparing (1.18.1) with (1.8.7), we conclude that the *quantum transport coefficient backward in time is the adjoint of the quantum transport coefficient forward in time.*

From the quantum mechanical fluctuation–dissipation relation (see Sect. 4.16), we know that the real part of the transport coefficient, which is related to damping, is negative. *This is not affected by the direction of time. However, the imaginary part, which is related to energy shift, is sensitive to the direction.*

1.19 The Master and Fokker–Planck Equations: Projection Techniques

In the following few sections we discuss how from the stochastic limit one can easily deduce the standard derivations of the master and Fokker–Planck equations.

In the notations of Sect. 1.9, both $U_t^{(\lambda)}$ and U_t ($j_t^{(\lambda)}$ and j_t) are solutions of Schrödinger equations (flow equations) in interaction representation either before or after the stochastic limit.

Suppose that in our quantum system one can distinguish two interacting parts, S and R corresponding respectively to the *slow* and *fast* degrees of freedom and that, correspondingly, the state space factorizes as the tensor product of two spaces: $\mathcal{H} = \mathcal{H}_S \otimes \mathcal{H}_R$. In the terminology of the quantum theory of open systems we can call *system* the slow degrees of freedom and *reservoir* the fast ones. Let the full Hamiltonian H_λ be given by

$$H_\lambda = H_S \otimes 1 + 1 \otimes H_R + \lambda H_{\rm I}.$$

One is interested in the long-term ($t \to \infty$) weak coupling ($\lambda \to 0$) behaviour of the Heisenberg evolution of an arbitrary system observable $b_S \in \mathcal{B}(\mathcal{H}_S)$, i.e.

$$b_S(t) := e^{iH_\lambda t}(b_S \otimes 1)e^{-iH_\lambda t}.$$

In the notations of Sect. 1.13 let $\xi_S \in \mathcal{H}_S$ be an arbitrary vector state of the system and consider the expectation value of

$$U_t^* b_S U_t =: j_t(b_S)$$

with respect to the state $\xi_S \otimes \Phi$, i.e.

$$\langle \xi_S \otimes \Phi, j_t(b_S)\xi_S \otimes \Phi\rangle. \tag{1.19.1}$$

Denoting ρ_S the density matrix $|\xi_S\rangle\langle\xi_S|$ and using the identity $\mathrm{Tr}(ab) = \mathrm{Tr}(ba)$, we can write the expectation value (1.19.1) as a trace over the $S+R$ system:

$$\mathrm{Tr}_{S,R}\{\rho_S \otimes |\Phi\rangle\langle\Phi|\; j_t(b_S)\} = \mathrm{Tr}_{S,R}\{j_t^*(\rho_S \otimes |\Phi\rangle\langle\Phi|)\; b_S\}, \tag{1.19.2}$$

where

$$j_t^*(z) := U_t z U_t^*. \tag{1.19.3}$$

By taking the partial trace over the reservoir space on the left-hand side of (1.19.2), we obtain

$$\mathrm{Tr}_S\{\rho_S P^t(b_S)\}, \tag{1.19.4}$$

where $P^t(b_S) =: b(t)$ is the partial expectation value

$$P^t(X) = \langle \Phi, j_t(X)\Phi\rangle = \langle \Phi, U_t^*(X \otimes 1)U_t\Phi\rangle. \tag{1.19.5}$$

Similarly, by taking the partial trace over the reservoir space on the right-hand side of (1.19.2), we obtain

$$\mathrm{Tr}_S\{P_*^t(\rho_S)X\}, \tag{1.19.6}$$

where $P_*^t(\rho_S)$ denotes the *effective density matrix* on the atomic system, i.e. P_*^t is the dual action of P^t on the density operators of S. More explicitly

$$\rho_S(t) = P_*^t(\rho_S) = \langle \Phi, j_t(\rho_S \otimes |\Phi\rangle\langle\Phi|)\Phi\rangle = \langle \Phi, U_t(\rho_S \otimes |\Phi\rangle\langle\Phi|)U_t^*\Phi\rangle. \tag{1.19.7}$$

As specified at the beginning, these manipulations can be done either before or after the stochastic limit. The basic difference between the two cases is that *before the stochastic limit the reduced evolution P^t is not a semigroup*. After the stochastic limit it is a semigroup, i.e. $P^t P^s = P^{t+s}$. If the reduced evolution is given by (1.19.5), by differentiating we obtain the *master equations*:

$$\frac{d}{dt}P^t(X) = L(P^t(X)), \tag{1.19.8}$$

where L is the infinitesimal generator whose typical form is given by (1.13.8) (a more general form is discussed in Sect. 4.14). By differentiating the dual reduced evolution given by (1.19.7), we obtain the *Fokker–Planck equation*:

$$\frac{d}{dt}P_*^t(\rho_S) = L^*\left(P_*^t(\rho_S)\right), \tag{1.19.9}$$

which is the dual (with respect to the trace) of the master equation.

1.20 The Master Equation in Open Systems: an Heuristic Derivation

In order to illustrate the difference between the master and the Fokker–Planck equations obtained before the stochastic limit and those obtained after, in this section we give a quick heuristic derivation of the master equation based on the *ansatz* that the *self-averaging property* (given by (1.20.8) below) is satisfied. The stochastic limit proves that this argument leads to the correct result. As a matter of fact the stochastic limit goes far beyond the master equation, because it describes the dynamics of the full composite system, not those only of the subsystem S.

Starting from the same Hamiltonian as in the previous section, we consider the Heisenberg equation:

$$\partial_t b(t) = \frac{i}{\hbar}\left[H_{\mathrm{I}}(t), b(t)\right] \tag{1.20.1}$$

under the initial condition

$$b(0) = b_S \otimes 1, \tag{1.20.2}$$

where b_S is a system observable. In integral form and after one iteration (1.20.1) becomes

$$\begin{aligned} b(t) &= b(0) + \frac{i}{\hbar}\,\lambda \int_0^t dt_1 [H_{\mathrm{I}}(t_1), b(0)] \\ &\quad - \frac{1}{\hbar^2}\,\lambda^2 \int_0^t dt_1 \int_0^{t_1} dt_2 \big[H_{\mathrm{I}}(t_1), [H_{\mathrm{I}}(t_2), b(t_2)]\big]. \end{aligned} \tag{1.20.3}$$

Introducing the time rescaling $t \mapsto t/\lambda^2$ and the change of variables $t_1 \mapsto \lambda^2 t_1$, $t_2 \mapsto \lambda^2 t_2$, we obtain

$$\begin{aligned} b(t/\lambda^2) &= b(0) + \frac{i}{\hbar} \int_0^t dt_1 \left[\frac{1}{\lambda} H_{\mathrm{I}}(t_1/\lambda^2), b(0)\right] \\ &\quad - \frac{1}{\hbar^2} \int_0^t dt_1 \int_0^{t_1} dt_2 \left[\frac{1}{\lambda} H_{\mathrm{I}}(t_1/\lambda^2), \left[\frac{1}{\lambda} H_{\mathrm{I}}(t_2/\lambda^2), b(t_2/\lambda^2)\right]\right]. \end{aligned} \tag{1.20.4}$$

We assume that the system and the reservoir are initially independent, i.e. the initial density matrix of the composite system has the form

$$\rho(0) = \rho_S \otimes \rho_R, \tag{1.20.5}$$

where $\rho_S \otimes \rho_R$ is invariant for the free evolution. Denoting $\langle\cdot\rangle_R$ the partial averaging with respect to the reservoir degrees of freedom, we suppose that the interaction Hamiltonian satisfies the mean zero condition:

$$\langle H_{\mathrm{I}}(t)\rangle_R = 0. \tag{1.20.6}$$

Taking $\langle \cdot \rangle_R$-expectation of both sides of (1.20.4), we find

$$\begin{aligned}&\langle b(t/\lambda^2)\rangle_R \\ &= b_S - \frac{1}{\hbar^2}\int_0^t dt_1 \int_0^{t_1} dt_2 \left\langle \left[\frac{1}{\lambda}H_{\mathrm{I}}(t_1/\lambda^2), \left[\frac{1}{\lambda}H_{\mathrm{I}}(t_2/\lambda^2), b(t_2/\lambda^2)\right]\right]\right\rangle_R .\end{aligned} \tag{1.20.7}$$

Now suppose that the following *self-averaging property* holds:

$$\begin{aligned}&\left\langle \left[\frac{1}{\lambda}H_{\mathrm{I}}(t_1/\lambda^2), \left[\frac{1}{\lambda}H_{\mathrm{I}}(t_2/\lambda^2), b(t_2/\lambda^2)\right]\right]\right\rangle_R \\ &= \left\langle \left[\frac{1}{\lambda}H_{\mathrm{I}}(t_1/\lambda^2), \left[\frac{1}{\lambda}H_{\mathrm{I}}(t_2/\lambda^2), \langle b(t_2/\lambda^2)\rangle_R\right]\right]\right\rangle_R ,\end{aligned} \tag{1.20.8}$$

where we identify $\langle b(t_2/\lambda^2)\rangle_R$ with $\langle b(t_2/\lambda^2)\rangle_R \otimes 1_R$. Then, because of the invariance of the expectation $\langle\cdot\rangle_R$ under the free evolution, (1.20.7) becomes

$$\begin{aligned}&\langle b(t/\lambda^2)\rangle_R \\ &= b_S - \frac{1}{\hbar^2}\int_0^t dt_1 \int_0^{t_1} d\tau_2 \left\langle \left[H_{\mathrm{I}}(0), \left[\frac{1}{\lambda^2}H_{\mathrm{I}}\left(\frac{t_2-t_1}{\lambda^2}\right), \langle b(t_2/\lambda^2)\rangle_R\right]\right]\right\rangle_R\end{aligned} \tag{1.20.9}$$

We see that, even assuming the self-averaging property, (1.20.9) is a complicated integro-differential equation. Let us show that as $\lambda \to 0$ this converges to a simple differential equation. Supposing that as $\lambda \to 0$ one has it that $\langle b(t/\lambda^2)\rangle_R$ converges to some operator $b_S(t)$ on the system space, then (1.20.9) will be well approximated by

$$b_S(t) = b_S - \frac{1}{\hbar^2}\int_0^t dt_1 \int_0^{t_1} dt_2 \left\langle \left[H_{\mathrm{I}}(0), \left[\frac{1}{\lambda^2}H_{\mathrm{I}}\left(\frac{t_2-t_1}{\lambda^2}\right), b_S(t_2)\right]\right]\right\rangle_R$$

and, with the same change of variables as in Lemma 1.9.1, this becomes

$$b_S(t) = b_S - \frac{1}{\hbar^2}\int_0^t dt_1 \int_{-t_1/\lambda^2}^0 d\tau \left\langle\left[H_{\mathrm{I}}(0), \left[H_{\mathrm{I}}(\tau), b_S(\lambda^2\tau + t_1)\right]\right]\right\rangle_R ,$$

which as $\lambda \to 0$ converges to (with $\hbar = 1$)

$$b_S(t) = b_S - \int_0^t dt_1 \int_{-\infty}^0 d\tau \left\langle\left[H_{\mathrm{I}}(0), \left[H_{\mathrm{I}}(\tau), b_S(t_1)\right]\right]\right\rangle_R$$

or, in differential form,

$$\partial_t b_S(t) = -\int_{-\infty}^0 d\tau \left\langle\left[H_{\mathrm{I}}(0), \left[H_{\mathrm{I}}(\tau), b_S(t)\right]\right]\right\rangle_R = L\big(b_S(t)\big),\ b_S(0) = b_S. \tag{1.20.10}$$

It can be proved (see Sect. 1.21 for an heuristic argument and Sect. 4.14 for a precise statement) that the operator L, on the right-hand side of (1.20.10) is the generator of a Markov semigroup. The solution of equation (1.20.10) depends linearly on b_S, so we can write

$$b_S(t) = P^t(b_S) = e^{tL} b_S(t) = \langle b(t) \rangle_R. \tag{1.20.11}$$

Therefore the 1-parameter family P^t defined by (1.20.11) is a quantum Markov semigroup whose generator L is given by

$$L(\cdot) = -\int_{-\infty}^{0} d\tau \left\langle \left[H_{\mathrm{I}}(0), [H_{\mathrm{I}}(\tau), \cdot\,] \right] \right\rangle_R. \tag{1.20.12}$$

The dual evolution is defined by

$$\mathrm{Tr}_S(\rho_S P^t(b_S)) = \mathrm{Tr}_S(P^t_*(\rho_S) b_S) = \mathrm{Tr}_S(\rho_S(t) b_S),$$

where Tr_S denotes the trace over the system and satisfies the master equation, i.e. the equation generated by the dual of L, denoted L^* and defined by

$$\mathrm{Tr}_S\big(\rho_S L(b_S)\big) = \mathrm{Tr}_S\big(L^*(\rho_S) b_S\big).$$

The explicit form of L^* is easy calculated using the identity

$$\langle X \rangle_R = \mathrm{Tr}(\rho_R X),$$

and this leads to *the master equation*:

$$\rho_S(t) = \rho_S(0) - \int_0^t dt_1 \int_{-\infty}^{0} d\tau \left\langle \left[H_{\mathrm{I}}(\tau), [H_{\mathrm{I}}(0), \rho_S(t_1) \otimes \rho_R] \right] \right\rangle_R.$$

Thus, given ρ_R and the interaction Hamiltonian, the explicit form of the master equation can be calculated. More on the master equation can be found in Sects. 1.15 and 4.14.

1.21 The Semiclassical Approximation for the Master Equation

By developing the double commutator in (1.20.9) and using the same change of variables, we find

$$\begin{aligned}
&\left\langle \left[H_{\mathrm{I}}(0), \left[H_{\mathrm{I}}(t_2), b(\lambda^2 t_2 + t_1) \right] \right] \right\rangle_R \\
&\quad = \langle H_{\mathrm{I}}(0) H_{\mathrm{I}}(t_2) b(\lambda^2 t_2 + t_1) \rangle_R - \langle H_{\mathrm{I}}(t_2) b(\lambda^2 t_2 + t_1) H_{\mathrm{I}}(0) \rangle_R \\
&\qquad - \langle H_{\mathrm{I}}(0) b(\lambda^2 t_2 + t_1) H_{\mathrm{I}}(t_2) \rangle_R + \langle b(\lambda^2 t_2 + t_1) H_{\mathrm{I}}(t_2) H_{\mathrm{I}}(0) \rangle_R.
\end{aligned}$$

Thus the self-averaging property is equivalent to the ansatz that, asymptotically as $\lambda \to 0$, is given by

$$\langle H_{\mathrm{I}}(0)H_{\mathrm{I}}(t_2)b(\lambda^2 t_2+t_1)\rangle_R = \langle H_{\mathrm{I}}(0)H_{\mathrm{I}}(t_2)\rangle_R \langle b(\lambda^2 t_2+t_1)\rangle_R$$
$$\langle H_{\mathrm{I}}(0)b(\lambda^2 t_2+t_1)H_{\mathrm{I}}(t_2)\rangle_R = \langle H_{\mathrm{I}}(0)\langle b(\lambda^2 t_2+t_1)\rangle_R H_{\mathrm{I}}(t_2)\rangle_R$$

(and the same relations for the adjoint).

In the case when $\langle \,\cdot\, \rangle_R$ is replaced by a *full average* (and not only a partial average over the R degrees of freedom) this ansatz is called, in quantum optics, a *semiclassical approximation*. As $\lambda \to 0$, as explained in the previous section, $\langle b(\lambda^2 t_2+t_1)\rangle_R$ converges to $\langle b(t_1)\rangle_R = b_S(t)$, and therefore the double commutator in (1.20.9) becomes

$$\langle H_{\mathrm{I}}(0)H_{\mathrm{I}}(\tau)\rangle_R b_S(t) + b_S(t)\langle H_{\mathrm{I}}(\tau)H_{\mathrm{I}}(0)\rangle_R \\ - \{\langle H_{\mathrm{I}}(\tau)b_S(t)H_{\mathrm{I}}(0)\rangle_R + \langle H_{\mathrm{I}}(0)b_S(t)H_{\mathrm{I}}(\tau)\rangle_R\},$$

where now $b_S(t)$ can be brought outside the $\langle \,\cdot\, \rangle_R$ average because $b_S(t)$ is a system operator and therefore it satisfies $\langle X b_S(t)\rangle_R = \langle X\rangle_R b_S(t)$.

Taking the τ-integral as in (1.20.10) and comparing this with the result of Sect. 1.8, we are led to introduce the *operator-valued transport coefficient*

$$\int_{-\infty}^{0} \langle H_{\mathrm{I}}(0)H_{\mathrm{I}}(\tau)\rangle_R d\tau =: Y^*, \quad \int_{-\infty}^{0} \langle H_{\mathrm{I}}(\tau)H_{\mathrm{I}}(0)\rangle_R d\tau = Y, \tag{1.21.1}$$

and the map

$$\kappa(b_S) := -\int_{-\infty}^{0} d\tau \{\langle H_{\mathrm{I}}(\tau)b_S H_{\mathrm{I}}(0)\rangle_R + \langle H_{\mathrm{I}}(0)b_S H_{\mathrm{I}}(\tau)\rangle_R\}. \tag{1.21.2}$$

Thus we find, for the generator L, the expression

$$L(b_S) = -b_S Y - Y^* b_S + \kappa(b_S). \tag{1.21.3}$$

This suggests the following *semiclassical approximation for the master equation: Given the time-dependent interaction Hamiltonian, $H_{\mathrm{I}}(t)$, calculate the operator-valued transport coefficients by means of* (1.21.1–2). *This gives the master equation by means of the rule* (1.21.3).

The above heuristic rule works well only in very special cases (the Fock case with a rotating wave approximation). The *stochastic golden rule*, to be discussed in Sect. 4.14 will give a fully rigorous rule which is applicable to a much more general context, explained as follows: (i) it applies not only to the master, but also to the stochastic Schrödinger equation; (ii) it allows general Gaussian states for the reservoir (in particular equilibrium states); and (iii) it does not require the rotating wave approximation.

The dual L^* of the operator L, with respect to the system trace duality, is defined by

$$\mathrm{Tr}_S\big(\rho L(X)\big) := \mathrm{Tr}_S\big(L^*(\rho)X\big)$$

and acts on the space of density operators ρ on $\mathcal{H}_S$ as follows:

$$L^*(\rho) = -Y\rho - \rho Y^+ + \kappa^*(\rho). \tag{1.21.4}$$

It gives the Fokker–Planck equation:

$$\partial_t \rho(t) = -Y\rho(t) - \rho(t)Y^+ + \kappa^*(\rho(t)). \tag{1.21.5}$$

1.22 Beyond the Master Equation

The vacuum transition amplitudes

$$\lim_{\lambda\to 0}\langle U^{(\lambda)}_{t/\lambda^2}\rangle = \langle U_t\rangle, \tag{1.22.1}$$

where we use the same symbol $\langle\ \cdot\ \rangle$ for the vacuum expectation of both the original field and the master field, are the simplest quantities that one can compute with the stochastic limit; however they already go beyond what one can do with the usual techniques, for example, of the master equation which only considers vacuum expectation values of observables of the system S evolved with the Heisenberg equation:

$$\langle\psi\otimes\Phi, U_t^*\,(X\otimes 1)\,U_t\psi\otimes\Phi\rangle = \langle\psi\langle\Phi, U_t^*\,(X\otimes 1)\,U_t\Phi\rangle\psi\rangle, \tag{1.22.2}$$

where X is an observable of the system. To compare this with (1.22.1), we consider, instead of the amplitude (1.22.1), the associated probability, i.e.

$$|\langle\psi\otimes\Phi, U_t\varphi\otimes\Phi\rangle|^2 = \langle\psi\otimes\Phi, U_t^*\,(|\varphi\rangle\langle\varphi|\otimes|\Phi\rangle\langle\Phi|)\,U_t\psi\otimes\Phi\rangle, \tag{1.22.3}$$

which cannot be reduced to the form of (1.22.2) because the former is the vacuum expectation of the identity preserving flow,

$$j_t(X) := U_t^*(X\otimes 1)U_t,$$

and the latter is the vacuum expectation of the flow,

$$j_t(X) := U_t^*(X\otimes|\Phi\rangle\langle\Phi|)U_t,$$

which is a paradigmatic example of a *flow with explosion*, i.e.

$$j_t(1)\neq 1.$$

The considerations above show that these flows arise naturally in physics.

The vacuum transition amplitudes (1.22.1) are only the simplest example of quantities of physical interest which can be computed by the stochastic limit, but not with the usual approximation techniques such as the master equation. More generally one can prove the existence of the limit, as $\lambda\to 0$, of any expression of the form

$$\langle\varphi_1\otimes\Phi_\lambda, U^{(\lambda)}_{t/\lambda^2}Z_\lambda U^{(\lambda)*}_{t/\lambda^2}\varphi_1'\otimes\Phi_\lambda'\rangle, \tag{1.22.4}$$

where

- Z_λ is a (finite) linear combination of operators of the form

$$X\otimes F_\lambda, \tag{1.22.5}$$

  X being a system operator and

$$F_\lambda = F\left(\frac{1}{\lambda}\,a_\lambda(t/\lambda^2,k), \frac{1}{\lambda}\,a^+(t/\lambda^2,k)\right) \tag{1.22.6}$$

  is a functional (e.g. a polynomial) of the rescaled field;

- φ_1, φ_1' are arbitrary states of the system;
- Φ_λ, Φ_λ' are reservoir states obtained by applying an arbitrary polynomial in the rescaled field to the reference vector Φ of the reservoir (e.g. the vacuum).

Moreover the limit of the matrix element (1.22.4) is equal to

$$\langle \varphi_1 \otimes \Phi_0, U_t Z U_t^* \varphi_1' \otimes \Phi_0' \rangle \tag{1.22.7}$$

where

- Z is a finite linear combination of operators of the form

$$X \otimes F_0,$$

X being as above and $F_0 = F\big(b(t,k), b^+(t,k)\big)$ is the same function F as above evaluated in the master field $b^\pm(t,k)$ (collective reservoir operators);
- φ_1, φ_1' are as above;
- Φ_0, Φ_0' are obtained by applying the some polynomials used to form Φ_λ, Φ_λ' to the reference vector Ψ of the master field (collective vectors).

The physical idea behind the consideration of observables of the form (1.22.5–6) is that, since the stochastic limit extracts the long-term cumulative behaviour of the interacting dynamics, we expect its effects to be best revealed on those reservoir observables or states which depend on this long-term cumulative dynamical behaviour.

The most general class of observables Z_λ for which the limit of (1.22.4) exists is not known at the moment, and its precise determination involves the problem of the estimate of the error in the stochastic limit. If the interaction Hamiltonian is linear in the reservoir fields, then this class can be considerably enlarged by including observables F_λ of the form

$$W(f_{S,T,\lambda}), \quad N(|f_{S,T,\lambda}\rangle\langle g_{S',T',\lambda}|), \quad \int_{S/\lambda^2}^{T/\lambda^2} A^+(S_t f) A(S_t f) dt,$$

where

- $W(h)$ is the Weyl operator with test function h (defined for boson fields), N is the number operator and f, g are test functions in the 1-particle reservoir space;
- S_t is an *effective 1-particle, free reservoir evolution* of the form (4.11.3);
- S, T, S', T' are real numbers and, if f is a test function, then $f_{S,T,\lambda}$ is defined by

$$f_{S,T,\lambda} := \int_{S/\lambda^2}^{T/\lambda^2} S_t f dt.$$

1.23 On the Meaning of the Decomposition $H = H_0 + H_{\mathrm{I}}$: Discrete Spectrum

The decomposition $H = H_0 + H_{\mathrm{I}}$ of an Hamiltonian into *free* and *interacting* parts does not have an absolute meaning because one could always add to H_{I} a function of H_0 and subtract from H_0 the same function. Intuitively what one would like to do is to subtract from the interaction Hamiltonian its *projection* on the diagonal algebra of the free Hamiltonian. The idea is that the diagonal part is responsible for the energy shifts and the off-diagonal part for the decays. This projection onto the diagonal algebra can be easily described in the case where the free Hamiltonian H_0 has a discrete spectrum, i.e.

$$H_0 = \sum_n \varepsilon_n P_n, \tag{1.23.1}$$

where ε_n are the eigenvalues and P_n the spectral projections. Then the decomposition between *free* and *interacting* part can be chosen in a canonical way as follows: since $\sum_n P_n = 1$, we can write

$$H_0 + H_{\mathrm{I}} = H_0 + \sum_{m,n} P_m H_{\mathrm{I}} P_n = H_0' + H_{\mathrm{I}}', \tag{1.23.2}$$

where

$$H_0' = H_0 + \sum_n P_n H_{\mathrm{I}} P_n = \sum_n (\varepsilon_n + P_n H_{\mathrm{I}} P_n) P_n, \tag{1.23.3}$$

$$H_{\mathrm{I}}' = \sum_{m \neq n} P_m H_{\mathrm{I}} P_n. \tag{1.23.4}$$

The term $\sum_n P_n H_{\mathrm{I}} P_n$ corresponds to the projection of the interaction in the diagonal algebra of the free Hamiltonian and its contribution to the interaction is just a shift in the free energy.

The off-diagonal part (1.23.4) is orthogonal to the new free Hamiltonian H_0' for the scalar product defined by the trace in the following very strong sense: for any (Borel) function $f : \mathbb{R} \to \mathbb{R}$, one has

$$\mathrm{Tr}(f(H_0') \cdot H_{\mathrm{I}}') = 0.$$

This is because H_0' commutes with H_0; so for any function f one has

$$f(H_0') = \sum_n P_n f(H_0') P_n,$$

and consequently

$$\mathrm{Tr}(f(H_0') H_{\mathrm{I}}') = \sum_{n \neq m} \mathrm{Tr}(P_n f(H_0') P_n H_{\mathrm{I}} P_m) = 0,$$

because $\mathrm{Tr} P_n X = \mathrm{Tr} X P_n$.

It is clear that the decomposition (1.23.2) in the diagonal ($P_n H_0' P_m = 0$ if $m \neq n$) and off-diagonal ($P_m H_{\rm I}' P_n = 0$ if $m = n$) parts is unique. In fact, if

$$H = H_0'' + H_{\rm I}''$$

is another decomposition of $H_0 + H_{\rm I}$ into diagonal and off-diagonal parts, then

$$H_0' - H_0'' = H_{\rm I}'' - H_{\rm I}'.$$

The left-hand side is diagonal and the right-hand side is off-diagonal only when $H_0' = H_0''$ and $H_{\rm I}' = H_{\rm I}''$. The unique decomposition (1.23.2) of the Hamiltonian $H_0 + H_{\rm I}$ into diagonal and off-diagonal parts is called the *canonical decomposition* of $H_0 + H_{\rm I}$. Writing

$$D_{mn} = P_m H_{\rm I} P_n, \quad m < n,$$

we can always suppose that $H_{\rm I}$ is purely off-diagonal (up to a change in H_0), i.e.

$$P_n H_{\rm I} P_n = 0, \quad \forall n.$$

Under this assumption

$$H_{\rm I} = \sum_{m<n} D_{mn} + \text{h.c.},$$

and, under free evolution, the operators D_{mn} evolve according to

$$e^{itH_0} D_{mn} e^{-itH_0} = e^{it(\varepsilon_m - \varepsilon_n)} D_{mn}.$$

This simple remark will be the starting point to put in canonical form Hamiltonians of a much more general type.

Remark 1.23.1. If we define the projection onto the diagonal algebra of H_0

$$E(X) := \sum_n P_n X P_n$$

(mathematically this is the conditional expectation onto the commutant of the algebra generated by H_0), then the off-diagonal part of $H_{\rm I}$ is defined by

$$H_{\rm I}' = H_{\rm I} - E(H_{\rm I}).$$

So the problem of extending the canonical decomposition of $H_0 + H_{\rm I}$ to the case in which H_0 has a continuous spectrum is equivalent to the problem of finding a good analogue, in this case, of the conditional expectation $E^{(1)}$.

1.24 Other Rescalings when the Time Correlations Are not Integrable

This section is added just as a remark pointing out possible developments in directions not pursued in the present book. The vacuum expectation of the second-order term of the iterated series is [see (1.8.8)]

$$\int_0^{\lambda^2 t} ds_1 \int_{-s_1/\lambda^2}^0 G(s_2)ds_2. \tag{1.24.1}$$

Now, suppose that for some nonzero function $\rho_\lambda(s)$ the limit

$$\lim_{\lambda\to 0} \frac{1}{\rho_\lambda(s)} \int_{-s/\lambda^2}^0 G(s_2)ds_2 =: F(s) \tag{1.24.2}$$

exists and write (1.24.1) in the form

$$\int_0^{\lambda^2 t} \rho_\lambda(s_1)G_\lambda(s_1)ds_1 = \int_0^{\lambda^2 t} G_\lambda(s_1)dR_\lambda(s_1), \tag{1.24.3}$$

where we have made

$$G_\lambda(s) := \frac{1}{\rho_\lambda(s)} \int_{-s/\lambda^2}^0 G(s_2)ds_2, \tag{1.24.4}$$

$$R_\lambda(s) = \int_0^s \rho_\lambda(s_1)ds_1. \tag{1.24.5}$$

Supposing that, for each λ, $R_\lambda(s_1)$ is an invertible function, in (1.24.3) one can make

$$R_\lambda(s_1) = \tau, \quad s_1 = 0 \Leftrightarrow \tau = 0, \quad s_1 = \lambda^2 t \Leftrightarrow \tau = R_\lambda(\lambda^2 t);$$

thus, (1.24.3) becomes

$$\int_0^{R_\lambda(\lambda^2 t)} G_\lambda(R_\lambda^{-1}(\tau))d\tau. \tag{1.24.6}$$

Suppose now that we can rescale time

$$t \mapsto t_\lambda \tag{1.24.7}$$

so that

$$R_\lambda(\lambda^2 t_\lambda) = t, \quad \forall\, \lambda. \tag{1.24.8}$$

This fixes t_λ to be

$$t_\lambda = \lambda^{-2} R_\lambda^{-1}(t). \tag{1.24.9}$$

So, after the rescaling (1.24.7), (1.24.1) becomes

$$\int_0^t G_\lambda \circ R_\lambda^{-1}(\tau)d\tau = \int_0^t \frac{1}{\rho_\lambda(R_\lambda^{-1}(\tau))} \int_{-R_\lambda^{-1}(\tau)/\lambda^2}^0 G(s_2)ds_2 d\tau. \quad (1.24.10)$$

Now, because of (1.24.2), if F is very regular, we would expect that (1.24.10) behaves like

$$\int_0^t F(R_\lambda^{-1}(\tau))d\tau, \quad (1.24.11)$$

and, in some favourable case, this might have a limit.

Example 1.24.1. Suppose that for large s, $G(s)$ is proportional $1/s$, for example,

$$\int_{-s}^0 G(|s_1|)ds_1 = \log s.$$

So we can choose

$$\rho_\lambda(s) = \log s/\lambda^2 = \rho(s/\lambda^2).$$

Now

$$R(s) = \int \log s ds = s \log s - s.$$

So

$$\begin{aligned} R_\lambda(s) &= \int_0^s \rho(s_1/\lambda^2)ds_1 = \lambda^2 \int_0^s \rho(s_1/\lambda^2)d(s_1/\lambda^2) \\ &= \lambda^2 \int_0^{s/\lambda^2} \rho(\tau)d\tau = \lambda^2 \cdot \left(\frac{s}{\lambda^2}\log\frac{s}{\lambda^2} - \frac{s}{\lambda^2}\right) \\ &= s\log\frac{s}{\lambda^2} - s. \end{aligned}$$

The derivative of $R_\lambda(s)$, i.e. $\log(s/\lambda^2) - 2$, is zero only for $s = e^2\lambda^2$. So, for $s > e^2\lambda^2$, $R_\lambda(s)$ is invertible. We are interested in $\lambda \to 0$. So, $\forall t > 0$, there exists λ_t such that, for $\lambda < \lambda_t$, $R_\lambda(s)$ is invertible for any $s \geq t$. In this case the function F is a constant c and therefore (1.24.11) would be of the form ct independent of λ.

1.25 Connections with the Classical Homogeneization Problem

The stochastic limit has a deep connection with the so-called *classical homogeneization problem*[(1)] which consists in the following: it is given a solution

$y(t)$ of a differential equation (ordinary or stochastic), and one wants to show that the rescaled process

$$y(t) \to \lambda y(t/\lambda^2) \tag{1.25.1}$$

converges to an independent increment process. (Most results up to now have been obtained when the limiting independent increment process is a classical Brownian motion.)

Typically one starts from a stochastic differential equation of the form

$$dy(t,x) = -\nabla U(y(x,t))dt + dw_t, \quad y(x,0) = x. \tag{1.25.2}$$

The associated diffusion has generator

$$L = \frac{1}{2}\Delta - \nabla U(y) \cdot \nabla.$$

Under general regularity conditions on the potential $U(y)$, the semigroup generated by L,

$$e^{tL} = P^t,$$

exists, and the (non-normalized) Gibbs measure with potential U, i.e.

$$\pi(dy) = e^{-2U(y)}dy,$$

is an invariant measure of it. If the semigroup P^t has a *mass gap*, i.e.

$$\lambda_0 := \inf\{\lambda > 0 : -\lambda \in \mathrm{spec}(L)\} > 0,$$

then it exhibits an *exponential tendence to equilibrium*, i.e.

$$\|P^t f - \langle \pi, f\rangle \cdot 1\|_\infty \le ce^{-t\lambda_0}\|f\|^2_{L^2},$$

and (dy) is called the *equilibrium measure* of it. Rescaling the solution of (1.25.2) as in (1.25.1), one has

$$y^\lambda(t) = \lambda y(\lambda^{-2}t) = -\lambda \int_0^{t/\lambda^2} \nabla U(y(s))ds + \lambda w_{t/\lambda^2}; \tag{1.25.3}$$

one expects that, as $\lambda \to 0$, the term $\lambda w_{t/\lambda^2}$ becomes negligible. Moreover, because of the strong mixing properties of the semigroup P^t, the random variables $\nabla U(y_s)$ are weakly dependent; thus one expects that the integral term in (1.25.3) converges to a Brownian motion by the effect of a functional central limit theorem.

1.26 Algebraic Formulation of the Stochastic Limit

The present section is not necessary for the understanding of the rest of the book. In it we formulate in an abstract algebraic setting the stochastic limit problem.

Let $\mathcal{A}_\lambda$ be a C^*-algebra (algebra of observables). We assume that $\mathcal{A}_\lambda \subseteq \mathcal{B}(\mathcal{H}_\lambda)$ where $\mathcal{H}_\lambda$ is a Hilbert space (always complex separable) and $\lambda \in \mathbf{R}$ is a parameter (in some cases $\lambda \in \mathbf{R}^2$, $\mathbf{R}^3, \dots$). Let there be given two 1-parameter automorphism groups of $\mathcal{A}_\lambda$

$$u_t^{0,\lambda}, \quad u_t^\lambda, \tag{1.26.1}$$

called respectively the *free evolution* (or free effective evolution) and the *interacting evolution.*

In the simplest cases $\mathcal{H}_\lambda = \mathcal{H}$ is independent of λ, and

$$u_t^{0,\lambda} = u_t^0; \tag{1.26.2}$$

but there are cases in which everything depends on λ (e.g. the low-density limit). We shall use the notation of (2.7.2) even in the general case.

The wave automorphism at time t is:

$$j_t^\lambda := u_t^\lambda \cdot u_{-t}^0 \tag{1.26.3}$$

is a u_{-t}^0-flow (also called an u_{-t}^0-1-*cocycle*) in the sense that its time shift

$$j_{[s,t]}^\lambda := u_s^0 \cdot j_{[s,t]}^\lambda \cdot u_{-s}^0 \tag{1.26.4}$$

satisfies the flow equation

$$j_{[s,t]}^\lambda \cdot j_{[r,s]}^\lambda = j_{[r,t]}^\lambda. \tag{1.26.5}$$

The time-scaling limit is introduced as follows: fix a λ-dependent time change

$$t \mapsto \tau(t,\lambda) \in \mathbf{R}, \quad \tau(t,\lambda) \to \infty \quad \text{as} \quad \lambda \to 0, \tag{1.26.6}$$

and consider the rescaled flow $j_{\tau(t,\lambda)}^\lambda$. The basic example is

$$\tau(t,\lambda) = t/\lambda^2. \tag{1.26.7}$$

Suppose that j_t^λ is a deterministic flow, i.e. it satisfies an ordinary differential equation of the form

$$\frac{dj_t^\lambda(a_0)}{dt} = j_t^\lambda(\delta_t(a_0)), \quad a_0 \in \mathcal{A}_0^\lambda, \tag{1.26.8}$$

where $\mathcal{A}_0^\lambda$ is a subspace (usually a subalgebra) of $\mathcal{A}_\lambda$ contained in the domain of the operators δ_t (usually unbounded derivations from $\mathcal{A}_0^\lambda$ to $\mathcal{A}_\lambda$).

The basic idea of the stochastic limit is that, in several interesting cases, one can prove that there exists a C^*-algebra $\mathcal{A}$, and, for a subalgebra $\mathcal{A}_0 \subseteq \mathcal{A}$, for each λ, a map

$$i_\lambda \ : \ a_0 \in \mathcal{A}_0 \mapsto i_\lambda(a_0) =: a_0^\lambda \in \mathcal{A}_0^\lambda$$

such that the limit

$$\lim_{\lambda\to 0} j^{\lambda}_{T(t,\lambda)}\big(i_\lambda(a_0)\big) =: j_t(a_0), \quad a_0 \in \mathcal{A}_0, \tag{1.26.9}$$

exists in a sense to be specified and satisfies a stochastic differential equation of the form

$$j_t(a_0) = 1 + \int_0^t j_s\big(\theta_\alpha(a_0)\big)dM_\alpha(s) + \int_0^t j_s\big(\theta_0(a_0)\big)ds, \quad \forall a_0 \in \mathcal{A}_0, \tag{1.26.10}$$

where the M_α are $\mathcal{A}$-valued finitely additive measures on R, θ_α are linear maps from $\mathcal{A}_0$ to $\mathcal{A}$ and (1.26.10) means that both sides are well defined and the identity holds.

One is interested in the following problems: given u_t^0 and u_t^λ, determine

(i) the structure of $\mathcal{A}$, $\mathcal{A}_0$ and of the $\mathcal{A}$–valued measures M_α;
(ii) the form of the stochastic equation (1.26.10) and its mathematical meaning;
(iii) the meaning of the limit (1.26.9).

In general the limit (1.26.9) does not exist in the usual operator topologies. The precise meaning of the limit is defined as follows:

(i) Fix a family $\mathcal{V}_\lambda \subseteq \mathcal{H}$ of vectors called *collective vectors* and depending on u_t^0, u_t^λ.
(ii) Prove that there exists a space $\mathcal{H}_{\text{noise}}$, a total family $\mathcal{V} \subseteq \mathcal{H}_{\text{noise}}$, and $\forall\lambda$ a one-to-one map

$$\Phi \in \mathcal{V} \to \Phi_\lambda \in \mathcal{V}_\lambda, \tag{1.26.11}$$

such that if Φ, Φ' are vectors in $\mathcal{V}$ whose images by (1.26.11) are respectively (Φ_λ) and (Φ'_λ) then

$$\lim_{\lambda\to 0}\langle \Phi_\lambda, \Phi'_\lambda\rangle = \langle \Phi, \Phi'\rangle. \tag{1.26.12}$$

Definition 1.26.1. *Let $\mathcal{A}_0$ be a subset of $\mathcal{A}$ (not necessarily an algebra). $j^\lambda_{T(t,\lambda)}$ is said to converge to j_t on $\mathcal{A}_0$ in the sense of the $\mathcal{V}_\lambda$-matrix elements if $\forall a_0 \in \mathcal{A}_0$.*

$$\lim_{\lambda\to 0}\langle \Phi_\lambda, j^\lambda_{T(t,\lambda)}(a_0)\Phi'_\lambda\rangle = \langle \Phi, j_t(a_0)\Phi'\rangle. \tag{1.26.13}$$

Note that, since $\mathcal{V}$ is total in $\mathcal{H}_{\text{noise}}$, the right-hand side of (1.26.13) defines a unique bounded operator $j_t(a_0) \in \mathcal{B}(\mathcal{H}_{\text{noise}})$, i.e. the notation (1.26.13) is consistent. In the most frequently considered examples, j_t^λ can be realized as inner automorphisms. In other words, if they are of the form

$$j_t^\lambda(a) = U_t^{(\lambda)*} a U_t^{(\lambda)}, \tag{1.26.14}$$

with $U_t^{(\lambda)}$ satisfying

$$\frac{\partial}{\partial t}U_t^{(\lambda)} = \frac{1}{i}H_{\mathrm{I}}^{(\lambda)}(t)U_t^{(\lambda)}, \tag{1.26.15}$$

then one expects that the same shall be true of its limit; similarly, for $a_0 \in A_0$

$$\lim_{\lambda\to 0}\langle \Phi_\lambda,\ j^\lambda_{t/\lambda^2}(a_0)\Phi'_\lambda\rangle = \lim_{\lambda\to 0}\langle \Phi_\lambda,\ U^{(\lambda)*}_{t/\lambda^2}\cdot a_0\cdot U^{(\lambda)}_{t/\lambda^2}\Phi'_\lambda\rangle = \langle \Phi,\ U_t^*\cdot a_0\cdot U_t\Phi'\rangle \tag{1.26.16}$$

for some unitary operator U_t. In this case, as a preliminary problem to (1.27.18), one studies the limit

$$\lim_{\lambda\to 0}\langle \Phi_\lambda, U^{(\lambda)}_{t/\lambda^2}\Phi'_\lambda\rangle = \langle \Phi,\ U_t\Phi'\rangle. \tag{1.26.17}$$

1.27 Notes

Section (1.1)

(1) By a Hilbert space we mean a complex separable Hilbert space and by a pre-Hilbert space we mean a complex vector space endowed with a (possibly degenerate) sesquilinear form whose induced topology is separable.

Section (1.6)

(1) Perhaps the first rigorous result, about the interaction of a system with a reservoir, where the role of a scaling limit involving $\lambda^2 t$ begun to emerge, is due to Bogoliubov [Bo45] who considered a classical system with Hamiltonian

$$H = H_{\mathrm{S}} + H_{\mathrm{R}} + H_{\mathrm{int}} \tag{1}$$

where

$$H_{\mathrm{S}} = \frac{1}{2}(p^2+\omega^2q^2)\,,\quad H_{\mathrm{R}} = \frac{1}{2}\sum_{n=1}^{N}(p_n^2+\omega_n^2q_n^2)\,,\quad H_{\mathrm{int}} = \lambda q\sum_{n=1}^{N}(\alpha_n q_n)\,. \tag{2}$$

He assumed that as $N\to\infty$

$$\sum_{\sigma<\omega_n}\frac{\alpha_n^2}{\omega_n^2} \to \int_\sigma^\infty J(\nu)d\nu \tag{3}$$

where $J(\nu)$ is a non-negative continuous function such that

$$\int_0^\infty J(\nu)d\nu < \infty\,. \tag{4}$$

Supposing that, at $t = 0$, the system is in the state $q = q_0, p = p_0$ and the state of the reservoir is a random variable with the distribution

$$\rho_R = e^{-H_R/T} \tag{5}$$

he proved that:

1) As $N \to \infty$ the limit distribution $\rho_S = \rho_S(t, p, q)$ of the random variables $p = p_t, q = q_t$, of the system, exists at any time $t > 0$.
2) If H_S is given by (2), define the function $\rho_S^0 = \rho_0(t, H_S) = \rho_0(t, p, q)$, by:

$$\rho_S^0(t, H_S) = \frac{\omega}{4\pi^2 T(1 - e^{-2a\lambda^2 t})} \int_0^{2\pi} d\phi \cdot \exp\{-\frac{H_S + E_0 e^{-2a\lambda^2 t} - 2\sqrt{H_S E_0} e^{-2a\lambda^2 t} \cos\phi}{T(1 - e^{-2a\lambda^2 t})}\},$$

where $a = \frac{\pi}{4} J(\omega)$, and E_0 is

$$\frac{1}{2}(p_0^2 + q_0^2),$$

where (p_0, q_0) is the initial state. (Notice that $\rho_S^0(t)$ depends only on $\lambda^2 t$.) Then for small λ the function ρ_S can be approximated by the function ρ_S^0, in the sense that, for any positive α, β as $\lambda \to 0$, one has, uniformly in the interval $\frac{\alpha}{\lambda^2} < t < \frac{\beta}{\lambda^2}$

$$\frac{1}{\Delta t_\lambda} \int_t^{t+\Delta t_\lambda} (\rho_S - \rho_S^0) dt \to 0$$

for any subsequence $\{\Delta t_\lambda\}$ such that

$$\lambda^2 \Delta t_\lambda \to 0 \quad , \qquad \Delta t_\lambda \to \infty .$$

Bogoliubov gave a rigorous proof of this result using a Volterra integro-differential equation.
His condition $\lambda^2 \Delta t_\lambda \to 0$ is, however, different from the one used in the stochastic limit ($\lambda^2 t \to$ constant).
Friedrichs, in the context of the now well-known Friedrichs model [Fri48], considered the scaling limit

$$\lambda \to 0 \ , \qquad t \to \infty \ , \qquad \lambda^2 t = \text{constant} \tag{6}$$

using second-order perturbation theory.
In the mid-1950s van Hove [vHo55] used this scaling as a device to extract the effects of a small perturbation of the global Hamiltonian of a composite system, on the reduced evolution of a subsystem, and to derive a Pauli-type master equation describing the irreversible time evolution of the observables of the subsystem (see Sect. (1.13) and the

Notes therein). Therefore, in the quantum theory of open systems, the limit (6) is known as the *van Hove* or the $\lambda^2 t$ limit.

Using the perturbative development of the dynamics in powers of λ, van Hove argued that the terms of order $2n$ should behave as $\lambda^{2n}t^n$ (in contrast with the rough estimate $\lambda^{2n}t^{2n}$). Furthermore, he thought it plausible that almost all the terms of order $2n$ should behave as $\lambda^{2n}t^{n-\varepsilon}$, for some $\varepsilon > 0$, and therefore vanish in the limit (6), while the remaining terms should sum to give a transport equation.

Notice that the $\lambda^{2n}t^n$-behaviour of the terms of order $2n$ of the iterated series is exactly of the same order of magnitude as the behaviour of the moments of order $2n$ of a Brownian motion with the time parameter rescaled as $\lambda^2 t$. Therefore a posteriori we can interpret van Hove's perturbative result as a first indication that the rescaling (6) should show a kind of *quantum Brownian motion* underlying the dynamics of the quantum system.

(2) Intuitively, the weak coupling–long time limit means that we are looking at times during which the electron has already weakly interacted many times with many impurities (long time cumulative effects). The net average effect of these interactions amounts to a loss of memory (Markovian approximation). Therefore in this limit we can expect to be able to approximate the microscopic time evolution, which contains complicated memory effects, with a simpler Markovian evolution.

Prigogine and Henin [PrHe57] and Prigogine and Balescu [PrBa59a], [PrBa59b] extended van Hove's results and, using the Liouvillian formulation of classical mechanics, first introduced the methods of perturbative quantum field theory and Feynman graphs in classical mechanics.

(3) The formulation of the stochastic limit program in the quantum Brownian motion formalism, the proof of the basic analytical estimates (see Part III), as well as the first proof that this program can be effectively realized are due to Accardi, Frigerio and Lu [AcFrLu87].

(4) The formulation of the stochastic limit approach in the formalism of distribution theory and quantum white noise on which this book was based is proposed by Accardi, Lu and Volovich [AcLuVo93].

Section (1.7)

(1) The relevance of the existence of two time scales for the possibility of a Markovian reduced description was first recognized by Bogoliubov [Bo45]. The derivation of kinetic equation is considered in [Bo48], [Bo84], [Sp80], [Zu71]. We refer to the phenomenological literature [WaBl53], [Abr61], [At73], [Ha72], [AlSt75] for a variety of examples of physical phenomena where the existence of two well-separated timescales can be easily recognized.

Nonrigorous derivations of a Markovian Master Equation from a Generalized Master Equation, based on the existence of two timescales, can be found, in addition to the already mentioned papers of van Hove and Prigogine, throughout the literature (see, for example, [Zw60], [Zw61], [Sew61], [ArKe64], [FaMa68], [Hak70], [Ag73]). Most derivations incorporate the Born approximation explicitly.

Section (1.8)

(1) As shown in Sect. 1.6 the first step of the method is to **rescale** some basic parameter of the model. This book shall be mainly devoted to the study of the time scaling $t \mapsto t/\lambda^2$ corresponding to the replacement

$$U_t^{(\lambda)} \mapsto U_{t/\lambda^2}^{(\lambda)} \tag{7}$$

in the evolution operator. However it is important to keep in mind that (7) is only one of the possible rescalings and different rescalings naturally enter in several physical models.
The model itself should suggest which is the natural parameter to be rescaled. The basic heuristic principle to guess the natural rescaling is that some measurable physical parameter should remain finite in the limit. There are models in which the rescaled quantity is *energy* (see [Ar94], [ArVo94]); others in which it is the *density of particles* (see Chap. 10) or *space–time* (see Sect. 3.13); in others, such as in the large N limit of QCD, it is the number of inner degrees of freedom (see [AcArVo95], [ArVo96]) .... All these types of rescaling give rise to different stochastic limits. Thus, in the stochastic limit, the *smallness* of the parameter λ is not given an absolute meaning, but only **relative to some other parameter**. In the rescaling (7), the parameter is time and the *smallness* of λ is relative to *long periods of time*, so long as the product $\lambda^2\tau = t$ remains constant as $\lambda \to 0$. Here τ denotes the *physical time* and t the *rescaled time*. Physically τ gives the timescale of the fast degrees of freedom and t that of the slow ones.

Section (1.9)

(1) To describe the situation properly, it is convenient to speak of the *square root* of the δ-function rather than the δ-function itself. In order to substantiate this last statement, let us consider a widely studied class of Hamiltonians corresponding to the situation in which

$$H_0^{(\lambda)} = H_0 \quad , \qquad H_{\mathrm{I}}^{(\lambda)}(t) = \lambda H_{\mathrm{I}}(t) \quad , \qquad H^{(\lambda)} = H_0 + \lambda H_{\mathrm{I}}(t) \, , \tag{8}$$

with both H_0 and H_{I} independent of λ. In these cases λ is called the *coupling constant* (examples in which both H_0 and H_{I} depend on λ, which is *not* a coupling constant, are discussed in Chap. 10, and in a book now in preparation which includes all the applications of the stochastic limit that were not included in this volume). In this notation the rescaled Hamiltonian becomes $H_{\mathrm{I}}(t/\lambda^2)/\lambda$. If $H_{\mathrm{I}}(s)$ is a square integrable scalar function, then not $H_{\mathrm{I}}(t/\lambda^2)/\lambda$ but its square would converge to a multiple of the δ-function. In this sense we say that asymptotically $H_{\mathrm{I}}^{(\lambda)}(t/\lambda^2)/\lambda^2$ behaves like the *square root* of the δ-function. This intuitive picture shall be substantiated in a precise mathematical phenomenon: *the white noise approach to the Ito table* (see [AcLuVo99]).

The equivalence between linear singular Hamiltonian equations driven by some form of the quantum white noise case and quantum stochastic differential equations was discovered by Accardi, Lu and Volovich (see [AcLuVo99] for history and references). It has led to a new approach to stochastic calculus: both the classical Ito calculus and all the existing quantum stochastic calculi are unified within this new approach and the possibility of a calculus with nonlinear white noise functionals emerges.

Even if apparently unrelated to the usual perturbation theory, the stochastic method has a deep connection with it. In typical perturbation theory one considers matrix elements (1.26.13) of the first few terms of the series (1.16.4) in increasing powers of λ and neglects the remaining ones. In the stochastic limit one distinguishes **inside each term** of the expansion (1.16.4) a relevant contribution and a negligible one. The remarkable new feature of this theory is that, for many models, the relevant contributions not only can be resummed but they also give rise to **a unitary evolution operator** U_t.

Section (1.13)

(1) In the first stage of development one only considered partial averages over the reservoir degrees of freedom, with respect to the Fock vacuum or to a thermal state of the reservoir fixed once and for all.
This limitation has the effect of sweeping away, in the limit, all the terms of the iterated series except those which, in the Wick ordering procedure, correspond to the scalar terms. Consequently, the structure of the underlying quantum noise cannot emerge from this very particular limit procedure.
On the other hand the quantum Feynman–Kac technique of [Ac78] shows that every quantum Markovian semigroup P^t can be obtained from a unitary evolution operator (Markovian cocycle) $U(t)$ by the formula

$$P^t(X) = E_0[U^+(t)(X \otimes 1)U(t)]\,, \tag{9}$$

where E_0 is the partial (conditional) expectation onto the system space, obtained by taking the partial average $\langle \cdot \rangle$ which only involves the reservoir degrees of freedom.
With the work of Hudson and Parthasarathy the notion of quantum Brownian motion was brought to its full power with the construction of the corresponding stochastic calculus [HuPa84] whose main result is a technique to construct, as solutions of quite general quantum stochastic

differential equations, unitary evolution operators (Markovian cocycles) $U(t)$ which satisfy all the properties required for application of the quantum Feynman–Kac formula.

This result suggested a more ambitious plan, with respect to the master equation approach to open systems, namely: *to deduce a limiting equation not only for the reduced evolution of the system (i.e. for the semigroup P^t), but for the whole coupled system including the reservoir (i.e. for the whole evolution operator $U(t)$).*

In other words, one wants to obtain the master equation as a corollary of the more general limit

$$\lim_{\lambda\to 0} U^{(\lambda)}_{t/\lambda^2} = U(t) \tag{10}$$

and, for all system observables X,

$$\lim U^{(\lambda)*}_{t/\lambda^2}(X\otimes 1)U^{(\lambda)}_{t/\lambda^2} = \lim_{\lambda\to 0} X^\lambda(t) = U^+(t)(X\otimes 1_{\mathrm{R}})U(t) =: X(t) \tag{11}$$

where $U(t)$ is the solution of the quantum stochastic differential equation of the form

$$dU(t) = \left\{\sum_{j=1}^n [D_j dA_j^+(t) - D_j^+ dA_j(t)] + K dt\right\} U(t) \quad , \qquad U(0)=1\,, \tag{12}$$

and the $A_j(t), A_j^+(t)$ are mutually independent quantum Brownian motions (see Chap. 5). In a series of papers starting from [AcFrLu87, AcFrLu90, AcFrLu91, AcLu91a, AcLu91b, AcLu91c, AcLu91d, AcLu91e] it has been shown that, under the scaling $t\to t/\lambda^2$:

(i) the quantum fields converge to quantum Brownian motions;
(ii) the limit (10) applies;
(iii) the limit propagator $U(t)$ satisfies an equation of the form (12) in particular it is unitary;
(iv) the limit (11) exists and $X(t)$ satisfies a quantum Langevin equation;
(v) the old results on the master equation are then recovered simply by applying the quantum Feynman–Kac formula (9).

Moreover, as shown in a multiplicity of quantum systems, involving the basic physical interactions, this is a rather universal phenomenon.

(2) That (1.13.8) is the most general form of the generator of a quantum Markovian semigroup was first proved by Gorini, Kossakowski and Sudarshan (1976) in the finite dimensional case and by Lindblad (1976) in the infinite dimensional case for bounded generators. This result has been extended to various classes of unbounded generators by several authors. We refer to [CheFa96], for the latest results in this direction. The theory of open systems and information dynamics are considered in [OhyPet93], [Ohy91]. For the bosonic and fermionic stochastic calculi see [Part92].

Section (1.17)

(1) A fact which was systematically applied by Prigogine in his quantum field theoretical approach to classical mechanics [Pr62].

Section (1.23)

(1) See also Antoniou and Suchanecki [AnSu98] and Accardi–Smolyanov [AcSmo93] for recent developments and for a discussion of the mathematical aspects of the problem. In Sect. 3.6 it is shown that, if H_0 is not discrete, but it has some discrete eigenvalues embedded into the continuous spectrum, then an extension of the canonical form is still possible.

Section (1.25)

(1) The homogenization problem in the classical case was studied by several authors. One of the first systematic expositions was the monograph by Bensoussan, Lions and Papanicolau [BeLiPa78], while a more recent survey is in [Ol94].

2. Quantum Fields

In this chapter we fix some notations which shall be used throughout the book. In particular some connections between probabilistic and quantum field theoretical notions are established. Since, in the stochastic limit, strongly nonlinear interactions break the validity of the standard commutation relations, leading to some *deformations* (even operator deformations) of them, in the following we shall need a notion of a free quantum field more general than the usual one [BoLoOkTo87], in the sense that it does not postulate a priori any commutation relation. For this reason we will introduce the notion of a Gaussian *quantum field* independently of the choice of specific commutation relations. On the other hand, it is known that *the standard (Bose or Fermi) commutation relations follow from the Gaussian statistics*, while the converse is true only in the Fock case. This suggests the general idea that statistics (i.e. the correlations) is a more basic physical notion than algebra (i.e. the commutation relations); in fact the former can be, at least in principle, compared with experiments, while the latter corresponds to an assumption directly on the mathematical model. Since the notion of a quantum field is a particular case of (generalized) quantum stochastic process, one can apply the reconstruction theorem of [AFL82] to conclude that the limit, in the sense of correlators, of a family of quantum fields is still a quantum field.

2.1 Creation and Annihilation Operators

Let the following be given: a set M (index set), a $*$-algebra $\mathcal{A}$ with identity 1 (we want a $*$-algebra so that the notions of unitarity, positivity, state, etc., are defined) and an expectation value or state $\langle \cdot \rangle$ on this algebra (i.e. a linear functional on $\mathcal{A}$ which is positive on positive elements and such that $\varphi(1) = 1$, see [Sak71] for these notions).

A *random field on* M is a map $k \in M \to a_k \in \mathcal{A}$. In the following we shall use indifferently the notation a_k or $a(k)$. The quantities a_k are called the *random variables* of the field. If the algebra $\mathcal{A}$ is commutative, the field is called *classical.* If $M = \mathbb{R}$, we speak of a *stochastic process.* The set M may have a complicated structure, i.e. a manifold, a loop space, etc. In this book we shall take $M = \mathbb{R}^d$ for some $d \geq 1$, and the variable k will be interpreted as *momentum*. Actually we will deal with generalized random fields on $\mathbb{R}^d$. In this case M will be a space of test functions on $\mathbb{R}^d$. We will often consider fields of the form $a(t,k)$, and in this case the variable t will be interpreted as *time*. The algebra $\mathcal{A}$ has an adjoint operation denoted $+$, and throughout this book we shall use the notation

$$X^\varepsilon = \begin{cases} X \text{ for } \varepsilon = 0\,, \\ X^+ \text{ for } \varepsilon = 1\,. \end{cases} \tag{2.1.1}$$

The quantity of basic physical interest are the *correlators* (or *correlation*, or *n-point functions*), defined by

$$\langle a_{p_1}^{\epsilon_1} \cdots a_{p_n}^{\epsilon_n} \rangle\,, \tag{2.1.2}$$

where n is any natural integer and, in the notation of (2.1.1), $\epsilon_1, \dots, \epsilon_n$ are either 0 or 1. In the classical case the correlators are also called *mixed moments*. Giving an expectation value is equivalent to giving a representation of the field algebra in an Hilbert space and a unit vector Φ in $\mathcal{H}$ such that the correlators are given by

$$\langle \Phi, a_{p_1}^{\epsilon_1} \cdots a_{p_n}^{\epsilon_n} \Phi \rangle\,. \tag{2.1.3}$$

The pair $a_k, a_{k'}^+$ is said to satisfy the q-commutation relations if

$$a_k a_{k'}^+ - q a_{k'}^+ a_k = \delta(k - k')\,. \tag{2.1.4}$$

In this case a_k are called *annihilators* and $a_{k'}^+$ *creators*.

We speak of a *boson field* if $q = +1$, of a *Fermi field* if $q = -1$, and of a *Boltzmann field* if $q = 0$. In many important examples (including the Bose and the Fermi case) the commutation relations can be deduced from the correlators. In this sense the algebra of the field is included in its statistics.

The commutation relations (2.1.4) show that, in general, the correlators (2.1.3) are distributions. Therefore, in order to produce numbers comparable with experiments, one has to integrate the correlators (2.1.3) against suitable test functions $\varphi(p_1 \dots p_n)$.

A field $a(t,k)$ is called *stationary* with respect to a given expectation value $\langle\,\cdot\,\rangle$ if its n-point correlation functions are time-translation invariant, i.e., for any $t_1, \dots, t_n, s \in \mathbb{R}$, $p_1, \dots, p_n \in \mathbb{R}^d$, one has

$$\langle a^{\epsilon_1}(t_1 + s, p_1) \cdots a^{\epsilon_n}(t_n + s, p_n) = \langle a^{\epsilon_1}(t_1, p_1) \cdots a^{\epsilon_n}(t_n, p_n)\rangle\,. \tag{2.1.5}$$

The Fourier transform of the 2-point correlation functions of a real-valued stationary process is called its *spectral function*. For a field we speak of its *spectral matrix* to mean the Fourier transform (with respect to the t variable) of its 2-point correlation matrix:

$$\begin{pmatrix} \langle a_s^+ a_{t+s} \rangle & \langle a_s a_{t+s} \rangle \\ \langle a_s^+ a_{t+s}^+ \rangle & \langle a_s a_{t+s}^+ \rangle \end{pmatrix}\,.$$

Let Ω be a set. A family of mean zero fields $a_\omega(t,k)$ with $\omega \in \Omega$ is called *independent* with respect to a given expectation value $\langle \cdot \rangle$ if their correlation functions, corresponding to different indices ω, vanish whenever there is a *singleton*, i.e. for any finite subset $F \subseteq \Omega$, and $\omega_1, \dots, \omega_n \in F$ (the possibility that $\omega_i = \omega_j$ for $i \neq j$ not being excluded.

One has

$$\left\langle a^{\varepsilon_1}_{\omega_1,k_1} \cdots a^{\varepsilon_n}_{\omega_n,k_n} \right\rangle = 0 \tag{2.1.6}$$

whenever there is an index $\alpha \in F$ such that $\omega_\alpha \neq \omega_\beta$ if $\alpha \neq \beta$.

Although a^ε_k are not bona fide operators but only operator-valued distributions, in all the situations discussed in the present book, it happens that for a large class of test functions there exist well-defined operators, defined on a common invariant domain $\mathcal{D}$ and satisfying the formal identity

$$A^+(g) = \int_{\mathbb{R}^d} g(k) a^+_k dk \quad , \qquad A(g) = \int_{\mathbb{R}^d} \overline{g}(k) a_k dk\,. \tag{2.1.7}$$

So the rigorous meaning of the *multiplication of distributions* a^ε_k is simply the multiplication of the corresponding operators $A^\varepsilon(g)$ on the common invariant domain $\mathcal{D}$. Formula (2.1.7) gives an automatic translation code from the distribution to the operator language and vice versa.

2.2 Gaussianity

The essence of Gaussianity consists in the fact that the expectation values of products of Gaussian random variables (correlators) are expressed as weighted sums, over a certain subset of pair partitions (all pair partitions, in the classical, boson and fermion cases), of the products of pair correlation functions over all the pairs of a single partition. By varying the weights and the subset of pair partitions, one obtains the various notions of Gaussianity. The main combinatorial formulae which allow one to work with (boson and fermion) Gaussian processes are discussed in Sect. 15.12.

Definition 2.2.1. *In the above notations, the field $a_k, a^+_{k'}$ together with the expectation value $\langle \cdot \rangle$ is called a* mean zero Gaussian field *if for each $n \in \mathbb{N}$, for each sequence of indices $k_1, \dots, k_n$ and each sequence $\epsilon_1, \dots, \epsilon_n$ in $\{0,1\}$ one has:*

$$\left\langle a^{\epsilon_1}_{k_1} \cdots a^{\epsilon_n}_{k_n} \right\rangle = 0 \quad , \text{ if } n \text{ is odd}\,, \tag{2.2.1}$$

$$\langle a^{\epsilon_1}_{k_1} \cdots a^{\epsilon_{2p}}_{k_{2p}} \rangle = \sum_{\mathcal{P}_0(2p)} \epsilon(i_1, j_1; i_2, j_2; \dots; i_p, j_p) \left\langle a^{\epsilon_{i_1}}_{k_{i_1}} a^{\epsilon_{j_1}}_{k_{j_1}} \right\rangle \cdots \left\langle a^{\epsilon_{i_p}}_{k_{i_p}} a^{\epsilon_{j_p}}_{k_{j_p}} \right\rangle \,, \tag{2.2.2}$$

where

- $\mathcal{P}_0(2p)$ *is a subset of the ordered partitions* $(i_1, j_1;\ i_2, j_2;\ \ldots\ ; i_p, j_p)$ *of the set* $(1, \ldots, 2p)$ *into pairs and (2.2.2) is extended to all pair partitions in* $\mathcal{P}_0(2p)$*;*
- *the term ordered above means that*

$$i_1 < \ldots < i_p \quad , \quad i_\alpha < j_\alpha \ , \ \forall \alpha = 1, \ldots, p\, ; \tag{2.2.3}$$

- *for each natural integer* n*,* $\epsilon(i_1, j_1;\ i_2, j_2;\ \ldots\ ; i_p, j_p)$ *is a complex number depending on the pair partition* $(i_1, j_1, \ldots, i_p, j_{2p})$*.*

Remark 2.2.1. A mean zero Gaussian field $a(t,k), a^+(t,k')$ is stationary if and only if its covariance is time-translation invariant:

$$\langle a^{\epsilon_1}(t,k) a^{\epsilon_2}(t',k')\rangle = \langle a^{\epsilon_1}(0,k) a^{\epsilon_2}(t'-t,k')\rangle \,. \tag{2.2.4}$$

Example 2.2.1: Generalized Wick theorem. Let Φ be the vector determining the expectation value, as in (2.1.3). If the field $a_k, a^+_{k'}$ satisfies the q-commutation relations and if for each $k \in \mathbb{R}^d$

$$a_k \Phi = 0\,, \tag{2.2.5}$$

then the field is the unique mean zero Gaussian with the following properties:

(i) The set $\mathcal{P}_0(2p)$ is the set of all the pair partitions of $(1, \ldots, 2p)$.
(ii) The factor $\epsilon(i_1, j_1;\ i_2, j_2;\ \ldots\ ; i_p, j_p)$ is equal to $q^{\partial\sigma}$, where σ is the permutation which brings $(1 \ldots 2p)$ into $(i_1, j_1;\ i_2, j_2;\ \ldots\ ; i_p, j_p)$ and $\partial\sigma$ is the minimum number of exchanges of σ.

In this case, i.e. if (2.2.5) holds, the representation of the field algebra is called the *Fock representation.*

Remark 2.2.2. The expectation value of a Gaussian field is called a Gaussian state. An immediate corollary of (2.2.1–2) is that *linear combinations and limits of Gaussian fields are still Gaussian fields.*

2.3 Types of Gaussian States: Gauge-Invariant, Squeezed, Fock and Anti-Fock

Definition 2.3.1. *Let* a_k*,* a^+_k *be a field in the sense of Definition 2.2.1. A* mean zero Gaussian state $\langle\ \cdot\ \rangle$ *on the field algebra is uniquely determined by its covariance:*

$$\begin{pmatrix} \langle a^+_k a_{k'}\rangle & \langle a_k a_{k'}\rangle \\ \langle a^+_k a^+_{k'}\rangle & \langle a_k a^+_{k'}\rangle \end{pmatrix} =: \begin{pmatrix} q_{11}(k,k') & q_{12}(k,k') \\ q_{21}(k,k') & q_{22}(k,k') \end{pmatrix} . \tag{2.3.1}$$

If the off-diagonal elements of the covariance are zero, the state $\langle \cdot \rangle$ *is called a* gauge-invariant *state because it is invariant under the so-called* gauge transformations of the first kind, *i.e. those transformations of the field operators of the form*

$$a_k \to e^{-i\alpha} a_k \quad , \qquad a_k^+ \to e^{i\alpha} a_k^+ .$$

where α *is an arbitrary number. If the off-diagonal elements of the covariance are nontrivial (nonzero), the state is called* squeezed. *A* Fock state *is a mean zero gauge-invariant Gaussian state with covariance*

$$\begin{pmatrix} 0 & 0 \\ 0 & \langle a_k a_{k'}^+ \rangle \end{pmatrix} , \tag{2.3.2}$$

meaning that only the correlation $\langle a_k a_{k'}^+ \rangle$ *is not zero. This is equivalent to the condition*

$$a_k \Phi = 0 \tag{2.3.3}$$

if Φ *is the vector implementing the state* $\langle \cdot \rangle$. *With the same notation, if the covariance has the form*

$$\begin{pmatrix} \langle a_{k'}^+ a_k \rangle & 0 \\ 0 & 0 \end{pmatrix} , \tag{2.3.4}$$

then we speak of an anti-Fock state.

A necessary condition for the right-hand side of (2.3.1) to be the covariance matrix of a Gaussian functional is that it is a matrix-valued distribution of positive type. For a general field the condition is not sufficient because the relations among a_k^+ and $a_{k'}$ impose some restrictions on the covariance. For example, the boson commutation relations imply

$$\langle a_k a_{k'}^+ \rangle = \delta(k - k') + \langle a_{k'}^+ a_k \rangle .$$

So no anti-Fock state can exist for bosons (such states exist for fermions).

2.4 Free Evolutions of Quantum Fields

Definition 2.4.1. An evolution *on any algebra is a 1-parameter automorphism group* u_t *of this algebra. This means that for any real numbers* s, t *and any elements* x, y *of the algebra, one has*

$$u_t(xy) = u_t(x)u_t(y) \ , \ u_t(x^+) = u_t(x)^+ \ , \ u_t(1) = 1 \ , \ u_{s+t} = u_s u_t . \tag{2.4.1}$$

The typical example is

$$u_t(x) := V_t^+ x V_t = e^{itH_0} x e^{-itH_0} , \tag{2.4.2}$$

where V_t is a 1-parameter group of unitaries in the algebra; in this case the automorphism u_t is called the *inner*, and H_0 is called the *Hamiltonian* of the field.

Because of (2.4.1), and in the notations of Sect. 2.1, an evolution on the polynomial algebra of the fields $a_k, a^+_{k'}$ is completely determined once one knows $u_t(a_k)$.

Definition 2.4.2. *If an evolution u_t^0 is such that there exists a real-valued function $\omega(k)$ such that, for all k,*

$$u_t(a_k) = e^{-it\omega(k)} a_k \,, \tag{2.4.3}$$

then u_t^0 (the Hamiltonian H_0, if (2.4.2) holds) is called a free evolution (a free Hamiltonian*). The function $\omega(k)$ is called the* free 1-particle Hamiltonian, *and H_0 is said to be the* second quantization *of $\omega(k)$; often (especially in the boson case) one uses the notation*

$$H_0 = d\Gamma\big(\omega(p)\big) \,.$$

If the fields satisfy a q-commutation relation, then one can prove that for any choice of the real-valued function $\omega(k)$ there exists a free evolution satisfying (2.4.3). Usually one wants $\omega(k)$ to be positive (if it is not then $d\Gamma(\omega(p))$ is unbounded from below even if $\omega(k)$ is bounded below) and, unless explicitly stated otherwise, we shall also assume that it vanishes nowhere (with the possible exception of sets of measure zero). Typical examples are

$$\omega(k) = \frac{k^2}{2m} \,, \tag{2.4.4}$$

$$\omega(k) = \sqrt{k^2 + m^2} \,, \tag{2.4.5}$$

$$\omega(k) = \mid k \mid \,, \tag{2.4.6}$$

m being a positive constant. In *position representation*, the momentum operator becomes

$$p = -i\partial_x \,, \tag{2.4.7}$$

and the 1-particle Hamiltonians H_1 corresponding to the choices (2.4.4–6) are respectively

$$H_1 = -\frac{1}{2m}\Delta = p^2/2m \,, \tag{2.4.8}$$

$$H_1 = \sqrt{-\Delta + m^2} \,, \tag{2.4.9}$$

$$H_1 = |p| \,. \tag{2.4.10}$$

2.5 States Invariant Under Free Evolutions

In this section we want to discuss the following problem: given a 1-particle free evolution, determined by a positive function of momentum ω (see Sect. 2.4) under which conditions a Gaussian state is invariant under the second quantized evolution associated with ω, i.e.

$$a_k \mapsto a(t,k) = e^{-it\omega_k} a_k \,. \tag{2.5.1}$$

The invariance under the evolution (2.5.1) is equivalent, for a Fock state, to the equality in the sense of distributions:

$$\langle a(t,k)a^+(t',k')\rangle = \langle a(0,k)a^+(t'-t,k')\rangle \,, \tag{2.5.2}$$

which, given (2.5.1) is the same as

$$e^{-it\omega_k} e^{it'\omega_{k'}} \langle a_k a^+_{k'}\rangle = e^{i(t'-t)\omega_{k'}} \langle a_k a^+_{k'}\rangle \,. \tag{2.5.3}$$

Writing, for simplicity

$$F(k,k') := \langle a_k a^+_{k'}\rangle \,, \tag{2.5.4}$$

(2.5.3) becomes

$$e^{-it\omega_k} F(k,k') = e^{-it\omega_{k'}} F(k,k') \,,$$

or equivalently

$$e^{it(\omega_{k'}-\omega_k)} F(k,k') = F(k,k') \,. \tag{2.5.5}$$

Now it is clear (see Lemma 2.5.1) that the identity (2.5.5) can hold if and only if the distribution support of F is in the set of all (k,k') such that $\omega_k = \omega_{k'}$. In particular, if we want the stationarity condition (2.5.3) to hold for all choices of ω, then the support of F must be on the diagonal of $\mathbb{R}^d \times \mathbb{R}^d$, and therefore, given the positivity of $\langle a_k a^+_k\rangle$, it must be a multiple of the δ-function.

Lemma 2.5.1. *Let ω be a smooth function, then (2.5.5) can hold if and only if*

$$\operatorname{supp} F \subseteq \{(k,k') \in \mathbb{R}^d \times \mathbb{R}^d : \omega_k = \omega_{k'}\} \,, \tag{2.5.6}$$

where the support is meant in the sense of distributions.

Proof 2.5.1. The identity (2.5.5) is equivalent to the following fact for any smooth function $\psi(k,k')$ with compact support:

$$\int\!\!\int dk dk' \psi(k,k')[e^{it(\omega_{k'}-\omega_k)} - 1] F(k,k') = 0 \,. \tag{2.5.7}$$

Choosing $\psi = 1$ in the ball B_R, of $\mathbb{R}^{2d}$ with center 0 and radius R, (2.5.7) implies that

$$B_R \cap \operatorname{supp} F \subseteq B_R \cap \{e^{it(\omega_{k'}-\omega_k)} = 1\} \,.$$

Letting $R \to +\infty$, we obtain (2.5.6).

Corollary 2.5.1. *A Fock state is invariant under arbitrary free evolutions, i.e. (2.5.3) holds for any choice of the function ω in (2.5.2), if and only if there exists a function $n(k)$ on $\mathbb{R}^d$ such that*

$$\langle a_k a_{k'}^+ \rangle = n(k)\delta(k-k') \, . \tag{2.5.8}$$

Proof 2.5.2. By Lemma 2.5.1, under our assumption, the support of $\langle a_k a_{k'}^+ \rangle$ must be contained in the intersection of all the sets $\{(k,k') \in \mathbb{R}^d \times \mathbb{R}^d : \omega_k = \omega_{k'}\}$ when ω varies among all possible smooth functions on $\mathbb{R}^d$. It is clear that this intersection contains only the diagonal subset of $\mathbb{R}^d \times \mathbb{R}^d$. Thus the support of $F(k,k')$ is contained in the diagonal, and this implies that it is a multiple of the δ-function.

Another corollary of our discussion is the following: if the (mean zero Gaussian) state instead of being a Fock state is gauge-invariant in the sense of Definition 2.3.1, i.e. has diagonal covariance

$$\begin{pmatrix} \langle a_k^+ a_{k'} \rangle & 0 \\ 0 & \langle a_k a_{k'}^+ \rangle \end{pmatrix} ,$$

then both distributions $\langle a_k^+ a_{k'} \rangle$, and $\langle a_k a_{k'}^+ \rangle$ must have support in the set (2.5.6). In particular, if the state is invariant for any choice of the dispersion function ω, then one must have

$$\langle a_k^+ a_{k'} \rangle = n(k)\delta(k-k') \tag{2.5.9}$$

$$\langle a_k a_{k'}^+ \rangle = m(k)\delta(k-k') \tag{2.5.10}$$

for some positive functions $m(k)$, $n(k)$. Finally, given the definition of free evolution (2.3.3), the time stationarity of the off-diagonal correlations,

$$\langle a^+(t,k)a^+(t',k') \rangle = \langle a^+(0,k)a^+(t'-t,k') \rangle \, , \tag{2.5.11}$$

leads to the identity

$$e^{it(\omega_k+\omega_{k'})} \langle a_k^+ a_{k'}^+ \rangle = \langle a_k^+ a_{k'}^+ \rangle \, , \tag{2.5.12}$$

which, if ω_k is positive, can only be satisfied in the set of k such that $\omega_k = 0$, corresponding to a trivial evolution. Summing up the above discussion:

Theorem 2.5.1. *A Gaussian state $\langle \cdot \rangle$ is invariant under all free evolutions (2.3.3) if and only if it has a diagonal correlation matrix (i.e. it is gauge invariant) and the diagonal correlations are supported in the diagonal [i.e. (2.5.9) and (2.5.10) hold].*

Definition 2.5.1. *A Gaussian state $\langle \cdot \rangle$ invariant under all free evolutions (2.3.3) is called a* free state. *General Gaussian states are also called quasi free states.*

Remark 2.5.1. The condition of being invariant under *all* free evolutions (2.3.3) is very restrictive, and it is easy, given a positive smooth function ω, to construct examples of states $\langle \cdot \rangle_\omega$ whose diagonal correlations are not supported in the diagonal, but identity (2.5.5) and positivity are satisfied.

2.6 Existence of Squeezed Stationary Fields

Starting from a Gaussian state on a field algebra, invariant under a given free evolution, and taking linear combinations of the field operators, one can define a new field algebra and look at the statistics of the new field with respect to the old state. In this section we will study some properties of these states.

Let $a_k, a^+_{k'}$ be a field and let $\langle \cdot \rangle$ be a mean zero gauge-invariant Gaussian state on the $a_k, a^+_{k'}$-algebra with covariance

$$\begin{pmatrix} \langle a^+_k a_{k'}\rangle & 0 \\ 0 & \langle a_k a^+_{k'}\rangle \end{pmatrix} = \begin{pmatrix} n(k) & 0 \\ 0 & m(k) \end{pmatrix} \delta(k-k') , \tag{2.6.1}$$

where $n(k)$, $m(k)$ are functions satisfying $m(k) = m(-k)$, $n(k) = n(-k)$.

Then, defining

$$b_k := \overline{c}_k a_k + s_k a^+_{-k} \quad , \quad b^+_k := c_k a^+_k + \overline{s}_k a_{-k} , \tag{2.6.2}$$

where c_k and s_k are smooth functions, the state $\langle \cdot \rangle$ also gives a mean zero gauge-invariant Gaussian state on the $b_k, b^+_{k'}$-algebra with covariance matrix

$$\begin{pmatrix} \langle b^+_k b_{k'}\rangle & \langle b_k b_{k'}\rangle \\ \langle b^+_k b^+_{k'}\rangle & \langle b_k b^+_{k'}\rangle \end{pmatrix} \tag{2.6.3}$$

$$= \begin{pmatrix} [|c_k|^2 n(k) + |s_k|^2 m(k)]\delta(k-k') & [c_k \overline{s}_{-k} n(k) + \overline{s}_k c_{-k} m(k)]\delta(k+k') \\ [\overline{c}_k s_{-k} n(k) + s_k \overline{c}_{-k} m(k)]\delta(k+k') & [|c_k|^2 m(k) + |s_k|^2 n(k)]\delta(k-k') \end{pmatrix}$$

Note the $\delta(k+k')$ in the off-diagonal elements. This means that in this state one can have simultaneous creation (or annihilation) of pairs, but momentum has to be conserved. The positivity of the state is guaranteed by construction. For example,

$$\begin{aligned} \langle b^+_k b^+_{k'}\rangle &= \langle (c_k a^+_k + \overline{s}_k a_{-k})(c_{k'} a^+_{k'} + \overline{s}_{k'} a_{-k'})\rangle \\ &= c_k \overline{s}_{k'} \langle a^+_k a_{-k'}\rangle + c_{k'} \overline{s}_k \langle a^+_{-k} a_{k'}\rangle \\ &= c_k \overline{s}_{k'} n(k)\delta(k+k') + \overline{s}_k c_{k'} m(k)\delta(-k-k') \\ &= (c_k \overline{s}_{-k} n(k) + \overline{s}_k c_{-k} m(k))\delta(k+k') , \\ \langle b_k b^+_{k'}\rangle &= \langle (\overline{c}_k a_k + s_k a^+_{-k})(c_{k'} a^+_{k'} + \overline{s}_{k'} a_{-k'})\rangle \\ &= \overline{c}_k c_{k'} \langle a_k a^+_{k'}\rangle + s_k \overline{s}_{k'} \langle a^+_{-k} a_{-k'}\rangle \\ &= |c_k|^2 m(k)\delta(k-k') + |s_k|^2 n(k)\delta(k-k') \\ &= (|c_k|^2 m(k) + |s_k|^2 n(k))\delta(k-k') . \end{aligned} \tag{2.6.4}$$

Now let us consider a free evolution associated with a positive function of the a_k field with dispersion function ω_k, i.e.

$$a(t,k) = e^{-it\omega_k} a_k \, . \tag{2.6.5}$$

Because of gauge invariance, the process $a(t,k)$ with the expectation value $\langle \cdot \rangle$ is a stationary (i.e. time-translation-invariant) process. But then also the process

$$b(t,k) := \bar{c}_k e^{-it\omega_k} a_k + s_k e^{it\omega_{-k}} a^+_{-k} \, , \quad b^+(t,k) := c_k e^{it\omega_k} a^+_k + \bar{s}_k e^{-it\omega_{-k}} a_{-k} \tag{2.6.6}$$

referred to the same expectation value is mean zero, Gaussian and stationary, because it is a linear combination of processes with these properties. The time-dependent field $b(t,k), b^+(t,k)$ is called a *squeezed stationary field.* In particular

$$\begin{pmatrix} \langle b^+(t,k)b(t',k')\rangle & \langle b(t,k)b(t',k')\rangle \\ \langle b^+(t,k)b^+(t',k')\rangle & \langle b(t,k)b^+(t',k')\rangle \end{pmatrix}$$

$$= \begin{pmatrix} \langle b^+(0,k)b(t'-t,k')\rangle & \langle b(0,k)b(t'-t,k')\rangle \\ \langle b^+(0,k)b^+(t'-t,k')\rangle & \langle b(0,k)b^+(t'-t,k')\rangle \end{pmatrix} .$$

Explicitly the diagonal part of the covariance is equal to

$$\begin{pmatrix} \langle b^+(0,k)b(t,k')\rangle & 0 \\ 0 & \langle b(0,k)b^+(t,k')\rangle \end{pmatrix} = \delta(k-k') \tag{2.6.7}$$

$$\cdot \begin{pmatrix} |c_k|^2 e^{-it\omega_{+k}} n(k) + |s_k|^2 e^{it\omega_{-k}} m(k) & 0 \\ 0 & |c_k|^2 e^{it\omega_{+k}} m(k) + |s_k|^2 e^{-it\omega_{-k}} n(k) \end{pmatrix} ,$$

while the off-diagonal part of the covariance is equal to

$$\begin{pmatrix} 0 & \langle b(0,k)b(t,k')\rangle \\ \langle b^+(0,k)b^+(t,k')\rangle & 0 \end{pmatrix} = \delta(k+k') \tag{2.6.8}$$

$$\cdot \begin{pmatrix} 0 & c_k \bar{s}_{-k} e^{-it\omega_k} n(k) + c_{-k} \bar{s}_k e^{it\omega_{-k}} m(k) \\ \bar{c}_k s_{-k} e^{it\omega_k} n(k) + \bar{c}_{-k} s_k e^{-it\omega_{-k}} m(k) & 0 \end{pmatrix} .$$

Remark 2.6.1. Instead of the b fields (2.6.2), one could also define the fields

$$d_k := \bar{c}_k a_k + s_k a^+_k \quad , \quad d^+_k := c_k a^+_k + \bar{s}_k a_k \, , \tag{2.6.9}$$

leading to the covariance

$$\begin{pmatrix} \langle d^+_k d_{k'}\rangle & \langle d_k d_{k'}\rangle \\ \langle d^+_k d^+_{k'}\rangle & \langle d_k d^+_{k'}\rangle \end{pmatrix}$$

$$= \begin{pmatrix} (|c_k|^2 n(k) + |s_k|^2 m(k)) & (c_k \bar{s}_k n(k) + \bar{s}_k c_k m(k)) \\ (\bar{c}_k s_k n(k) + s_k \bar{c}_k m(k)) & (|c_k|^2 m(k) + |s_k|^2 n(k)) \end{pmatrix} \delta(k-k') \, , \tag{2.6.10}$$

and also in this case, by the same argument as above, one would obtain a mean zero, Gaussian, time-translation-invariant field. However it is not clear if these states can be physically realized since for them simultaneous pair creation (or annihilation) would take place without conservation of momentum.

Remark 2.6.2. A field $b(t,k), b^+(t,k)$ which satisfies (2.6.6), or the analogous formula deduced from (2.6.7), can be said to have a *quasi-free evolution.* From a mathematical point of view the discussion of free and quasi-free evolutions can be unified using the formalism of second quantization and test functions (it is easy to verify that a quasi-free evolution is essentially a free evolution in a different 1-particle space). However we shall see that from the physical point of view their behaviour is quite different therefore we prefer to keep these two notions distinguished.

Remark 2.6.3. Since a gauge-invariant state is invariant under all free evolutions, it follows that nothing changes in the above discussion if, instead of (2.6.6), the time-dependent field $b(t,k)$ is defined by

$$b(t,k) := \bar{c}_k e^{-it\omega_k} a_k + s_k e^{it\omega'_{-k}} a^+_{-k} , \tag{2.6.11}$$

where ω_k and ω'_k are arbitrary real functions. In particular the state induced on the algebra of the b field will still be mean zero, Gaussian and time-translation invariant. This remark will have important consequences in what follows (see Chap. 3).

2.7 Positivity of the Covariance

It remains to be proved that the covariance, given by (2.6.7–8) is positive definite. This follows from the following lemma.

Lemma 2.7.1. *Let $(\sigma_{jk}(k,k'))$ be a positive definite matrix-valued smooth function. Then the matrix-valued distribution*

$$\begin{pmatrix} \sigma_{11}(k,k')\delta(k-k') & \sigma_{12}(k,k')\delta(k+k') \\ \sigma_{21}(k,k')\delta(k+k') & \sigma_{22}(k,k')\delta(k-k') \end{pmatrix} \tag{2.7.1}$$

is positive definite.

Proof 2.7.1. By Schur's lemma it is sufficient to prove the theorem in the case in which $\sigma_{ij} = 1$ for $i,j = 1,2$. In this case, using Fourier transform, (2.7.1) becomes

$$\int_{\mathbb{R}} dt \begin{pmatrix} e^{itk}e^{-itk'} & e^{itk}e^{itk'} \\ e^{-itk}e^{-itk'} & e^{-itk}e^{itk'} \end{pmatrix} = \int_{\mathbb{R}} dt \begin{pmatrix} e^{itk} \\ e^{-itk} \end{pmatrix} \begin{pmatrix} e^{-itk'} & e^{itk'} \end{pmatrix} \tag{2.7.2}$$

and the thesis follows because the right-hand side of (2.7.2) is an integral of positive definite matrices.

2.8 Dynamical Systems in Equilibrium: the KMS Condition

By extending to the quantum case Boltzmann's variational argument, von Neumann [vonN33] proved that, for a finite system with a state space that is an Hilbert space $\mathcal{H}$, the equilibrium states at *inverse temperature* $\beta = 1/kT > 0$ (T is interpreted as temperature and k is the Boltzmann constant) have the form

$$\langle x \rangle = \mathrm{Tr}(e^{-\beta H} x)/\mathrm{Tr}(e^{-\beta H}), \tag{2.8.1}$$

where $H = H^*$ is a self-adjoint operator on $\mathcal{H}$ such that

$$\mathrm{Tr}(e^{-\beta H}) < +\infty \tag{2.8.2}$$

The state $\langle \cdot \rangle$ is called the Gibbs state at inverse temperature $\beta > 0$, and its density matrix is

$$W = e^{-\beta H}/\mathrm{Tr}(e^{-\beta H}) \quad , \quad 0 \le W \le 1 \, , \quad \mathrm{Tr}W = 1 \, . \tag{2.8.3}$$

Now let

$$x \to x(t) := e^{itH} x e^{-itH} \quad , \tag{2.8.4}$$

x being an observable, be the Heisenberg dynamics associated with H.

Theorem 2.8.1. *In the above notations the following are equivalent:*

(i) The state $\langle \cdot \rangle$ has the form (2.8.1).
(ii) For any pair of observables x, y the map $t \mapsto \langle xy(t) \rangle$ can be analytically continued and the state $\langle \cdot \rangle$ satisfies the following condition, called the KMS condition at inverse temperature $\beta > 0$:

$$\langle xy(i\beta) \rangle = \langle yx \rangle \, . \tag{2.8.5}$$

Proof 2.8.1. (i) implies (ii). If the state $\langle \cdot \rangle$ has the form (2.8.1) then, denoting $N = 1/\mathrm{Tr}(e^{-\beta H})$, one has

$$\langle xy(i\beta) \rangle = N\mathrm{Tr}(e^{-\beta H} x e^{-\beta H} y e^{\beta H}) = N\mathrm{Tr}(x e^{-\beta H} y) = \langle yx \rangle \, .$$

(ii) implies (i). If (2.8.5) holds then, denoting W the density matrix of $\langle \cdot \rangle$, we can write it in the form

$$\mathrm{Tr}(W x e^{-\beta H} y e^{\beta H}) = \mathrm{Tr}(W y x) \, . \tag{2.8.6}$$

Making $x = 1$, we deduce $e^{\beta H} W e^{-\beta H} = W$, i.e. $e^{\beta H}$ and W commute and (2.8.6) becomes

$$\mathrm{Tr}(W e^{\beta H} x e^{-\beta H} y) = \mathrm{Tr}(x W y) \, .$$

Since y is arbitrary, this is equivalent to

$$We^{\beta H} x e^{-\beta H} = xW\,,$$

or

$$We^{\beta H} x = xWe^{\beta H}\,.$$

Since also x is arbitrary, this implies that $We^{\beta H}$ is a constant. The normalization condition $\mathrm{Tr}(W) = 1$ fixes this constant to be $1/\mathrm{Tr}(e^{-\beta H})$.

Thus for a finite system the Gibbs formula for the equilibrium state is equivalent to the KMS property. For an infinite system there are no density matrices so one usually takes the KMS condition as the definition of the equilibrium states. There are several indications that this is a reasonable definition in the sense that it describes the physics of equilibrium systems.

2.9 Equilibrium States: the KMS Condition

For an Heisenberg evolution u_t, the KMS condition is

$$\langle x u_{t+i\beta}(y)\rangle = \langle u_t(y)x\rangle\ , \quad \forall\, x, y \in \mathcal{A}\,. \tag{2.9.1}$$

Taking the Fourier transform of both sides, this becomes

$$\int_{-\infty}^{+\infty} \langle x u_{t+i\beta}(y)\rangle e^{it\omega} dt = \int_{-\infty+i\beta}^{+\infty+i\beta} \langle x u_\tau(y)\rangle e^{i(\tau - i\beta)\omega} d\tau = \int_{-\infty}^{+\infty} \langle u_t(y)x\rangle e^{it\omega} dt\,. \tag{2.9.2}$$

Equivalently

$$e^{\beta\omega} \int_{-\infty}^{+\infty} \langle x u_t(y)\rangle e^{it\omega} dt = \int_{-\infty}^{+\infty} \langle u_t(y)x\rangle e^{it\omega} dt\,. \tag{2.9.3}$$

So the KMS condition is equivalent to a symmetry property of the right and the left spectral functions associated with the dynamical system.

Remark 2.9.1. Since the state φ is u_t invariant, it follows that the map $\big((x, s), (y, t)\big) \mapsto \varphi\big(u_s(x^*)u_t(y)\big) = \varphi\big(x^* u_{t-s}(y)\big)$ is of positive type, and therefore its Fourier transform

$$(x \mid y)(\omega) := \int_{-\infty}^{+\infty} e^{i\omega t} \varphi\big(x^* u_t(y)\big) dt$$

must be a pre-scalar product for any ω. The KMS condition then is

$$(x \mid y)(\omega) = e^{-\beta\omega t}(y^* \mid x^*)(-\omega)\,.$$

To understand the relation between the two scalar products, one has to perform the stochastic limit backward in time.

2.10 q-Gaussian Equilibrium States

In this section we shall consider a field $a_k, a^+_{k'}$ satisfying the q-commutation relations

$$a_k a^+_{k'} - q a^+_{k'} a_k = \delta(k-k')\,, \tag{2.10.1}$$

and we want to know when a Gaussian state $\langle\,\cdot\,\rangle$ is an equilibrium state, i.e. it satisfies the KMS condition, for a given free dynamics, determined by the 1-particle evolution $e^{it\omega_k}$. Such a state must be necessarily invariant for the dynamics. In this section we shall assume that the correlation matrix of the state $\langle\,\cdot\,\rangle$ is diagonal and has the form (2.5.9–10).

Theorem 2.10.1. *In the notations and assumptions stated above, the Gaussian state $\langle\,\cdot\,\rangle$ is an equilibrium state, for the free dynamics, determined by $e^{it\omega_k}$ at inverse temperature β, if and only if its covariance has the form*

$$\begin{pmatrix} \langle a^+_k a_{k'}\rangle & 0 \\ 0 & \langle a_k a^+_{k'}\rangle \end{pmatrix} = \begin{pmatrix} n(k) & 0 \\ 0 & m(k) \end{pmatrix} \delta(k-k') = \begin{pmatrix} \frac{1}{e^{\beta\omega_k}-q} & 0 \\ 0 & \frac{e^{\beta\omega_k}}{e^{\beta\omega_k}-q} \end{pmatrix} \delta(k-k')\,. \tag{2.10.2}$$

Proof 2.10.1. If $\langle\,\cdot\,\rangle$ is a Gaussian state, then it is sufficient to check the KMS condition for the two-point function. This gives in particular

$$\langle a(0,k) a^+(t+i\beta,k')\rangle = \langle a^+(t,k') a(0,k)\rangle\,. \tag{2.10.3}$$

If the evolution is free, this means

$$e^{i(t+i\beta)\omega_{k'}} \langle a_k a^+_{k'}\rangle = e^{it\omega_{k'}} \langle a^+_{k'} a_k\rangle\,,$$

or equivalently

$$e^{-\beta\omega_{k'}} \langle a_k a^+_{k'}\rangle = \langle a^+_{k'} a_k\rangle\,. \tag{2.10.4}$$

Therefore (2.10.4) becomes

$$e^{-\beta\omega_{k'}} m(k)\delta(k-k') = n(k)\delta(k-k')\,,$$

or equivalently

$$e^{-\beta\omega_k} m(k) = n(k)\,.$$

Any pair $m(k), n(k)$ with this property can be written in the form

$$m(k) = \frac{\alpha_k}{1-e^{-\beta\omega_k}}\,, \quad n(k) = \frac{\alpha_k}{e^{\beta\omega_k}-1}\,,$$

where α_k is a numerical function which includes additional physical information on the state (q-deformed Gaussian, equilibrium states, etc.).

If the Gaussian field satisfies q-commutation relations (2.10.1), then the expectation value of (2.10.1) gives

$$\langle a_k a^+_{k'} \rangle - q\langle a^+_{k'} a_k \rangle = \big(m(k) - qn(k)\big)\delta(k-k') = \delta(k-k')\,,$$

or equivalently

$$m(k) - qn(k) = 1\,.$$

So, if the equilibrium condition (2.10.4) also holds, this gives

$$\alpha_k = \frac{e^{\beta\omega_k} - 1}{e^{\beta\omega_k} - q}\,,$$

so that

$$n(k) = \frac{1}{e^{\beta\omega_k} - q} \quad , \quad m(k) = \frac{e^{\beta\omega_k}}{e^{\beta\omega_k} - q} = n(k) + \frac{1}{q}\,. \tag{2.10.5}$$

Conversely, if $\langle\,\cdot\,\rangle$ is a Gaussian state with covariance (2.10.2), then the KMS condition is easily verified.

Remark 2.10.1. Note that if $q \geq 0$ the positivity of $n(k), m(k)$ imposes some restrictions on the existence of these states (e.g. if the minimum of ω_k is zero, then q must be ≤ 1).

Remark 2.10.2. Note that in the limit $\beta \to +\infty$ if ω_k is a strictly positive function then a finite temperature state reduces to the Fock state.

Remark 2.10.3. In some important cases (e.g. Bose condensation, see Sect. 2.17) $n(k)$ is a distribution and (2.10.2) is understood in the sense of convolution of distributions.

2.11 Boson Gaussianity

Here we use the notations of Definition 2.2.1, and we consider the boson case. We write $E(b) = \langle b \rangle$ for the expectation value. The following considerations hold whenever the range of E is commutative (in the present book this range is the complex numbers). Then the identity (2.2.2) becomes

$$E(b_1 \dots b_{2p}) = \sum_{\substack{i_1 < \dots < i_p \\ i_\alpha < j_\alpha}} E(b_{i_1} b_{j_1}) \dots E(b_{i_p} b_{j_p})\,. \tag{2.11.1}$$

If in (2.11.1) we choose

$$b_1 = b_2 = \dots = b_{2p} =: b\,, \tag{2.11.2}$$

then we obtain

$$E(b^{2p}) = \frac{(2p)!}{2^p p!} E(b^2)^p\,, \qquad \forall p \in \mathbf{N}\,, \tag{2.11.3}$$

where the combinatorial factor $(2p)!/(2^p p!)$ counts the number of terms in (2.11.1).

An element $b \in \mathcal{B}$ satisfying (2.11.3) and

$$E(b^{2p+1}) = 0 \quad , \quad \forall p \in \mathbf{N}, \tag{2.11.4}$$

will be called the *mean zero* (or centered) *E-Gaussian element.* Note that if E is a $*$-map [i.e. $E(b)^* = E(b^*)$], then b is E-Gaussian if and only if b^* is.

Corollary 2.11.1. *Let $\mathcal{B}$ be a topological $*$-algebra; let $E : \mathcal{B} \to \mathbf{C}$ be a continuous linear map; and let $b \in \mathcal{B}$ be a mean zero, boson E-Gaussian element. Then for every complex number z one has*

$$\sum_{n=0}^{\infty} \frac{z^n}{n!} E(b^n) = e^{\frac{1}{2}z^2 E(b^2)} . \tag{2.11.5}$$

Proof 2.11.1. From (2.11.3–4) one has

$$\sum_{n=0}^{\infty} \frac{z^n}{n!} E(b^n) = \sum_{p=0}^{\infty} z^{2p} \frac{E(b^{2p})}{(2p)!} = \sum_{p=0}^{\infty} z^{2p} \frac{E(b^2)^p}{2^p p!} = e^{\frac{1}{2}z^2 E(b^2)} . \tag{2.11.6}$$

Remark 2.11.1. Note that formally (2.11.5) is equivalent to the well-known identity for Gaussian measures:

$$E(e^{zb}) = e^{\frac{1}{2}z^2 E(b^2)} . \tag{2.11.7}$$

This formal argument becomes rigorous if the algebras $\mathcal{B}$ and $\mathcal{C}$ have a topology which allows one to speak of the convergence of the exponential series and the linear functional E is continuous for this topology, so to allow the exchange of the summation and the expectation in (2.11.7).

One can show that (2.11.7) characterizes boson Gaussianity.

2.12 Boson Fock Fields

In this section we discuss the *boson Fock fields*, the description of which includes some of the most famous classical stochastic processes, such as Brownian motion and the Poisson process. We shall describe these fields in the distribution language. The rule explained in Sect. 2.1 allows to translate this description in the test function language.

Definition 2.12.1. *Let G be a positive distribution on $\mathbb{R}^d$. A free* scalar boson Fock field *on $\mathbb{R}^d$ with* covariance *G is defined by*

- *an Hilbert space $\mathcal{H}$, called the* Fock space*;*
- *pair of operator-valued distributions $a(k), a^+(k)$ ($k \in \mathbb{R}^d$) called respectively* creation *and* annihilation *densities;*

- *a unit vector* Φ *in* $\mathcal{H}$, *called* the vacuum vector *and satisfying*

$$a(k)\Phi = 0\,, \tag{2.12.1}$$

$$\Phi \in \mathrm{Dom}\,(a^+(k)) \qquad , \ \forall\, k \in \mathbb{R}^d\,. \tag{2.12.2}$$

Moreover the n-particle *(or* number*) vectors*

$$\{(a^+(k_1)\dots a^+(k_n))\Phi \ :\ n \in \mathbb{N}\} \tag{2.12.3}$$

are total in $\mathcal{H}$ *and belong to the domain of* $a_{k_1}\dots a_{k_n}$ *for any* n *and for any* $k_1\dots k_n \in \mathbb{R}^d$ *and on the linear span of these vectors they satisfy*

$$\langle \xi, a(k)\eta\rangle = \langle a^+(k)\xi, \eta\rangle \tag{2.12.4}$$

and the so-called canonical commutation relations (CCR)

$$[a(k), a^+(k')] = G(k)\delta(k-k')\,, \tag{2.12.5a}$$

$$[a(k), a(k')] = 0\,. \tag{2.12.5b}$$

Remark 2.12.1. The reason why the function G is called the *covariance* of the field is clarified by the following:

Theorem 2.12.1. *The random variables* $a(k), a^+(k')$ *are mean zero boson Gaussian variables with respect to the vacuum vector* Φ, *with* 2*-point functions (or covariance)*

$$\begin{pmatrix} \langle a_k^+ a_{k'}\rangle & \langle a_k a_{k'}\rangle \\ \langle a_k^+ a_{k'}^+\rangle & \langle a_k a_{k'}^+\rangle \end{pmatrix} = \begin{pmatrix} 0 & 0 \\ 0 & G(k) \end{pmatrix} \delta(k-k')\,. \tag{2.12.6}$$

Conversely, if $a(k), a^+(k')$ *are random variables with these properties, then they satisfy the commutation relations (2.12.5a,b) (so they are a free boson Fock field).*

Proof 2.12.1. The identities (2.12.6) are immediate consequences of (2.12.1) and (2.12.5a,b). The boson Gaussianity is proved by induction on the n-point correlators of $a(k), a^+(k')$: if the first (from left) term of the correlator is a creation or the last an annihilation, then the correlator is zero. Suppose the first creator is at the kth place, with $k > 1$. Commute it with all the annihilators on its left so as to bring it to the first place. Because of (2.12.5a,b) each commutation gives a 2-point function. The induction assumption plus a little combinatorial argument show that, starting from all the pair partitions on $2p-2$ points, one arrives at all the pair partitions on $2p$ points, completing the proof.

To prove the converse, one expands the generic matrix element of the commutators (2.12.5a) in two n-particle vectors ξ, η and checks that all the terms in which $a(k)$ and $a^+(k')$ are not paired between themselves vanish, while the terms in which they are paired between themselves give rise to $G(k)\delta(k-k')\langle\xi,\eta\rangle$. A similar proof can be made with (2.12.5b).

Remark 2.12.2. From (2.12.5a,b) and (2.12.6) one sees that for boson Fock fields the 2-point function coincides with the commutator. This is also true for Fermi Fock fields with anticommutator replacing the commutator, but in general is not true for nonFock fields.

From Theorem 2.12.1 one deduces that two boson (or fermion) Fock fields with the same covariance are equivalent, i.e. they have the same correlators. Moreover, limits of Fock fields are still Fock fields.

If $G(k) \equiv 1$, i.e. if instead of the commutation relations (2.1.2) $a(k), a^+(k')$ satisfy

$$[a(k), a^+(k')] = \delta(k - k'), \tag{2.12.7}$$

we simply speak of the *scalar boson Fock field* on $\mathbb{R}^d$.

Passing from the distribution to the operator picture, i.e. integrating the above identities with respect to test functions $f_1 \dots f_n$ one obtains

$$A^+(f) = \int a^+(k) f(k) dk, \tag{2.12.8}$$

and for the number vectors

$$\psi_N = A^+(f_1) \dots A^+(f_N) \Phi, \tag{2.12.9}$$

where Φ is the Fock vacuum. In particular (2.12.5a,b) become, in this picture,

$$[A(f), A^+(g)] = \langle f, g \rangle \quad , \quad [A(f), A(g)] = 0, \tag{2.12.10}$$

where the scalar product is given by (2.12.5a,b). The relations (2.12.10) are called *the Heisenberg commutation relations* or the canonical commutation relations in unbounded form.

Remark 2.12.3. An equivalent (up to technical conditions on the domains) formulation of (2.12.10) can be given in terms of the *Weyl operators*. Such a formulation is convenient for a translation of several aspects of quantum field theory uniquely in terms of bounded operators, but requires a much lengthier preparation before entering in the heart of the problems, i.e. interacting fields and several of the fundamental observables of the theory, which are unbounded, are expressed only in an indirect way in terms of bounded objects.

Remark 2.12.4. In (2.12.5a) one might replace the term $G(k)\delta(k - k')$ by a positive definite distribution $G(k, k')$, thus obtaining a slightly more general class of fields.

Remark 2.12.5. It is not known at the moment if the GNS representation of every Gaussian state give rise to some kind of commutation relation. The main difficulty in answering this question lies in the fact that an explicit classification of all the Gaussian states in the sense of Definition 2.4.1 is not available. This would require a classification of the possible weights and the possible subsets of the set of all pair partitions that can give rise to a state on the noncommutative polynomial algebra generated by b_j.

2.13 Free Hamiltonians for Boson Fock Fields

Lemma 2.13.1. *If $a^+(k)$ and $a(k)$ are boson creation and annihilation operators satisfying the usual commutation relations*

$$[a(k), a^+(k')] = \delta(k - k') \quad , \; k, k' \in \mathbf{R}^3 \, , \tag{2.13.1}$$

for any 1-particle free evolution $S_t = e^{it\omega(k)}$, the associated free Hamiltonian has the form

$$H_0 = \int \omega(k) a^+(k) a(k) dk \, . \tag{2.13.2}$$

In the sense that the commutator, with H_0, of any polynomial in the field operators coincides with the commutator of the same polynomial with the right-hand side of (2.13.2).

Proof 2.13.1. Since the free evolution is multiplicative, i.e. in the notations (1.19.5),

$$u_t^0(xy) = u_t^0(x) u_t^0(y) \quad , \quad \forall x, y \, , \tag{2.13.3}$$

it follows that any polynomial in the field operators is in the domain of the commutator with H_0. Again by multiplicativity, in order to check the identity of the commutator of such a polynomial with H_0 and with the right-hand side of (2.13.3), it is sufficient to check it on $a(k)$ and $a^+(k)$. This follows from the commutation relations

$$[a_k^+, a_h^+ a_h] = a_h^+ [a_k^+, a_h] = -a_h^+ \delta(k - h) \, ,$$

from which one deduces

$$[H_0, a_k^+] = \int \omega(h) [a_h^+ a_h, a_k^+] dh = \omega(k) a_k^+ \, ,$$

and therefore

$$\left[\int \omega(h) a_h^+ a_h dh, a_k \right] = -\omega(k) a_k \, ;$$

hence Definition 2.4.2 implies that

$$\partial_t a_k = -i\omega(k) a_k \quad , \quad a_k(0) = a_k \, ,$$

$$a_k(t) = e^{-it\omega(k)} a_k \quad , \quad a_k^+(t) = e^{it\omega(k)} a_k \, .$$

So the evolution generated by the Hamiltonian (2.13.2) is indeed a free evolution in the sense of Definition 2.4.2. The action of the free evolution (2.13.2) on the Weyl operators is

$$e^{itH_0} W(g) e^{-itH_0} = W\left(e^{it\omega(p)} g \right) \, . \tag{2.13.4}$$

2.14 White Noises

Definition 2.14.1. *A quantum field $b^{\pm}(t,k)$ $(t \in \mathbb{R}, k \in \mathbb{R}^d)$ with expectation value $\langle \cdot \rangle$, is called a* white noise *if it is a mean zero Gaussian field with covariance*

$$\langle b^{\varepsilon}(t,k) b^{\varepsilon'}(t',k')\rangle = \delta(t-t') G_{\varepsilon,\varepsilon'}(k,k'),$$

where $G_{\varepsilon,\varepsilon'}(k,k')$ is a positive definite distribution (as a function of the variables $\varepsilon, k, \varepsilon', k'$).

2.15 Boson Fock White Noises and Classical White Noises

A basic feature common to all the models to which the stochastic limit approximation can be applied is that, in the stochastic limit, the rescaled fields (see Sect. 3.5 for a discussion of this notion) converge, in the sense of correlators, to a *quantum (boson, Fermi, free, etc.) white noise* (see Theorem 3.7.1 below for the simplest occurrence of this phenomenon). The prototype of such noises is the *boson Fock white noise.*

Definition 2.15.1. *In the notations of Definition 2.12.1, a* boson Fock white noise *on $\mathbb{R}^d$ is a boson Fock field on $\mathbb{R}^{d+1}$ with vacuum vector Φ and commutator of the form*

$$[b(t,k), b^+(t',k')] = \delta(t-t') G(k) \delta(k'-k), \tag{2.15.1a}$$

$$[b(t,k), b(t',k')] = 0, \tag{2.15.1b}$$

$$b(t,k)\Phi = 0. \tag{2.15.1c}$$

Definition 2.15.2. *A* classical white noise *on $\mathbb{R}^{d+1}$, with covariance G, is a classical self-adjoint stochastic process $X(t,k)$ indexed by $\mathbb{R}^{d+1}$ with a cyclic vector Φ such that*

(i) the family $X(t,k)$ is Φ-Gaussian; and
(ii) the $X(t,k)$ are δ-correlated in time, i.e.

$$\langle \Phi, X(t,k) X(t',k')\Phi\rangle = \delta(t-t') G(k) \delta(k'-k). \tag{2.15.2}$$

Definition 2.15.1 is justified by the following remark: defining the two stochastic processes

$$q(t,k) := b(t,k) + b^+(t,k) \tag{2.15.3}$$

$$p(t,k) := \frac{1}{i}\left[b(t,k) - b^+(t,k)\right] \tag{2.15.4}$$

one has

Lemma 2.15.1. *Both stochastic processes $\{q(t,k)\}$ and $\{p(t,k)\}$ are isomorphic to the classical white noise on $\mathbb{R}^{d+1}$. Moreover, if b satisfies (2.15.1a,b), then*

$$[q(t,k), p(t',k')] = 2i\delta(t-t')G(k)\delta(k'-k)\,. \tag{2.15.5}$$

Proof 2.15.1. Both processes are Gaussian as they are a linear combination of Gaussian processes. That they are classical, i.e. all their random variables commute, follows from the parity of the δ-function and the commutator identity (2.15.1a,b). Therefore to prove that each of the processes $\{q(t,k)\}$ and $\{p(t,k)\}$ is isomorphic to the classical white noise on $\mathbb{R}^{d+1}$ with propagator G, it suffices to show that their vacuum 2-point functions coincide. Using (2.15.1c), this is an easy computation.

2.16 Boson Fock White Noises and Classical Wiener Processes

The operators

$$B_t(g) = \int_0^t d\tau \int_{\mathbb{R}^d} dk\bar{g}(k)b(\tau,k) \quad , \quad B_t^+(g) = \int_0^t d\tau \int_{\mathbb{R}^d} dkg(k)b^+(\tau,k) \tag{2.16.1}$$

define a *quantum Brownian motion.* The self-adjoint operators (for varying t and fixed g)

$$P_t(g) := \frac{1}{i}\left[B_t(g) - B_t^+(g)\right] \tag{2.16.2}$$

form a commuting family of classical random variables whose statistics in the *vacuum vector* Φ is completely determined by the relation

$$\left\langle \Phi, e^{iP_t(g)}\Phi \right\rangle = e^{-\frac{t}{2}\|g\|^2} = \exp\left(-\frac{t}{2}\int_{\mathbb{R}^d} dk\,|g(k)|^2\right)\,. \tag{2.16.3}$$

Therefore, if the dk integral is equal to 1, the family $\big(P_t(g)\big)$ can be identified to a standard classical Wiener process.

2.17 Boson Thermal Statistics and Thermal White Noises

In general a thermal equilibrium state for a system of conserved particles is determined by two parameters, β and μ, so that the *energy per unit volume* and the *number of particles per unit volume* (density) may have independent arbitrary values. When the interaction Hamiltonian does not preserve the number of particles, it is natural to consider only the energy per unit volume as given so that the number of particles is no longer an independent parameter and in this case one puts $\mu = 0$ in (2.17.4), thus obtaining (2.17.3). This is usually done for photons (Planck's formula) and for phonons or, more generally for thermal fields.

From Theorem 2.10.1 we deduce that a Gaussian *equilibrium state*, at inverse temperature β, for the Bose field $a(k)$ with standard commutation relations

$$[a(k), a^+(k')] = \delta(k - k') \tag{2.17.1}$$

is a Gaussian state on the polynomial algebra generated by the creation and annihilation operators (2.17.1) with covariance

$$\begin{pmatrix} \langle a_k^+ a_{k'} \rangle & \langle a_k a_{k'} \rangle \\ \langle a_k^+ a_{k'}^+ \rangle & \langle a_k a_{k'}^+ \rangle \end{pmatrix} = \begin{pmatrix} n(k) & 0 \\ 0 & 1 + n(k) \end{pmatrix} \delta(k - k') , \tag{2.17.2}$$

where the factor $1 + n(k)$ arises from the commutation relations, $n(k)$ is the Bose–Einstein density

$$n(k) := \frac{e^{-\beta\omega_1(k)}}{1 - e^{-\beta\omega_1(k)}} = \frac{1}{e^{\beta\omega_1(k)} - 1} , \tag{2.17.3}$$

and $H_R^1 = \omega_1(p)$ is the free 1-particle Hamiltonian in momentum representation [see (2.4.3)]. The passage to the *grand canonical* state at inverse temperature β, *chemical potential* $\mu \in \mathbb{R}$ and *fugacity* $z = e^{\beta\mu}$ is realized by replacing $\omega_1(k)$ in (2.17.3) by $\omega_1(k) - \mu$. This leads to the density

$$n(k) = \frac{e^{-\beta[\omega_1(k) - \mu]}}{1 - e^{-\beta[\omega_1(k) - \mu]}} = \frac{z e^{-\beta\omega_1(k)}}{1 - z e^{-\beta\omega_1(k)}} . \tag{2.17.4}$$

This representation can be realized on an Hilbert space $\mathcal{H}$ with a cyclic vector Φ; we shall use such a realization throughout. Note however that in this representation it is no longer true that the cyclic vector Φ is annihilated by the annihilator. As shown in Sect. 2.18 a concrete form of this representation can be realized by taking the tensor product of a Fock representation with an anti-Fock one.

Remark 2.17.1. Note that in this representation it is no longer true that the commutator coincides with the 2-point function.

Remark 2.17.2. For fermions the expression of the density (2.17.4) is replaced by the Fermi–Dirac density:

$$n(k) = \frac{1}{e^{\beta[\omega_1(k) - \mu]} + 1} .$$

Thus for fermions the sign of μ can be arbitrary, while for bosons, in order to guarantee the positivity of the correlations (2.17.2) for an arbitrary dispersion function $\omega_1(k)$, one can only have $\mu \leq 0$. For bosons the density, as a function of μ (≤ 0) increases, with a maximum value at $\mu = 0$. In the case $\mu = 0$ the zero-momentum states ($= 0$) might give an additional, singular contribution to the density, described by a multiple of the δ-function, so that the density becomes

$$n(k) = \frac{1}{e^{\beta\omega_1(k)} - 1} + c\delta(k) ,$$

where c is a constant. This additional term accounts for the so-called *Bose–Einstein condensation.*

Definition 2.17.1. *Let γ and γ_Q be positive real numbers. The* boson, finite temperature white noise *with commutator γ and covariance γ_Q is the mean zero Gaussian field b_t, b_t^+ $(t \in \mathbf{R})$ with commutator*

$$[b_t, b_s^+] = \gamma\delta(t-s) \quad , \quad [b_t, b_s] = 0 \tag{2.17.5}$$

and 2-point correlations

$$\langle b_t^+ b_s \rangle = \gamma_Q \delta(t-s) \quad , \quad \langle b_t b_s \rangle = 0 . \tag{2.17.6}$$

2.18 Canonical Representation of the Boson Thermal States

It is often convenient to represent an equilibrium state, in the sense of Sect. 2.10, in terms of Fock states. To this goal, one introduces a pair of independent creation and annihilation operators $c_1(k)$, $c_1^+(k)$, $c_2(k)$, $c_2^+(k)$, with commutation relations

$$[c_i(k), c_j^+(k')] = \delta_{ij}\delta(k-k') \tag{2.18.1}$$

acting in a Fock space $\mathcal{H}_0$ with vacuum vector Φ_0

$$c_i(k)\Phi_0 = 0 \ , \quad i = 1, 2 , \tag{2.18.2}$$

and defines

$$a(k) = \sqrt{n(k)+1}c_1(k) + \sqrt{n(k)}c_2^+(k) \tag{2.18.3}$$

so that

$$a^+(k) = \sqrt{n(k)+1}c_1^+(k) + \sqrt{n(k)}c_2(k) . \tag{2.18.4}$$

Then from (2.18.1, 3, 4) one obtains the commutation relations (2.17.1) for $a(k)$, $a^+(k)$. A concrete way to realize the independent pair $c_1^\varepsilon(k)$, $c_2^\varepsilon(k)$ is to *double* the Fock space by considering $\mathcal{H}_0 \otimes \mathcal{H}_0$ with the corresponding vacuum vector $\Phi = \Phi_0 \otimes \Phi_0$ and to identify $c_1^\varepsilon(k)$ with $c_1^\varepsilon(k) \otimes 1$ and $c_2^\varepsilon(k)$ with $1 \otimes c_2^\varepsilon(k)$.

With these notations one can easily check that the vacuum 2-point function of the operators $a(k')$ and $a^+(k)$ given by (2.10.2, 3, 6). The representation of the CCR algebra (2.10.1) shall be called *the canonical representation of the thermal Gaussian state of Sect. 2.10.*

Notice however that if one wants the evolution of the $a(k)$ operator to be free in the sense of Definition 2.4.2 then one must have

$$a^+(t,k) = e^{it\omega(k)}a^+(k) = \sqrt{n(k)+1}e^{it\omega_k}c_1^+(k) + \sqrt{n(k)}e^{it\omega_k}c_2(k)\,, \tag{2.18.5}$$

so we see that $c_2(k)$ must evolve not like a free annihilator but like a free creator, i.e.

$$c_2(t,k) = e^{it\omega_k}c_2(k)$$

this is equivalent to replacing ω_k by $-\omega_k$ (negative energy) in the c_1, c_2 representation. Such an evolution is easily realized by considering not the 1-particle space itself (*ket vectors*) but its conjugate space (*bra vectors*).

Remark 2.18.1. Note that if one wants to use (2.18.3) to pass from the representation in terms of operator-valued distributions a_k, a_k^+ to a representation in terms of *bona fide operators* then one has to introduce test functions. In the case of the annihilator (2.18.3) gives

$$A(g) = \int dk\overline{g}(k)a_k d_k = \int dk\overline{g}(k)\sqrt{n(k)+1}c_1(k) + \int dk\overline{g}(k)\sqrt{n(k)}c_2^+(k)\,. \tag{2.18.6}$$

Note that on the right-hand side of (2.18.6) *the creator $c_2^+(k)$ goes with the conjugate of the test function g.* This means that if we take elements $g \in L^2(\mathbb{R}^d)$ as test functions of the c_1 field we must take their complex conjugates $\overline{g}$ as test functions of the c_2 field. This is equivalent to replacing the usual scalar product in $L^2(\mathbb{R}^d)$ with its complex conjugate:

$$\langle f, g\rangle = \int \overline{f}(k)g(k)dk \mapsto \langle g, f\rangle = \int f(k)\overline{g}(k)dk\,.$$

This fact is expressed by saying that one takes the Fock representation for the c_1 field and *the anti-Fock representation* for the c_2 field.

Remark 2.18.2. Note that in the above discussion the form (2.10.5) of the functions $m(k), n(k)$ plays no role at all. Therefore the present construction is applicable to all gauge-invariant states. This representation plays an important role in the deduction of the explicit form of the stochastic limit in the nonFock case.

Complementary Material

2.19 Spectral Representation of Quantum White Noise

Given a quantum white noise $b(t)$, its inverse Fourier transform

$$b(\omega) : = \frac{1}{\sqrt{2\pi}} \int dt \quad e^{-i\omega t}\, b(t)$$

is a Bose field in the frequency domain satisfying the relations

$$[b(\omega), b^{+}(\omega')] \;=\; \delta(\omega - \omega')\,, \tag{2.19.1}$$

so operators at different frequencies commute. In this approach the frequency appears as the conjugate variable of time in the Fourier transform, in particular it must be one-dimensional. Moreover, since ω can be any real number ($\omega \in \mathbb{R}$), this means that one is introducing negative frequencies. The operator

$$\int_{\omega}^{\omega+d\omega} b(\omega')d\omega' = b(\chi_{[\omega,\omega+d\omega]})$$

is interpreted as an annihilator of quanta in the bandwidth $[\omega, \omega + d\omega]$, i.e. $b(\omega)$ annihilates one quantum at the frequency ω. The white noise is recovered as the Fourier transform of the field $b(\omega)$:

$$b(t) : = \frac{1}{\sqrt{2\pi}} \int d\omega\; e^{-i\omega t}\, b(\omega)\,. \tag{2.19.2}$$

The representation (2.19.2) of quantum white noise is called *the spectral representation* and is widely used in quantum optics. The emergence of an *energy* unbounded below is not surprising due to the fact that this is an *effective energy.* In Section 4.11 one of the possible mechanisms through which a perfectly legal, bounded-below Hamiltonian can give rise to an effective Hamiltonian which is unbounded below shall be described.

Conversely, if one starts from a Bose field in the frequency domain satisfying the relations (2.19.1) and defines the operators $b(t)$ by (2.19.2), then the commutation relations are δ-like in time:

$$[b(t), b^{+}(t')] \;= \delta(t - t')\,,$$

as required by the definition of white noise. So one can define a white noise starting from its spectral representation. In order to complete this definition one has to assign a (boson) Gaussian expectation value, and this is uniquely determined by the matrix

$$\begin{pmatrix} \langle b^{+}(\omega)\; b(\omega')\rangle & \langle b(\omega)\; b(\omega')\rangle \\ \langle b^{+}(\omega)\; b^{+}(\omega)\rangle & \langle b(\omega')\; b^{+}(\omega')\rangle \end{pmatrix}.$$

The requirement that different modes are uncorrelated is expressed by

$$\langle b^+(\omega)\, b(\omega')\rangle \;=\; \overline{N}(\omega)\, \delta(\omega-\omega')\,.$$

By the commutation relation (2.19.1) this uniquely determines the correlation

$$\langle b(\omega)\, b^+(\omega')\rangle \;=\; [1+\overline{N}(\omega)]\, \delta(\omega-\omega')$$

If we want the processes b_s, b_t^+ to be δ-correlated, then $\overline{N}(\omega)$ must be independent of ω, since the quantity

$$\int_{\Omega}^{\Omega+d\Omega} a^+(\omega)a(\omega)\, d\omega$$

denotes the number of quanta in the bandwidth $[\Omega, \Omega+d\Omega]$ (a number one would like to be finite). This means that in a white-noise ensemble the number of quanta per unit bandwidth is constant, i.e.

$$\langle b^+(\omega)\, b(\omega')\rangle_{\mathrm{WN}} := \overline{N}\, \delta(\omega-\omega')\,.$$

This is not the case, for example, in a thermal ensemble at temperature T, where the average number of quanta per unit bandwith is not constant, being given by the Planck formula:

$$\langle b^+(\omega)\, b(\omega)\rangle_T \;=:\; \overline{N}_T(\omega) \;=\; 1/\left(e^{\frac{\hbar\omega}{kT}}-1\right)\,.$$

To give the full vacuum correlations, we must also give the *squeezing correlations* $\langle b(\omega)\, b(\omega')\rangle$. If we want $\langle b_s b_t\rangle = c\delta(t-s)$, these correlations cannot be arbitrary. In fact

$$\langle b_s b_t\rangle = \int\!\!\int e^{-i\omega s}\langle b(\omega)b(\omega')\rangle e^{-i\omega' t} d\omega d\omega'\,. \tag{2.19.3}$$

So if we choose $\langle b(\omega)b(\omega')\rangle = \delta(\omega-\omega')$, then (2.19.2) is equal to

$$\int e^{-i\omega(t+s)} d\omega = 2\pi\delta(t+s)\,,$$

which gives the wrong time correlations. The choice which gives the correct time correlations

$$\langle b_s b_t\rangle = \sigma_{--}\delta(s-t) = \sigma_{--}\delta(t-s)\,,$$

where σ_{--} is a constant, is

$$\langle b_\omega b_{\omega'}\rangle = \delta(\omega+\omega')\,, \tag{2.19.4}$$

i.e. *it is possible to create or annihilate two quanta, but only at opposite frequencies.* Conversely, starting from a quantum white noise, one easily verifies that the associated frequency field satisfies (2.19.3). In this case the correlation matrix is then

$$\langle b_{\omega}^{\varepsilon_1} b_{\omega'}^{\varepsilon_2} \rangle = \sigma_{\varepsilon_1 \varepsilon_2} \delta(\omega - \varepsilon_1 \varepsilon_2 \omega') ,$$

where $\sigma_{\varepsilon_1 \varepsilon_2}$ is a constant matrix,

$$\varepsilon = \pm 1 \ , \quad b^{(-1)} = b \ , \quad b^{(+1)} = b^+ \ .$$

The matrix

$$\begin{pmatrix} \langle b^+(\omega)\, b(\omega)\rangle & \langle b(\omega)\, b(\omega)\rangle \\ \langle b^+(\omega)\, b^+(\omega)\rangle & \langle b(\omega)\, b^+(\omega)\rangle \end{pmatrix} = \begin{pmatrix} \overline{N} & -c \\ -\bar{c} & \overline{N}+1 \end{pmatrix} = \begin{pmatrix} \sigma_{+-} & \sigma_{--} \\ \sigma_{++} & \sigma_{-+} \end{pmatrix}$$

must be positive definite, i.e.

$$1 + 2\overline{N} \geq 0 \qquad , \qquad \overline{N}(\overline{N}+1) \geq |c|^2 \, .$$

The case $c = \overline{N} = 0$ characterizes the Fock (or vacuum) white noise, while the squeezed white noise is characterized by the condition $c \neq 0$.

2.20 Locality of Quantum Fields and Ultralocality of Quantum White Noises

The basic difference between ordinary quantum fields and quantum white noises is that quantum fields satisfy the property of *locality*, but quantum white noises enjoy a property that could be called *ultralocality*. The difference between the two is explained in the following.

The *field density* is expressed, in terms of the creation and annihilation density, by

$$\begin{aligned} \varphi(x^0, x) &= \frac{1}{(2\pi)^{3/2}} \int_{\mathbf{R}^3} e^{ix^o\omega(k) - ikx} \, a^+(k) \frac{dk}{\sqrt{2\omega(k)}} + \text{h.c.} \\ &= \frac{1}{(2\pi)^{3/2}} \int_{\mathbf{R}^3} e^{-ikx} \, a^+(x^0, k) \frac{dk}{\sqrt{2\omega(k)}} + \text{h.c.} \end{aligned} \tag{2.20.1}$$

The density is understood in the sense of distributions, i.e. integrated in the variables x^0, x with appropriate test functions. Let $M = \{\mathcal{H}, \mathcal{A}\}$ be a model of QFT, where the field algebra $\mathcal{A}$ is generated by the set of operators (fields) $\{\varphi_\alpha(f)\}$ and $f \in \mathcal{S}(\mathbb{R}^4)$ is a test function. The locality property is [BoLoOkTo87]

$$[\varphi_\alpha(f_1), \varphi_\beta(f_2)]_{\varepsilon_{\alpha\beta}} = 0 \tag{2.20.2}$$

if $\operatorname{supp}(f_1)$ is space similar to $\operatorname{supp}(f_2)$. Here $[\cdot,\cdot]_-$ denotes the commutator, $[\cdot,\cdot]_+$ the anti-commutator and

$$\varepsilon_{\alpha\beta} = \begin{cases} -1 \text{ if } \varphi_\alpha \text{ or } \varphi_\beta \text{ is a Bose field}, \\ +1 \text{ if } \varphi_\alpha \text{ and } \varphi_\beta \text{ are Fermi fields}. \end{cases}$$

In the above notations, the *ultralocality* condition means that the commutation relations (2.20.1) hold whenever

$$\text{supp}(f_1) \cap \text{supp}(f_2) = \emptyset . \tag{2.20.3}$$

In this case the fields $\varphi_\alpha(f)$ are called *Bose or Fermi multidimensional quantum white noises*. If condition (2.20.3) is replaced by the weaker condition

$$\text{time} - \text{supp}(f_1) \cap \text{time} - \text{supp}(f_2) = \emptyset , \tag{2.20.4}$$

then one speaks of a *Bose or Fermi quantum white noises*.

Example 2.20.1. Free (Gaussian) quantum fields and free (Gaussian) quantum white noises.

Example 2.20.2. Equal time canonical commutation relations for quantum fields:

$$[\varphi_\alpha(t,x), \pi_\beta(t,y)]_{\varepsilon_{\alpha\beta}} = i\delta_{\alpha\beta}\delta(x-y) . \tag{2.20.5}$$

Canonical commutation relations for quantum white noises:

$$[\psi_\alpha(t,x), p_\beta(\tau,y)]_{\varepsilon_{\alpha\beta}} = i\delta_{\alpha\beta}\delta(t-\tau)\delta(x-y) . \tag{2.20.6}$$

Note that φ_α and ψ_α are not necessarily Gaussian.

Remark 2.20.1. The condition of ultralocality (2.20.4) is more restrictive than the condition of locality (2.20.2). Therefore quantum white noises form a subclass of quantum fields.

3. Those Kinds of Fields We Call Noises

In this chapter we discuss one of the main theses of the stochastic limit approach, namely the fact that *in the stochastic limit the quantum fields become white noises,* in the simplest possible situation, i.e. for mean zero Gaussian (free) fields. In these cases one has only to prove the convergence of the covariance (2-point) function and therefore the problem is reduced to the convergence of a d-dimensional integral.

3.1 Convergence of Fields in the Sense of Correlators

There are many notions of convergence of random fields. In this section we introduce the notion of *convergence in the sense of correlators,* which shall be constantly used in the following. The idea is that $a_\lambda \to a$ if the corresponding correlators converge. More precisely:

Definition 3.1.1. *Let $a^{\pm}_{\lambda}(k)$ be a family of fields with cyclic vector Φ_λ, parameterized by a real number λ, and let $a^{\pm}_{k}$ be another field with cyclic vector Φ. $a^{\pm}_{\lambda}(k)$ is said to converge to $a^{\pm}_{k}$ in the sense of correlators as $\lambda \to \lambda_0$ (λ_0 a real number or $+\infty$) if, for any natural integer n, for any $k_1 \dots k_n \in \mathbb{R}^d$, and any choice of $\epsilon_1 \dots \epsilon_n \in \{0,1\}$, one has*

$$\lim_{\lambda\to\lambda_0} \langle \Phi_\lambda, a^{\epsilon_1}_{\lambda}(k_1)\dots a^{\epsilon_n}_{\lambda}(k_n)\Phi_\lambda\rangle = \langle \Phi, a^{\epsilon_1}(k_1)\dots a^{\epsilon_n}(k_n)\Phi\rangle, \tag{3.1.1}$$

where the limit is meant in the sense of distributions.

Definition 3.1.1 requires a priori knowledge of the limit field, $\{a^{\pm}_{k}, \mathcal{H}, \Phi\}$ which, in concrete situations, is hardly ever guessable. The following theorem shows that this knowledge is not necessary, i.e. from the limit of the correlators one can reconstruct, in an abstract sense, the limit field. However a *good* convergence theorem should always be completed with the explicit construction of the limit field.

Theorem 3.1.1. *In the notations of Definition 3.1.1, suppose that, as $\lambda \to 0$, the correlators*

$$\langle \Phi_\lambda, a_\lambda^{\epsilon_1}(k_1)\dots a_\lambda^{\epsilon_n}(k_n)\Phi_\lambda\rangle$$

converge, in the sense of distributions, to a distribution $F(k_1 \dots k_n)$. Then there exists a field $\{a_k^\pm, \mathcal{H}, \Phi\}$ such that

$$F(k_1 \dots k_n) = \langle \Phi, a^{\epsilon_1}(k_1)\dots a^{\epsilon_n}(k_n)\Phi\rangle$$

Proof 3.1.1. The limit preserves the compatibility conditions for correlation kernels (correlators) described in [AFL82]. Therefore the thesis follows from the general reconstruction theorem proved in that paper.

Remark 3.1.1. From Definition 2.2.1 of Gaussianity it immediately follows that

(i) the limit, in the sense of correlators, of a family of Gaussian processes is Gaussian; and
(ii) a family of Gaussian processes converges, in the sense of correlators, if and only if its covariance matrix converges in the sense of distributions.

Thus, for Gaussian processes, checking the convergence, in the sense of correlators, is equivalent to checking the convergence of the 2-point function. In this chapter we shall exploit this fact.

Remark 3.1.2. It is important to keep in mind the fact that the limit in the sense of correlators may destroy the eventual commutation relations among the $a_\lambda^\pm(k)$. The phenomenon of *stochastic bosonization* (see Sect. 11.10), and more generally the *block principle* of Sects. 11.7 and 15.4, provides an example of this statement.

3.2 Generalized White Noises as the Stochastic Limit of Gaussian Fields

In this section we begin to show how quantum white noises arise as stochastic limits of free fields. The result described by Theorem 3.2.1, and its oscillating generalization given by Theorem 3.7.1, are simple examples of *quantum central limit theorems* and will be the first step of the stochastic limit technique. They correspond to the control of the limit of the 0-order term in the iterated series.

Theorem 3.2.1. *Let $a(t,k)$, $a^+(t,k)\,(t \in \mathbb{R}, k \in \mathbb{R}^d)$ be a mean zero Gaussian quantum field in the sense of Definition 2.14.1 with respect to a given expectation value $\langle \cdot \rangle$. In the notations* (2.1.1) *and for $\varepsilon, \varepsilon' \in \{0,1\}$ suppose that the expectation value $\langle \cdot \rangle$ is stationary, i.e.*

$$\langle a^{\varepsilon_1}(t_1+s,k_1)\dots a^{\varepsilon_n}(t_n+s,k_n)\rangle = \langle a^{\varepsilon_1}(t_1,k_1)\dots a^{\varepsilon_n}(t_n,k_n)\rangle, \tag{3.2.1}$$

and that the time correlations $\langle a^{\varepsilon}(0,k)a^{\varepsilon'}(t,k')\rangle$ *are integrable in the sense of distributions i.e.*

$$\int_{-\infty}^{+\infty} dt \left| \int_{\mathbb{R}^d}\int_{\mathbb{R}^d} dk dk' \overline{g}_1(k) g_2(k') \langle a^{\varepsilon}(0,k)a^{\varepsilon'}(t,k')\rangle \right| < \infty \tag{3.2.2}$$

for any pair of test functions g_1, g_2 *in* $\mathcal{S}(\mathbb{R}^d)$, *i.e. the Schwartz functions. Then the limit, in the sense of distribution correlators,*

$$\lim_{\lambda\to 0}\frac{1}{\lambda}\, a(t/\lambda^2,k) = b_t(k) = b(t,k), \tag{3.2.3}$$

exists and is the white noise $b_t(k)$ [*of the same Gaussian type as* $a(t,k)$] *with covariance*

$$\langle b_t^{\varepsilon}(k) b_{t'}^{\varepsilon'}(k')\rangle = \delta(t-t')\hat{G}_{\varepsilon,\varepsilon'}(0,k,k') = \delta(t-t')\int_{-\infty}^{+\infty} d\tau \langle a^{\varepsilon}(0,k)a^{\varepsilon'}(\tau,k')\rangle, \tag{3.2.4}$$

where $\hat{G}_{\varepsilon,\varepsilon'}(0,k,k')$ *is the spectral matrix of the field* $a(t,k)$ *evaluated at zero.*

Remark 3.2.1. When we say that $b_t(k)$ is a white noise *of the same Gaussian type* as $a(t,k)$, we mean that if we start with a field satisfying some commutation relations (Bose, Fermi, q-deformed, etc.) then the limit white noise satisfies the same type of relations. We shall see in Part II that this is not true for the nonlinear extension of the present theorem.

Remark 3.2.2. More explicitly, the covariance of the white noise $b_t(k)$ is

$$\begin{pmatrix} \langle b_t^+(k) b_{t'}(k')\rangle & \langle b_t(k) b_{t'}(k')\rangle \\ \langle b_t^+(k) b_{t'}^+(k')\rangle & \langle b_t(k) b_{t'}^+(k')\rangle \end{pmatrix} = \begin{pmatrix} \hat{G}_{+,-}(0,k,k') & \hat{G}_{-,-}(0,k,k') \\ \hat{G}_{+,+}(0,k,k') & \hat{G}_{-,+}(0,k,k') \end{pmatrix} \delta(t-t'). \tag{3.2.5a}$$

Remark 3.2.3. The limit (3.2.3) has to be understood in the sense of correlators for the associated operator-valued distributions, i.e.

$$\lim_{\lambda\to+\infty} \lambda \int_{-\infty}^{+\infty} dt \int d^3k a(\lambda^2 t,k)\varphi(t)g(k) = \int_{-\infty}^{+\infty} dt \int b(t,k)\varphi(t)g(k)d^3k, \tag{3.2.5b}$$

where $\varphi(t)$ and $g(k)$ are test functions.

Proof 3.2.1. Because of mean zero, stationarity (3.2.1) and Gaussianity we have only to prove the convergence of the covariance $\frac{1}{\lambda^2}\langle a^{\varepsilon}(0,k)a^{\varepsilon'}(t/\lambda^2,k')\rangle$ in the sense of distributions. If for any pair of test functions g_1, g_2 one defines

$$F(t) := \int_{\mathbb{R}^d} dk \int_{\mathbb{R}^d} dk' \overline{g}_1(k) g_2(k') \langle a^{\varepsilon}(0,k)a^{\varepsilon'}(t,k')\rangle,$$

then the thesis follows by applying Lemma 1.9.1 to the function F, which is possible due to assumption (3.2.2).

Remark 3.2.4. The integrals (3.2.4) (for $\varepsilon, \varepsilon' = 0, 1$) can be explicitly calculated in terms of test functions. Here we illustrate this calculation for the aa^+ correlation. Because of Gaussianity, to prove (3.2.3) it is enough to consider only the vacuum 2-point function which has the form

$$I := \langle \Phi, \lambda^{-1} \int\int a(\lambda^{-2}t, k) g(t,k) dt dk \cdot \lambda^{-1} \int a^+(\lambda^{-2}s, p) f(s,p) ds dp \Phi \rangle$$

$$= \lambda^{-2} \int dt \int ds \int dk \; e^{i\lambda^{-2}(s-t)\omega(k)} g(t,k) f(s,k), \tag{3.2.6}$$

where $f(t,k)$ and $g(t,k)$ are test functions on $\mathbb{R}^{d+1}$ and $\omega(k)$ is a nondegenerate almost everywhere smooth function. Introducing the function

$$G(t, s', s) := \int dk \; e^{-it\omega(k)} g(s,k) f(s',k), \tag{3.2.7}$$

one can rewrite the right-hand side of (3.2.6) in the form

$$I = \lambda^{-2} \int dt \; ds \; G\;(\lambda^{-2}(t-s), s, t). \tag{3.2.8}$$

Making the change of variables

$$(t-s)/\lambda^2 = \sigma, \qquad s = \tau, \tag{3.2.9}$$

one finds

$$I = \int\int d\sigma \; d\tau G\;(\sigma, \tau, \tau + \lambda^2 \sigma). \tag{3.2.10}$$

Our assumptions imply that $G(t, s', s)$ is bounded over the third argument and belongs to $L^1(\mathbf{R}^2)$ with respect to t and s'. By dominated convergence one can pass to the limit $\lambda \to 0$ under the integral, obtaining

$$\lim_{\lambda \to 0} I = \int d\sigma \int d\tau \; G\;(\sigma, \tau, \tau), \tag{3.2.11}$$

which, using the definition of the function G, can be rewritten in the form

$$\int d\sigma \; d\tau \int dk \; e^{i\omega(k)\sigma} g(\tau,k) f(\tau,k). \tag{3.2.12}$$

This, in the sense of distribution theory, is equal to

$$2\pi \int d\tau \int dk \delta(\omega(k)) g(\tau,k) f(\tau,k), \tag{3.2.13}$$

and, from another part, we know that this is equal to

$$\langle \Psi, \int d\tau \int dk \; b(\tau,k) g(\tau,k) \int d\tau' \int dk' b^*(\tau',k') f(\tau',k') \Psi \rangle, \tag{3.2.14}$$

which is the 2-point function of a white noise on $\mathbb{R}^d$ with covariance

$$G(k-k') = \delta(\omega(k)) \delta(k-k'). \tag{3.2.15}$$

This proves the limit relation (3.2.3).

In concrete physical situations the operators $a^+(k)$ and $a(k)$ are boson or fermion creation and annihilation operators [see (2.12.1)],

$$[a(k), a^+(k')]_\pm = \delta(k-k') \ , \quad k, k' \in \mathbb{R}^3; \tag{3.2.16}$$

let their free dynamical evolutions be given by [see (2.5.1)]

$$a(t,k) = e^{-it\omega(k)}a(k) \ , \quad a^+(t,k) = e^{it\omega(k)}a^+(k), \tag{3.2.17}$$

where $\omega(k)$ is a nondegenerate bounded below real function such that the function $t \mapsto \int_{\mathbb{R}^d} e^{it\omega_k} g_k dk$ is integrable for any Schwartz test function. In this case the following modification of Lemma 1.9.1 is useful:

Corollary 3.2.1. *Let ω be a real-valued function on $\mathbb{R}^d$ such that for any test function g, the map*

$$t \in \mathbb{R} \mapsto \int_{\mathbb{R}^d} dk g(k) e^{it\omega(k)} \tag{3.2.18}$$

is integrable (we say in this case that $t \mapsto e^{it\omega(k)}$ is integrable in the sense of distributions). Then, in the sense of distributions on $\mathbb{R}^{d+1}$,

$$\lim_{\lambda\to 0} \frac{1}{\lambda^2} e^{i\omega(k)(t-t')/\lambda^2} = \delta(t-t')2\pi\delta(\omega(k)).$$

Proof 3.2.2. Because of Lemma 1.9.1 we have only to prove that for any test function g

$$\int_{-\infty}^{+\infty} d\sigma \int_{\mathbb{R}^d} dk g(k) e^{i\sigma\omega(k)} = 2\pi \int_{\mathbb{R}^d} \delta(\omega(k)) g(k) dk.$$

This follows by exchanging the $d\sigma$ and dk integrals, which can be done because g is a Schwartz test function and the map (3.2.18) is integrable.

Theorem 3.2.2. *If the fields $a(k)$ and $a^+(k)$ are mean zero Gaussian and satisfy the q-deformed relations*

$$a(k)a^+(k') - qa^+(k')a(k) = \delta(k-k'), \tag{3.2.19}$$

then the rescaled fields

$$\frac{1}{\lambda} a\left(\frac{t}{\lambda^2}, k\right) := \frac{1}{\lambda} e^{-\frac{i\omega(k)}{\lambda^2}t} a(k)$$

converge in the sense of distribution correlators to a q-deformed white noise; more precisely,

$$\lim_{\lambda\to 0} \frac{1}{\lambda} e^{-\frac{i\omega(k)}{\lambda^2}t} a(k) = b_t(k) \quad , \quad \lim_{\lambda\to 0} \frac{1}{\lambda} e^{\frac{i\omega(k)}{\lambda^2}t} a^+(k) = b_t^+(k), \tag{3.2.20}$$

where $b_t(k)$ is the q-deformed white noise satisfying

$$b_t(k)b_{t'}^+(k') - qb_{t'}^+(k')b_t(k) = \delta(t-t')2\pi\delta(\omega(k))\delta(k-k'). \tag{3.2.21}$$

Proof 3.2.3. By Gaussianity one reduces the n-point correlation function to a two-point one, and this has the form

$$\left\langle \Phi, \frac{1}{\lambda} a\left(\frac{t}{\lambda^2}, k\right) \frac{1}{\lambda} a^+\left(\frac{t'}{\lambda^2}, k'\right) \Phi \right\rangle = \frac{1}{\lambda^2} e^{\frac{i}{\lambda^2}\omega(k)(t'-t)} \delta(k-k').$$

The result then follows from Theorem 3.2.1 and Corollary 3.2.1.

3.3 Existence of Fock, Temperature and Squeezed White Noises

In the notations of the previous section, let us suppose that the field $a(\tau,k)$ is in the Fock state and suppose that the free evolution is given by (2.4.3) with

$$\omega(k) = \omega_1(k) - \omega_0, \tag{3.3.1}$$

where ω_0 is an *additional frequency* whose physical meaning is explained in Chap. 4. The Bose or Fermi case here does not matter since we are only interested in the covariance which, in both cases, is

$$\begin{pmatrix} \langle a^+(0,k)a(\tau,k')\rangle & \langle a(0,k)a(\tau,k')\rangle \\ \langle a^+(0,k)a^+(\tau,k')\rangle & \langle a(0,k)a^+(\tau,k')\rangle \end{pmatrix} = \begin{pmatrix} 0 & 0 \\ 0 & e^{i\tau[\omega_1(k)-\omega_0]} \end{pmatrix} \delta(k-k'). \tag{3.3.2}$$

Therefore the covariance of the associated Fock white noise is

$$\begin{pmatrix} \langle b^+(0,k)b(\tau,k')\rangle & \langle b(0,k)b(\tau,k')\rangle \\ \langle b^+(0,k)b^+(\tau,k')\rangle & \langle b(0,k)b^+(\tau,k')\rangle \end{pmatrix} = \begin{pmatrix} 0 & 0 \\ 0 & \delta(\omega_1(k)-\omega_0) \end{pmatrix} \delta(\tau)\delta(k-k'). \tag{3.3.3}$$

If the state of the a field is the grand canonical boson state at inverse temperature β and chemical potential $\mu \in \mathbb{R}$, then its covariance is

$$\begin{pmatrix} \langle a^+(0,k)a(\tau,k')\rangle & \langle a(0,k)a(\tau,k')\rangle \\ \langle a^+(0,k)a^+(\tau,k')\rangle & \langle a(0,k)a^+(\tau,k')\rangle \end{pmatrix}$$
$$= \begin{pmatrix} e^{-i\tau[\omega_1(k)-\omega_0]} \frac{e^{-\beta[\omega_1(k)-\mu]}}{1-qe^{-\beta[\omega_1(k)-\mu]}} & 0 \\ 0 & e^{i\tau[\omega_1(k)-\omega_0]} \frac{1}{1-qe^{-\beta[\omega_1(k)-\mu]}} \end{pmatrix} \delta(k-k'), \tag{3.3.4}$$

with $q=+1$ in the Bose case and $q=-1$ in the Fermi case. Therefore the covariance of the associated thermal white noise with the same parameters is

$$\begin{pmatrix} \langle b^+(0,k)b(\tau,k')\rangle & \langle b(0,k)b(\tau,k')\rangle \\ \langle b^+(0,k)b^+(\tau,k')\rangle & \langle b(0,k)b^+(\tau,k')\rangle \end{pmatrix}$$

$$= \begin{pmatrix} \frac{e^{-\beta[\omega_1(k)-\mu]}}{1-qe^{-\beta[\omega_1(k)-\mu]}} & 0 \\ 0 & \frac{1}{1-qe^{-\beta[\omega_1(k)-\mu]}} \end{pmatrix} \delta(\tau)\delta(k-k')\delta(\omega_1(k)-\omega_0). \quad (3.3.5)$$

Now suppose that the field is in the squeezed state described in Sect. 2.6 but with quasi-free evolution given not by (2.6.6) but by

$$a(t,k) := \overline{c}_k e^{-it(\omega_k-\omega_0)} a_k + s_k e^{it(\omega_{-k}+\omega_0)} a^+_{-k}. \quad (3.3.6)$$

Notice that this evolution is obtained simply by multiplying a field $b(t,k)$ which evolves according to the evolution (2.6.6) by the scalar factor $e^{it\omega_0}$. In Sect. 4.10 we shall see how this factor naturally arises from the interaction. The fact that the state induced by the original gauge invariant state (2.6.1) on the algebra generated by the b fields is still mean zero, Gaussian and time-translation invariant follows from Remark 2.2.2. Under these assumptions the diagonal part of the covariance is equal to

$$\begin{pmatrix} \langle a^+(0,k)a(t,k')\rangle & 0 \\ 0 & \langle a(0,k)a^+(t,k')\rangle \end{pmatrix} = \delta(k-k')$$

$$\cdot \begin{pmatrix} |c_k|^2 e^{-it(\omega_k-\omega_0)} n_k \\ +|s_k|^2 e^{it(\omega_{-k}+\omega_0)} m_k & 0 \\ 0 & |c_k|^2 e^{it(\omega_k-\omega_0)} m_k \\ & +|s_k|^2 e^{-it(\omega_{-k}+\omega_0)} n_k \end{pmatrix}, \quad (3.3.7)$$

while the off-diagonal part of the covariance is equal to

$$\begin{pmatrix} 0 & \langle a(0,k)a(t,k')\rangle \\ \langle a^+(0,k)a^+(t,k')\rangle & 0 \end{pmatrix} = \delta(k+k')$$

$$\cdot \begin{pmatrix} 0 & \overline{c}_k s_{-k} e^{it(\omega_{-k}+\omega_0)} m_k \\ & +s_k \overline{c}_{-k} e^{-it(\omega_k-\omega_0)} n_k \\ c_k \overline{s}_{-k} e^{-it(\omega_{-k}+\omega_0)} m_k & 0 \\ +\overline{s}_k c_{-k} e^{it(\omega_k-\omega_0)} n_k & \end{pmatrix}, \quad (3.3.8)$$

with ω_k given by (3.3.1). Then the associated white noise is squeezing with the diagonal part of the covariance equal to

$$\begin{pmatrix} \langle b^+(0,k)b(t,k')\rangle & 0 \\ 0 & \langle b(0,k)b^+(t,k')\rangle \end{pmatrix} = \delta(t)\delta(k-k')$$

$$\cdot \begin{pmatrix} \delta(\omega_k-\omega_0)|c_k|^2 n_k \\ +|s_k|^2 m_k \delta(\omega_{-k}+\omega_0) & 0 \\ 0 & |c_k|^2 m_k \delta(\omega_k-\omega_0) \\ & +|s_k|^2 n_k \delta(\omega_{-k}+\omega_0) \end{pmatrix}, \quad (3.3.9)$$

while the off-diagonal part of the covariance is equal to

$$\begin{pmatrix} 0 & \langle b(0,k)b(t,k')\rangle \\ \langle b^+(0,k)b^+(t,k')\rangle & 0 \end{pmatrix} = \delta(t)\delta(k+k')$$

$$\cdot \begin{pmatrix} 0 & \begin{matrix} \overline{c}_k s_{-k} m_k \delta(\omega_{-k}+\omega_0) \\ +s_k \overline{c}_{-k} n_k \delta(\omega_k - \omega_0) \end{matrix} \\ \begin{matrix} \overline{c}_k s_{-k} n_k \delta(\omega_k - \omega_0) \\ +\overline{c}_{-k} s_k m_k \delta(\omega_{-k} - \omega_0) \end{matrix} & 0 \end{pmatrix} . \tag{3.3.10}$$

3.4 Convergence of the Field Operator to a Classical White Noise

Since the field density is given by

$$\begin{aligned} \varphi(x^0, x) &= \frac{1}{(2\pi)^{3/2}} \int_{\mathbf{R}^3} e^{ix^0\omega(k)-ikx} \, a^+(k) \frac{dk}{\sqrt{2\omega(k)}} + \text{h.c.} \\ &= \frac{1}{(2\pi)^{3/2}} \int_{\mathbf{R}^3} e^{-ikx} \, a^+(x^0, k) \frac{dk}{\sqrt{2\omega(k)}} + \text{h.c.}, \end{aligned} \tag{3.4.1}$$

the rescaling of the annihilation density

$$a(t,k) \to \frac{1}{\lambda} a(t/\lambda^2, k), \tag{3.4.2}$$

is equivalent to the following rescaling of the field density:

$$\varphi(x^0, x) \to \frac{1}{\lambda}\varphi(x^0/\lambda^2, x). \tag{3.4.3}$$

The limit (3.2.3) implies

$$\lim_{\lambda\to 0} \frac{1}{\lambda}\varphi\left(\frac{t}{\lambda^2}, \; x\right) = w(t,x), \tag{3.4.4}$$

where the vacuum correlations of $w(t,x)$ coincide with those of a *classical white noise*.

Complementary Material

3.5 Beyond the Master Equation: The Master Field

The master field (white noise) associated with a given interaction strongly depends on the interaction itself, on the free evolution and also on the reference state. In order to determine its form, one first writes the rescaled interaction Hamiltonian in (1.9.7) as a sum of terms of the form

$$\frac{1}{\lambda} H_{\mathrm{I}}^{(\lambda)}\left(\frac{t}{\lambda^2}\right) = a_{t/\lambda^2}^{(\lambda)+} + a_{t/\lambda^2}^{(\lambda)}, \tag{3.5.1}$$

where the operators $a_t^{(\lambda)+}$, $a_t^{(\lambda)}$ are not necessarily creation and annihilation operators, but certain functions of them (depending on the specific form of the interaction) called *rescaled fields* whose explicit form is strongly model dependent and cannot be given in general (see Sect. 4.9.1 for the simplest example). However an easily applicable heuristic rule which allows one to deduce their correct form is the following: *the form of the rescaled fields is given by the first-order term of the time-rescaled iterated series.* By a quantum central limit theorem effect, these rescaled fields converge, as $\lambda \to 0$, to some new field operators b, called the *master field* or *quantum noise*, acting on some new Hilbert space. Symbolically

$$\frac{1}{\lambda} a_\lambda(t/\lambda^2) \to b(t). \tag{3.5.2}$$

The convergence of the rescaled fields should be interpreted in the sense of convergence of correlation functions (also called *Wightman functions* in quantum field theory or *mixed moments* in probability theory), that is, for any $n \in \mathbb{N}$ and $t_1 \dots t_n \in \mathbb{R}$, there exists the limit

$$\lim_{\lambda\to 0} \left\langle \Phi_\lambda, \frac{1}{\lambda} A_\lambda^{\varepsilon_1}(t_1) \dots \frac{1}{\lambda} A_\lambda^{\varepsilon_n}(t_n) \Phi_\lambda \right\rangle =: G(t_1 \dots t_n). \tag{3.5.3}$$

One then explicitly constructs the following:

- An Hilbert space $\mathcal{H}$ and a vector $\Psi \in \mathcal{H}$.
- For any $t \in \mathbb{R}$, a field operator $b(t)$ (typically a quantum white noise (see Sect. 2.15), on $\mathcal{H}$, such that $G(t_1 \dots t_n)$ is equal to the Ψ-expectation value of the product of b operators; more precisely, the limit (3.5.3) can be written

$$\lim_{\lambda\to 0} \left\langle \Phi, \frac{1}{\lambda} A_\lambda^{\varepsilon_1}(t_1) \dots \frac{1}{\lambda} A_\lambda^{\varepsilon_n}(t_n) \Phi \right\rangle = \langle \Psi, b^{\varepsilon_1}(t_1) \dots b^{\varepsilon_n}(t_n) \Psi \rangle. \tag{3.5.4}$$

It may happen that, even if one starts from a single scalar field, the master field is a field with many (possibly an infinite number) components (see

Sect. 6.2). The explicit form of the master interaction Hamiltonian, and therefore the white noise equation, cannot be given in general, because it depends strongly on the specific model.

3.6 Discrete Spectrum Embedded in the Continuum

This section extends the discussion on the *canonical form of the interaction Hamiltonian* begun in Sect. 1.23. For a system–reservoir Hamiltonian of the form

$$H = H_S + H_R + (D \otimes F^+ + D^+ \otimes F), \tag{3.6.1}$$

if the free system Hamiltonian H_S has a discrete spectrum, i.e.

$$H_S = \sum_n \varepsilon_n P_n, \tag{3.6.2}$$

where ε_n are the eigenvalues and P_n the spectral projections, then we can apply the same argument as in Sect. 1.23. Let us illustrate this procedure in the case in which H_0 has a discrete nondegenerate spectrum:

$$H_S = \sum_j \epsilon |\epsilon_j\rangle\langle\epsilon_j|. \tag{3.6.3}$$

Denoting

$$F_j := F\langle\epsilon_j, D\epsilon_j\rangle, \tag{3.6.4}$$

$$D_{ij} := \langle\epsilon_i, D\epsilon_j\rangle \mid \epsilon_i\rangle\langle\epsilon_j \mid, \tag{3.6.5}$$

one can write

$$\begin{aligned} H &= H_S + H_R + \sum_j (\langle\epsilon_j, D\epsilon_j\rangle F^+ + \langle\epsilon_j, D^+\epsilon_j\rangle F)|\epsilon_j\rangle\langle\epsilon_j| \\ &\quad + \sum_{i\neq j} (D_{ij} \otimes F^+ + D_{ij}^+ \otimes F) \\ &= H_S + H_R + \sum |\epsilon_j\rangle\langle\epsilon_j| \otimes \left(F_j^+ + F_j\right) + \sum_{i\neq j} (D_{ij} \otimes F^+ + \text{ h.c.}). \end{aligned} \tag{3.6.6}$$

In the case of a linear interaction, i.e.

$$F = A(g) = \int \bar{g}(k) a_k dk, \tag{3.6.7}$$

one has

$$H_S + H_R + [D \otimes A^+(g) + D^+ \otimes A(g)] \tag{3.6.8}$$

or, denoting

$$g_j(k) := \langle \varepsilon_j, D\varepsilon_j \rangle g(k), \tag{3.6.9}$$

the diagonal part of (3.6.1) can be rewritten as

$$H_S + \int \omega_k a_k^+ a_k dk + \sum \int |\epsilon_j\rangle\langle\epsilon_j| \otimes [g_j(k)a_k^+ + g_j^-(k)a_k]dk. \tag{3.6.10}$$

Using the identities (for constant c)

$$\omega a^+ a + ca^+ + \bar{c}a = \omega \left(a + \frac{c}{\omega}\right)^+ \left(a + \frac{c}{\omega}\right) - \frac{|c|^2}{\omega},$$

one can rewrite the diagonal part of (3.6.1) in the form

$$H_S + \sum \int \omega_k |\epsilon_j\rangle\langle\epsilon_j| \otimes \left(a_k + \frac{\bar{g}_j(k)}{\omega_k}\right)^+ \left(a_k + \frac{\bar{g}_j(k)}{\omega_k}\right) dk$$

$$- \sum |\epsilon_j\rangle\langle\epsilon_j| \int \frac{|g_j(k)|^2}{\omega_k} dk.$$

Thus, introducing the new creation and annihilation operators

$$b_{k,j}^+ = a_k^+ + \frac{g_j(k)}{\omega_k} \quad , \quad b_{k,j} = a_k + \frac{\bar{g}_j(k)}{\omega_k},$$

the new free system Hamiltonian,

$$H_S(g) := \sum_j \left(\epsilon_j + \int \frac{|g_j(k)|^2}{\omega_k} dk\right) |\epsilon_j\rangle\langle\epsilon_j|,$$

and the new free system–field Hamiltonian,

$$\sum_j |\epsilon_j\rangle\langle\epsilon_j| \otimes \int \omega_k b_{k,j}^+ b_{k,j} dk,$$

the full Hamiltonian (3.6.1) assumes the following canonical form:

$$H_S(g) + \sum_j \int dk\omega_k |\epsilon_j\rangle\langle\epsilon_j| \otimes b_{k,j}^+ b_{k,j} + \sum_{i \neq j} [D_{ij} \otimes A^+(g) + \text{ h.c.}]$$

$$=: H_S + H_{SR} + H_{\mathrm{I}}, \tag{3.6.11}$$

and the whole Hilbert space splits into a sum

$$\mathcal{H} = \oplus \mathcal{H}_j$$

on each of whose factors the Hamiltonian $H_S + H_{SR}$ acts as a usual free Hamiltonian, i.e.

$$H_j = c_j + \int b_{kj}^+ b_{kj} dk,$$

with c_j being a constant. Note that

$$[H_S, H_{RS}] = 0.$$

Now consider the commutators

$$[H_{SR}, |\varepsilon_\mu\rangle\langle\varepsilon_\nu|] \quad \text{or equivalently} \quad [H_{SR}, D_{\mu,\nu}^+].$$

One has

$$[H_{SR}, |\varepsilon_\mu\rangle\langle\varepsilon_\nu|] = \sum_j \int dk\omega_k (|\varepsilon_j\rangle\langle\varepsilon_j|, |\varepsilon_\mu\rangle\langle\varepsilon_\nu|) b_{k,j}^+ b_{k,j}$$

$$= \sum_j \int dk\omega_k (\delta_{j\mu}|\varepsilon_j\rangle\langle\varepsilon_\nu| - \delta_{\nu j}|\varepsilon_\mu\rangle\langle\varepsilon_j|) b_{k,j}^+ b_{k,j}$$

$$= \int dk\omega_k (|\varepsilon_\mu\rangle\langle\varepsilon_\nu| - |\varepsilon_\mu\rangle\langle\varepsilon_\nu|) b_{k,j}^+ b_{k,j} = 0,$$

$$[H_{SR}, b_{k',\mu}^+] = \sum_j \int dk\omega_k |\varepsilon_j\rangle\langle\varepsilon_j| [b_{k,j}^+ b_{k,j}, a_{k'}^+]$$

$$= \sum_j \int dk\omega_k |\varepsilon_j\rangle\langle\varepsilon_j| [a_k^+ a_k, a_{k'}^+] = \omega_{k'} a_{k'}^+.$$

However, for $\mu \neq \nu$,

$$[H_{SR}, |\varepsilon_\mu\rangle\langle\varepsilon_\nu| a_{k'}^+] = [H_{SR}, |\varepsilon_\mu\rangle\langle\varepsilon_\nu|] a_{k'}^+ + |\varepsilon_\mu\rangle\langle\varepsilon_\nu| [H_{SR}, a_{k'}^+]$$

$$= |\varepsilon_\mu\rangle\langle\varepsilon_\nu| [H_{SR}, a_{k'}^+] = \omega_{k'} |\varepsilon_\mu\rangle\langle\varepsilon_\nu| a_{k'}^+.$$

So that

$$e^{itH_{SR}} |\varepsilon_\mu\rangle\langle\varepsilon_\nu| a_{k'}^+ e^{-itH_{SR}} = e^{-it\omega_{k'}} |\varepsilon_\mu\rangle\langle\varepsilon_\nu| a_{k'}^+.$$

This means that the contribution of H_{SR} to the free evolution of the interaction Hamiltonian H_{I}, defined by (3.6.11), is the same as that of H_R.

3.7 The Stochastic Limit of a Classical Gaussian Random Field

Let us now consider a classical analogue of the stochastic limit of free (Gaussian) quantum fields.

Let $u(t, x)$ be a mean zero real-valued classical *Gaussian random field* on $\mathbb{R}^{d+1}$ defined on the probability space $(\Omega, \mathcal{F}, P)$ and with covariance

$$E(u(t_1, x_1)u(t_2, x_2)) = D(t_1 - t_2, x_1 - x_2) \quad , \qquad t_i \in \mathbb{R},\ x_i \in \mathbb{R}^d. \quad (3.7.1)$$

The covariance (3.7.1) implies that $u(t, x)$ is stationary in time. Thus denoting for any $t \in \mathbb{R}$ and $f, g \in \mathcal{S}(\mathbb{R}^d)$

$$D(t, f, g) := \int\int dx_1 dx_2 \overline{f}(x_1) D(t, x_1 - x_2) g(x_2) \qquad (3.7.2)$$

and defining

$$u(t, f) := \int dx u(t, x) f(x),$$

if we can exchange the E expectation with the dx integral, then the process $\{u(t, f) : t \in \mathbb{R},\ f \in \mathcal{S}(\mathbb{R}^d)\}$ is Gaussian stationary in time with covariance

$$E(u(t_1, f)^* u(t_2, g)) = D(t_1 - t_2, f, g). \qquad (3.7.3)$$

More generally, one might start from a mean zero Gaussian field on $\mathbb{R} \times \mathcal{S}(\mathbb{R}^d)$ defined by (3.7.3). In this case the right-hand side of (3.7.1) is the integral kernel (possibly a distribution) of (3.7.3).

Theorem 3.7.1. *Suppose that for any* $f, g \in \mathcal{S}(\mathbb{R}^d)$

$$\int_{\mathbb{R}} dt |D(t, f, g)| < +\infty. \qquad (3.7.4)$$

Then the rescaled field

$$u_\lambda(t, x) := \frac{1}{\lambda} u\left(\frac{t}{\lambda^2}, x\right) \qquad (3.7.5)$$

converges to the white noise $\xi(t, x)$ *with covariance*

$$E(\xi(t, f)\xi(s, g)) = \delta(t - s) \int_{\mathbb{R}} d\tau D(\tau, f, g), \qquad (3.7.6)$$

where the limit is understood in the sense of correlations and in the distribution sense.

Proof 3.7.1. Because of Gaussianity it is enough to consider

$$I = \lim_{\lambda \to 0} \int \frac{1}{\lambda^2} D\left(\frac{t_2 - t_1}{\lambda^2}, x, y\right) \varphi(t_1)\psi(t_2)f(x)g(y)dt_1dt_2dxdy, \qquad (3.7.7)$$

where $g \in \mathcal{S}(\mathbb{R}^d)$ and φ, ψ are test functions. With the change of variables $(t_2 - t_1)/\lambda^2 = s$, $t_1 = t$, the integral (3.7.7) becomes

$$\int\int dsdt\varphi(\lambda^2 s + t)\psi(t) \int D(s, x, y)f(x)g(y)dxdy$$

$$= \int\int dsdt\varphi(\lambda^2 s + t)\psi(t)D(s, f, g).$$

The thesis now follows by dominated convergence.

Remark 3.7.1. We express the content of Theorem 3.7.1 by the statement

$$\lim_{\lambda \to 0} u_\lambda(t, x) = \lim_{\lambda \to 0} \frac{1}{\lambda} u\left(\frac{t}{\lambda^2}, x\right) = \xi(t, x), \qquad (3.7.8)$$

where $\xi(t, x)$ is the classical white noise random field with covariance (3.7.6).

Remark 3.7.2. By Bochner's theorem, $D(t, x)$ is the Fourier transform of a positive measure on $\mathbb{R}^{d+1}$. Assuming that this measure has a density, i.e.

$$D(t, x) = \int_{\mathbb{R}^{d+1}} e^{itk_0 + ixk} \mu(k_0, k)dkdk_0, \qquad (3.7.9)$$

where $\mu(k_0, k)$ is a sufficiently regular function, we have

$$\int_{\mathbb{R}} dtD(t, x) = \int_{\mathbb{R}} dt \int_{\mathbb{R}} dk_0 \int_{\mathbb{R}^d} dke^{itk_0}e^{ixk}\mu(k_0, k)$$

$$= 2\pi \int_{\mathbb{R}} dk_0\delta(k_0) \int_{\mathbb{R}^d} dke^{ixk}\mu(k_0, k)$$

$$= 2\pi \int_{\mathbb{R}^d} dke^{ixk}\mu(0, k) = D(0, x),$$

which is the classical analogue of the result established in the quantum case in Theorem 3.2.1.

3.8 Semiclassical Versus Semiquantum Approximation

The limit relation (3.2.3), i.e.

$$\lim_{\lambda \to 0} \frac{1}{\lambda} a\left(\frac{t}{\lambda^2}, k\right) = b(t, k), \qquad (3.8.1)$$

is a direct relation between stochastic processes and quantum theory *in real time*, i.e. without any need to introduce imaginary time.

Recalling the expression (2.4.3) for $a(t, k)$ and the expression (2.20.1) for the field, one realizes that in the corresponding limit for the field λ can also be interpreted as the square root of Planck's constant. This interpretation is particularly clear in the functional integral approach of Chap. 9 and is illustrated in the finite dimensional case by the discussion in Sect. 1.2.

In the usual semiclassical expansion the leading term of the asymptotic expansion of the quantum dynamics is the classical solution. Our leading approximation to quantum theory is not a smooth classical trajectory or field configuration, but rather a *quantum stochastic theory*, with some simplified features with respect to the original one.

For this reason it might be more appropriate to call this new type of expansion a *semiquantum approximation* rather than a semiclassical one.

The root of this different behaviour has to be looked for in the fact that the usual semiclassical approach can be applied to a class of initial states with a very special dependence upon Planck's constant, more precisely one considers initial states of the form (WKB ansatz)

$$\psi(x) = \varphi(x) e^{\frac{i}{\hbar} S(x)}.$$

This special choice of the initial data is carried over also in the functional integral extension of the WKB method to the space of quantum histories.

In our approach (3.8.1) has to be meant in the sense of the finite correlators with respect to a quantum Gaussian state, for example, the Fock vacuum or a finite temperature state or a ground state, etc.

The fact that the Wiener measure of smooth trajectories is zero, and yet the smooth trajectories are the leading approximation in the standard semiclassical expansion, has counterparts in several standard asymptotic techniques of classical probability theory (the functional form of the law of iterated logarithm, action estimates, large deviations, asymptotics of supports of diffusion processes, etc.), where the same thing happens.

Therefore, since the quantum white noise that we obtain in the limit can be thought as a pair of (noncommuting) classical white noises (see Sect. 2.15), it is clear that, by restricting one's attention to these classical subprocesses of the quantum process, the present method provides a natural frame to bring these powerful tools of classical probability to the service of quantum field theory, without the loss of insight which accompanies the passage to imaginary time.

The existence of several classical subprocesses *underlying* the full quantum process and in some sense *driving it* shall be a characteristic of the stochastic limit and establishes a deep relation between classical probability and quantum theory *in real time* (see Sect. 4.15).

3.9 An Historical Example: The Damped Harmonic Oscillator

Historically the notion of quantum white noise and its connections with quantum dissipation arose in connection with a basic physical example: the damped harmonic oscillator.One considers an atom in interaction with an electromagnetic field. After introducing the *dipole* and the *rotating wave* approximations (see Section 4.9), the Hamiltonian of the composite system becomes

$$H = \omega_0 c^+ c + \int dk \omega(k) a_k^+ a_k + \lambda \int dk [a_k^+ \otimes c + a_k \otimes c^+] g(k), \tag{3.9.1}$$

where $\omega(k), g(k)$ are suitable cutoff functions, c, c^+ are the annihilation and creation operators of the atom (described as an harmonic oscillator), a_k, a_k^+ are the annihilation and creation operators corresponding to the kth mode of the field in the momentum representation. Denoting

$$c_t := e^{itH} c e^{-itH} \ , \ a_k(t) = e^{itH} a_k e^{-itH}, \tag{3.9.2}$$

the Heisenberg equations of motion for c_t and $a_k(t)$ become

$$\begin{aligned} \dot{c}_t &= -i\omega_0 c_t - i\lambda \int dk a_k(t) g(k), \\ \dot{a}_k(t) &= -i\omega(k) a_k(t) - i\lambda g(k) c_t. \end{aligned} \tag{3.9.3}$$

Solving for $a_k(t)$ one finds

$$a_k(t) = a_k e^{-it\omega(k)} - i\lambda g(k) e^{-it\omega(k)} \int_0^t ds e^{is\omega(k)} c_s.$$

Inserting $a_k(t)$ into the equation for c_t, one obtains

$$\dot{c}_t = -i\omega_0 c_t - i\lambda \int dk g(k) a_k e^{-it\omega(k)} - \lambda^2 \int_0^t ds c_s \int dk g^2(k) e^{-i(t-s)\omega(k)}. \tag{3.9.4}$$

Introducing the notations

$$\lambda^2 \int dk g^2(k) e^{-i(t-s)\omega(k)} =: \phi_\lambda(t-s), \tag{3.9.5a}$$

$$-i\lambda \int dk g(k) a_k e^{-it\omega(k)} =: \xi_t^{(\lambda)}, \tag{3.9.5b}$$

the reduced equation for c_t becomes

$$\dot{c}_t = -i\omega c_t - \int_0^t ds \phi_\lambda(t-s) c_s + \xi_t^{(\lambda)}. \tag{3.9.6}$$

Note that up to now *no approximations* have been made. Equation (3.9.6) is usually called a *quantum Langevin equation* driven by the quantum *random force* ξ_t, also called *colored quantum noise.* Note that from (3.9.5b) it follows that the quantum noise acts *exclusively on the field degrees of freedom.* Now let Ψ denote the Fock vacuum for $a_k^{\pm}$ (i.e. for the field) and let us consider the Ψ statistics of the random force ξ_t.

Because of Gaussianity the statistics are uniquely determined by the 2-point (correlation) functions:

$$\langle \Psi, \xi_t^{(\lambda)} \Psi \rangle = \left\langle \Psi, \left(\xi_t^{(\lambda)} + \xi_s^{(\lambda)^+} \right) \Psi \right\rangle = \left\langle \Psi, \xi_t^{(\lambda)} \xi_s^{(\lambda)} \Psi \right\rangle = 0 \; ,$$

$$\left\langle \Psi, \xi_t^{(\lambda)} \xi_s^{(\lambda)^+} \Psi \right\rangle = \phi_\lambda(t-s), \tag{3.9.7}$$

where the function $\phi_\lambda(t-s)$, which is the same one as that which appears in (3.9.6), is defined by (3.9.5a) and it is smooth (provided that the functions g and ω are regular enough). This justifies the term *coloured noise* for the Gaussian noise defined by (3.9.7). However, in the usual (classical) Langevin equation, the random force is a *white noise*, i.e. a mean zero, δ-correlated random process. The stochastic limit explains the dynamical origins of the δ correlation. In fact, if we make the rescaling $t \to t/\lambda^2$, then with the notation

$$c_\lambda(t) := c\left(\frac{t}{\lambda^2}\right) \tag{3.9.8}$$

(3.9.4) becomes

$$\dot{c}_\lambda(t) = -i\omega_0 c_\lambda(t) - \int_0^t d\sigma c_\lambda(\sigma) \frac{1}{\lambda^2} \int dk |g_k|^2 e^{i\omega_k \frac{(t-\sigma)}{\lambda^2}}$$
$$+ \frac{i}{\lambda} \int a_k g_k e^{i\omega_k \frac{t}{\lambda^2}} dk. \tag{3.9.9}$$

From Theorem 3.2.1 we know that the last term in (3.9.9) converges as $\lambda \to 0$ to the quantum white noise $b_t(g)$ with variance

$$\gamma = 2\pi \int |g_k|^2 \delta(\omega_k) dk.$$

Similarly from Proposition 1.2.1 we know that

$$\frac{1}{\lambda^2} e^{i\omega(k)\frac{t}{\lambda^2}} \to 2\pi\delta(\omega(k))\delta(t)$$

in the sense of distributions. This means that the formal limit as $\lambda \to 0$ of (3.9.9) is the white noise equation

$$\dot{c}(t) = (-i\omega_0 - \gamma)c(t) + ib_t(g),$$

which is a *true Langevin equation* in the sense that the random force b_t now has the usual white noise statistics. The basic advantage of the stochastic limit is that *it gives with a single limit theorem, all the possible Langevin equations of interest*, thus making useless the individual deduction of each single equation.

3.10 Emergence of the White Noise: A Traditional Derivation

It is interesting to compare the stochastic limit approach with a standard argument used in the physical literature to justify the emergence of the quantum white noise and of the associated stochastic equations. This is based on the remark that, in the case of a one-dimensional space ($d = 1$) and with the choices

$$\omega(k) = k, \tag{3.10.1}$$

$$g(k) = g_0 = \text{ const. } \in \mathbb{R}, \tag{3.10.2}$$

the function $\phi_\lambda(t-s)$, defined by (3.9.5a) becomes a positive multiple of the δ-function, say $\kappa\delta(t-s)$, with

$$\kappa = 2\pi g_0^2 \lambda^2, \tag{3.10.3}$$

and the random force takes the form

$$b_t := -i g_0 \lambda \int_{\mathbb{R}} dk a_k e^{-itk}. \tag{3.10.4}$$

Note that the ansatz (3.10.1) makes sense only if the momentum space is assumed to be one-dimensional, which is rarely the case in interesting models. Moreover, even in the one-dimensional case, (3.10.1) means that the free energy is unbounded below, which is again an unphysical assumption. The situation with a stochastic limit is much more satisfactory: it is not bound to the one–dimensional case and the flat spectrum has a natural physical interpretation not as energy, but as *energy fluctuations* around the resonant levels. Moreover, as shown in this section and in the following one, the arguments used in the nonrigorous derivations are not at all simpler than those used in the stochastic limit to obtain a rigorous proof.

From (3.10.4) it follows that the Ψ-statistics of b_t is Gaussian with mean zero and correlations

$$\langle \Psi, b_t \Psi \rangle = \langle \Psi, b_s b_t \Psi \rangle = \langle \Psi, b_s^+ b_t^+ \Psi \rangle = 0, \tag{3.10.5}$$

$$\langle \Psi, b_s b_t^+ \Psi \rangle = \kappa \delta(t-s). \tag{3.10.6}$$

Moreover the commutation relations for b_s^+, b_t are

$$[b_s, b_t] = [b_s^+, b_t^+] = 0 \qquad , \forall s, t, \tag{3.10.7}$$

$$[b_t, b_s^+] = \lambda^2 \int e^{-i(t-s)k} g_0^2 dk = \kappa \cdot \delta(t-s). \tag{3.10.8}$$

Thus b_t is the Fock white noise with variance κ. With the choices (3.10.1–2) and in the above notations, the equation for a_t becomes

$$\dot{a}_t = (-i\omega - \kappa) a_t + b_t. \tag{3.10.9}$$

3.11 Heuristic Origins of Quantum White Noise

In this section we discuss some standard arguments used to justify the assumptions introduced to account for the emergence of quantum white noise. The starting point of these arguments is a discrete family of one-dimensional boson fields depending on a one-dimensional parameter ω,

$$[b_i(\omega), b_j^+(\omega')] = \delta_{ij}\ \delta(\omega - \omega') \quad , \quad [b_i(\omega), b_j(\omega')] = 0,$$

linearly interacting with a system with Hamiltonian

$$H_{\mathrm{I}}(\omega) = \sum_j k_j(\omega)\ [R_j \otimes b_j^+(\omega) - R_j^+ \otimes b_j(\omega)],$$

where $k_j(\omega)$ are cutoff functions describing the strength of the coupling of the system to the field and R_j are system operators. The total interaction is then

$$H_{\mathrm{I}} = i \sum_j \frac{1}{\sqrt{2\pi}} \int_{\Omega_j - \vartheta_j}^{\Omega_j + \vartheta_j} k_j(\omega)\ [R_j \otimes b_j^+(\omega) - R_j^+ \otimes b_j(\omega)]\ d\omega. \quad (3.11.1)$$

Where Ω_j is called the *carrier frequency* and ϑ_j the *half-bandwidth*. The free dynamics of the fields is given by

$$b_j(\omega) \mapsto e^{-i\omega t}\ b_j(\omega).$$

In the interaction representation *(with trivial dynamics for the system)* we obtain

$$H_{\mathrm{I}}(t) = i \sum_j \frac{1}{\sqrt{2\pi}} \left\{ R_j \otimes \left(\int_{\Omega_j - \vartheta_j}^{\Omega_j + \vartheta_j} k_j(\omega) d\omega\ e^{+i\omega t} b_j^+(\omega) \right) - \text{h.c.} \right\}.$$

Within this scheme, the emergence of the quantum white noise is justified by the following three ansatzen, often called *approximations* (see [Bar94]). The term *ansatz* seems to be more appropriate than the term *approximation* because in the three following points it is not clear *what approximates what* and in which sense.

- The *1st Markov ansatz*, which corrresponds to the formal replacement

$$k_j(\omega) \to 1,$$

  i.e. to a flat spectrum of the frequency field.
- The *2nd Markov ansatz*, which corrresponds to the formal replacement

$$\vartheta_j \to +\infty,$$

  i.e. to an infinite bandwidth.

- The *3rd Markov ansatz*, which corresponds to the formal replacement
$$\langle b_j(\omega,t)b^+_{j'}(\omega',t')\rangle - \langle b_j(\omega,t)\rangle\langle b^+_{j'}(\omega',t')\rangle = \delta_{j,j'}\bar{N}_j\delta(\omega-\omega'),$$
i.e. δ-correlated frequency modes.

Using the three ansatzen above the interaction Hamiltonian (3.11.1) is replaced by
$$H_{\mathrm{I}}(t) = i\sum_j \frac{1}{\sqrt{2\pi}}\left\{R_j \otimes b_t^{(j)+} - \text{h.c.}\right\},$$
where $b_t^{(j)+}$ are given by (see Section 2.19)
$$\int_{-\infty}^{+\infty} d\omega\ e^{i\omega t} b_j^+(\omega) =: b_t^+$$
and satisfy
$$\langle b_j(t)b^+_{j'}(t')\rangle - \langle b_j(t)\rangle\langle b^+_{j'}(t')\rangle = \delta_{j,j'}\bar{N}_j\delta(t-t'),$$
i.e. they are independent finite temperature quantum white noise.

Notice that, independently on the one-dimensional limitation, this is not a physical justification for the origin of quantum white noise, but rather an ad hoc mathematical ansatz. In fact the infinite bandwidth assumption would be physically justified in the case $k_j(\omega) \to 0$ as $\omega \to \infty$; however this is contradictory with the flat spectrum assumption, which requires the replacement $k_j(\omega) \to 1$.

In summary, it is not clear why postulating these three ansatzen gives more insight than directly postulating that the field is approximated by a quantum white noise of the required type. The proof of Theorem 3.2.2 is not only mathematically simpler (and rigorous) but it also provides a much more satisfactory physical picture: the white noise naturally emerges simply by looking at the fast time scale of the standard quantum dynamics.

3.12 Relativistic Quantum White Noises

In this section, which is not essential for what follows, we investigate some possible forms of relativistic quantum noises.

By analogy with the one-dimensional boson Fock white noise, consider a pair of operator-valued distributions $\xi(x), \eta(x)$ acting in an Hilbert space $\mathcal{H}$ over the Minkowski space with coordinates $x = (x^0, x^1, x^2, x^3)$. The commutator of ξ and η should be a c-number,
$$[\xi(x), \eta(y)] = \Delta(x, y). \tag{3.12.1}$$
Let fields ξ and η be scalars, i.e. transform as

$$U(g)\xi(x)U(g)^{-1} = \xi(gx), \tag{3.12.2}$$

where $U(g)$ is a unitary representation of the Poincaré group in the Hilbert space $\mathcal{H}$. Applying the transformation $U(g)$ to (3.12.1) one obtains

$$\Delta(gx, gy) = \Delta(x, y). \tag{3.12.3}$$

Here $g = (a, \Lambda)$, where a is a vector translation and Λ a Lorentz rotation. From (3.12.3) it follows that Δ depends only on the difference between x and y,

$$\Delta(x, y) = \Delta(x - y), \tag{3.12.4}$$

and is invariant under Lorentz rotations:

$$\Delta(\Lambda z) = \Delta(z). \tag{3.12.5}$$

Therefore, if $\Delta(z)$ is a function, then it could depend only on the Lorentz square:

$$z^2 = (z^0)^2 - (z^1)^2 - (z^2)^2 - (z^3)^2.$$

If it is a distribution, then also δ-functions and their derivatives could arise. So the simplest possibility for a relativistic generalization of the 1-dimensional quantum white noise is

$$[\xi(x), \eta(y)] = i\delta(x - y), \tag{3.12.6}$$

where $\delta(x - y)$ is the 4-dimensional δ-function and we assume that the fields ξ and η are Hermitian. The relation (3.12.6) is understood in the usual distribution sense, i.e.

$$[\xi(f), \eta(g)] = i\langle f, g\rangle, \tag{3.12.7}$$

where f and g are test functions, on a dense set in $\mathcal{H}$.

One can realize the commutation relation (3.12.6) by introducing the boson creation and annhilation operators

$$[a(k), a^+(k')] = \delta^{(4)}(k - k') \tag{3.12.8}$$

in the Fock space $\Gamma(L_2(\mathbb{R}^4))$ and then defining

$$\xi(x) = \frac{1}{(2\pi)^2}\int \left[e^{-ik\cdot x}a(k) + e^{ikx}a^+(k)\right] d^{(4)}k,$$

$$\eta(x) = \frac{i}{(2\pi)^2}\int \left[e^{ik\cdot x}a^+(k) - e^{-ikx}a(k)\right] d^{(4)}k.$$

Similarly, starting from the operators $b(t), b(t)^+$, obtained in the stochastic limit, we can form a scalar relativistic quantum white noise by

$$\xi(x) = b(x^0) \otimes a(\bar{x}) + b^+(x^0) \otimes a^+(\bar{x}) \, ,$$
$$\eta(x) = \frac{1}{i} \left[b(x^0) \otimes a(\bar{x}) - b^+(x^0) \otimes a^+(\bar{x}) \right] .$$

In order to compare the relativistic quantum white noise with the usual free scalar relativistic quantum field $\varphi(x)$, let us remind ourselves that the latter, expressed in terms of creation and annihilation operators, has the form

$$\varphi(x) = \frac{1}{\sqrt{(2\pi)^3}} \int \frac{d^3k}{\sqrt{2\omega(k)}} \left[a^+(k) e^{i\omega(k)x^0 - i\bar{k}\bar{x}} + a(k) e^{-i\omega(k)x^0 + i\bar{k}\bar{x}} \right] ,$$

where $\omega(k) = \sqrt{k^2 + m^2}$ and $m \geq 0$ is a mass.

One can also introduce quantum white noises for different spins. For example, a vector quantum white noise is a pair $\xi_\mu(x), \eta_\mu(x)$ with the commutator

$$[\xi_\mu(x), \eta_\gamma(y)] = k_{\mu\gamma}\delta(x-y).$$

Here $\mu, \gamma = 0, 1, 2, 3$ and the tensor $k_{\mu\gamma}$ in the simplest case has the form

$$k_{\mu\gamma} = a \; g_{\mu\gamma},$$

where a is a constant and $g_{\mu\gamma}$ the Minkowski tensor: $g_{\mu\gamma} = \text{diag}(-1, 1, 1, 1)$. In more complex cases $k_{\mu\gamma}$ could be a differential operator, e.g.

$$k_{\mu\gamma} = a \; g_{\mu\gamma} + \partial_\mu \partial_\gamma + \dots ,$$

where $\partial_\mu = \partial/\partial x_\mu$.

A spinor quantum white noise is a pair of spinor fields $\psi_x(\alpha), \eta_x(\beta)$ with the following anticommuting relations:

$$\{\psi_\alpha(x), \eta_\beta(y)\} = t_{\alpha\beta}\delta(x-y),$$

where

$$T_{\alpha\beta} = a\delta_{\alpha\beta} + b\gamma^\mu_{\alpha\beta}\partial_\mu,$$

a, b are constants, α, β are spinor indices, $\delta_{\alpha,\beta}$ is the Kronecker symbol, and γ^μ are the Dirac matrices.

3.13 Space–Time Rescalings: Multidimensional White Noises

In this section we discuss the extension of the result of Sect. 3.4 to the multidimensional relativistic case, so as to include space–time rescalings; that is we consider the relativistic scalar field and we rescale not only time but also space variables. In the limit one would expect to obtain a multidimensional relativistic white noise; but we shall see that the situation is not so straightforward and that a meaningful stochastic limit can be obtained not for the convergence of arbitrary correlators but only for the convergence of the *chronological correlators*, i.e. expectations of chronologically ordered products of field operators. We will show that in the stochastic limit in the sense of chronological order we obtain a relativistic white noise.

Consider a boson Fock field $\varphi(x^0, x)$ as in Sect. 3.4,

$$\varphi(x) = \frac{1}{(2\pi)^{3/2}} \int \frac{d\overline{k}}{\sqrt{2\omega(\overline{k})}} \left[e^{i\omega(\overline{k})x^0 - i\overline{k}\overline{x}} a^+(\overline{k}) + e^{-i\omega(\overline{k})x^0 + i\overline{k}\overline{x}} a(\overline{k}) \right],$$

and define the rescaled field as

$$\varphi_\lambda(x) = \frac{1}{\lambda^d} \varphi(x/\lambda^2) = \frac{1}{\lambda^d} \varphi(x^0/\lambda^2, \overline{x}/\lambda^2) =: \varphi_\lambda(x^0, \overline{x}). \tag{3.13.1}$$

By repeating, in a purely formal way, for this rescaling the arguments of Sect. 3.4 and using Gaussianity to reduce the convergence, in the sense of vacuum correlators, to the convergence of the vacuum 2-point function, we see that, after the rescaling (3.13.1) this 2-point function becomes

$$\begin{aligned} \langle \varphi_\lambda(x^0, x)\varphi_\lambda(y^0, y)\rangle &= \int_{\mathbf{R}^d} dk_1 \int_{\mathbf{R}^d} dk_1 \langle a(x^0, k_1) a^+(x^0, k_2)\rangle \\ &\quad \cdot e^{ik_1 x/\lambda^2} e^{-ik_2 y/\lambda^2} \\ &= \frac{1}{(2\pi)^d} \frac{1}{\lambda^{2d}} \int dk_1 \int dk_2 \frac{e^{-i(x^0/\lambda^2)\omega(k_1)} e^{i(y^0/\lambda^2)\omega(k_2)}}{2\omega(k_1)} \\ &\quad \cdot e^{ik_1 x/\lambda^2} e^{-ik_2 y/\lambda^2} \delta(k_1 - k_2) \\ &= \frac{1}{(2\pi)^d} \frac{1}{\lambda^{2d}} \int dk \frac{e^{-i(x^0 - y^0)\omega(k)/\lambda^2}}{2\omega(k)} e^{ik(x-y)/\lambda^2}. \end{aligned} \tag{3.13.2}$$

Introducing test functions f and g in $\mathbb{R}^{d+1}$,

$$\frac{1}{\lambda^{2d}} \int dx^0 dx \int dy^0 dy f(x^0, x) g(y^0, y) \int dk \frac{e^{-i(x^0 - y^0)\omega(k)/\lambda^2}}{2\omega(k)} e^{i(x-y)/\lambda^2}. \tag{3.13.3}$$

Setting

$$x^0 - y^0 = z^0 \ , \quad x - y = z,$$

(3.13.3) becomes

$$\frac{1}{\lambda^{2d}} \int dz^0 dz \int dy^0 dy f(z^0 + y^0, z + y) g(y^0, y) \int dk \frac{e^{-i(z^0/\lambda^2)\omega(k)}}{2\omega(k)} e^{ikz/\lambda^2}. \tag{3.13.4}$$

Now making

$$\lambda^{-2}z^0 = x^0 \ , \quad \lambda^{-2}z = x,$$

we obtain

$$\int dx^0 dx \int dy^0 dy f(\lambda^2 x^0 + y^0, \lambda^2 x + y) g(y^0, y) \int \frac{e^{-ix^0\omega(k)}}{2\omega(k)} e^{ikx} dk.$$

As $\lambda \to 0$ this converges, in the sense of distributions, to

$$\langle f, g\rangle_{L^2(\mathbb{R}^{d+1})} \int dx^0 dx \int e^{-ix^0\omega(k)} e^{ikx} dk = \langle f, g\rangle \delta(\omega(0)) \frac{(2\pi)^{d+1}}{2\omega(0)}. \quad (3.13.5)$$

This should correspond to the limit relation

$$\lim_{\lambda\to 0} \varphi_\lambda(x^0, x) = b(x^0, x), \quad (3.13.6)$$

i.e. to the fact that as $\lambda \to 0$ the rescaled field (3.13.1) converges, in the sense of vacuum correlators, to a *boson Fock white noise* $b(x^0, x)$ on $\mathbb{R}^{d+1}$ with variance

$$\langle b(x^0, x) b(y^0, y)\rangle = \delta(\omega(0)) \delta(x^0 - x) \delta(y^0 - y).$$

The above expression is, however, meaningless because of the divergent expression

$$\delta\left(\omega(0)\right) = \frac{1}{(2\pi)^{d+1}} \int_{\mathbb{R}^{d+1}} dx^0 dx \int_{\mathbb{R}^d} e^{ix^0\omega(k)} e^{-ikx} dk. \quad (3.13.7)$$

In order to deal with the divergent constant $\delta\left(\omega(0)\right)$, one has to introduce a momentum cutoff and a renormalization. This will be discussed in the following section.

3.14 The Chronological Stochastic Limit

The stochastic limit is a singular limit, and the result depends on the exact prescription of the limiting procedure. The discussion in the previous section shows that there are (at least) two natural prescriptions for the stochastic limit: *one can use the convergence in the sense of Wightman functions or in the sense of Green functions* (see Sect. 14.5 for additional discussion).

In the first approach one considers the correlators

$$\lim_{\lambda\to 0} \langle a_\lambda^{\varepsilon_1}(t_1, f_1) \dots U^{(\lambda)}(t/\lambda^2) \dots a_\lambda^{\varepsilon_w}(t_n, f_n)\rangle, \quad (3.14.1)$$

and in the second one the time-ordered correlators

$$\lim_{\lambda\to 0} \langle T(a_\lambda^{\varepsilon_1}(t_1, f_1) \dots U^{(\lambda)}(t/\lambda^2) \dots a_\lambda^{\varepsilon_n}(t_n, f_n)\rangle, \quad (3.14.2)$$

where T is the operation of the chronological product and in both cases

$$a_\lambda^\varepsilon(t,f) = \frac{1}{\lambda}\int dk a^\varepsilon(k) e^{i\varepsilon t\omega(k)/\lambda^2} f(k). \tag{3.14.3}$$

The results of the two limits are different for a free evolution.

Theorem 3.14.1. *For $\omega(k)=\sqrt{k^2+m^2}$ and $m>0$, we have*

$$\lim_{\lambda\to 0}\langle a_\lambda(t_1,f_1)a_\lambda^+(t_2,f_2)\rangle = 0, \tag{3.14.4}$$

$$\lim_{\lambda\to 0}\left\langle T\left(a_\lambda(t_1,f_1)a_\lambda^+(t_2,f_2)\right)\right\rangle = -i\delta(t_1-t_2)\cdot\int \frac{dk}{\sqrt{k^2+m^2}}\overline{f}_1(k)f_2(k). \tag{3.14.5}$$

Proof 3.14.1. To prove (3.14.4–5), we use the relations

$$T\left(A(t_1)B(t_2)\right) = \theta(t_1-t_2)A(t_1)B(t_2) + \theta(t_2-t_1)B(t_2)A(t_1),$$

$$\theta(t_1-t_2) = \frac{1}{2\pi i}\int_{\mathbb{R}} \frac{e^{i\alpha(t_1-t_2)}}{\alpha - i0}\,d\alpha.$$

Then

$$\left\langle T\left(a_\lambda(t_1,f_1)a_\lambda^+(t_2,f_2)\right)\right\rangle = \frac{1}{2\pi i}\int_{-\infty}^{\infty}\frac{e^{i\alpha(t_1-t_2)/\lambda^2}}{\alpha-i0}\,d\alpha$$

$$\cdot\int\frac{1}{\lambda^2}e^{-i(t_1-t_2)\omega(k)/\lambda^2}\overline{f}_1(k)f_2(k)dk$$

$$= \frac{1}{2\pi i}\int dk\overline{f}_1(k)f_2(k)\int_{-\infty}^{\infty} d\alpha\frac{1}{\alpha-i0}\frac{1}{\lambda^2}e^{\frac{i(t_1-t_2)}{\lambda^2}[\alpha-\omega(k)]}.$$

When $\lambda\to 0$ this becomes

$$\frac{1}{2\pi i}\int dk\overline{f}_1(k)f_2(k)\int_{-\infty}^{\infty} d\alpha\frac{1}{\alpha-i0}2\pi\delta(\alpha-\omega(k))\delta(t_1-t_2)$$

$$= i\delta(t_1-t_2)\int\frac{dk\overline{f}_1(k)}{\omega(k)}f_2(k).$$

Thus, the theorem is proved.

More generally, for any two-point correlation function we have

$$\lim_{\lambda\to 0}\langle a_\lambda(t_1,f_1)a_\lambda^+(t_2,f_2)\rangle = 2\pi\delta(t_1-t_2)\int dk\delta\big(\omega(k)\big)\cdot\overline{f}_1(k)f_2(k) \tag{3.14.6}$$

and

$$\lim_{\lambda\to 0}\left\langle T\left(a_\lambda(t_1,f_1)a_\lambda^+(t_2,f_2)\right)\right\rangle = -i\delta(t_1-t_2)\cdot\int\frac{dk}{\omega(k)}\overline{f}_1(k)f_2(k). \tag{3.14.7}$$

In all of the above cases one cannot identify the limiting object with the ordinary (quantum) white noise because its *chronological covariance*, being equal to $i\delta(t_1 - t_2)$, is purely imaginary. Similar results are obtained when dealing with the functional integral approach to the stochastic limit (Chap. 9). The emergence of a *complex covariance* has the same origin as the causal commutator, discussed in Sect. 7.4.

Theorem 3.14.2. *Let φ be the same quantum scalar field as in Sect. 3.13 with $\omega(\overline{k}) = \sqrt{\overline{k}^2 + m^2}$, $m > 0$, $\overline{k} \in \mathbb{R}^{d-1}$ and let the rescaled field be given by (3.14.1). Then one has the relation*

$$\lim_{\lambda\to 0} \langle T\left(\varphi_\lambda(x)\varphi_\lambda(y)\right)\rangle = \frac{-i}{m^2}\,\delta(x-y), \tag{3.14.8}$$

where T denotes the chronological order (see Sect. 8.1).

Proof 3.14.2.

$$\begin{aligned}\langle T\left(\varphi_\lambda(x)\varphi_\lambda(y)\right)\rangle &= \frac{i}{(2\pi)^d\lambda^{2d}}\int dk \frac{e^{ik(x-y)/\lambda^2}}{k^2 - m^2 - i0}\\ &= \frac{i}{(2\pi)^d}\int dp \frac{e^{ip(x-y)}}{\lambda^4 p^2 - m^2 - i0}.\end{aligned}$$

Since, as $\lambda \to 0$ this converges to $-\frac{i}{m^2}\,\delta(x-y)$, the theorem is proved.

Remark 3.14.1. We have proved the existence of the stochastic limit in the sense of the chronological order. Before we had seen that for the space–time rescaling there is no meaningful stochastic limit in the sense of matrix elements.

Remark 3.14.2. Finally let us notice that in the Euclidean formulation we obtain in the stochastic limit for the free fields the classical white noise. In fact

$$\begin{aligned}\langle\phi_\lambda(t_1,x_1)\phi_\lambda(t_2,x_2)\rangle &= \frac{1}{(2\pi)^4}\frac{1}{\lambda^2}\int_{\mathbb{R}^4}\frac{e^{ik_0(t_1-t_2)/\lambda^2 + i\overline{k}(\overline{x}_1-\overline{x}_2)}dk_0 d\overline{k}}{k_0^2 + \overline{k}^2 + m^2}\\ &= \frac{1}{(2\pi)^4}\int_{\mathbb{R}^4}\frac{e^{ip_0(t_1-t_2)+i\overline{k}(\overline{k}_1-\overline{x}_2)}dp_0 d\overline{k}}{\lambda^4 p_0^2 + \overline{k}^2 + m^2},\end{aligned}$$

which tends to

$$\frac{1}{(2\pi)^3}\,\delta(t_1-t_2)\int_{\mathbb{R}^3}\frac{e^{i\overline{k}(\overline{x}_1-\overline{x}_2)}d\overline{k}}{\overline{k}^2 + m^2},$$

i.e. the covariance of a classical white noise.

3.15 Notes

The distribution approach to the stochastic limit and its interpretation as a new semiclassical expansion was introduced in [AcLuVo93].

The stochastic limit for a relativistic model is considered in [FaPa99].

The chronological stochastic limit is presented here for the first time.

4. Open Systems

In this chapter we begin the study of the stochastic limit of the simplest interacting systems by considering the open system scheme, described in Sect. 1.13.

We formulate the problem for a general reservoir and then, starting from Sect. 4.12 we specialize to the case in which the reservoir is a boson system. The modifications needed to include fermions are discussed in Chaps. 15 and 16.

In Sect. 4.1, as a paradigmatic example of open system interaction, we describe the nonrelativistic quantum electrodynamics (QED) Hamiltonian. This is an important case not only in itself, but also because a great multiplicity of model Hamiltonians can be obtained from it by introducing simplifying assumptions (e.g. dipole or rotating wave approximation or consideration of only a finite number of atomic levels) or variations. After introducing notations we discuss the dipole and the rotating-wave approximations. The stochastic limit for the multiplicative (dipole-type) interaction Hamiltonian is derived in Sects. 4.11 and 4.12. Then we explain how to read physics from the stochastic dynamical equations.

Motivated by the example of the QED Hamiltonian, we introduce in Sect. 4.8 the general class of interactions which shall be studied in Part I. The crucial difference between Parts I and II is not so much in the form of the interactions considered, but in the fact that in Part I the system Hamiltonian has a discrete spectrum. We shall see in Part II that removing this assumption leads to some qualitative new features such as the vanishing of the crossing diagrams in the stochastic limit and the subsequent emergence of new, non-Gaussian, statistics.

4.1 The Nonrelativistic QED Hamiltonian

The nonrelativistic QED Hamiltonian describes an atom in interaction with the electromagnetic field. The quantized electromagnetic field is confined in a box with periodic boundary conditions (so that the opposite faces are identified, giving a torus) and with an ultraviolet (i.e. high-momentum) cutoff, so that also the momenta are in a box Λ rather than in the infinite lattice, dual to the torus. It interacts, through so-called *minimal coupling*, with N quantum spinless particles (atoms, molecules, electrons, etc.). In the nonrelativistic approximation, the Hamiltonian of this system is given by

$$H_\Lambda = \sum_{j=1}^{N} \left[\frac{1}{2m_j} \left(p_j - \frac{e}{c} A_\Lambda(q_j) \right)^2 + V(q_j) \right] + \sum_{\sigma=1,2} \sum_{k \in \Lambda} |k| a_{k,\sigma}^+ a_{k,\sigma} + \sum_{i \neq j} V_{ij}(q_i - q_j), \tag{4.1.1}$$

where p_j, q_j denote momentum and position of the jth particle in the 3-dimensional space, i.e. $p_j = (p_{j1}, p_{j2}, p_{j3})$, $q_j = (q_{j1}, q_{j2}, q_{j3})$. $a(k, \sigma)$ are Boson annihilators, $\sigma = 1, 2$ is a polarization index and the expansion of A_Λ in Fourier series is

$$A_\Lambda(q) = (\text{const.}) \sum_\sigma \frac{1}{|\Lambda|^{1/2}} \sum_k \frac{1}{|k|^{1/2}} e_{k,\sigma} \left(a_{k,\sigma} e^{ik \cdot q} + \text{h.c.} \right), \tag{4.1.2}$$

where the square roots in the denominator arise from the normalization of the functions $e^{ik \cdot x}$ on the torus and $e(k,1)$, $e(k,2)$ are orthonormal vectors in $\mathbf{R}^3$ (polarization vectors)satisfying

$$k \cdot e_{k,\sigma} = 0 \quad , \quad \sigma = 1, 2. \tag{4.1.3}$$

Expanding the square in (4.1.1), neglecting for the moment the A^2 term and introducing the N-body system Hamiltonian

$$H_S = \sum_{j=1}^{N} \frac{1}{2m_j} p_j^2 + \sum_{j=1}^{N} V(q_j) + \sum_{i \neq j} V_{ij}(q_i - q_j), \tag{4.1.4}$$

one is reduced to the Hamiltonian

$$\begin{aligned} H_\Lambda &= H_S + \sum_{k,\sigma} |k| a_{k,\sigma}^+ a_{k,\sigma} - \sum_{j=1}^{N} \frac{e}{2m_j c} \left[p_j \cdot A_\Lambda(q_j) + A_\Lambda(q_j) \cdot p_j \right] \\ &=: H_0 + H_{\mathrm{I}}, \end{aligned} \tag{4.1.5}$$

where, using (4.1.2) for $A_\Lambda(q_j)$, the interaction Hamiltonian can be written as follows:

$$\begin{aligned} H_{\mathrm{I}} = &\frac{1}{|\Lambda|^{1/2}} \sum_{k,\sigma} \sum_{j=1}^{N} \frac{e}{2m_j c} \frac{1}{|k|^{1/2}} \\ &\cdot \left[\left(e_{k\sigma} \cdot p_j e^{ik \cdot q_j} + e^{ik \cdot q_j} e_{k\sigma} \cdot p_j \right) a_{k\sigma} + \left(e_{k\sigma} \cdot p_j e^{-ik \cdot q_j} + e^{-ik \cdot q_j} e_{k\sigma} \cdot p_j \right) a_{k\sigma}^+ \right]. \end{aligned} \tag{4.1.6}$$

Note that $e_{k\sigma} \cdot p_j$ is the component of p_j along $e_{k\sigma}$, and therefore, due to the relation

$$[q_j \cdot u, p_j \cdot v] = iu \cdot v \ , \quad u, v \in \mathbf{R}^3, \tag{4.1.7}$$

it commutes with $k \cdot q_j$ because of (4.1.3). Therefore

$$H_{\mathrm{I}} = \frac{1}{|\Lambda|^{1/2}} \sum_{k,\sigma} \sum_{j=1}^{N} \frac{e}{m_j c} \frac{1}{|k|^{1/2}} \left(e_{k\sigma} \cdot p_j e^{ik\cdot q_j} a_{k\sigma} + e_{k\sigma} \cdot p_j e^{-ik\cdot q_j} a^+_{k\sigma} \right)$$
$$= \frac{1}{|\Lambda|^{1/2}} \sum_{k,\sigma} \sum_{j=1}^{N} \frac{e}{m_j c} \frac{1}{|k|^{1/2}} \left(e^{ik\cdot q_j} e_{k\sigma} \cdot p_j a_{k\sigma} + e^{-ik\cdot q_j} e_{k\sigma} \cdot p_j a^+_{k\sigma} \right). \tag{4.1.8}$$

Summing over the polarization index and introducing the *total annihilator*

$$\sum_{\sigma} e_{k\sigma} a_{k\sigma} =: a_k, \tag{4.1.9}$$

one obtains

$$H_{\mathrm{I}} = \frac{1}{|\Lambda|^{1/2}} \sum_{k} \sum_{j=1}^{N} \frac{e}{m_j c} \frac{1}{|k|^{1/2}} \left(e^{ik\cdot q_j} a_k + e^{-ik\cdot q_j} a^+_k \right) \cdot p_j. \tag{4.1.10}$$

Introducing the cutoff function

$$g(k) := \frac{1}{|k|^{1/2}}, \tag{4.1.11}$$

(in fact a truncated version of it) and the module notation

$$A(ge^{i(\cdot)\cdot q_j}) := \sum_{k\in\Lambda} g(k) e^{ik\cdot q_j} a_k \quad , \quad A^+(ge^{i(\cdot)\cdot q_j}) := \sum_{k\in\Lambda} g(k) e^{-ik\cdot q_j} a^+_k, \tag{4.1.12}$$

we finally write the interaction Hamiltonian in the form

$$H_{\mathrm{I}} = \sum_{j=1}^{N} \frac{e}{m_j c} \frac{1}{|\Lambda|^{1/2}} \left(A(ge^{i(\cdot)\cdot q_j}) + A^+(ge^{i(\cdot)\cdot q_j}) \right) \cdot p_j, \tag{4.1.13}$$

which we can write in the symmetrized form

$$H_{\mathrm{I}} = \left(A(ge^{i(\cdot)\cdot q_N}) + A^+(ge^{i(\cdot)\cdot q_N}) \right) \cdot \frac{p_N}{|\Lambda|^{1/2}}, \tag{4.1.14}$$

where we identify the operators $e^{-ik\cdot q_j}$ to the components of a column vector denoted symbolically $e^{-ik\cdot q_N}$ and the operators p_j to the components of a row vector denoted symbolically p_N (the *collective atomic variables*[(1)]).

Replacing in (4.1.10) the sum over momenta $|\Lambda|^{-1/2} \sum_{k\in\Lambda}$ by a dk integral one gets,

$$A(ge^{i(\cdot)\cdot q}) = \int d^3k \; \left[\overline{g}(k) a_k e^{ik\cdot q} + h.c. \right],$$

where we have used the usual notation

$$A(g) = \int d^3k \overline{g}(k) a(k). \tag{4.1.15}$$

Taking all the masses to be equal, one can write the interaction Hamiltonian in the form

$$H_{\mathrm{I}} = -\frac{e}{mc} \sum_{\sigma=1,2} \int d^3k \, \left[a_\sigma^+(k) e^{-ik\cdot q} + a_\sigma(k) e^{ik\cdot q}\right] \, g_\sigma(k) \cdot p, \tag{4.1.16}$$

where the explicit form of $g_\sigma(k)$ is obtained by modifying (4.1.11) both near zero and at infinity so as to guarantee not only integrability but also the validity of condition (4.5.4). In more compact notations,

$$H_{\mathrm{I}} = \frac{e}{mc} \sum_{j=1}^{N} \left(A(g e^{i(\cdot)\cdot q_j}) + A^+(g e^{i(\cdot)\cdot q_j}) \right) \cdot p_j. \tag{4.1.17}$$

The *polaron Hamiltonian* (see Sect. 12.15) is obtained simply by omitting from (4.1.16) the scalar product with the momentum operator p.

4.2 The Dipole Approximation

The dipole approximation consists in replacing by zero the quantity $k \cdot q$ in the expression (4.1.16) for the interaction Hamiltonian[(1)]. Under this approximation the potential no longer depends on the position of the atom and one has, using the notation (4.1.15)

$$A(q) = \int d^3k \, \left[\overline{g}(k) a_k e^{ik\cdot q} + \text{h.c.}\right] \sim \int d^3k \, \left[\overline{g}(k) a_k + \text{h.c.}\right]$$

$$= A(g) + A^+(g) = Q(g). \tag{4.2.4}$$

With these notations the QED Hamiltonian in the dipole approximation becomes

$$H_{\mathrm{I}} = \frac{e}{mc} \left[A(g) + A^+(g)\right] \cdot p_N = \frac{e}{mc} Q(g) \cdot p_N. \tag{4.2.5}$$

If we put

$$\frac{e}{mc} =: \lambda$$

in (4.1.17), in the dipole approximation, it becomes[(4)]

$$H_{\mathrm{I}} = \lambda Q(g) \cdot p = -i\lambda \, (A_g + A_g^+)(a - a^+). \tag{4.2.6}$$

Remark 4.2.1. Usually the Hamiltonian describing the interaction of the electromagnetic field with an atom in the dipole approximation is defined by

$$H_{\mathrm{I}} = -E(q,t) \cdot d, \tag{4.2.7}$$

where $E(q,t)$ denotes the electric field and d the dipole moment of the atom (whose strength is the coupling constant). Recall that, according to standard notations,

$$E \text{ electric field} = P(g) = \frac{c}{i}\,[A(g) - A^+(g)], \tag{4.2.8}$$

$$H \text{ magnetic field} = Q(g) = [A(g) + A^+(g)] \tag{4.2.9}$$

where c is a constant proportional to e. Thus, according to (4.2.7) one should have

$$H_{\mathrm{dip}} = -E(g) \cdot qc = -P(g) \cdot qc. \tag{4.2.10}$$

In order to understand the difference between these two conventions, note that by making $c = 1$ one has

$$Q(g) \cdot p = \frac{1}{i}\,(A_g + A_g^+)(a - a^+) = \frac{1}{i}\,(A_g a - A_g^+ a^+) + \frac{1}{i}\,(A_g^+ a - A_g a^+), \tag{4.2.11}$$

$$-P(g) \cdot q = \frac{1}{i}\,(A_g - A_g^+)(a + a^+) = +\frac{1}{i}\,(A_g a - A_g^+ a^+) + \frac{1}{i}\,(A_g a^+ - A_g^+ a), \tag{4.2.12}$$

i.e. the difference between the two terms is the sign of the term $A_g a^+ - A_g^+ a$, which is opposite in the two cases.[(5)]

4.3 The Rotating-Wave Approximation

Often, beyond the dipole approximation, one also introduces the *rotating-wave approximation* which consists in dropping from the Hamiltonian (4.2.11–9) the terms aA_g, $a^+A_g^+$, corresponding to creation of one photon in the field and a simultaneous jump of one level up in the atom (and the converse process). After this approximation the interaction Hamiltonians (4.2.11–9) become effectively equal and (4.2.6) becomes

$$H_{\mathrm{I}} = -i\lambda(aA_g^+ - a^+A_g). \tag{4.3.1}$$

We shall see that in some cases (see Remark 5.5.1) this additional approximation is not needed, in the sense that the contribution of the pair creation terms vanishes automatically in the stochastic limit. Replacing, in the Hamiltonian (4.3.1), the atomic annihilation operator a by a more or less arbitrary atomic operator D, one obtains the so-called *spin–boson Hamiltonian*

$$H_{\mathrm{I}} = -i\lambda(DA_g^+ - D^+A_g), \tag{4.3.2}$$

which is the prototype of a large class of model Hamiltonians [see (4.9.2)]. In Sect. 4.2 it was shown how the spin–boson Hamiltonians is deduced from the standard nonrelativistic QED Hamiltonian by introducing the dipole approximation.[(6)]

4.4 Composite Systems

Motivated by the example of the QED Hamiltonian let us consider the following general situation:

Definition 4.4.1. *A composite system of two given quantum dynamical systems*

$$S \equiv \{\mathcal{H}_S, H_S\}, \quad R \equiv \{\mathcal{H}_R, H_R\}, \tag{4.4.1}$$

is another quantum dynamical system of the form[(7)]

$$\{\mathcal{H}_S \otimes \mathcal{H}_R, H_{SR}\}, \tag{4.4.2}$$

$$H_{SR} = H_S \otimes 1_R + 1_S \otimes H_R + H_{\mathrm{I}}, \tag{4.4.3}$$

where H_I contains all the new physics, with respect to the isolated systems, and is called the interaction Hamiltonian, while H_S (H_R) is called the free Hamiltonian of S (R).

When no confusion is possible, we shall leave implicit the symbol $\otimes$, so, for example, instead of (4.4.3) we often write

$$H_{SR} = H_S + H_R + H_{\mathrm{I}}. \tag{4.4.4}$$

In fact we shall consider a whole family of such systems

$$(S, R)_\lambda \equiv \{\mathcal{H}_S \otimes \mathcal{H}_R, H_S + H_R + \lambda H_{\mathrm{I}}\}$$

parameterized by a positive number λ called *the coupling constant.* For any such system one introduces the *free Hamiltonian*

$$H_0 = H_S + H_R, \tag{4.4.5}$$

the *total Hamiltonian*

$$H^{(\lambda)} = H_S + H_R + \lambda H_{\mathrm{I}} \tag{4.4.6}$$

and the following unitary operators:

$$\text{the free evolution } V_t^0 = e^{-itH_0}, \tag{4.4.7}$$

$$\text{the total evolution } V_t^\lambda = e^{-itH^{(\lambda)}}, \tag{4.4.8}$$

$$\text{the interacting representation evolution } U_t^{(\lambda)} = (V_t^0)^+ V_t^\lambda. \tag{4.4.9}$$

$U_t^{(\lambda)}$ (also called the *wave operator* at time t) is the solution of the following *Schrödinger equation in the interaction picture*:

$$\partial_t U_t^{(\lambda)} = -i\lambda H_{\rm I}(t)\, U_t^{(\lambda)}, \quad U_0^{(\lambda)} = 1, \tag{4.4.10}$$

where

$$H_{\rm I}(t) = (V_t^0)^+ H_{\rm I} V_t^0. \tag{4.4.11}$$

Up to now the two systems have entered in a completely symmetric way, and there is no reason to call one of them *system* and the other one *reservoir*. Let us now introduce those assumptions which justify why S is called the system and R the reservoir.

4.5 Assumptions on the Environment (Field, Gas, Reservoir, etc.)

Throughout the present book we shall assume that the reservoir R is a *Gaussian quantum field* a_k^+, a_k over $\mathbb{R}^d$, with

$$d \geq 3$$

and satisfying the q-commutation relations

$$a_k a_{k'}^+ - q a_{k'}^+ a_k = [a_k, a_{k'}^+]_q = \delta(k - k'), \quad k, k' \in \mathbb{R}^d, \tag{4.5.1}$$

for some complex number q. In fact we shall be mainly interested in the Bose ($q = 1$) or Fermi ($q = -1$) case, but we want to emphasize that our method works for arbitrary q. We recall that the Gaussianity assumption means that there exists a unit vector $\Phi \in \mathcal{H}_R$ (vacuum, equilibrium, ground, squeezing, etc.), defining the Gaussian statistics of the field in the sense of Definition 2.2.1. We also assume that Φ is invariant under the free reservoir evolution (in the gauge-invariant, δ-correlated case this always happens; see Sect. 2.5). It is moreover required that the time correlations of a_k^+ and a_k with respect to Φ decay sufficiently rapidly to be integrable in the sense of distributions (see Theorem 3.2.1), formally

$$\int_{-\infty}^{+\infty} | \langle a^{\varepsilon_1}(0,k) a^{\varepsilon_2}(t,k') \rangle | \, dt < +\infty, \tag{4.5.2}$$

where condition (4.5.2) is meant in the same sense as condition (3.2.2), a^ε denotes either a^+ or a_k and

$$a(t,k) = e^{itH_0} a_k e^{-itH_0} = e^{-it\omega_1(k)} a_k. \tag{4.5.3}$$

The suffix 1 in $\omega_1(k)$ denotes 1-particle evolution and sometimes will be omitted.

Condition (4.5.2), which in particular requires a continuous spectrum, will be taken as the main characteristic of a *reservoir* and breaks the symmetry between systems R and S. Throughout this book the 1-particle space of the reservoir will be the space $L^2(\mathbb{R}^d)$ of square integrable complex-valued functions defined on $\mathbb{R}^d$; we shall denote $\mathcal{K}$ a subspace of $L^2(\mathbb{R}^d)$ such that any pair of functions $f, g \in \mathcal{K}$ satisfy the integrability condition

$$\int_{-\infty}^{+\infty} ds \mid \langle f, e^{is\omega(p)} g \rangle \mid < +\infty, \tag{4.5.4}$$

which will be called *the basic analytical condition* and in particular implies that the 1-particle reservoir Hamiltonian H_1 has a continuous spectrum. Since the map $s \mapsto \langle g, e^{is\omega(p)} g \rangle$ is positive definite, it follows from Bochner's theorem [Yosh66] that the sesquilinear form

$$(g|g) := \int_{-\infty}^{+\infty} ds \langle g, e^{is\omega(p)} g \rangle = 2\pi \langle g, \delta(\omega(p)) g \rangle \tag{4.5.5}$$

is a pre-scalar product. Endowed with this pre-scalar product, $\mathcal{K}$ becomes a pre-Hilbert space and, when no confusion can arise, we shall use the same symbol to denote this space and its completion (for the meaning of the δ-function $\delta\big(\omega(p)\big)$ of an operator see Definition 1.2.1).

4.6 Assumptions on the System Hamiltonian

We shall assume that the system S is *discrete* in the sense that its free Hamiltonian H_S has a discrete spectrum

$$H_S = \sum_n \varepsilon_n P_n, \tag{4.6.1}$$

where ε_n are the eigenvalues and P_n the spectral projections. More generally, $\mathcal{H}_S$ can be decomposed into orthogonal subspaces corresponding to the discrete, the absolutely continuous and the singular parts of the spectrum of H_S. The vectors in the discrete spectrum subspace of $\mathcal{H}_S$ are called the *bound states* of the system S. To restrict one's attention to this subspace means that the system is initially prepared in a bound state. In the case of the hydrogen atom this means that we ignore the effect of the ionized states. Often, in atomic physics, one is interested only in the behaviour of bound states, so several interesting cases are included in the Hamiltonians discussed the following. The continuous part of the spectrum shall be dealt with in Part II, and we shall see that qualitatively new phenomena arise in this case.

4.7 The Free Hamiltonian

The free reservoir Hamiltonian H_R will usually have an absolutely continuous spectrum coinciding with the positive half line $\mathbb{R}_+$, with infinite multiplicity. Thus, in the assumption (4.6.1), the Hamiltonian H_0, defined by (4.4.5), has the following spectrum:

$$\begin{aligned} \sigma_{ac}(H_0) &= \mathbb{R}_+, \\ \sigma_{sc}(H_0) &= \emptyset, \\ \sigma_{pp}(H_0) &= \{\varepsilon_1 \dots \varepsilon_M\}. \end{aligned} \tag{4.7.1}$$

Denote by ψ_j an eigenvector of H_S corresponding to the eigenvalue ε_j. Clearly $\Psi_j = \psi_j \otimes \Phi$ is an eigenfunction of H_0 corresponding to the same eigenvalue. So these eigenvalues of H_0 are embedded into the continuous spectrum.

4.8 Multiplicative (Dipole-Type) Interactions: Canonical Form

Definition 4.8.1. *H_{I} is called a multiplicative or dipole-type interaction Hamiltonian if it has the form*

$$H_{\mathrm{I}} = \int d^3k \quad \{D(k) \otimes a^+(k) + D^+(k) \otimes a(k)\}, \tag{4.8.1}$$

where $\{D(k) : k \in \mathbb{R}^3\}$ is a family of system operators (i.e. acting on $\mathcal{H}_S$) called the response terms or currents and containing local information about the interaction.

This is most general dipole-type interactions between a system (the D field) and an environment (the a field). One can think, for example, of two interacting fields the spectrum of one which is discretized by introducing a cutoff (see Chap. 14). Many interactions encountered both in the theory of open systems and in quantum field theory have the form (4.8.1) for an appropriate choice of $D(k)$; in particular this form includes the QED Hamiltonian (4.1.16) and many other examples that we shall discuss in the following.

Let H_S, (ε_n), and (P_n) be as in Sect. 4.6, so that (4.6.1) holds. Then, arguing as in Sect. 3.6, we can write the interaction Hamiltonian in the form

$$H_{\mathrm{I}} = \int dk \sum_{m,n} P_m D(k) P_n \otimes a_k^+ + \text{h.c.}, \tag{4.8.2}$$

so that the time-evolved interaction Hamiltonian becomes

$$H_{\mathrm{I}}(t) = \int dk \sum_{m,n} P_m D(k) P_n \otimes e^{it(\omega_k + \varepsilon_m - \varepsilon_n)} a_k^+ + \text{h.c.} \tag{4.8.3}$$

Now let us introduce the *Bohr frequencies*

$$\omega = \varepsilon_n - \varepsilon_m, \tag{4.8.4}$$

and let us label them by an index q so that, by construction

$$\omega_q \neq \omega_p \quad \text{if} \quad q \neq p. \tag{4.8.5}$$

Then we introduce the operators

$$D_q(k) := \sum_{\varepsilon_n - \varepsilon_m = \omega_q} P_m D(k) P_n, \tag{4.8.6}$$

and (4.8.3) becomes

$$H_{\mathrm{I}}(t) = \int dk \sum_q D_q(k) \otimes e^{it(\omega_k - \omega_q)} a_k^+ + \text{ h.c.}, \tag{4.8.7}$$

where the system operators $D_q(k)$ satisfy

$$e^{itH_S} D_q(k) e^{-itH_S} = e^{-it\omega_q} D_q(k). \tag{4.8.8}$$

The rescaled Schrödinger equation, in the interaction representation, associated to (4.8.7) is

$$\partial_t U_{t/\lambda^2}^{(\lambda)} = -\frac{i}{\lambda} H_{\mathrm{I}}(t/\lambda^2) U_{t/\lambda^2}^{(\lambda)}. \tag{4.8.9}$$

4.9 Approximations of the Multiplicative Hamiltonian

Instead of discussing immediately the stochastic limit of the evolution $U_{t/\lambda^2}^{(\lambda)}$ associated to the Hamiltonian (4.8.7), we follow the historical developement of the subject by starting from some very strong simplifying assumptions on the interaction Hamiltonian H_{I} and then showing how, each time these assumptions are weakened, some new qualitative phenomena arise. This approach not only helps to put in the correct perspective the role of the single assumptions, but also allows one to understand the basic ideas of the stochastic limit in the simplest possible situations, without unnecessary complications. We shall proceed in three steps, in increasing order of complexity.

4.9.1 Rotating-Wave Approximation Hamiltonians

The first and simplest class of Hamiltonians we are going to consider are obtained from (4.8.7) through the following approximations:

(i) *Generalized dipole approximation:*

$$D_q(k) \hookrightarrow D_q^{\text{dipole}}(k) = g_q(k) D_q, \tag{4.9.1}$$

where $g_q(k)$ is a test function (cutoff or form factor) and D_q is a fixed system operator.

(ii) *Only one Bohr frequency*, i.e. the sum over q in (4.8.7) is replaced by a single term.

Approximation (i) is valid whenever the response does not vary appreciably for those values of the wavelength (i.e. the argument k) of the reservoir particles which are of interest for the problem considered (see Sect. 4.2). The band of variation of these wavelengths are delimited by the form factors g_q. In QED this approximation is not appropriate for large momenta.

Approximation (ii) means that we concentrate our attention on only two levels of the system S.

Under these assumptions, (4.8.7) is replaced by the approximate Hamiltonian

$$H_{\mathrm{I}} = D^+ \otimes A(g) + D \otimes A^+(g) = \int_{\mathbb{R}^d} dk [D^+ \otimes \overline{g}(k) a_k + D \otimes g(k) a_k^+], \tag{4.9.2}$$

where g is a test function (*form factor*) in the space $\mathcal{K}$, defined at the end of Sect. 4.5, describing the strength of the interaction of the system with the environment D is an operator on the system space satisfying the following condition: there exists a dense subspace $\mathcal{D}_S \subset \mathcal{H}_S$, and for any pair of vectors $\xi, \eta \in \mathcal{D}_S$, there exists an $\alpha > 0$ such that

$$\sum_{n=1}^{\infty} \frac{|\langle \xi, D^n \eta \rangle|}{[n/\alpha]!} < \infty, \tag{4.9.3}$$

where $[n/\alpha]$ denotes the integer part of n/α. This condition is obviously satisfied if D is bounded. If also $\mathcal{H}_S$ is a second quantization space then, choosing $\mathcal{D}_S$ to be the linear span of the finite particle (or coherent) vectors, one can choose D to be any polynomial in the creation or annihilation operators.

4.9.2 No Rotating-Wave Approximation Hamiltonians with Cutoff

The interaction Hamiltonian H_{I} is a (possibly infinite) sum of terms of the form (4.9.2), i.e.

$$H_{\mathrm{I}} = \sum_{q=1}^{N} \left(D_q^+ \otimes A(g_q) + D_q \otimes A^+(g_q) \right), \tag{4.9.4}$$

and for any time t

$$e^{itH_S} D_q e^{-itH_S} = e^{-it\omega_q} D_q. \tag{4.9.5}$$

4.9.3 Neither Dipole nor Rotating-Wave Approximation Hamiltonians Without Cutoff

This is the original multiplicative coupling Hamiltonian (4.8.1) without any approximation. Here quantities relative to the system act as cutoff on the field through a mechanism whose implications deserves further investigation because it clarifies the physical meaning of the cutoff functions (or form factors) which are present in all the interaction Hamiltonians considered in this book. Let us illustrate the situation in the case in which the system Hamiltonian H_S has a purely nondegenerate spectrum. In this case, in the notation of (4.6.1) and denoting $|\varepsilon_n\rangle$ the eigenvector of H_S belonging to the eigenvalue ε_n, one has

$$H_S = \sum_n \varepsilon_n |\varepsilon_n\rangle\langle\varepsilon_n|, \tag{4.9.6}$$

and the operator (4.8.6) becomes

$$D_q(k) = \sum_{\varepsilon_n - \varepsilon_m = \omega_q} P_m D(k) P_n = \sum_{\varepsilon_n - \varepsilon_m = \omega_q} |\varepsilon_m\rangle\langle\varepsilon_n| \langle\varepsilon_m, D(k)\varepsilon_n\rangle. \tag{4.9.7}$$

Thus, if we introduce the notation

$$g_{m,n}^{(q)}(k) := \langle\varepsilon_m, D(k)\varepsilon_n\rangle, \tag{4.9.8}$$

(4.8.7) takes the form

$$\begin{aligned} H_{\mathrm{I}}(t) &= \sum_q \sum_{\varepsilon_n - \varepsilon_m = \omega_q} |\varepsilon_m\rangle\langle\varepsilon_n| \otimes \int dk\, g_{m,n}^{(q)}(k) a_k^+ e^{it(\omega_k - \omega_q)} + \text{h.c.} \\ &= \sum_q \sum_{\varepsilon_n - \varepsilon_m = \omega_q} |\varepsilon_m\rangle\langle\varepsilon_n| \otimes A\left(S_t^{(q)} g_{m,n}^{(q)}\right) + \text{h.c.} \end{aligned} \tag{4.9.9}$$

From (4.9.8–9) it is clear that the cutoff functions of the a field are matrix elements operators of the system. This shows that, at least in all cases in which the matrix elements (4.9.8) have a sufficiently rapid decay, the introduction of cutoff functions is not necessarily something *introduced by hand* and external to the theory, but may be something canonically deduced from the interaction (see Sect. 5.18 for further discussion).

In Chap. 11 we will show that, as a consequence of the *block principle* (see Sect. 11.7), the qualitative picture does not change very much if one enlarges the class of multiplicative couplings by replacing, in (4.9.2), the linear powers of the field operators $A(g), A^+(g)$ with arbitrary polynomials.

This shows that the essential restriction of the multiplicative coupling considered in this chapter is not so much the bilinearity in the system–reservoir coupling, as the fact that, in the time-evolved interaction Hamiltonian (4.9.9), the frequency ω_q of the system operator D does not depend on k. This restriction will be removed in Chap. 12, and we shall see that this removal introduces substantial changes in the overall picture after the stochastic limit.

4.10 The Generalized Rotating-Wave Approximation

Definition 4.10.1. *The interaction Hamiltonian H_I is said to satisfy the generalized rotating-wave approximation condition if there exists a fixed strictly positive real number,*

$$\omega_0 > 0, \tag{4.10.1}$$

such that for any $t \in \mathbb{R}$ one has

$$e^{itH_S} D e^{-itH_S} = e^{-i\omega_0 t} D. \tag{4.10.2}$$

Example 4.10.1. Take the system space $\mathcal{H}_S$ to be the Fock space of a 1-mode field (harmonic oscillator) with creation and annihilation operators a, a^+ and

$$H_S = \omega_0 a^+ a, \quad D = a. \tag{4.10.3}$$

Clearly, if $D = a$, (4.10.2) is satisfied, and also the analytical condition (4.9.3) is satisfied if one chooses $\mathcal{D}_S$ to be the linear span of the number vectors. Replacing a by any of its powers a^m simply amounts to replacing ω_0 in (4.10.2) by $m\omega_0$; condition (4.9.3) is still satisfied. Another example, based on 2-level systems and the Pauli matrix, is discussed in Sect. 5.9.

Remark 4.10.1. In some sense the two examples mentioned above are the *building blocks* of all operators satisfying condition (4.10.2) above. In fact, suppose that $|\varepsilon\rangle$ is an eigenvector of H_S corresponding to the eigenvalue ε, i.e.

$$H_S|\varepsilon\rangle = \varepsilon|\varepsilon\rangle,$$

then

$$e^{itH_S} D|\varepsilon\rangle = \left(e^{itH_S} D e^{-itH_S}\right) e^{itH_S}|\varepsilon\rangle = e^{it(\varepsilon-\omega_0)} D|\varepsilon\rangle,$$

i.e. $D|\varepsilon\rangle$ is an eigenvector corresponding to the eigenvalue $\varepsilon - \omega_0$. This means that the operator D *reduces* the energy of the system; therefore, since in the interaction Hamiltonian (4.9.4) it is paired to a creation operator, it behaves as an annihilation operator. Thus the interaction Hamiltonian (4.9.2) describes an exchange of single quanta between the system and the reservoir. In the optical interpretation, ω_0 is the proper frequency of the laser.

Iterating the above argument we see that for any $k \in \mathbb{N}$ there are only two possibilities:

(i) $D^k|\varepsilon\rangle = 0$.
(ii) $\varepsilon - k\omega$ is an eigenvalue of H_S. However, since H_S is bounded below, for each of its eigenvectors $|\varepsilon\rangle$ there must exist some $k = k_\varepsilon \in \mathbb{N}$ such that condition (i) is verified (see [AcKo00b] for more detailed analysis).

4.11 The Stochastic Limit of the Multiplicative Interaction

Under condition (4.10.2) the interaction Hamiltonian (4.9.2) evolved with the free evolution is as follows:

$$H_{\mathrm{I}}(t) := \int dk \left[D^+ \otimes \bar{g}_k e^{-it\omega(k)} a_k + D \otimes g_k e^{it\omega(k)} a_k^+\right] =: D^+ a_t + D a_t^+, \tag{4.11.1}$$

where we omit the symbol $\otimes$ (often the case in the following) and where we have introduced the shorthand notations:

$$a_t := \int dk \bar{g}_k e^{-it\omega(k)} a_k = A(S_t g), \tag{4.11.2}$$

$$S_t := S_t^1 e^{-it\omega_0} = e^{it(\omega_1(k)-\omega_0)} =: e^{it\omega(k)}, \tag{4.11.3}$$

$$\omega(k) := \omega_1(k) - \omega_0 \tag{4.11.4}$$

($\omega_1(k)$ is as in Sect. 4.5). Thus we see that *as far as the interaction H_{I} is concerned*, under the assumption (4.10.2), one can assume that the system Hamiltonian is absent (i.e. it is zero) at the cost of replacing the 1-particle reservoir Hamiltonian by the *effective* 1-particle Hamiltonian, given by (4.11.4), which also depends on the atomic frequency ω_0.

In these notations and assumptions and with a_t given by (4.11.2), the Schrödinger equation in the interaction representation takes the form

$$\partial_t U_t^{(\lambda)} = -i\lambda \left(D^+ a_t + D a_t^+\right) U_t^{(\lambda)}. \tag{4.11.5}$$

The rescaled evolution operator satisfies the rescaled Schrödinger equation:

$$\partial_t U_{t/\lambda^2}^{(\lambda)} = -i H_{\mathrm{I}}^{(\lambda)}(t) U_{t/\lambda^2}^{(\lambda)}, \tag{4.11.6}$$

where the *rescaled interaction Hamiltonian* is defined by

$$H_{\mathrm{I}}^{(\lambda)}(t) := \frac{1}{\lambda} H_{\mathrm{I}}(t/\lambda^2) = D^+ \left(\frac{1}{\lambda} a_{t/\lambda^2}\right) + D\left(\frac{1}{\lambda} a_{t/\lambda^2}^+\right) =: D^+ a_t^{(\lambda)} + D a_t^{(\lambda)+} \tag{4.11.7}$$

and the *rescaled fields* are

$$a_t^{(\lambda)} = \frac{1}{\lambda} a_{t/\lambda^2} = \frac{1}{\lambda} \int_{\mathbb{R}^d} dk \bar{g}_k e^{-i(t/\lambda^2)\omega(k)} a_k. \tag{4.11.8}$$

Now we apply the general scheme, described in Sect. 3.2, to deduce the convergence of the rescaled fields and of (4.11.5). We know from Theorem 3.2.1 that the rescaled fields (4.11.8) converge in the sense of correlators to a quantum white noise b_t, b_t^+ whose covariance is determined by the reference state of the a field as explained in Sect. 3.2. In symbols

$$a_t^{(\lambda)} = \frac{1}{\lambda} a_{t/\lambda^2} \to b_t, \quad a_t^{(\lambda)+} = \frac{1}{\lambda} a^+_{t/\lambda^2} \to b_t^+, \tag{4.11.9}$$

$$\lim_{\lambda\to 0} \frac{1}{\lambda} H_{\mathrm{I}} \left(\frac{t}{\lambda^2}\right) = H_t = D^+ b_t + D b_t^+, \tag{4.11.10}$$

and the white noise Hamiltonian equation associated with the right-hand side of (4.11.7) is

$$\partial_t U_t = -iH_t U_t = -i(D^+ b_t + D b_t^+) U_t. \tag{4.11.11}$$

Theorem 4.11.1. *Equation (4.11.11) can be given a meaning and, under the assumption (4.9.3), gives a unique unitary solution U_t. Moreover,*

$$\lim_{\lambda\to 0} U^{(\lambda)}_{t/\lambda^2} = U_t,$$

where convergence is meant in the sense of the matrix elements with respect to the collective vectors (see Sect. 1.26 for a general definition of these vectors).

Proof 4.11.1. The meaning of equation (4.11.11) is explained in Sect. 4.12. The proof is performed using the estimates developed in Part III. We shall not discuss the details here.

4.12 The Normal Form of the White Noise Hamiltonian Equation

Equation (4.11.11) is only apparently a usual Hamiltonian equation. In fact it is very singular because b_t, b_t^+ is a quantum white noise. In order to give it meaning, one has to bring it to its normally ordered form, i.e. instead of the term $b_t U_t$ we would like to have the term $U_t b_t$. Rewriting the equation in the form

$$\partial_t U_t = -iD b_t^+ U_t - iD^+ U_t b_t - iD^+ [b_t, U_t], \tag{4.12.1}$$

we see that the normal form of the equation involves the computation of the commutator $[b_t, U_t]$. This is done through the following lemma, whose proof shall be given in Sect. 7.9.

Lemma 4.12.1. *Let* U_t *denote the unique unitary solution of* (4.11.11). *Then the following identities hold:*

$$[b_t, U_t] = b_t U_t - U_t b_t = -i\gamma_- D U_t, \tag{4.12.2}$$

$$[U_t^*, b_t^+] = U_t^* b_t^+ - b_t^+ U_t^* = i\overline{\gamma}_- U_t^* D^+, \tag{4.12.3}$$

$$[b_t, U_t^*] = i\gamma_- U_t^* D, \tag{4.12.4}$$

where γ_- *is given by*

$$\gamma_- := \int_{-\infty}^{0} \langle g, S_s g\rangle ds = \int_{-\infty}^{0} dt \int_{\mathbb{R}^d} e^{-it[\omega_1(k)-\omega_0]} \mid g(k) \mid^2 dk$$
$$= \int_{-\infty}^{0} ds\, e^{+is\omega_0} G_{-+}(s). \tag{4.12.5}$$

Applying this lemma, it is possible to give a precise meaning to (4.11.11) by bringing it to normal order, as described by the following theorem:

Theorem 4.12.1. *If the reference state of the a field is the Fock vacuum, then the normally ordered form of the white noise Hamiltonian* (4.11.11) *is the following equation:*

$$\partial_t U_t = -i(D^+ U_t b_t + D b_t^+ U_t) - Y U_t, \tag{4.12.6}$$

where

$$Y = \gamma_- D^+ D \tag{4.12.7}$$

is the operator transport coefficient defined in Sect. 1.8 with respect to the vacuum expectation value.

Proof 4.12.1. The term $D b_t^+ U_t$ is already in normally ordered form. Thus we have only to consider the term

$$D^+ b_t U_t = D^+ [b_t, U_t] + D^+ U_t b_t.$$

From Lemma 4.12.1 we know that $[b_t, U_t] = -i\gamma_- D U_t$, and from this the thesis follows.

Definition 4.12.1. *Given a white noise Hamiltonian (4.11.10), the operator* Y, *the coefficient of* U_t *in the term not including noise operators in (4.12.6), is called the drift coefficient or the Ito correction term..*

4.13 Invariance of the Ito Correction Term Under Free System Evolution

The additional term $-YU_t$, appearing in the white noise equation (4.12.6) as a consequence of the normal order, is called the *Ito correction term.* As already remarked, this term is nothing but the *operator-valued transport coefficient* introduced in Sect. 1.8.

A corollary of the generalized rotating-wave approximation (4.10.2) is that *the coefficient $-\gamma_- D^+ D$ of the Ito correction term commutes with the system Hamiltonian H_S*. In fact, using the differential form of (4.10.2), we find

$$[H_S, Y] = [H_S, D^+ D] = [H_S, D^+]D + D^+[H_S, D] = \omega_0 D^+ D - \omega_0 D^+ D = 0. \tag{4.13.1}$$

In terms of the representation of γ, discussed in Lemma 4.20.1, the Ito correction term corresponds to a non-self-adjoint correction to the system Hamiltonian,

$$iH_S \to iH_S - \gamma_- D^+ D = i(H_S + \mathrm{Im}(\gamma_-)D^+ D) - \frac{1}{2}\gamma D^+ D. \tag{4.13.2}$$

Since H_S commutes separately with the real and imaginary part, we see that the effect of the imaginary part of the Ito correction is a *global (operator) shift of amount* Im $\gamma_- D^+ D$ *in the spectrum of the system Hamiltonian.*

We shall see that the commutation relation (4.13.1) is a quite universal phenomenon in the stochastic limit and has an important consequence, namely: *the energy levels of the free system Hamiltonian H_S, after the stochastic limit, undergo a classical stochastic process.*

This gives a theoretical foundation to the empirical successes achieved through the use of classical stochastic processes in the description of quantum systems.

4.14 The Stochastic Golden Rule: Langevin and Master Equations

Given the normally ordered form of the equation, one can calculate all the quantities of physical interest. Therefore it is useful to formulate a simple rule which allows one to write down immediately the white noise equation for the most general multiplicative interactions (4.8.1) written in canonical form, i.e. in the notations of Sect. 4.9:

$$\begin{aligned} H_\mathrm{I}(t) &= \sum_q \sum_{\varepsilon_n - \varepsilon_m = \omega_q} |\varepsilon_m\rangle\langle\varepsilon_n| \otimes \int dk\, g^{(q)}_{m,n}(k) a^+_k e^{it(\omega_k - \omega_q)} + \text{h.c.} \\ &= \sum_q H_\mathrm{I}^{(q)}(t), \end{aligned}$$

where, for $q \neq q'$, $\omega_q \neq \omega_{q'}$ and each Hamiltonian $H_\mathrm{I}^{(q)}(t)$ satisfies condition (4.10.2) with ω_0 replaced by ω_q (notice that ω_q might be negative), more precisely:

$$H_{\mathrm{I}}^{(q)}(t) := \sum_{\varepsilon_n-\varepsilon_m=\omega_q} |\varepsilon_m\rangle\langle\varepsilon_n| \otimes \int dk\; g_{m,n}^{(q)}(k) a_k^+ e^{it(\omega_k-\omega_q)} + \text{h.c.}$$
$$= \sum_{\varepsilon_n-\varepsilon_m=\omega_q} |\varepsilon_m\rangle\langle\varepsilon_n| \otimes A\left(S_t^{(q)} g_{m,n}^{(q)}\right) + \text{h.c.}$$

We call the following set of prescriptions *the stochastic golden rule* because it is a generalization of the *Fermi golden rule* (how to deduce the latter from the former is discussed in Sect. 5.8). This rule allows one just by inspection of the given interaction Hamiltonian:

- to define the *rescaled fields* (hence to choose the collective vectors);
- to determine the limit space on which the limit of $U_{t/\lambda^2}^{(\lambda)}$, $U_{t/\lambda^2}^{(\lambda)} X U_{t/\lambda^2}^{*(\lambda)}$ and $U_{t/\lambda^2}^{*(\lambda)} X U_{t/\lambda^2}^{(\lambda)}$ live (the *master space*);
- to write down the white noise equations (or the equivalent quantum stochastic differential form of these equations) satisfied by
  (i) the limit evolution operator U_t (stochastic Schrödinger equation),
  (ii) the limit interacting Heisenberg evolutions $U_t^*(1 \otimes X)U_t$ forward (inner) Langevin equation,
  (iii) the limit interacting Heisenberg evolutions $U_t(1 \otimes X)U_t^*$ backward (outer) Langevin equation,
- to write down the backward (for observables) and forward (for density matrices) master equations, canonically associated to the corresponding Langevin equations (as explained in Sect. 1.13).

For the Hamiltonian above and for an arbitrary mean zero gauge-invariant Gaussian state of the reservoir (thus including equilibrium states), the rule is the following:

(i) Write, as in (4.11.7), the rescaled interaction Hamiltonians $(1/\lambda) \cdot H_{\mathrm{I}}^{(q)}(t/\lambda^2)$ in terms of the rescaled fields:

$$a_\lambda^{+(q)}(t/\lambda^2) := \frac{1}{\lambda} a_k^+ e^{i(\omega_k-\omega_q)t/\lambda^2}. \tag{4.14.1}$$

(ii) Use Theorem 3.2.1 to find the master field as the limit of the rescaled fields $\left(a_{t/\lambda^2}^{(\lambda)\pm} \to b_t^\pm\right)$, in the sense of correlators with respect to the reference state of the field, and write down the white noise Hamiltonian equation as the formal limit of the rescaled equation (4.11.6):

$$\partial_t U_t = -iH_t U_t. \tag{4.14.2}$$

(iii) Compute the operator-valued transport coefficient Y, i.e. the limit of the expectation of the second-order term of the iterated series (at time $t = 1$) with respect to the reference state of the field (see Lemma 1.8.1):

$$\lim_{\lambda\to 0}\lambda^2\int_0^{1/\lambda^2}dt_1\int_0^{t_1}dt_2\langle H_{\mathrm{I}}(t_1)H_{\mathrm{I}}(t_2)\rangle$$
$$=\sum_q\int_{-\infty}^0 ds\left\langle H_{\mathrm{I}}^{(q)}(0)H_{\mathrm{I}}^{(q)}(s)\right\rangle = Y. \qquad (4.14.3)$$

The normally ordered form of the white noise Hamiltonian equation is then

$$\partial_t U_t = -i : H_t U_t : \ - YU_t, \qquad (4.14.4)$$

where $: H_t U_t :$ is obtained from $H_t U_t$ simply by bringing all the time annihilators contained in H_t to the right of U_t. If the state is nonFock, one expresses it in terms of Fock and anti-Fock states as described in Sect. 2.18, and the annihilators have to be meant in the sense of this realization. Using equation (4.14.4), it is easy, following the method explained in Sect. 4.17, to compute the matrix elements of U_t in many different types of states (vacuum, nonzero temperature, coherent, n-particle, etc.).

(iv) Write the white noise Langevin equation for the Heisenberg evolution of an observable X (in interaction representation) simply by differentiating the product $X_t = U_t X U_t^*$, replacing $\partial_t U_t$ by its normally ordered form, and bringing the resulting product to normal order using Lemma 4.12.1. This gives rise to an expression of the form

$$\partial_t X_t = \theta_+(X_t)b_t + b_t^+\theta_-(X_t) + \theta_0(X_t) - YX_t - X_tY^* \qquad (4.14.5)$$

or to a sum of such expressions. Here $\theta_+,\theta_-,\theta_0$ are linear operators depending only on the system variables, i.e. such that for any expectation value $\langle\,\cdot\,\rangle$, involving only the master field $b_t^{\pm}$, one has

$$\langle\theta(Z)\rangle = \theta(\langle Z\rangle). \qquad (4.14.6)$$

This step allows one to estimate all the relevant physical quantities and is an essential step beyond the master equation; for example, in the classical case the master equation corresponds only to the semigroup equation of a Markov process, but the Langevin equation corresponds to the full stochastic differential equation.

The explicit form of the maps $\theta_\pm$, θ_0 strongly depends on the following:

- the interaction Hamiltonian H_{I};
- the free evolution of the (S, R) system;
- the reference state of the R (reservoir) system.

Therefore a single formula encompassing all possible cases does not exist. Several explicit examples are discussed in the following sections.

(v) The master equation for $\overline{X}_t := \langle X_t \rangle$, is the vacuum expectation value (in the sense of (iii) above and therefore includes non zero temperature expectation values) of the white noise Langevin equation. Thus, using (4.14.5–6):

$$\partial_t \overline{X}_t = \theta_0(\overline{X}_t) - Y\overline{X}_t - \overline{X}_t Y^*. \tag{4.14.7}$$

θ_0 and Y depend on the reference state of the reservoir. However in all cases (4.14.7) has the general form (1.13.10), corresponding to the GKSL generator [GKS76], [Li76]. The master equation for the density operator is the dual of (4.14.7).

All the results of the stochastic limit are obtained as an application of the stochastic golden rule. We shall see that, with minor adjustments, this rule is valid also in the Fermi case and for a very large class of interaction Hamiltonians, free evolutions (of both the system and the reservoir) and reference states of the field.

4.15 Classical Stochastic Processes Underlying Quantum Processes

We illustrate the way to use the stochastic golden rule in the simplest case in which the interaction Hamiltonian is (4.11.1), the a field is boson and its reference state is the Fock vacuum. In this case we have shown in Sect. 4.12 that (4.14.4) becomes

$$\partial_t U_t = -ib_t^+ D U_t - iD^+ U_t b_t - \gamma_- D^+ D U_t, \tag{4.15.1}$$

with γ_- given by (4.12.5). If we fix a system observable X and denote $X_t = U_t X U_t^*$ then, since $[b_t^{\pm}, X] = 0$, the white noise Langevin equation for X_t is given by

$$\begin{aligned}
\partial_t X_t &= (\partial_t U_t) X U_t^* + U_t X (\partial_t U_t^*) \\
&= (-ib_t^+ D U_t - iD^+ U_t b_t - \gamma_- D^+ D U_t) X U_t^* \\
&\quad + U_t X (iU_t^* b_t D^+ + ib_t^+ U_t^* D - \overline{\gamma}_- U_t^* D^+ D) \\
&= -ib_t^+ D X_t - iD^+ U_t X [b_t, U_t^*] - iD^+ X_t b_t - \gamma_- D^+ D X_t \\
&\quad + iX_t b_t D^+ + i[U_t, b_t^+] X U_t^* D + ib_t^+ X_t D - \overline{\gamma}_- X_t D^+ D.
\end{aligned}$$

Using Lemma 4.12.1 we explicitly compute the commutators and find

$$\begin{aligned}
&-ib_t^+ D X_t + \gamma_- D^+ X_t D - iD^+ X_t b_t - \gamma_- D^+ D X_t \\
&\qquad + iX_t D^+ b_t + \overline{\gamma} D^+ X_t D + ib_t^+ X_t D - \overline{\gamma}_- X_t D^+ D.
\end{aligned}$$

Thus we finally arrive at the normally ordered Langevin white noise equation,

$$\partial_t X_t = - ib_t^+[D, X_t] + i[X_t, D^+]b_t + 2\,\mathrm{Re}\,(\gamma_-)D^+X_tD \\ - \gamma_-D^+DX_t - \overline{\gamma}_-X_tD^+D, \tag{4.15.2}$$

and from this, taking the vacuum expectation value, we get the master equation for $\overline{X}_t := \langle X_t \rangle$, i.e.

$$\partial_t\overline{X}_t = 2\,\mathrm{Re}\,(\gamma_-)D^+\overline{X}_tD - \gamma_-D^+D\overline{X}_t - \overline{\gamma}_-\overline{X}_tD^+D. \tag{4.15.3}$$

The generator of the Markov semigroup associated with the master equation (4.15.3) is then

$$L(Z) = 2\,\mathrm{Re}\,(\gamma_-)D^+ZD - \gamma_-D^+DZ - \overline{\gamma}_-ZD^+D. \tag{4.15.4}$$

Remark 4.15.1. As a corollary of the generalized rotating-wave approximation condition (4.10.2) we deduce (see Sect. 4.13) that *the generator L commutes with the free system evolution:*

$$L\left(e^{itH_S}Ze^{-itH_S}\right) = e^{itH_S}L(Z)e^{-itH_S}. \tag{4.15.5}$$

This is an important property because it implies that *when restricted to the abelian algebra generated by H_S, the generator L becomes the generator of a classical jump Markov process among the energy levels of H_S.* We shall see that this property is of general validity in the stochastic limit, i.e. it continues to hold also when the generalized rotating-wave approximation condition (4.10.2) is dropped.

Note that this property does not remain true for the Langevin equation. In Sect. 4.27 we shall see that there exists a natural generalization of the commutation relation (4.15.5) to the Langevin equation which gives rise to an interesting family of classical jump processes canonically associated with the initial quantum process.

We shall also prove (Sect. 4.27) that these commutation relations are a universal property of the stochastic limit.

4.16 The Fluctuation–Dissipation Relation

In order to understand the physical meaning of the Ito correction term (4.13.2), we distinguish in it two parts:

(i) an Hamiltonian part: $i(H_S + \mathrm{Im}(\gamma_-)D^+D)$;
(ii) a dissipative part: $-\frac{1}{2}\,\gamma D^+D = -\ \mathrm{Re}\,(\gamma_-)D^+D$.

The Hamiltonian part generates a usual reversible dynamics, while the evolution generated by $-(1/2)\gamma D^+D$ is a contraction, in general nonunitary, i.e. dissipative. On the other hand γ is the covariance of the noise, which is a measure of the strength of the fluctuations. For this reason (4.12.7) can be considered as the prototype of *the quantum mechanical fluctuation–dissipation relation.* More precisely, recalling (4.12.5) and the facts that:

- $Y = -\gamma_- D^+D$ is the drift of (4.12.6) (see Definition 4.12.1);
- D^+b_t, Db_t^+ are the coefficients of the white noise Hamiltonian (4.11.11) (*noise terms*);
- $\langle \cdot \rangle$ denotes vacuum expectation; we can rewrite (ii) above in the form

$$-\langle D^+b_t Db_t^+\rangle = -\langle b_t b_t^+\rangle D^+D = -\gamma D^+D = \frac{1}{2}(Y + Y^*)$$

or, in other words,

$$-\text{ covariance of the noise terms } = \text{ real part of the drift.} \tag{4.16.1}$$

(4.16.1) is a universal formulation of *the quantum mechanical fluctuation–dissipation relation.* In fact we shall prove that it is equivalent (up to the regularity conditions) to the unitarity of the solution of the white noise Hamiltonian (4.11.11).

In summary, *not only does the dissipative term in* (4.12.7) *not break the unitarity of the evolution, but the reason why a dissipative term must be present is precisely to guarantee the unitarity condition.*

Starting from the following section, we shall now illustrate, in a multiplicity of examples, the meaning of the energy shift (i) above and of the dissipative addition (ii).

4.17 Vacuum Transition Amplitudes

Using the normally ordered equation (4.12.6) for U_t one can easily calculate the matrix elements $\langle \psi, U_t \phi\rangle$ for a large class of states ψ, ϕ in the limit space (number or coherent vectors, etc.). As an illustration we compute some vacuum-to-vacuum amplitude of the evolution operator U_t (we mean the vacuum for the operators b_t, b_t^+). In Sect. 5.13 we shall show the connections of this result with the *Fermi golden rule.*

Now let us consider a special type of states, i.e. states of the form $\psi \otimes \Phi$ where ψ is an arbitrary state of the system and Φ is the vacuum state of the reservoir. The probability amplitude that the system will be measured in state ψ at time t if it initially was in the state φ, while the reservoir remains in its vacuum state Φ, is

$$\langle \psi \otimes \Phi, U_t(\varphi \otimes \Phi)\rangle = \langle \psi, \langle U_t\rangle \varphi\rangle. \tag{4.17.1}$$

The quantities are called *vacuum transition amplitudes* (see Sect. 1.22). The quantity

$$\langle U_t\rangle := \langle \Phi, U_t \Phi\rangle \tag{4.17.2}$$

is the partial expectation with respect to the vacuum state. Note that $\langle U_t\rangle$ is not a scalar but an operator on the system space. The reason why it is preferable to work with partial expectations is that, as we shall see in a moment, one can easily deduce an equation satisfied in full generality by the partial expectation value (4.17.2), while only for very special choices of the state ψ one can deduce an equation for the full expectation value (4.17.1). In fact the transition rate, corresponding to the amplitude (4.17.4) is

$$\frac{d}{dt}\langle \psi \otimes \Phi, U_t(\varphi \otimes \Phi)\rangle = \left\langle \psi, \left(\frac{d}{dt}\langle U_t\rangle\right)\varphi\right\rangle ,$$

and the time derivative in parentheses can be easily computed using the normal form (4.12.6) of the equation for U_t which gives:

$$\frac{d}{dt}\langle U_t\rangle = -\gamma_- D^+ D\langle U_t\rangle, \tag{4.17.3}$$

with γ_- given by (4.12.5). The solution of (4.17.3) with $U_t|_{t=0} = 1$ is

$$\langle U_t\rangle = \exp(-t\gamma_- D^+ D). \tag{4.17.4}$$

Therefore the explicit form of the expectation value (4.17.1) is

$$\langle \psi, e^{-t\gamma_- D^+ D}\psi\rangle = \langle \psi \otimes \Phi, U_t(\psi \otimes \Phi)\rangle. \tag{4.17.5}$$

Since both the operator D^+D and the real part of γ_- are positive, we see that the generic (see Sect. 4.18) behaviour one should expect from the amplitude (4.17.5) is of an *exponential decay driven by the real part of* γ_- *and modulated by its imaginary part.* The operator $\gamma_- D^+ D$ could be interpreted as an *operator inverse lifetime of the vacuum in the limit space* in the sense that, on each spectral subspace of D^+D, the inverse lifetime is uniquely defined.

Remark 4.17.1. From physical considerations we expect the real part of γ_- to be positive, in fact, if the scalar product $\langle \psi \otimes \Phi, U_t(\varphi \otimes \Phi)\rangle$ has to be interpreted as a transition amplitude, then its square is a probability; hence, it is less than or equal to 1 for any unit vector ψ. This means that the operator $\langle U_t\rangle$ should be a contraction for each t, which is not the case if the real part of γ_- is strictly less than 0. We shall see in the following section that the real part of γ_- *is always positive*, and therefore it indeed describes an exponential decay. There are interesting physical examples in which this real part is zero. In these cases the behaviour of the vacuum shall depend on the operator D^+D and purely oscillatory behaviours are possible. An example shall be given in Sects. 5.14 and 5.15.

Definition 4.17.1. *A state α is said to undergo an exponential decay under the evolution operator U_t if, for large t, the transition amplitude from the state α to itself in the interval $(0, t)$ decays exponentially, i.e.*

$$\langle \alpha, U_t\alpha\rangle \sim e^{it\Delta\varepsilon_\alpha/\hbar}, \tag{4.17.6}$$

where $\Delta\varepsilon_\alpha$ *is a complex number with a* positive imaginary part*:*

$$\Delta\varepsilon_\alpha = -\frac{1}{2i}\Gamma_\alpha + \Omega_\alpha, \quad \Gamma_\alpha > 0, \quad \Omega_\alpha \in \mathbb{R}. \tag{4.17.7}$$

More generally one speaks of exponential decay of a transition amplitude,

$$\langle \beta, U_t\alpha\rangle \sim e^{it\Delta\varepsilon_{\alpha,\beta}/\hbar}. \tag{4.17.8}$$

4.18 Mass Gap of D^+D and Speed of Decay

In the notations of the previous section, let us first consider the case in which 0 is an eigenvalue of D^+D and ψ is a corresponding normalized eigenvector:

$$D^+D\psi = 0. \tag{4.18.1}$$

Then (4.17.5) implies that the amplitude $\langle \psi\otimes\Phi, U_t\psi\otimes\Phi\rangle$ has the constant value 1. More generally:

Definition 4.18.1. *We say that a positive self-adjoint operator* A *acting on an Hilbert space* $\mathcal{H}$ *has a gap at the zero eigenvalue, if* $\mathcal{H}$ *can be decomposed as an orthogonal sum*

$$\mathcal{H} = \mathcal{H}_0 \oplus \mathcal{H}_0^\perp,$$

with

$$A \equiv 0 \quad \text{on} \quad \mathcal{H}_0,$$
$$A \geq \lambda_1 1 \quad \text{on} \quad \mathcal{H}_0^\perp$$

for some real constant $\lambda_1 > 0$.

Remark 4.18.1. Clearly if A is a positive operator with a discrete spectrum in which 0 is not an accumulation point, then there is a constant c_A such that $A - c_A 1$ has a gap at the zero eigenvalue.

Theorem 4.18.1. *Suppose that* D^+D *has a gap* c *at the zero eigenvalue then, denoting* $\mathcal{H}_0$ *the zero eigenspace of* D^+D*, if* ψ *is any vector in* $\mathcal{H}_S$ *and* ψ_0 *denotes its projection in* $\mathcal{H}_0$*, then for any vector* $\xi \in \mathcal{H}_S$ *one has*

$$|\langle \xi\otimes\Phi, U_t\psi\otimes\Phi\rangle - \langle \xi\otimes\Phi, \psi_0\otimes\Phi\rangle| \leq \|\xi\|\cdot\|\psi_0^\perp\|e^{-\lambda_1 t\gamma}, \tag{4.18.2}$$

where λ_1 *is the gap and* $\gamma = 2\,\mathrm{Re}\,\gamma_-$.

Proof 4.18.1. Writing $\psi = \psi_0 + \psi_0^\perp$ with $\psi_0 \in \mathcal{H}_0$ and $\psi_0^\perp \in \mathcal{H}_0^\perp$, one has

$$\langle \xi\otimes\Phi, U_t\psi\otimes\Phi\rangle = \langle \xi, e^{-t\gamma_- D^+D}\psi\rangle = \langle \xi, \psi_0\rangle + \langle \xi, e^{-t\gamma_- D^+D}\psi_0^\perp\rangle.$$

But

$$\|e^{-t\gamma_- D^+D}\psi_0^\perp\|^2 = \langle \psi_0^\perp, e^{-t(\gamma_- + \overline{\gamma}_-)D^+D}\psi_0^\perp\rangle \leq e^{-2t\,\mathrm{Re}\,\gamma_- c}\|\psi_0^\perp\|^2$$

and therefore the left-hand side of (4.18.2) is less than or equal to $\|\xi\|\cdot\|\psi_0^\perp\|e^{-ct\gamma}$, which is the thesis.

If D^+D does not have a gap, then nonexponential decays are mathematically possible; it is interesting to find physically meaningful cases of such behaviours. An example illustrating this phenomenon in the polaron case is given in Sect. 12.15.

4.19 How to Avoid Decoherence

The interaction Hamiltonian in the white noise equation (4.11.11),

$$H_{\mathrm{I}}(t) = -i(D^+b_t + Db_t^+), \tag{4.19.1}$$

is very singular, so it does not make sense to speak of its eigenvalues and eigenvectors. In the previous section we have seen that the vacuum behaves *as if* it were an eigenvector with a complex eigenvalue. It is therefore natural to ask oneself if there exist states which behave *as if* they were eigenvectors with *real* eigenvalues, i.e. whose transition amplitudes have an oscillating behaviour in time. Here we outline a method which can be used to produce examples of these *pseudo-eigenstates.*

Equation (4.17.5) shows that vectors of the form $\psi \otimes \Phi$, where Φ, the vacuum of the field, can be a pseudo-eigenstate only if either the real part of γ_- is equal to zero or the system state ψ is in the zero eigenspace of D^+D. We shall consider next the case of an exponential vector $\psi(f)$ with test function f and recall that such a state is characterized by the condition

$$b_t\psi(f) = f(t)\psi(f).$$

This implies that the equation for the transition amplitude between two such vectors, say $\psi(f)$ and $\psi(g)$, becomes

$$\frac{d}{dt}\langle\psi(f), U_t\psi(g)\rangle = (-iD^+g(t) - iD\overline{f}(t) - \gamma_-D^+D)\langle\psi(f), U_t\psi(g)\rangle.$$

Since f, g are square integrable (this is necessary if the vectors $\psi(f)$, $\psi(g)$ have to be in the Fock space), they must tend to 0 as $t \to +\infty$ (with the possible exception of sets of smaller and smaller measure). Therefore the decay rate of the amplitude $\langle\psi(f), U_t\psi(g)\rangle$ shall be eventually (i.e. for long times) the same as for the vacuum. This shows that if we want to prevent exponential decay then we have to consider exponential "vectors" with test functions which do not vanish at infinity. The simplest example would be the case of a test function that is constant everywhere (plane waves). Such objects are not vectors in Fock space, but one can give rigorous meaning to the expectation value with respect to them as linear functionals on the polynomial algebra generated by the field operators with test functions say, with compact support, which are a dense subspace of the 1-particle space.

For the moment let us discuss only an approximate result.

To simplify the discussion let us choose the system state ξ to be also a complex eigenvector of D with complex eigenvalue z (say D is an annihilation operator and $\xi = |z\rangle$, a coherent vector), then denoting

$$\beta := \xi \otimes \psi(f), \quad \alpha = \xi \otimes \psi(g),$$

we obtain

$$\frac{d}{dt}\langle \beta, U_t\alpha\rangle = \left(-i\overline{z}g(t) - iz\overline{f}(t) - \gamma_-|z|^2\right)\langle \beta, U_t\alpha\rangle .$$

So, if we could choose f, g and z so that

$$\mathrm{Im}(\overline{z}g(t) + z\overline{f}(t)) = (\mathrm{Re}\,\gamma_-)|z|^2,$$

we would obtain a purely oscillatory behaviour of the amplitude $\langle \beta, U_t\alpha\rangle$. As already explained, such a choice can not be made for any time t; however we can take f and g to be constant in an arbitrarily large interval of time, say $[0, T]$, and in this interval the transition amplitude $\langle \beta, U_t\alpha\rangle$ shall exhibit a purely oscillatory behaviour.

Finally note that, since U_t is localized on the space up to time t (adapted), denoting $\alpha_T = \xi \otimes \psi(\chi_{[0,T]}g)$ one can use the prescription

$$\lim_{T\to\infty} \frac{\langle \beta_T, U_t\alpha_T\rangle}{\langle \beta_T, \alpha_T\rangle}$$

to give meaning to the transition amplitude $\langle \beta_T, U_t\alpha_T\rangle$ even when f, g are only locally square integrable. So, for example, one could give meaning to transitions between *coherent vectors with constant test functions* or plane waves.

4.20 The Energy Shell Scalar Product: Linewidths

The Ito correction term, i.e. the additional term which arises in (4.12.6) with respect to the Hamiltonian white noise equation (4.11.11), is given, in the Fock case, by $-D^+D\gamma_-$, where γ_- is given by the integral

$$\gamma_- := (g \mid g)_- = \int_{-\infty}^{0} \langle g, S_s g\rangle ds = \int_{-\infty}^{0} dt \int_{\mathbb{R}^d} e^{-it(\omega_1(k)-\omega_0)} \mid g(k) \mid^2 dk. \tag{4.20.1}$$

On the other hand in Sect. 4.12 we have seen that the covariance of the white noise obtained as limit of the original quantum field is given by the full integral

$$\begin{aligned}\gamma := (g \mid g) &= \int_{-\infty}^{+\infty} \langle g, S_s g\rangle ds = \int_{-\infty}^{+\infty} dt \int_{\mathbb{R}^d} e^{it(\omega_1(k)-\omega_0)} \mid g(k) \mid^2 dk \\ &= \int_{-\infty}^{+\infty} ds\, e^{-is\omega_0} G_{-+}(s). \end{aligned} \tag{4.20.2}$$

From this we see that

$$\gamma = \hat{G}_{-+}(\omega_0), \tag{4.20.3}$$

where as usual ^ denotes the Fourier transform.

In the present section we investigate the physical meaning of these quantities. To this goal recall that in the notation (4.20.2) one has

$$G_{-+}(t) = \langle g, S_t^1 g\rangle = \int_{\mathbb{R}^d} dk \mid g_k \mid^2 e^{it\omega(k)} = \langle a_0 a_t^+\rangle = \langle A(g)A^+(S_t g)\rangle, \tag{4.20.4}$$

i.e. $\langle g, S_t^1 g\rangle$ is the Fock time-correlation function of the stationary process a_t^+ (see Sect. 2.1) with respect to the Fock vacuum. According to a standard terminology in the theory of stationary processes, the Fourier transform of the time-correlation function of such a process is called its *spectral function* (see Sect. 2.1). Since the time-correlation function of a process is a positive-definite function, by Bochner's theorem it follows that its Fourier transform is positive. This justifies the fact that sometimes we shall call the sesquilinear form $(g \mid g)$ the *scalar product* in the state space of the quantum white noise and $(g \mid g)_-$ the *semiscalar product.* These quantities are well defined whenever the function g (*cutoff function* or *form factor*) is a rapidly decreasing test function (see Sect. 4.1). In these notations we have the basic rule: *the covariance of the white noise obtained as the limit of the original quantum field is the spectral function of the original quantum field evaluated at the resonant frequency* ω_0.

Lemma 4.20.1. *The real part of the semiscalar product is given by*

$$\operatorname{Re}\gamma_- = \operatorname{Re}(g|g)_- = \frac{1}{2}(g|g) = \frac{1}{2}\gamma. \tag{4.20.5}$$

In particular the real part of γ_- *is positive because the scalar product is positive.*

Proof 4.20.1. This follows from the relation $G(t)^* = G(-t)$. Explicitly,

$$\begin{aligned}\operatorname{Re}(g|g)_- &= \frac{1}{2}[(g|g)_- + \overline{(g|g)_-}] = \frac{1}{2}\left(\int_{-\infty}^0 \langle g, S_t g\rangle dt + \int_{-\infty}^0 \langle g, S_{-t} g\rangle dt\right)\\ &= \frac{1}{2}\int_{-\infty}^{+\infty} \langle g, S_t g\rangle dt = \frac{1}{2}(g \mid g).\end{aligned}$$

Recalling that $S_t = S_t^1 e^{-it\omega_0}$ (4.11.3), we conclude that *the real part of the transport coefficient is nothing but the power spectrum function*, i.e. the Fourier transform of the 1-particle time-correlation function at the resonant frequency ω_0:

$$\Gamma_g(\omega_0) := (g \mid g) = (g \mid g)(\omega_0) = \int_{-\infty}^{+\infty} e^{-it\omega_0}\langle g, S_t^1 g\rangle dt, \tag{4.20.6}$$

i.e. the scalar product $(g \mid g)$ is the Fourier transform of the (time) vacuum correlation function of the free creator process, evaluated at the resonant frequency ω_0. The explicit form of the spectral function can be found using the relation

$$\int_{-\infty}^{\infty} dt\, e^{-ixt} = 2\pi\delta(x),$$

which gives

$$\begin{aligned}\Gamma_g(\omega_0) := (g \mid g) = (g \mid g)(\omega_0) &= \int_{-\infty}^{+\infty} \left\langle g, e^{-it(H_1-\omega_0)} g \right\rangle dt \\ &= 2\pi \langle g, \delta(H_1 - \omega_0) g\rangle ;\end{aligned}$$

the δ-function of an operator is discussed in Sect. 1.2. Keeping in mind that $H_1 = \omega_1(p)$, we see that

$$\Gamma_g(\omega_0) := (g \mid g) = (g \mid g)(\omega_0) = \int dk \mid g(k) \mid^2 \delta\big(\omega_1(k) - \omega_0\big), \qquad (4.20.7)$$

which shows that, in the stochastic limit of the simplest model defined by condition (4.10.2), the interaction is concentrated on the *energy surface*

$$\omega_1(k) = \omega_0$$

defined by the resonant frequency ω_0. This surface is a generalization of the *Fermi surface* widely used in solid-state physics. For this reason the scalar product $(g \mid g)$ is called *energy shell scalar product*. From (4.17.5) and (4.20.5) we conclude that *the energy shell scalar product $(g \mid g)$ (i.e. the covariance of the master field) gives the expression of the decay rates.*

Remark 4.20.1. Note that since we are assuming that $\omega_1(k) \geq 0$ it follows that the stochastic limit becomes trivial if $\omega_0 < 0$ (and in several cases also when $\omega_0 = 0$). The physical implications of this fact shall be discussed in Sect. 4.25.

4.21 Dispersion Relations and the Ito Correction Term

The connection (4.20.5) between the scalar and the semiscalar product can be made explicit by using the identity (see Sect. 4.12):

$$\int_{-\infty}^{0} e^{-it\omega} dt = \pi\delta(\omega) - i\text{P.P.}\frac{1}{\omega}, \qquad (4.21.1)$$

where $\text{P.P.}\frac{1}{\omega}$ denotes the principal part distribution. Recalling that $\omega(k) = \omega_1(k) - \omega_0$ or, in operator form,

$$\omega(p) = H_1 - \omega_0 = \omega_1(p) - \omega_0, \qquad (4.21.2)$$

where H_1 is the 1-particle Hamiltonian of the field, we obtain

$$(g \mid g)_-(\omega_0) = \pi\langle g, \delta(H_1 - \omega_0)g\rangle - i\left\langle g, \text{P.P.}\frac{1}{H_1 - \omega_0}g\right\rangle . \tag{4.21.3}$$

In a more explicit form:

$$(g|g)_-(\omega_0) = \pi \int_{\mathbb{R}^d} |g(k)|^2 \delta(\omega_1(k) - \omega_0) dk - i\text{P.P.} \int_{\mathbb{R}^d} dk \frac{|g(k)|^2}{\omega_1(k) - \omega_0} . \tag{4.21.4}$$

The δ-function in (4.21.4) can be explicitly calculated for a rather general form of the function $\omega_1(k)$ (see [Ske96]). In the following section we shall consider some particular cases and show that the connection between the real and imaginary parts of $(g \mid g)_-(\omega_0)$ is a universal phenomenon of the stochastic limit which relates it to the well-known dispersion relations.

4.22 The Case: $H_1 = |k^2|$

It is instructive to evaluate explicitly the semiscalar product $(g \mid g)_-$ in the case when the 1-particle field Hamiltonian is the usual Laplacian, i.e. in momentum representation,

$$H_1 = \omega_1(k) =\mid k \mid^2 . \tag{4.22.1}$$

Changing to polar coordinates, $k = \sqrt{\omega}\sigma$ in momentum space, and defining $\tilde{\Gamma}_g(\omega)$ by

$$\tilde{\Gamma}_g(\omega) := \omega^{\frac{d-2}{2}} \int_{S^{(d-1)}} \mid g\left(\sqrt{\omega}\sigma\right) \mid^2 d\sigma_{d-1}, \tag{4.22.2}$$

where $S^{(d-1)} \subseteq \mathbb{R}^d$ is the unit sphere and $d\sigma_{d-1}$ the normalized measure on it, the right-hand side of (4.21.3) becomes

$$\begin{aligned}(g|g)_-(\omega_0) &= \frac{1}{2}\int_{\mathbb{R}} \tilde{\Gamma}_g(\omega)\delta(\omega - \omega_0)d\omega - i\text{P.P.} \int_{\mathbb{R}} \frac{1}{2(\omega^2 - \omega_0)} \tilde{\Gamma}_g(\omega)d\omega \\ &=: \frac{1}{2}\Gamma(\omega_0) - i\Omega(\omega_0). \end{aligned} \tag{4.22.3}$$

Using the formula

$$\delta\left(x^2 - c^2\right) = \frac{1}{2c}\left[\delta(x - c) + \delta(x + c)\right], \quad c > 0,$$

and the fact that $\omega_0 > 0$, we see that the first integral, which represents the decay rate, is equal to $\tilde{\Gamma}_g(\omega_0)$, while the the second integral, which represents the energy shift, is its Hilbert transform evaluated at the point ω_0, i.e.

$$\Omega(\omega_0) := \text{P.P.} \int_{\mathbb{R}} \frac{1}{2(\omega^2 - \omega_0)} \tilde{\Gamma}_g(\omega)d\omega. \tag{4.22.4}$$

In summary, *the energy shift $\Omega(\omega_0)$ is the Hilbert transform of the spectral function $\tilde{\Gamma}_g(\omega)$ (decay rate) evaluated at the resonant frequency ω_0.*

This Hilbert transform connection is a typical example of a Kramers–Krönig dispersion relation expressing causality and is not specific to this example, but has general validity. More precisely, if instead of (4.22.1), H_1 has the form

$$H_1 = \omega_1(k) = \rho(|k|), \tag{4.22.5}$$

i.e. if ω_1 depends only on $|k|$, then one can repeat the above calculation with minor modifications.

4.23 Multiplicative Coupling with the Rotating Wave Approximation: Arbitrary Gaussian State

Consider again the multiplicative interaction Hamiltonian of Sect. 4.8:

$$H_{\mathrm{I}} = D^+ \int dka(k)\overline{g}(k) + D \int dka^+(k)g(k) = D^+A(g) + DA^+(g); \tag{4.23.1}$$

however, now the reference state of the field operators $a(k), a^+(k)$ is a general mean zero Gaussian state, as described in Sect. 2.2. In this section the fields can be either Fermi or Bose. We suppose that the generalized rotating-wave approximation condition (4.10.2) holds, i.e.

$$e^{itH_S} De^{-itH_S} = e^{-it\omega_0} D. \tag{4.23.2}$$

The equation for the rescaled evolution operator $U^{(\lambda)}_{t/\lambda^2}$ is

$$\partial_t U^{(\lambda)}_{t/\lambda^2} = -\frac{i}{\lambda} H_{\mathrm{I}}(t/\lambda^2) U^{(\lambda)}_{t/\lambda^2} = -i\left(\frac{1}{\lambda}\, a^{(\lambda)+}(t/\lambda^2) + \text{h.c.}\right) U_{t/\lambda^2}, \tag{4.23.3}$$

where $a^{(\lambda)}(t), a^{(\lambda)+}(t)$ are the rescaled operators, i.e.

$$a_t^{(\lambda)} = \int dk\overline{g}(k)\frac{1}{\lambda}\, e^{-it\omega(k)/\lambda^2} a(k). \tag{4.23.3b}$$

From Theorem 3.2.1 we know that the rescaled fields $e^{i(t/\lambda^2)(\omega_k-\omega_0)}a_k^+/\lambda$ converge to a white noise $b^+(t,k)$, whose covariance depends on the state of the a field as described in (3.2.4). Therefore the rescaled fields $a^+(t/\lambda^2)/\lambda$ will converge to the white noise $b^+(t)$ given by

$$b(t) = \int dk\overline{g}(k)b(t,k), \tag{4.23.4}$$

and (4.23.3) will converge to the limit white noise equation

$$\partial_t U_t = -i\big(b^+(t) + b(t)\big)U_t = -i\left(D \otimes \int dk\, g(k)b^+(t,k) \;+\; \text{h.c.}\right) U_t$$
$$= -i\left(D \otimes b^+(t) \;+\; \text{h.c.}\right) U_t. \tag{4.23.5}$$

The normal form of (4.23.5), according to the stochastic golden rule, is

$$\partial_t U_t = -i \int dk : \Big(g(k)D \otimes b^+(t,k) + \;\text{h.c.}\Big)U_t \;: -YU_t, \tag{4.23.6}$$

where the normal order : : is meant in the sense of Sect. 4.14. Y is the operator transport coefficient:

$$Y = \int_{-\infty}^{0} dt\langle H_{\rm I}(0)H_{\rm I}(t)\rangle, \tag{4.23.7}$$

where $\langle\,\cdot\,\rangle$ denotes expectation with respect to the reference state of the a field; the explicit expression of this coefficient for an arbitrary mean zero Gaussian state can be easily calculated. In the following sections we shall give some examples.

4.24 Multiplicative Coupling with RWA: Gauge Invariant State

If the reference state of the field operators $a(k), a^+(k)$ is an arbitrary gauge-invariant state, in particular a thermal state, as described in Sect. 2.10, its covariance is

$$\begin{pmatrix} \left\langle a_t^{(\lambda)+} a_\tau^{(\lambda)} \right\rangle & 0 \\ 0 & \left\langle a_t^{(\lambda)} a_\tau^{(\lambda)+} \right\rangle \end{pmatrix} = \begin{pmatrix} \frac{1}{\lambda^2}\, G^{+-}_{\beta,\mu}\left(\frac{\tau-t}{\lambda^2}\right) & 0 \\ 0 & \frac{1}{\lambda^2}\, G^{-+}_{\beta,\mu}\left(\frac{\tau-t}{\lambda^2}\right) \end{pmatrix} \delta(k-k'), \tag{4.24.1}$$

where, expressed in terms of the Planck factors $n(k), m(k)$ (defined by (2.10.5)) and the shifted 1-particle dynamics defined by (4.11.3), the functions $G^{\varepsilon\varepsilon'}_{\beta,\mu}$ are

$$G^{+-}_{\beta,\mu}(t) = \int e^{-it\omega(k)}|g(k)|^2 n(k)dk, \quad G^{-+}_{\beta,\mu}(t) = \int e^{-it\omega(k)}|g(k)|^2 m(k)dk. \tag{4.24.2}$$

For a thermal state, the difference between the Bose case and the Fermi case, at this level, is only in the form of the factors $n(k), m(k)$. For finite temperature states beyond the basic analytical assumption (4.5.2), we must also require the integrability of the finite temperature correlations:

$$\int_{-\infty}^{+\infty} |G^{-+}_{\beta,\mu}(t)|dt < +\infty, \quad \int_{-\infty}^{+\infty} |G^{+-}_{\beta,\mu}(t)|dt < +\infty, \tag{4.24.3}$$

and similarly for G^{+-}. Denoting

$$\gamma^{+-}_{\beta,\mu} := \int_{-\infty}^{\infty} d\sigma G^{+-}_{\beta,\mu}(\sigma), \quad \gamma^{-+}_{\beta,\mu} := \int_{-\infty}^{\infty} d\sigma G^{-+}_{\beta,\mu}(\sigma), \tag{4.24.4}$$

we know from Theorem 3.2.1 that as $\lambda \to 0$ the rescaled fields (4.23.3b) converge, in the sense of correlators to the finite temperature white noise with covariance

$$\begin{pmatrix} \langle b^+(t)b(\tau)\rangle & \langle b(t)b(\tau)\rangle \\ \langle b^+(t)b^+(\tau)\rangle & \langle b(t)b^+(\tau)\rangle \end{pmatrix} = \begin{pmatrix} \gamma^{+-}_{\beta,\mu} & 0 \\ 0 & \gamma^{-+}_{\beta,\mu} \end{pmatrix} \delta(\tau - t). \tag{4.24.5}$$

Given the form (4.11.1) of the interaction Hamiltonian and the diagonal form of the covariance, Y is easily computed from (4.23.7) to be

$$\begin{aligned} Y &= DD^+ \int_{-\infty}^{0} ds \langle A^+(g)A(g_s)\rangle + D^+D \int_{-\infty}^{0} ds \langle A(g)A^+(g_s)\rangle \\ &= DD^+ \int_{-\infty}^{0} ds \langle g_s, ng\rangle + D^+D \int_{-\infty}^{0} ds \langle mg, g_s\rangle, \end{aligned}$$

where

$$g_s(k) = e^{is\omega(k)} g(k), \tag{4.24.6}$$

$mg(k) = m(k)g(k)$ and $ng(k) = n(k)g(k)$. Since $\omega(k), m(k), n(k)$ are real and commute, this is equivalent to

$$Y = DD^+ \int_{-\infty}^{0} dt \langle ng, g_{-t}\rangle + D^+D \int_{-\infty}^{0} dt \langle mg, g_t\rangle. \tag{4.24.7}$$

Now in the drift term we distinguish two contributions: a Fock one, corresponding to the m term, and an anti-Fock one, corresponding to the n term. The difference in the time sign in the two integrals means that time flows from the past ($t = -\infty$) to the present ($t = 0$) in the Fock part and from the present ($t = 0$) to the future ($t = +\infty$) in the anti-Fock one.

4.25 Red Shifts and Blue Shifts

The difference between the Fock and the anti-Fock parts in the drift term (4.24.7) has an interesting physical consequence. Explicit computation of the integrals in (4.24.7) gives for the Fock part

$$\int_{-\infty}^{0} dt \langle mg, g_t\rangle = \int_{\mathbb{R}^d} |g(k)|^2 m(k) \delta(\omega_1(k) - \omega_0) dk - i\text{P.P.} \int dk \frac{|g(k)|^2 m(k)}{\omega_1(k) - \omega_0}, \tag{4.25.1}$$

and for the anti-Fock part

$$\int_{-\infty}^{0} dt\langle mg, g_{-t}\rangle = \int_{\mathbb{R}^d} |g(k)|^2 n(k)\delta(\omega_1(k)-\omega_0)dk + i\text{P.P.}\int dk \frac{|g(k)|^2 n(k)}{\omega_1(k)-\omega_0}. \tag{4.25.2}$$

Thus, as in the Fock case, in the stochastic limit only the master fields corresponding to positive Bohr frequencies survive. However, differently from the Fock case, now the negative Bohr frequencies contribute to the energy with a shift which is red in the Fock case, but blue in the anti-Fock case. In particular, in the thermal case, we know from Sect. 2.10 that the factors m, n have respectively the form

$$m(k) = \frac{1}{1-qe^{-\beta[\omega_1(k)-\omega_0]}}, \quad n(k) = \frac{e^{-\beta[\omega_1(k)-\omega_0]}}{1-qe^{-\beta[\omega_1(k)-\omega_0]}}, \tag{4.25.3}$$

with $q = \pm 1$ corresponding to the Bose and Fermi cases respectively. So, especially at high frequencies, we expect the anti-Fock contribution to be smaller than the Fock one and the overall shift due to the negative frequencies to still be red. As in the Fock case we expect the highest contribution to the shift to come from the region of momenta near the positive resonant frequencies ω_0. However, at high temperature (i.e. small β), the shift is larger for bosons and smaller for fermions than in the Fock case.

4.26 The Free Evolution of the Master Field

The equation for U_t (j_t) has been obtained as limit of the Schrödinger (Heisenberg) equation *in the interaction representation.* For this reason U_t (j_t) is not a quantum evolution, i.e. a 1-parameter unitary group (a 1-parameter group of $*$-automorphisms) but only a localized evolution operator, or a *Markovian cocycle* (a quantum flow). If we want a bona fide quantum dynamics, then we have to put back the free evolution which has been subtracted to pass to the interaction representation. However, after the stochastic limit, while the free atomic evolution remains unchanged, the free evolution of the a field does not, and therefore the following problem arises: *Which is the natural candidate for the free evolution in the master space?*

In other words we have to find a *free master Hamiltonian* $\tilde{H}_0$ such that the 1-parameter unitary family

$$e^{-it\tilde{H}_0}U_t =: V_t$$

is a quantum evolution (i.e. $V_{t+s} = V_tV_s$). From the general discussion in Sect. 1.3 we know that this is equivalent to the fact that U_t satisfies the cocycle condition (1.3.11) with respect to $\exp\{-it\tilde{H}_0\}$.

In the literature on quantum white noise it is usually assumed, more or less implicitly, that the free evolution of the a field becomes the second quantization $\tilde{S}_t$ of *the time shift* S_t in the white noise space, and consequently the Schrödinger evolution after the stochastic limit is defined by

$$V_t = (e^{itH_S} \otimes \tilde{S}_t)U_t. \tag{4.26.1}$$

However this choice is not good both for mathematical and physical reasons. From the mathematical point of view we should notice that the 1-parameter family V_t is not bona fide quantum dynamics, because it does not satisfy the condition

$$V_{s+t} = V_s V_t. \tag{4.26.2}$$

This is shown as follows: since U_t is a $(1 \otimes \tilde{S}_t)$ cocycle, we know from Sect. 1.2 that the family

$$V_t^0 := (1 \otimes \tilde{S}_t)U_t \tag{4.26.3}$$

is a 1-parameter unitary group. From this it easily follows that V_t, as defined by (4.26.1), satisfies (4.26.2) if and only if

$$e^{isH_S} V_t^0 e^{-isH_S} = V_t^0, \quad \forall\, s, t \in \mathbb{R}. \tag{4.26.4}$$

But one easily sees that this is impossible. In fact the normally ordered form of the white noise Hamiltonian equation satisfied by V_t^0 is

$$\partial_t V_t^0 = -i \left[Db^+(t) V_t^0 + D^+ V_t^0 b(t) \right] - \left(Y + \frac{i}{\hbar} K_0 \right) V_t^0, \tag{4.26.5}$$

where K_0 denotes the generator of $\tilde{S}_t$. Now, if (4.26.4) were true, (4.26.5) should also be satisfied by $e^{isH_S} V_t^0 e^{-isH_S} := \tilde{V}_t^0$ (for any choice of s). But because of the rotating-wave approximation (4.10.2) and of the discussion in Sect. 4.13, the equation satisfied by this is easily seen to be

$$\partial_t \tilde{V}_t^0 = -i \left[e^{-is\omega_0} Db^+(t) \tilde{V}_t^0 + e^{is\omega_0} D^+ \tilde{V}_t^0 b(t) \right] - \left(Y + \frac{i}{\hbar} K_0 \right) \tilde{V}_t^0. \tag{4.26.6}$$

The weak point of this argument is clear: before the limit, in order to pass to the interaction representation, we have subtracted the free evolution both of the system and of the field, while, using (4.26.1), we are putting back the free evolution of the system but not that of the field. Now notice that the master fields $b(t, k)$ can be intuitively thought to be of the form

$$b(t, k) = b(t) \otimes b(k), \tag{4.26.7}$$

where $b(t)$ is a white noise on $\mathbb{R}$ and $b(k)$ an independent field on $\mathbb{R}^d$ (in reality this intuitive representation is literally true only when the master field is Boltzmannian).

The representation (4.26.7) and the factor $\delta(\omega_k - \omega_0)$ appearing in the covariance of the $b(t,k)$ field suggest that the $b(k)$ field should live on the *level surface* in $\mathbb{R}^d$ defined by

$$\omega(k) = \omega_0. \tag{4.26.8}$$

This implies that the free evolution of the a field acts on the master field $b(k)$ as follows:

$$e^{-is\tilde{\omega}}b(k)e^{is\tilde{\omega}} = e^{-is\omega_0}b(k), \quad \forall s \in \mathbb{R}\ ,\ \forall k \in \mathbb{R}^d; \tag{4.26.9}$$

this means that the correct free evolution of the master field is not $\tilde{S}_t \otimes 1$, but

$$R_t := \tilde{S}_t \otimes e^{-it\tilde{\omega}}, \tag{4.26.10}$$

where the tensor product refers to the decomposition (4.26.7) and $e^{it\tilde{\omega}}$ is the free evolution characterized by the property

$$e^{-it\tilde{\omega}}b(k)e^{-it\tilde{\omega}} = e^{-it\omega_0}b(k). \tag{4.26.11}$$

If we now define

$$V_t = \left(e^{itH_S} \otimes R_t\right) U_t = \left(e^{itH_S} \otimes \tilde{S}_t \otimes e^{it\tilde{\omega}}\right) U_t, \tag{4.26.12}$$

then it is easy to verify that condition (4.26.4) is satisfied and we have bona fide quantum dynamics. In fact, defining V_t^0 as in (4.26.3) and V_t as in (4.26.12), condition (4.26.2) is equivalent to

$$\left(e^{isH_S} \otimes e^{is\tilde{\omega}}\right) V_t^0 \left(e^{-isH_S} \otimes e^{-is\tilde{\omega}}\right) = V_t^0, \quad \forall\, s,t \in \mathbb{R}. \tag{4.26.13}$$

This identity is true because, due to the compensating gauge factor (4.26.11), both sides of (4.26.13) satisfy (4.26.5) with the same initial condition. In conclusion let us underline that the separation (4.26.7) of the t and the k variables in $b(t,k)$ is not necessary, and the evolution R_t can be abstractly defined by the condition

$$\rho_s\big(b(t,k)\big) = R_s b(t,k) R_s^+ = e^{-is\omega_0}b(t+s,k).$$

4.27 Algebras Invariant Under the Flow

Lemma 4.27.1. *The flow*

$$j_t(X) = X_t = U_t X U_t^*$$

commutes with the free evolution, i.e. for any s, $t \in \mathbb{R}$,

$$V_s^{0*} j_t(X) V_s^0 = j_t\left(V_s^{0*} X V_s^0\right). \tag{4.27.1}$$

In particular, if T is any set of operators which is left invariant by the free evolution, then also $j_t(T)$ has this property.

Proof 4.27.1. Fix $s \in \mathbb{R}$ and denote

$$\tilde{X}_t = V_s^{0*} X_t V_s^0 = V_s^{0*} j_t(X) V_s^0,$$

then $\tilde{X}_t$ satisfies

$$\begin{aligned}\partial_t \tilde{X}_t = V_s^{0*}(\partial_t X_t) V_s^0 &= -ie^{it\omega}[e^{-it\omega}D, \tilde{X}_t] - ie^{-it\omega} b_t^+ [e^{it\omega}D^+, \tilde{X}_t] b_t \\ &+ 2\,\mathrm{Re}\,\gamma_- D^+ \tilde{X}_t D - \gamma_- D^+ D \tilde{X}_t - \overline{\gamma}_- \tilde{X}_t D^+ D\end{aligned}$$

with the initial condition

$$\tilde{X}_0 = V_s^{0*} X V_s^0.$$

But this is the same equation satisfied by

$$j_t\left(V_s^{0*} X V_s^0\right)$$

with the same initial condition. Therefore (4.27.1) holds.

In the previous section we have seen that the free evolution for the master field can be intuitively represented by

$$e^{itH_S} \otimes \tilde{S}_t \otimes e^{it\tilde{\omega}} =: V_t^0; \tag{4.27.2}$$

its generator is

$$H_S + K_0 + \tilde{\omega}, \tag{4.27.3}$$

where K_0 is the generator of the time shift. Consider now the vectors of the form

$$|\varepsilon_m\rangle \otimes |b^+(t_n, k_n) \ldots b^+(t_1, k_1)\Phi\rangle. \tag{4.27.4}$$

The action of the unitary operator (4.27.2) on a vector of the form (4.27.4) gives

$$e^{it(\varepsilon_m + n\omega)}|\varepsilon_m\rangle \otimes |b^+(t_n + t, k_n) \ldots b^+(t_1 + t, k_1)\Phi\rangle,$$

and this gives a natural candidate for the abelian algebra left invariant by the flow.

4.28 Notes

Section (4.1)

(1) From (4.1.13), (4.1.14) it is clear that for QED the natural choice for the coupling constant is $\lambda = e/mc$.
In many cases the rescaling parameter λ has an obvious interpretation in terms of coupling constant, density, energy, etc. but there are other possible interpretations. In fact, starting from the interaction Hamiltonian (4.1.14), one can define a number $\lambda > 0$ by the prescription

$$\frac{N}{|\Lambda|} = \text{density of molecules in } \Lambda =: \lambda^2. \tag{4.1.18}$$

Under these assumptions and notations, and with the further notation

$$\lambda g =: g_\lambda \tag{4.1.19}$$

the interaction Hamiltonian (4.1.4), or equivalently (4.1.14), becomes

$$H_{\mathrm{I}} = \left(A(g_\lambda e^{i(\cdot)\cdot q_N}) + A^+(g_\lambda e^{i(\cdot)\cdot q_N})\right) \cdot \frac{P_N}{N^{1/2}}. \tag{4.1.20}$$

Section (4.2)

A direct study of the interaction Hamiltonian (4.1.10) is very complicated. The simple notation (4.1.1) for this Hamiltonian has been chosen in order to give intuition of the various simplifying approximations which have been introduced in the physical literature (single or in combination) to obtain at least a qualitative understanding of the QED interaction. In the following a list of the main approximations is given.

(1) The dipole approximation is widely used in quantum optics. Its usual justification is that at optical wavelengths the magnitude of the wave vector is of the order of

$$|k| \sim 2\pi/4 \cdot 10^5 \text{ cm}^{-1}, \tag{4.2.1}$$

while the typical atomic radius is about

$$q_{at} \sim 10^{-8} \text{ cm} \tag{4.2.2}$$

(see [Lou73], § 4.2, Eberly and Milionni (1987), I.C.). Within these ranges one has

$$\exp ik \cdot q \sim \exp i(10)^{-3} \sim 1 + i10^{-3} + O(10^{-6}), \tag{4.2.3}$$

i.e. the dipole approximation is justified up to an order of 10^{-3}. It is clear from these arguments that one should expect the dipole approximation to fail when the high frequencies play a major role.

(2) In the literature (see [Lou73], Chap 4.21) there are other heuristic justifications of the replacement of $Q(g)$ with $P(g)$ based on the identity

$$H_{at} = \frac{1}{2m}p^2 + V(q) \qquad [q, H_{at}] = \frac{i}{2m}\,p,$$

which implies, denoting φ_m the eigenvectors of H_{at} with eigenvalues λ_m ($H\varphi_m = \lambda_m\varphi_m$), that

$$\langle\varphi_k, p\varphi_{k'}\rangle = \frac{2m}{i}\,\langle\varphi_k, [q, H]\varphi_{k'}\rangle = 2m(\lambda_{k'} - \lambda_k)\langle\varphi_k, q\varphi_{k'}\rangle,$$

$$\langle\psi, p_j\psi'\rangle = (im/\hbar)(\lambda_k - \lambda_{k'})\langle\psi, q_j\psi'\rangle,$$

so the matrix elements of the two coincide up to a constant.

Section (4.3)

(1) A frequently used approximation of the QED Hamiltonian consists in simplifying drastically the operators $A^{\pm}(g_\lambda)$ and the atomic operators P_N while keeping λ, N, and t (time) fixed. A typical example is the *Dicke maser Hamiltonian*, obtained from (4.1.20) by introducing the following approximations:

(i) The field is reduced to a 1–mode field by neglecting all modes except, say, the k_0 mode with $|k_0| = 1$.

(ii) The atomic variables are replaced by 2–level system variables according to the following rules:

$$\frac{1}{2m_j} p_j^2 \to \varepsilon \sigma_3^{(j)} \quad ; \quad \varepsilon > 0, \tag{4.3.3}$$

$$k \cdot q_j \to k_0 \cdot x_j \quad ; \; x_j \in \mathbf{R}^3 \tag{4.3.4}$$

(this means that position is not quantized),

$$\frac{e}{m_j c} p_j \to \sigma_+^{(j)} \tag{4.3.5}$$

with the replacements

$$\frac{1}{N^{1/2}} \sum_{j=i}^{N} \frac{e}{2m_j c}(p_j e^{ik\cdot q_j} + e^{ik\cdot q_j} p_j) \to \frac{1}{N^{1/2}} \sum_{j=1}^{N} e^{ik_0\cdot x_j} \sigma_+^{(j)}; \tag{4.3.6}$$

the Hamiltonian (4.1.20) becomes:

$$H_N = a^+ a + \varepsilon \sum_{j=1}^{N} \sigma_3^{(j)} + \frac{\lambda}{N^{1/2}} \sum_{j=1}^{N} (a \otimes \sigma_+^{(j)} e^{ik_0\cdot x_j} + \text{h.c.}). \tag{4.3.7}$$

By absorbing the scalar phases into the definition of $\sigma_\pm^{(j)}$, which changes neither the commutation relations nor $\sigma_3^{(j)}$, we arrive at the form

$$H_N = a^+ a + \varepsilon S_N^{(\varepsilon)} + \lambda \left(a \otimes \frac{P_N}{N^{1/2}} + a^+ \otimes \frac{P_N^+}{N^{1/2}} \right). \tag{4.3.8}$$

Starting from the Dicke Hamiltonian and letting

$$N \to +\infty \qquad |\Lambda| \to +\infty, \tag{4.3.9}$$

so that (4.1.18) remains valid, this limit gives rise to the quantum *laws of large numbers* and *central limit theorems*.

Other limits:

(i) Combining dipole and rotating-wave approximations leads to the *spin–boson Hamiltonian.*

(ii) In (4.1.20), with N fixed, replace g_λ by some function such that

$$\lim_{\lambda \to 0} g_\lambda \to \delta \; = \; \text{delta function} \tag{4.3.10}$$

in some distribution sense, but keep t fixed (singular coupling limit) (see [GoFri76], [HeLi75]). This limit also gives rise to quantum Brownian motion.

(iii) Start from the full Hamiltonian (4.1.20) (possibly with the dipole or rotating-wave approximation) and let

$$t \to +\infty \qquad \lambda \to 0 \qquad \lambda^2 t = \tau = \text{const}, \tag{4.3.11}$$

then we obtain the stochastic limit.

(iv) After the weak coupling limit, perform the limit (4.3.9).

The stochastic limit for models not of system–reservoir type (this essentially means that condition (iv) is not fullfilled) is discussed in Part II. It is technically more difficult and gives rise to totally new phenomena, for example the breaking of the usual Bose–Einstein (or Fermi–Dirac) statistics and of the Bose or Fermi commutation relations.

5. Spin–Boson Systems

The generalized rotating-wave approximation condition (4.10.2) is too restrictive. In the present chapter we begin to realize the second step of the program described in Sect. 4.9, i.e. we shall generalize the multiplicative Hamiltonian (4.8.1) by dropping this condition and allowing the system operators to be arbitrary [modulo the analytical condition (4.9.3)]. We shall see that this leads to a new phenomenon, namely: *even if we start with a single scalar boson Fock field, in the limit we shall have not a single, but an infinity of independent quantum noises* – one for each Bohr frequency [see (4.8.4) for this notion] of the system; this is the stochastic resonance principle. In this chapter the stochastic golden rule is applied to the investigation of the stochastic limit for the general spin–boson Hamiltonian, describing a discrete system coupled with a boson field. The spin–boson Hamiltonian is widely used in physics [4], in studying quantum computing [Vol99], in studying stochastic resonance[(1)] [AcKoVo97, Gam98, Gri98, ImYuOh99], etc.

The stochastic limit for the spin–boson Hamiltonian is derived in Sect. 5.9 Then a more general spin–boson Hamiltonian is defined. In Sect. 5.16 we derive the Langevin equation for this model and discuss a general and easily verifiable criterion of existence for the solution of this equation (the quantum flow) which is also applicable to infinite systems (e.g. spin systems). In Sect. 5.17 we derive the corresponding master equation for the density matrix and show that it describes the collapse of the density matrix to a classical Gibbs distribution (convergence to equilibrium). Macroscopic quantum effects (conservation of coherence) for this model are found and the problem of the control of these effects using the parameters of the model is investigated. In Sect. 5.18 we give an effective criterium for the quantum decoherence generalizing the approach of Sects. 5.10 and 5.11: quantum coherence will decay exponentially if the values of the following quantities are nonzero,

$$\mathrm{Re}\,(g|g)^{+}_{\sigma\sigma'} = \int dk\,|g(k)|^2 2\pi\delta(\omega(k)+E(\sigma')-E(\sigma))n(k),$$

(see (5.20.15) for the definition of these quantities). Using this characterization of quantum decoherence, we find that for this model the quantum decoherence is controllable by the state of the reservoir (which can be controlled by filtering). As a particular case of the above construction, in Sect. 5.19 we derive the Glauber dynamics (and generalizations thereof) for a system of spins (see [AcKo99a, AcKo00b]).

5.1 Dropping the Rotating-Wave Approximation

Following the scheme of Sect. 4.9.2 our starting point is the evolution equation

$$\partial_t U_t^{(\lambda)} = -i\lambda H_{\mathrm{I}}(t) U_t^{(\lambda)} , \tag{5.1.1}$$

where $H_{\mathrm{I}}(t)$ has the canonical form (4.9.4) i.e. in the notations (4.11.3–4)

$$H_{\mathrm{I}}(t) = e^{itH_o} H_{\mathrm{I}} e^{-itH_o} = \sum_{q=1}^{N} \{D_q^+ \otimes a_q(t) + D_q \otimes a_q^+(t)\} , \tag{5.1.2}$$

$$a_q(t) := \int e^{-i(\omega(k)-\omega_q)t} g(k) a_k dk = A\left(e^{-it\omega_q} S_t^1 g\right) , \tag{5.1.3}$$

and where with D_q defined by (4.8.6) we can assume that the *nondegeneracy condition* (4.8.5), i.e.

$$\omega_q \neq \omega_{q'} \qquad \text{for} \qquad q \neq q' \tag{5.1.4}$$

is satisfied. Therefore the time-rescaled evolution operator satisfies

$$\begin{aligned} \partial_t U_{t/\lambda^2}^{(\lambda)} &= -\frac{i}{\lambda} H_{\mathrm{I}} \left(\frac{t}{\lambda^2}\right) U_{t/\lambda^2}^{(\lambda)} \\ &= -i \sum_{q=1}^{N} \{D_q^+ \otimes a_q^{(\lambda)}(t) + D_q \otimes a_q^{(\lambda)+}(t)\} U_{t/\lambda^2}^{(\lambda)} , \end{aligned} \tag{5.1.5}$$

where $a_q^{(\lambda)}(t), a_q^{(\lambda)+}(t)$ are the rescaled fields associated with (5.1.3):

$$a_q^{(\lambda)}(t) := \frac{1}{\lambda} a_q \left(\frac{t}{\lambda^2}\right) , \quad a_q^{(\lambda)+}(t) := \frac{1}{\lambda} a_q^+ \left(\frac{t}{\lambda^2}\right) . \tag{5.1.6}$$

According to the stochastic golden rule, the first step is to find the the limit of these rescaled fields, i.e. the master field.

5.2 The Master Field

The limit of the single rescaled operators $a_q^{(\lambda)\pm}(t)$ is known from Theorem 3.2.1, so because of Gaussianity, the only thing we need to know is the limit of the correlations among fields with different indices q. The rescaled operators $a_q^{(\lambda)\pm}(t)$ in (5.1.6) differ among themselves only by the phase factor $e^{i\omega_q t}$. We shall prove in this section that, by the Riemann–Lebesgue lemma, this phase factor leads, in the stochastic limit, to the independence of the quantum white noises corresponding to different indices. The time correlation between $a_q(t)$ and $a_q^+(t)$ is

$$\langle a_p(t)a_q^+(\tau)\rangle = e^{-i\tau\omega_q + it\omega_p} G(\tau - t) , \tag{5.2.1}$$

where $G(t)$ is the usual (Fock) time correlation, i.e.

$$G(t) = \langle g, S_t g\rangle = \int e^{i\omega_1(k)t}|g(k)|^2 dk = \langle A^+(g), A^+(S_t^1 g)\rangle ; \tag{5.2.2}$$

consequently the time correlation between $a_q^{(\lambda)}(t)$ and $a_q^{(\lambda)+}(t)$ is

$$\left\langle a_p^{(\lambda)}(t)a_q^{(\lambda)+}(\tilde{\tau})\right\rangle = \frac{1}{\lambda^2} e^{-i\omega_q\tau/\lambda^2 + i\omega_p t/\lambda^2} G\left(\frac{\tau - t}{\lambda^2}\right) , \tag{5.2.3}$$

and all the other ones vanish (Fock case). We know that the stochastic limit of the covariance (5.2.2) is, in the sense of distributions,

$$\lim_{\lambda\to 0} \frac{1}{\lambda^2} G\left(\frac{\tau - t}{\lambda^2}\right) = \delta(\tau - t) \int G(\sigma) d\sigma . \tag{5.2.4}$$

However for the covariance (5.2.3) one has the following

Proposition 5.2.1. *For an integrable function G on $\mathbb{R}$, the following relation holds in the distribution sense:*

$$\begin{aligned} \lim_{\lambda\to 0} \frac{1}{\lambda^2} e^{-i\frac{\tau}{\lambda^2}\omega_q + i\frac{t}{\lambda^2}\omega_p} G\left(\frac{\tau - t}{\lambda^2}\right) &= \delta_{p,q}\gamma^{(p)}\delta(\tau - t) \\ &= \begin{cases} \gamma^{(p)}\delta(\tau - t) & \text{if } p = q , \\ 0 & \text{if } p \neq q , \end{cases} \end{aligned} \tag{5.2.5}$$

where, denoting $\hat{G}$ the Fourier transform of G,

$$\gamma^{(p)} = \int_{-\infty}^{+\infty} d\sigma e^{-i\sigma\omega_p} G(\sigma) = \hat{G}(\omega_p) . \tag{5.2.6}$$

Proof 5.2.1. After integration over test functions φ_1 and φ_2, with compact support, the left-hand side of (5.2.5) becomes

$$\int dt \varphi_1(t) \int d\tau \varphi_2(\tau) e^{-i\omega_q\tau/\lambda^2 + i\omega_p t/\lambda^2} \frac{1}{\lambda^2} G\left(\frac{\tau - t}{\lambda^2}\right) . \tag{5.2.7}$$

After the change of variables $\tau \to \sigma = (\tau - t)/\lambda^2$, it becomes

$$\int dt \varphi_1(t) e^{-i(\omega_q - \omega_p)t/\lambda^2} \int d\sigma \varphi_2(t + \lambda^2\sigma) e^{-i\omega_q\sigma} G(\sigma) . \tag{5.2.8}$$

Now, when $\lambda \to 0$, the integral over $d\sigma$ tends to

$$\varphi_2(t) \int d\sigma e^{-i\omega_q \sigma} G(\sigma) \,, \tag{5.2.9}$$

and the convergence is dominated because G is integrable and

$$| \varphi_2(t + \lambda^2 \sigma) e^{-i\omega_q \sigma} G(\sigma) | \leq | G(\sigma) | \operatorname{supp}_t(| \varphi_2(t) |)$$

thus we can replace the second integral in (5.2.8) by (5.2.9) in the limit $\lambda \to 0$. Therefore, the integral over dt in (5.2.9) tends to zero if $\omega_q \neq \omega_p$ by the Riemann–Lebesgue lemma. If $\omega_q = \omega_p$ (5.2.8) tends to

$$\int dt \varphi_1(t) \varphi_2(t) \int d\sigma e^{i\omega_q \sigma} G(\sigma) = \hat{G}(\omega_q) \int dt \varphi_1(t) \varphi_2(t) \,, \tag{5.2.10}$$

and this ends the proof.

We now formulate a generalization of Theorem 3.2.1.

Theorem 5.2.1. *Let $a(t,k)$, $a^+(t,k)$ $(t \in \mathbb{R}, k \in \mathbb{R}^d)$ be a mean zero Gaussian quantum field in the sense of Definition 2.2.1 with respect to a given expectation value $\langle \cdot \rangle$. Suppose that the process $a^\pm(t,k)$ is stationary with respect to $\langle \cdot \rangle$, in the sense of (2.2.4), and that its covariance matrix, with respect to the time variable, is integrable in the sense of distributions, i.e. in the notations (2.1.1) and for $\varepsilon, \varepsilon' = 0, 1$:*

$$\int_{-\infty}^{+\infty} dt \left| \int_{\mathbb{R}^d} \int_{\mathbb{R}^d} dk dk' \overline{g}_1(k)_2(k') \langle a^\varepsilon(0,k) a^{\varepsilon'}(t,k') \rangle \right| < \infty \tag{5.2.11}$$

for any pair of test functions g_1, g_2 in the Schwartz space. For each real number ω, define the new process

$$a_\omega(t,k) := e^{-it\omega} a(t,k) \,.$$

Then the limit, in the sense of distribution correlators,

$$\lim_{\lambda \to 0} \frac{1}{\lambda} a_\omega(t/\lambda^2, k) = b_\omega(t,k) \,, \tag{5.2.12}$$

exists and is the white noise $b_\omega(t,k)$ [of the same Gaussian type as $a(t,k)$] with covariance

$$\begin{aligned} \langle b_\omega^\varepsilon(t,k) b_\omega^{\varepsilon'}(t',k') \rangle &= \delta(t-t') \int_{-\infty}^{+\infty} d\tau e^{-i\tau\omega} \langle a^\varepsilon(0,k) a^{\varepsilon'}(\tau,k') \rangle \\ &= \delta(t-t') \hat{G}_{\varepsilon,\varepsilon'}(\omega,k,k') \,, \end{aligned} \tag{5.2.13}$$

where $\hat{G}_{\varepsilon,\varepsilon'}(\omega,k,k')$ is the (time) spectral matrix of the process $a^\pm(t,k)$ (see Sect. 2.1). Moreover the quantum white noises $\{b_\omega(t,k) : \omega \in \mathbb{R}\}$ are mutually independent in the sense of (2.1.7).

Remark 5.2.1. In the boson Fock case, this means that

$$[b_\omega(t,k), b^+_{\omega'}(t',k')] = \delta_{\omega,\omega'}\delta(t-t')\hat{G}_{-,+}(\omega,k,k') .$$

Proof 5.2.2. The first statement follows from Theorem 3.2.1. Since the family $\{b_\omega : \omega \in \mathbb{R}\}$ is Gaussian as a limit of Gaussian families, to prove independence it is sufficient to prove that for $\omega \neq \omega'$ the pair correlations between $b_\omega(t,k)$ and $b_{\omega'}(t,k)$ vanish. But this is an immediate consequence of Proposition 5.2.1.

Remark 5.2.2. The physical meaning of Theorem 5.2.1 becomes clear by looking at the explicit form of the interaction Hamiltonian in (5.1.2–3) and remembering that the index q, in that formula, parameterizes the mutually different Bohr frequencies of the system. The decomposition (5.1.5) of the interaction Hamiltonian suggests that, before the limit, the original field *splits into a family of effective fields, each of which is interacting with its own Bohr frequency* and with the other effective fields. Proposition 5.2.1 says that the mutual interaction of these effective fields becomes negligible in the stochastic limit, and after the limit this mutual independence becomes exact. In Sect. 11.8 the stochastic resonance principle shall be extended to nonlinear interactions.

Remark 5.2.3. Note that, if the initial reservoir state is gauge invariant, then so is the limit white noise.

5.3 The White Noise Hamiltonian Equation

With the expression of the master fields $b_q(t), b^+_q(t)$, given by Theorem 5.2.1, the stochastic golden rule gives the white noise Hamiltonian equation

$$\partial_t U_t = -i\,H(t)U_t = -i\sum_{q=1}^{N}\{D^+_q \otimes b_q(t) + D_q \otimes b^+_q(t)\}U_t \tag{5.3.1}$$

as well as its normally ordered form

$$\partial_t U_t = -i\sum_{q=1}^{N} D^+_q U_t b_q(t) + D_q b^+_q(t) U_t - Y U_t\,, \tag{5.3.2}$$

where Y is the operator transport coefficient

$$Y = \lim \lambda^2 \int_0^{1/\lambda^2} dt_1 \int_0^{t_1} dt_2 \langle H_{\mathrm{I}}(t_1) H_{\mathrm{I}}(t_2)\rangle \tag{5.3.3}$$

and $\langle\,\cdot\,\rangle$ denotes expectation with respect to the reference state of the original a field.

5.4 The Operator Transport Coefficient: no Rotating-Wave Approximation, Arbitrary Gaussian Reference State

In order to calculate the operator transport coefficient without introducing the rotating-wave approximation, we have to generalize Lemma 1.8.1. In fact the multiplicity of the Bohr frequencies ω_q breaks the stationarity condition so that the basic assumption of this lemma is not satisfied.

Lemma 5.4.1. *Let $f \in L^1(\mathbb{R})$ and let ω be a real number. Then for any $t \geq 0$,*

$$\lim_{\lambda \to 0} \lambda^2 \int_0^{t/\lambda^2} dt_1 \int_0^{t_1} dt_2 e^{it_1\omega} f(t_2 - t_1) = 0 \tag{5.4.1}$$

if $\omega \neq 0$, and if $\omega = 0$,

$$\lim_{\lambda \to 0} \lambda^2 \int_0^{t/\lambda^2} dt_1 \int_0^{t_1} dt_2 f(t_2 - t_1) = t \int_{-\infty}^0 f(\sigma) d\sigma \,. \tag{5.4.2}$$

Proof 5.4.1. One has the identity

$$\lambda^2 \int_0^{t/\lambda^2} dt_1 \int_0^{t_1} dt_2 e^{it_1\omega} f(t_2 - t_1) = \int_0^t ds_1 e^{is_1\omega/\lambda^2} \int_0^{s_1/\lambda^2} dt_2 f(t_2 - s_1/\lambda^2)$$

$$= \int_0^t ds_1 e^{is_1\omega/\lambda^2} \int_{-s_1/\lambda^2}^0 d\sigma f(\sigma) = \int_0^t e^{is_1\omega/\lambda^2} ds_1 \int_{-s_1/\lambda^2}^0 f(\sigma) d\sigma \,.$$

Therefore (5.4.1) follows from the double integral lemma of Sect. 15.8 and (5.4.2) from dominated convergence.

Proposition 5.4.1. *The operator transport coefficient associated with the interaction Hamiltonian (5.1.2–3) and with an arbitrary Gaussian state of the field a_k, a_k^+, with integrable time correlations, is given by*

$$\sum_n [D_\mu^+ D_\mu \hat{G}_{-+}^{-1}(-\omega_\mu) + D_\mu D_\mu^+ \hat{G}_{+-}^{(-)}(\omega_\mu) + D_{\omega_\mu}^+ D_{-\omega_\mu}^+ \hat{G}_{--}^{(-)}(\omega_\mu)$$

$$+ D_{-\omega_\mu} D_{\omega_\mu} \hat{G}_{++}^{(-)}(-\omega_\mu)] \,, \tag{5.4.3}$$

where $D_{\omega_\mu} = D_\mu$; $G_{\varepsilon\varepsilon'}$ is the spectral function of the field $e^{-it\omega_k} a_k$ (see Sect. 2.1) and $\hat{G}^{(-)}(\omega)$ is the causal Fourier transform of $G(t)$, i.e. the Fourier transform of $\chi_{(-\infty,0]} G(t)$.

Proof 5.4.2. Given the form (5.1.2–3) of the interaction Hamiltonian, we have

$$\langle H_{\rm I}(t_1) H_{\rm I}(t_2) \rangle = \sum_{\mu\nu} \sum_{\varepsilon,\varepsilon'=0,1} D_\mu^{1-\varepsilon} D_\nu^{1-\varepsilon'} e^{(-1)^\varepsilon it_1\omega_\mu} e^{(-1)^{\varepsilon'} it_2\omega_\nu} \langle a^\varepsilon(t_1) a^{\varepsilon'}(t_2) \rangle \,.$$

For a gauge-invariant state only the coefficients $\langle a(t_1)a^+(t_2)\rangle$ and $\langle a^+(t_1)a(t_2)\rangle$ are different from zero. For the Fock term we have

$$\sum_{\mu\nu} D_\mu^+ D_\nu e^{-it_1\omega_\mu} e^{it_2\omega_\nu} \langle a(t_1)a^+(t_2)\rangle \,,$$

$$\sum_{\mu,\nu} D_\nu^+ D_\nu e^{i(t_2-t_1)\omega_\nu} e^{it_1(\omega_\mu-\omega_\nu)} \langle a(0)a^+(t_2-t_1)\rangle \,.$$

Since the function $e^{it\omega_\nu}\langle a(0)a^+(t)\rangle$ is integrable by assumption, we can apply Lemma 5.4.1 to it and conclude that for $\omega_\mu \neq \omega_\nu$ the contribution of this term to the transport coefficient is zero. If $\omega_\mu = \omega_\nu$ then we apply (5.4.2). The same argument apply to the anti-Fock term. For the off-diagonal terms one has (recalling that $\varepsilon = 0$ corresponds to creator and $\varepsilon = 1$ to annihilator):

$$\sum_{\mu,\nu} D_\mu^+ D_\nu^+ e^{-it_1\omega_\mu} e^{-it_2\omega_\nu} \langle a(0)a(t_2-t_1)\rangle$$

$$= \sum_{\mu,\nu} D_\mu^+ D_\nu^+ e^{-t_1(\omega_\mu+\omega_\nu)} e^{-i(t_2-t_1)\omega_\nu} \langle a(0)a(t_2-t_1)\rangle \,,$$

and, again by Lemma 5.4.1, the contribution of this term to the operator transport coefficient is zero if $\omega_\mu + \omega_\nu \neq 0$, but if $\omega_\mu + \omega_\nu = 0$, it is given by

$$\sum D_{\omega_\mu}^+ D_{-\omega_\mu}^+ \int_{-\infty}^0 e^{-i\sigma\omega_\nu} \langle a(0)a(\sigma)\rangle d\sigma = \sum D_{\omega_\mu}^+ D_{-\omega_\mu}^+ \hat{G}_{--}^{(-)}(\omega_\mu) \,. \quad (5.4.4)$$

Similarly, for the DD term one finds

$$\sum_\mu D_{\omega_\mu} D_{-\omega\omega_\mu} \hat{G}_{++}^{(-)}(-\omega_\mu) \,. \quad (5.4.5)$$

5.5 Different Roles of the Positive and Negative Bohr Frequencies

A priori all the Bohr frequencies enter on the same footing in (5.4.2–3). However the stochastic limit shows that in fact the positive and negative Bohr frequencies play different roles. Let us initially consider the Fock case in which only the correlations $\langle b_\omega(t,k)b_\omega^+(t',k')\rangle$ are nontrivial, so in the transport coefficient (5.4.3) only one spectral function for each Bohr frequency appears:

$$\hat{G}_{-+}(\omega_p) = \int_{-\infty}^{+\infty} d\sigma e^{-i\sigma\omega_p} G_{-+}(\sigma) \,.$$

In this case, according to Proposition 5.4.1, the transport coefficient is

$$Y = \sum_{q=1}^{N} \gamma_-^{(q)} D_q^+ D_q \,, \quad (5.5.1)$$

where $\gamma_-^{(q)}$ are defined by

$$\gamma_-^{(q)} := \int_{-\infty}^0 dt G_{-+}(t) = \int_{-\infty}^0 dt \left\langle g, S_t^{(q)} g \right\rangle$$
$$= \int_{-\infty}^0 dt \int_{\mathbb{R}^d} e^{-it(\omega_1(k)-\omega_q)} |g(k)|^2 . \tag{5.5.2}$$

Therefore the normal form of the white noise equation (5.3.1) is

$$\partial_t U_t = -i \sum_{q=1}^N [D_q^+ U_t b_q(t) + D_q b_q^+(t) U_t] - \left(\sum_{q=1}^N \gamma_-^{(q)} D_q^+ D_q \right) U_t . \tag{5.5.3}$$

Using (4.21.1) we write explicitly the coefficient $\gamma_-^{(q)}$ in a form which separates the damping (i.e. real) part from the energy shift (i.e. the imaginary part):

$$\gamma_-^{(q)} = \pi \int_{\mathbb{R}^d} |g(k)|^2 \delta(\omega_1(k) - \omega_q) dk - i\text{P.P.} \int dk \frac{|g(k)|^2}{\omega_1(k) - \omega_q} . \tag{5.5.4}$$

Remark 5.5.1. Assuming that $\omega_1(k) > 0$ almost everywhere, we see that the argument of the δ-function can be zero only if $\omega_q > 0$. On the other hand, from the fluctuation–dissipation theorem of Sect. 4.16, we know that the real part of the Ito correction term is the covariance of the noise; so the fact that the contribution relative to a given frequency vanishes means that there is no master field with that characteristic frequency. In particular, with an interaction Hamiltonian of the form (4.9.2) and under the assumption (4.10.2) with $\omega_o > 0$, we see that if we add to the Hamiltonian a term of the form $D \otimes a_k$ (e.g. pair annihilation) then the master field corresponding to this term will be identically zero. Summing up we conclude that: *in the stochastic limit only the master fields corresponding to positive Bohr frequencies survive.*

But the explicit form (5.5.4) also shows that this is not the case for the imaginary part; in other words: *in the stochastic limit the negative Bohr frequencies contribute with an overall red shift to the energy.* Recalling from (4.8.4) that $\omega_q = \varepsilon_n - \varepsilon_m$ for some pair of energy levels ε_n, ε_m, we see that the positivity of ω_q means that in the transition $\varepsilon_n \to \varepsilon_m$ the energy of the system decreases. So we can expect that if the field is initially prepared in the vacuum state then in the interaction (5.1.2), even *before the limit the dominating contributions should come from those interactions which decrease the energy of the system.*

Since a priori all the Bohr frequencies enter the evolution equation (5.1.1) symmetrically and after the stochastic limit this symmetry is destroyed, it is natural to call this a stochastic *symmetry breaking* phenomenon. This phenomenon[(1)] has observable physical effects because it means that even if those terms of the Hamiltonian (5.1.2) corresponding to negative Bohr frequencies do not contribute to the damping, they leave a physically observable track in the form of a shift in the system energy; thus something of their role survives, like the grin of the Cheshire cat. For this reason this phenomenon was called the *Cheshire cat effect.*(2) In Sects. 5.9–5.11 we will illustrate this effect in the SQUID Hamiltonian considered in [LCDFGZ87].

5.6 No Rotating-Wave Approximation with Cutoff: Gauge Invariant States

We discuss here the general case in which the reference state of the field operators $a(k)$, $a^+(k)$ is an arbitrary mean zero gauge-invariant Gaussian state with covariance (2.6.1). Thermal states correspond to the choice (2.10.2) of the coefficient m, n of the covariance.

In this case, according to Proposition 5.4.1 the drift term in equation (5.3.2) is split into Fock and anti-Fock parts as follows:

$$Y = \sum_{q=1}^{N} \gamma_-^{(q,m)} D_q^+ D_q + \sum_{q=1}^{N} \gamma_-^{(q,n)} D_q D_q^+ \,, \tag{5.6.1}$$

where $\gamma_-^{(q)}$ are defined by

$$\gamma_-^{(q,m)} := \int_{-\infty}^{0} dt \int_{\mathbb{R}^d} e^{-it(\omega_1(k)-\omega_q)} |g(k)|^2 m(k) \,, \tag{5.6.2}$$

$$\gamma_-^{(q,n)} := \int_{-\infty}^{0} dt \int_{\mathbb{R}^d} e^{it(\omega_1(k)-\omega_q)} |g(k)| n(k) \,, \tag{5.6.3}$$

and each term of the series has the form (4.8.6). Thus the qualitative picture is different from the Fock case only in the fact that now the contributions of the negative frequencies is not only *a red shift* (corresponding to the Fock term) but also includes a *blue* component (corresponding to the anti-Fock term).

However the fact that negative Bohr frequencies do not contribute to the damping also remains true for this class of reservoir states.

5.7 No Rotating-Wave Approximation with Cutoff: Squeezing States

In this section we show that, in contrast to what happens for the Fock and equilibrium (more generally, gauge-invariant) states, in the case of squeezing states the negative Bohr frequencies contribute to the definition of the master fields.

Since the Fock and anti-Fock contributions have already been calculated in Sect. 5.6, we have only to calculate the contribution of the off-diagonal terms to the transport coefficient (5.4.3) which are given by (5.4.4–5). Using (4.21.1) in these identities, one finds

$$\hat{G}^{(-)}_{--}(\omega_q) = \pi \int_{\mathbb{R}^d} dk |g(k)|^2 \delta(\omega_1(k) + \omega_q) - i\text{P.P.} \int_{\mathbb{R}^d} dk \frac{|g(k)|^2}{\omega_1(k) + \omega_q} ,$$

$$\hat{G}^{(-)}_{++}(-\omega_q) = \pi \int_{\mathbb{R}^d} dk |g(k)|^2 \delta(\omega_1(k) + \omega_q) + i\text{P.P.} \int dk \frac{|g(k)|^2}{\omega_1(k) + \omega_q} ,$$

which show that now *also the negative frequencies contribute to the real part of the drift and therefore to the damping.* So a less-symmetric (non-gauge-invariant) state re-establishes a symmetry between positive and negative frequencies which is broken by a more-symmetric (gauge-invariant) state.

5.8 The Stochastic Golden Rule for Dipole Type Interactions and Gauge-Invariant States

As explained at the end of Sect. 4.14, the explicit form of the operators defining the Langevin equation (structure maps) strongly depends on the free evolution, the interaction Hamiltonian and the reference state of the reservoir. In this section we prove that, if the interaction Hamiltonian is of dipole (generalized multiplicative) type and the reference state of the field operators $a(k)$, $a^+(k)$ is an arbitrary mean zero gauge-invariant Gaussian state with covariance (2.6.1) [e.g. a thermal state see (2.10.2)], then the Langevin equation has a fairly explicit form. The stochastic golden rule (for this type of interaction and state) allows one to immediately write the correct Langevin (hence master) equation, given the Hamiltonian system.

Since we restrict ourselves to the boson case, the coefficient m of the covariance in (2.6.1) is equal to $n + 1$. This generalization is relevant because it gives several examples of stationary nonequilibrium states, so that we can effectively speak of *nonequilibrium quantum field theory.* The total Hamiltonian has the usual form

$$H = H_0 + \lambda H_{\text{I}} = H_S + H_R + \lambda H_{\text{I}} \tag{5.8.1}$$

and acts on the space $\mathcal{H}_S \otimes \mathcal{F}$. In this chapter we will only consider interactions of the following type.

Definition 5.8.1. *A dipole interaction Hamiltonian is a sum of operators of the form*

$$H_{\text{I}} = \sum_j \left\{ D_j^* \otimes A(g_j) + D_j \otimes A^*(g_j) \right\} , \tag{5.8.2}$$

where D_j are system operators and $A^{\pm}(g_j)$ are creation and annihilation operators of a quantum field. In the case of a bosonic quantum field, (5.8.2) defines the generalized spin–boson model. We will assume that (5.8.2) satisfies the following additional conditions:

(i) *the index j runs a finite set (finite volume Hamiltonian). For particular models, such as models of spin systems, the index j in (5.8.2) labels the particles in the system. Assuming that the number of particles in the system is proportional to the volume occupied by the system, we can conclude that the interaction is proportional to the volume.*

(ii) *The operators D_j are bounded.*

(iii) *The cutoff functions g_j are Schwartz functions (in most of the applications all of them can be taken equal to a single function g).*

(iv) *The free evolution of the boson field is $a_k^+ \mapsto e^{it\omega_k} a_k^+$, and the dispersion function $\omega(k)$ and the cutoff functions g_j are related by the following basic analytical condition:*

$$\int_{\mathbb{R}} dt |\langle g_i, e^{it\omega(p)} g_j \rangle| = \int_{\mathbb{R}} dt \left| \int_{\mathbb{R}^d} e^{it\omega(k)} \overline{g}_i(k) g_j(k) dk \right| < +\infty \,. \tag{5.8.3}$$

If $\omega(k) = k^2 = \sum_{i=1}^d k_j^2$, then (5.8.3) is satisfied if $d \geq 3$.

(v) $\omega(k) \geq 0, \qquad \forall k.$

(vi) *The $(d-1)$-dimensional Lebesgue measure of the surface $\{k : \omega(k) = 0\}$ is equal to zero [this implies, in particular, $\delta(\omega(k)) = 0$] [for example, $\omega(k) = k^2 + m$ with $m \geq 0$ or $\omega(k) = |k|$].*

Often we will simplify the notations by omitting the symbol $\otimes$. So, for example, (5.8.2) will simply be written

$$H_{\mathrm{I}} = \sum_j \{D_j^* A(g_j) + D_j A^*(g_j)\} \,. \tag{5.8.4}$$

All these assumptions [in particular, condition (i)] can be relaxed in various ways, but we shall not discuss these generalizations here. Under these assumptions the iterated series of the (5.8.2) converges weakly on some subspace of the space $\mathcal{H}_S \otimes \mathcal{F}$, for example, the space $\mathcal{H}_S \otimes \mathcal{D}$, where $\mathcal{D}$ is the algebraic linear span of the exponential and number vectors. To perform the stochastic limit, one needs to calculate the free evolution of the interaction Hamiltonian:

$$H_{\mathrm{I}}(t) = e^{itH_0} H_{\mathrm{I}} e^{-itH_0} \,.$$

Using the identity (see (4.6.1))

$$1 = \sum_r P_{\varepsilon_r} \,,$$

we write the interaction Hamiltonian in the form

$$H_{\mathrm{I}} = \sum_j \sum_{r,r'} P_{\varepsilon_r} D_j^* P_{\varepsilon'_r} \int dk \overline{g}_j(k) a(k) + \mathrm{h.c.} \tag{5.8.5}$$

Let us introduce the set of energy differences (*Bohr frequencies*)

$$F := \{\omega = \varepsilon_r - \varepsilon_{r'} : \varepsilon_r, \varepsilon_{r'} \in \operatorname{Spec} H_S\}, \tag{5.8.6}$$

and for each $\omega \in F$, the set

$$\begin{aligned} F_\omega := &\{\varepsilon_r \in \operatorname{Spec} H_S : \varepsilon_r - \omega \in \operatorname{Spec} H_S\} \\ = &\{\varepsilon_r \in \operatorname{Spec} H_S : \exists \varepsilon_r' \in \operatorname{Spec} H_S, \varepsilon_r - \varepsilon_r' = \omega\}. \end{aligned} \tag{5.8.7}$$

With these notations we rewrite (5.8.5) in the *canonical form:*

$$\begin{aligned} H_{\mathrm{I}} &= \sum_j \sum_{\omega \in F} \sum_{\varepsilon_r \in F_\omega} P_{\varepsilon_r} D_j^* P_{\varepsilon_r - \omega} \int dk \overline{g}_j(k) a(k) + \text{h.c.} \\ &= \sum_j \sum_{\omega \in F} E_\omega^* (D_j) \int dk \overline{g}_j(k) a(k) + h.c.\,, \end{aligned} \tag{5.8.8}$$

where we have introduced the operators

$$E_\omega(X) := \sum_{\varepsilon_r \in F_\omega} P_{\varepsilon_r - \omega} X P_{\varepsilon_r}\,. \tag{5.8.9}$$

It is easy to see that for any system operator X the free evolution of $E_\omega(X)$ is

$$e^{itH_S} E_\omega(X) e^{-itH_S} = e^{-it\omega} E_\omega(X)\,. \tag{5.8.10}$$

Using this and (2.4.3) for the free evolution of boson fields, we obtain for the free evolution of the interaction Hamiltonian:

$$H_{\mathrm{I}}(t) = \sum_j \sum_{\omega \in F} E_\omega^* (D_j) \int dk \overline{g}_j(k) e^{-it[\omega(k) - \omega]} a(k) + \text{h.c.} \tag{5.8.11}$$

5.9 The Stochastic Golden Rule

Note that in the canonical form of (5.8.11) the whole time dependence of the interaction Hamiltonian has been shifted to the phase factor multiplying the field operators. This is the advantage of the canonical form. Using this, we rewrite the rescaled interaction Hamiltonian in the form

$$\begin{aligned} \frac{1}{\lambda} H_{\mathrm{I}}(t/\lambda^2) &= \sum_j \sum_{\omega \in F} E_\omega^* (D_j) \int dk \overline{g}_j(k) \frac{e^{-i(t/\lambda^2)[\omega(k) - \omega]}}{\lambda} a(k) + \text{h.c.} \\ &= \sum_j \sum_{\omega \in F} E_\omega^* (D_j) \int dk \overline{g}_j(k) a_{\lambda,\omega}(t,k) + \text{h.c.}\,, \end{aligned}$$

i.e. as a function of the *rescaled creation and annihilation operators*, defined by

$$a_{\lambda,\omega}(t,k) := \frac{1}{\lambda}\, e^{-i\frac{t}{\lambda^2}\,[\omega(k)-\omega]}a(k) \qquad , \ \omega \in F\,. \tag{5.9.1}$$

The first step in investigating the limit of $(1/\lambda)H_{\mathrm{I}}(t/\lambda^2)$ consists in looking at the limit of the rescaled creation and annihilation operators. The result is expressed by the stochastic golden rule whose formulation we describe in the following two theorems.

Theorem 5.9.1. *Suppose the reference state of the field is mean zero and Gaussian, with a correlation matrix given by (2.6.1). Then each rescaled annihilation (creation) operator (5.9.1), corresponding to a nonpositive frequency*

$$\omega = \varepsilon_r - \varepsilon_{r'} \leq 0$$

converges, in the sense of correlators, to zero. While, if $\omega > 0$, one has

$$\lim_{\lambda\to 0} a_{\lambda,\omega}(t,k) = \lim_{\lambda\to 0}\frac{1}{\lambda}\, e^{-i\frac{t}{\lambda^2}\,[\omega(k)-\omega]}a(k) =: b_\omega(t,k)\,, \tag{5.9.2}$$

where $b_\omega(t,k)$ is a quantum white noise or master field satisfying the commutation relations

$$[b_\omega(t,k), b^*_{\omega'}(t',k')] = \delta_{\omega,\omega'}2\pi\delta(t-t')\delta(\omega(k)-\omega)\delta(k-k') \tag{5.9.3}$$

with a reference vector Ψ which defines the mean zero gauge-invariant white noise with correlations

$$\langle b^*_\omega(t,k) b_{\omega'}(t',k')\rangle = \delta_{\omega,\omega'}2\pi\delta(t-t')\delta(\omega(k)-\omega)\delta(k-k')n(k)\,, \tag{5.9.4a}$$

$$\langle b_\omega(t,k) b^*_{\omega'}(t',k')\rangle = \delta_{\omega,\omega'}2\pi\delta(t-t')\delta(\omega(k)-\omega)\delta(k-k')(n(k)+1)\,. \tag{5.9.4b}$$

In particular white noises corresponding to different frequencies ω are independent.

Proof 5.9.1. Since the rescaled operators are Gaussian, their limit in the sense of correlators is uniquely determined by the limit of their pair correlation functions, e.g.

$$\langle a_{\lambda,\omega}(t,k) a^*_{\lambda,\omega'}(t',k')\rangle = \frac{1}{\lambda^2}\, e^{-i\frac{t}{\lambda^2}\,[\omega(k)-\omega]} e^{i\frac{t'}{\lambda^2}\,[\omega(k')-\omega']}\delta(k-k')$$

$$= \frac{1}{\lambda^2}\, e^{-i\frac{(t-t')}{\lambda^2}\,[\omega(k)-\omega]} e^{i\frac{t'}{\lambda^2}\,(\omega-\omega')}\delta(k-k')\,.$$

To find these limits we use

$$\lim_{\lambda\to 0}\frac{1}{\lambda^2}\exp\left(\frac{it}{\lambda^2}\, f(k)\right) e^{i\frac{t'}{\lambda^2}\,(\omega-\omega')} = \delta_{\omega,\omega'}2\pi\delta(t)\delta(f(k))\,, \tag{5.9.5}$$

which shows that the term $\delta(f(k))$ in (5.9.5) is not identically equal to zero only if $\omega = \omega'$ and $f(k) = 0$ for some k in a set of nonzero $(d-1)$-dimensional

Lebesgue measure [this motivates condition (vi) of Definition 5.8.1, on the dispersion $\omega(k)$ of the boson field].

Finally note that since $\omega(k)$ is supposed to be ≥ 0 from (5.9.4a,b) it follows that if $\omega \leq 0$ then the above correlations vanish identically, which means that $b_\omega(t,k) \equiv 0$.

Theorem 5.9.1 implies that the stochastic limit of the rescaled interaction Hamiltonian is equal to

$$h(t) = \sum_j \sum_{\omega \in F} E_\omega^* (D_j) \int dk \overline{g}_j(k) b_\omega(t,k) + \text{h.c.} \tag{5.9.6}$$

$$= \sum_{\omega \in F} \int dk \left(\sum_j E_\omega^*(D_j) \overline{g}_j(k) \right) b_\omega(t,k) + \text{h.c.}$$

The stochastic golden rule for this class of interactions and states is obtained by combining Theorem 5.9.1 with the following result:

Theorem 5.9.2. *The stochastic limit of the evolution equation in the interaction picture for our model is the (singular) white noise equation*

$$\frac{d}{dt} U_t = -ih(t) U_t \,, \tag{5.9.7}$$

and its (causally) normally ordered form is the stochastic Schrödinger equation

$$dU_t = [-idH(t) - Gdt] U_t \qquad , \quad t > 0 \,, \tag{5.9.8}$$

with the initial condition $U_0 = 1$ and where

(i) $h(t)$ is the white noise Hamiltonian (5.9.6) and $dH(t)$, called the martingale term, is the stochastic differential,

$$dH(t) = \int_t^{t+dt} h(s) ds = \sum_j \sum_{\omega \in F} \left[E_\omega^* (D_j) \, dB_{j\omega}(t) + E_\omega (D_j) \, dB_{j\omega}^*(t) \right] \tag{5.9.9}$$

driven by the quantum Brownian motions

$$dB_{j\omega}(t) := \int_t^{t+dt} \int dk \overline{g}_j(k) b_\omega(\tau,k) d\tau =: \int_t^{t+dt} b_\omega(\tau, g_j) d\tau \,; \tag{5.9.10}$$

(ii) The operator G, called the drift, is given by

$$G = \sum_{ij} \sum_{\omega \in F} \left[(g_i|g_j)_\omega^- E_\omega^* (D_i) \, E_\omega (D_j) + \overline{(g_i|g_j)_\omega^+} E_\omega (D_i) \, E_\omega^* (D_j) \right] \,, \tag{5.9.11}$$

where the explicit form of the constants $(g_i|g_j)^{\pm}_{\omega}$ *is*

$$(g_i|g_j)^-_\omega = \int dk\, \overline{g_i(k)} g_j(k) \frac{-i[n(k)+1]}{\omega(k)-\omega-i0} \tag{5.9.12}$$

$$= \pi \int dk\, \overline{g_i(k)} g_j(k) [n(k)+1] \delta(\omega(k)-\omega) - i\,\text{P.P.} \int dk\, \overline{g_i(k)} g_j(k) \frac{n(k)+1}{\omega(k)-\omega},$$

$$(g_i|g_j)^+_\omega = \int dk\, \overline{g_i(k)} g_j(k) \frac{-in(k)}{\omega(k)-\omega-i0} \tag{5.9.13}$$

$$= \pi \int dk\, \overline{g_i(k)} g_j(k) n(k) \delta(\omega(k)-\omega) - i\,\text{P.P.} \int dk\, \overline{g_i(k)} g_j(k) \frac{n(k)}{\omega(k)-\omega}.$$

Remark 5.9.1. The constants $(g_i|g_j)^{\pm}_{\omega}$ are called *generalized susceptivities* and have an important physical meaning. In some sense they contain all the physical information on the original Hamiltonian system. An important class of physical states is the one in which $n(k)$ itself is a function of $\omega(k)$, i.e.

$$n(k) = n(\omega(k)). \tag{5.9.14}$$

In this case the generalized susceptivities become

$$\text{Re}\,(g_i|g_j)^-_\omega = [n(\omega)+1]\gamma_{ij}(\omega)$$
$$\text{Re}\,(g_i|g_j)^+_\omega = n(\omega)\gamma_{ij}(\omega),$$

where

$$\gamma_{ij}(\omega) = \pi \int dk\, \overline{g_i(k)} g_j(k) \delta(\omega(k)-\omega).$$

Therefore

$$\frac{\text{Re}\,(g_i|g_j)^-_\omega}{\text{Re}\,(g_i|g_j)^+_\omega} = \frac{n(\omega)+1}{n(\omega)}. \tag{5.9.15}$$

Recalling that $n(\omega)$ is the density of field quanta (photon, phonons, etc.) at frequency ω, and comparing (5.9.15) with the well-known formula of radiation theory (the Einstein formula)

$$\frac{W_{\text{emission}}}{W_{\text{absorption}}} = \frac{\overline{n}_\omega + 1}{\overline{n}_\omega}, \tag{5.9.16}$$

which gives the ratio of the probability of emission and absorption of a light quantum by an atom [see [Heit54], Chap. V, §17, formula (18)], we gain some physical intuition of the meaning of the generalized susceptivities. In fact the ratio (5.9.16) "is just that which is necessary to preserve the correct thermal equilibrium of the radiation with the gas" ([Heit54], p. 180). This suggests that the ratios (5.9.15) may play a similar role for some stationary nonequilibrium states.

Remark 5.9.2. In (5.9.10) we introduced the *smeared master fields*:

$$b_\omega(t,g) := \int_{\mathbb{R}^d} dk\overline{g}(k)b_\omega(t,k)\,, \tag{5.9.17}$$

with covariance

$$\langle b_\omega(t,g_i)b_\omega^*(t',g_j)\rangle = \delta(t-t')2\,\mathrm{Re}(g_i|g_j)_\omega^-\,, \tag{5.9.18}$$

$$\langle b_\omega^*(t,g_j)b_\omega(t',g_j)\rangle = \delta(t-t')2\,\mathrm{Re}(g_i|g_j)_\omega^+\,. \tag{5.9.19}$$

Moreover we know that $\mathrm{Re}(g_i|g_j)_\omega^\pm$ can be $\neq 0$ only if $\omega > 0$.

Proof 5.9.2. The limit dynamical equation is obtained as follows: the first term $dH(t)$ in (5.9.8) is just the limit of the iterated series solution for the rescaled Schrödinger equation in the interaction picture

$$\lim_{\lambda\to 0}\frac{1}{\lambda}\int_t^{t+dt} H_{\mathrm{I}}\left(\frac{\tau}{\lambda^2}\right)d\tau\,.$$

The second term Gdt, i.e. the *drift*, is equal to the limit of the expectation value in the reservoir state of the second term in the same iterated series

$$\begin{aligned}&\lim_{\lambda\to 0}\frac{1}{\lambda^2}\int_t^{t+dt} dt_1\int_t^{t_1} dt_2\left\langle H_{\mathrm{I}}\left(\frac{t_1}{\lambda^2}\right)H_{\mathrm{I}}\left(\frac{t_2}{\lambda^2}\right)\right\rangle\\ &=\lim_{\lambda\to 0}\frac{1}{\lambda^2}\int_t^{t+dt} dt_1\int_t^{t_1} dt_2\left\langle H_{\mathrm{I}}(0)H_{\mathrm{I}}\left(\frac{t_2-t_1}{\lambda_2}\right)\right\rangle\end{aligned}$$

Making the change of variables $\tau = (t_2-t_1)/\lambda^2$, we obtain

$$\lim_{\lambda\to 0}\int_t^{t+dt} dt_1\int_{(t-t_1)/\lambda^2}^0 d\tau\langle H_{\mathrm{I}}(0)H_{\mathrm{I}}(\tau)\rangle = dt\int_{-\infty}^0 d\tau\langle H_{\mathrm{I}}(0)H_{\mathrm{I}}(\tau)\rangle\,. \tag{5.9.20}$$

Computing the expectation value and using the fact that the limits of oscillating factors of the form $\lim_{\lambda\to 0}\exp(ict_1/\lambda^2)$ (with $t_1 \neq 0$) vanish unless the constant c is equal to zero; we see that we can have a nonzero limit only when all the oscillating factors in (5.9.20) cancel. In conclusion we obtain

$$\begin{aligned}G=&\sum_{ij}\sum_{\omega\in F}\int_{-\infty}^0 d\tau\Big(\int dk\,\overline{g_i(k)}g_j(k)e^{i\tau[\omega(k)-\omega]}\left(n(k)+1\right)E_\omega^*\left(D_i\right)E_\omega\left(D_j\right)\\ &+\int dk\,g_i(k)\overline{g_j(k)}e^{-i\tau[\omega(k)-\omega]}n(k)E_\omega\left(D_i\right)E_\omega^*\left(D_j\right)\Big)\,;\end{aligned} \tag{5.9.21}$$

therefore, from the known formula of distribution theory

$$\int_{-\infty}^0 e^{it\omega}dt = \frac{-i}{\omega-i0} = \pi\delta(\omega) - i\,\mathrm{P.P.}\frac{1}{\omega}\,, \tag{5.9.22}$$

we obtain

$$G = \sum_{ij} \sum_{\omega \in F} \left(\int dk \, \overline{g_i(k)} g_j(k) \frac{-i[n(k)+1]}{\omega(k) - \omega - i0} E_\omega^* (D_i) \, E_\omega (D_j) \right. \tag{5.9.23}$$
$$\left. + \int dk \, g_i(k) \overline{g_j(k)} \frac{in(k)}{\omega(k) - \omega + i0} E_\omega (D_i) \, E_\omega^* (D_j) \right),$$

which, again using (5.9.22), is seen to coincide with the expression (5.9.11) for the drift G.

Remark 5.9.3. We know that $n(k) \geq 0$, therefore (5.9.12–13) imply that, for each $\omega \in F$, the matrices

$$\left(\mathrm{Re}(g_i|g_j)_\omega^\pm \right)_{i,j} \tag{5.9.24}$$

are positive. Therefore the operator

$$G = \mathrm{Re} G + i \, \mathrm{Im}\, G = \frac{1}{2} (G + G^*) + i \frac{1}{2i} (G - G^*)$$
$$= \sum_{i,j} \sum_{\omega \in F} [\mathrm{Re}(g_i|g_j)_\omega^- E_\omega^*(D_i) E_\omega(D_j) + \mathrm{Re}(g_i|g_j)_\omega^+ E_\omega(D_i) E_\omega^*(D_j)]$$
$$+ i \sum_{ij} \sum_{\omega \in F} [\mathrm{Im}(g_i|g_j)_\omega^- E_\omega^*(D_i) E_\omega(D_j) - \mathrm{Im}(g_i|g_j)_\omega^+ E_\omega(D_i) E_\omega^*(D_j)]$$

has a positive real part.

Remark 5.9.4. Note that, from (5.9.11) and from the explicit form of the generalized susceptivities (5.9.12–13), it follows that it is perfectly possible for

$$\mathrm{Re}(g|g)_\omega^\pm = 0 \tag{5.9.25}$$

but $\mathrm{Im}(g|g)_\omega^\pm \neq 0$. In particular, if (5.9.25) takes place for all ω, the stochastic (white noise) Schrödinger equation (5.9.8) reduces to a usual Schrödinger equation,

$$\partial_t U_t = -i(\mathrm{Im} G) U_t \quad , \quad t > 0 , \tag{5.9.26}$$

and there is no damping at all. This has been called a *Cheshire cat effect* in [AcKoVo97] because it means that in the stochastic limit the field disappears, but a trace of its interaction with the system remains in the form of a shift in the original Hamiltonian of the latter: just like the grin of the Cheshire cat. More generally we can say that, in any case, even if the negative Bohr frequencies ω do not generate quantum noises, hence they do not contribute to damping, the corresponding values $\mathrm{Im}(g|g)_\omega^\pm$ in (5.9.11) might be nonzero, hence they produce an energy shift in the system. Recently this fact has been experimentally proved in cavity QED. From a physical point of view this is a quite natural fact.

If the rescaled annihilator $a_{\lambda,\omega}(t,k)$ (5.9.1) corresponds to a strictly positive Bohr frequency $\omega > 0$, then it describes the absorption of a boson with energy $\omega(k)$ and the simultaneous transition of the system from the lower energy level ε' to the upper energy level $\varepsilon = \varepsilon' + \omega$. The creator $a^*_{\lambda,\omega}(t,k)$ describes the symmetric process.

If $\omega \le 0$ then $a_{\lambda,\omega}(t,k)$ describes the absorption of a boson with energy $\omega(k)$ and the simultaneous transition of the system from the upper energy level ε to the lower energy level $\varepsilon' = \varepsilon + \omega$ (recall that $\omega \le 0$). The creator describes the corresponding absorption.

For a better intuition one can imagine that the system itself is a quantum field and transition with positive frequency corresponding to emission of a quantum. Intuitively we expect that the most probable processes should be those in which one quantum (reservoir) is absorbed, another (system) is emitted, and the two have the same frequency. On the contrary, we expect that processes involving absorption of a quantum with some energy and emission of a quantum with another energy, or simultaneous absorption (or emission) of two quanta, should be much less probable. The stochastic golden rule shows that this intuition is correct: processes without conservation of energy do not contribute in the stochastic limit.

Note however, while we can immediately read this simple fact from the stochastic equation (5.9.8), it is quite hidden in the original Hamiltonian equation (5.9.7).

Remark 5.9.5. The independence of the noises corresponding to different Bohr frequencies is the essence of *the stochastic resonance principle.* According to this principle a system can absorb a quantum by jumping from energy level 1 to energy level 2 only if the energy $\omega(k)$ of the quantum is equal to the difference of energies of levels 2 and 1

$$E(2) - E(1) = \varepsilon = \omega(k)\,.$$

Correspondingly, when the system jumps from energy level 2 to energy level 1 it emits a quantum with the energy $\omega(k)$.

5.10 The Langevin Equation

Using the stochastic golden rule we can find the Langevin equation, which is the limit of the Heisenberg evolution in the interaction representation of any observable $X = X_S \otimes 1_{\text{res}}$ of the system. The Langevin equation is the equation satisfied by the stochastic flow j_t, defined by

$$j_t(X) = U_t^* X U_t\,, \tag{5.10.1}$$

where U_t satisfies (5.9.8), i.e. in the notations (5.9.9–10):

$$dU_t = [-idH(t) - Gdt]U_t\,. \tag{5.10.2}$$

To derive the Langevin equation, we may apply the stochastic golden rule or, alternatively, we can apply to the stochastic differential

$$dj_t(X) = j_{t+dt}(X) - j_t(X) = dU_t^* X U_t + U_t^* X dU_t + dU_t^* X dU_t\,, \quad (5.10.3)$$

the *causal Ito table*, according to which the only nonvanishing products among the quantum stochastic differentials are

$$dB_{i\omega}(t)dB_{j\omega}^*(t) = 2\mathrm{Re}\,(g_i|g_j)_\omega^- dt, \quad dB_{i\omega}^*(t)dB_{j\omega}(t) = 2\mathrm{Re}\,(g_i|g_j)_\omega^+ dt\,, \quad (5.10.4)$$

where the generalized susceptivities $(g_i|g_j)_\omega^\pm$ are given by (5.9.12–13).

Remark 5.10.1. The usual Ito table is obtained from the causal one by replacing the constants $(g_i|g_j)_\omega^\pm$ with $(g_i|g_j)_\omega$.

Combining the terms in (5.10.3) and using (5.9.10–11) and (5.10.2, 4), we obtain the Langevin equation,

$$\begin{aligned} dj_t(X) &= \sum_\alpha j_t \circ \theta_\alpha(X) dM^\alpha(t) \qquad (5.10.5)\\ &= \sum_{n=-1,1;j\omega} j_t \circ \theta_{nj\omega}(X) dM^{nj\omega}(t) + j_t \circ \theta_0(X) dt\,, \end{aligned}$$

where the maps θ_α are called *structure maps* and

$$dM^{-1,j\omega}(t) = dB_{j\omega}(t), \quad \theta_{-1,j\omega}(X) = -i[X, E_\omega^*(D_j)]\,, \quad (5.10.6)$$

$$dM^{1,j\omega}(t) = dB_{j\omega}^*(t), \quad \theta_{1,j\omega}(X) = -i[X, E_\omega(D_j)]\,, \quad (5.10.7)$$

$$\begin{aligned} \theta_0(X) = \sum_{ij}\sum_{\omega\in F} \Big[&-i\,\mathrm{Im}\,(g_i|g_j)_\omega^- [X, E_\omega^*(D_i) E_\omega(D_j)] \qquad (5.10.8)\\ &+ i\,\mathrm{Im}\,(g_i|g_j)_\omega^+ [X, E_\omega(D_i) E_\omega^*(D_j)]\\ &+ 2\mathrm{Re}\,(g_i|g_j)_\omega^- \left(E_\omega^*(D_i) X E_\omega(D_j) - \frac{1}{2}\{X, E_\omega^*(D_i) E_\omega(D_j)\}\right)\\ &+ 2\mathrm{Re}\,(g_i|g_j)_\omega^+ \left(E_\omega(D_i) X E_\omega^*(D_j) - \frac{1}{2}\{X, E_\omega(D_i) E_\omega^*(D_j)\}\right)\Big] \end{aligned}$$

is a quantum Markovian generator. The structure map $\theta_0(X)$ has the standard form of the generator of a master equation [GKS76, Lind76]:

$$\theta_0(X) = \Psi(X) - \frac{1}{2}\{\Psi(1), X\} + i[H, X]\,,$$

where Ψ is a completely positive map and H is self-adjoint. In our case $\Psi(X)$ is a linear combination of terms of the type

$$E_\omega^*(D_i) X E_\omega(D_j)\,.$$

The structure maps θ_α in (5.10.5) satisfy the following stochastic Leibnitz rule, see [AcKo99a].

Theorem 5.10.1. *For any pair of operators in the system algebra X, Y, the structure maps in the Langevin equation (5.10.5) satisfy*

$$\theta_\alpha(XY) = \theta_\alpha(X)Y + X\theta_\alpha(Y) + \sum_{\beta,\gamma} c_\alpha^{\beta\gamma}\theta_\beta(X)\theta_\gamma(Y),$$

where the structure constants $c_\alpha^{\beta\gamma}$ are given by the causal Ito table [see (5.10.4)]

$$dM^\beta(t)dM^\gamma(t) = \sum_\alpha c_\alpha^{\beta\gamma} dM^\alpha(t).$$

The conjugation rules of $dM^\alpha(t)$ and θ_α are connected in such a way that (5.10.5) defines a $*$-flow ($* \circ j_t = j_t \circ *$).

Proof 5.10.1. The proof of this theorem is by direct calculation using (5.10.6–8).

5.11 Subalgebras Invariant Under the Generator

In this section we prove that the generators of the Markovian semigroups, arising in the stochastic limit, have a very special structure which have important consequences. This structure is a consequence of the following, well-known, elementary fact.

Lemma 5.11.1. *Let $\mathcal{A}$ be an abelian subalgebra of an algebra $\mathcal{B}$ generated by the partition of the identity (P_n) ($P_nP_m = \delta_{m,n}P_n, \sum_n P_n = 1$). An element x is in the commutant of $\mathcal{A}$ (the set of elements $y \in \mathcal{B}$ such that ($xy = yx$ for any $x \in \mathcal{A}$) if and only if*

$$x = \sum_n P_n x P_n .$$

Proof 5.11.1. It is clear that any element of the form $\sum_n P_n x P_n$ is in the commutant of $\mathcal{A}$. Conversely, if x is in the commutant of $\mathcal{A}$, then $x = (\sum_n P_n^2)x = \sum_n P_n x P_n$.

Corollary 5.11.1. *The Markovian generator θ_0, defined by (5.10.8), and therefore also the associated Markov semigroup $P^t = \exp(t\theta_0)$, maps the commutant $L^\infty(H_S)'$ of the abelian algebra $L^\infty(H_S)$, generated by the spectral projections of H_S, into itself. Moreover:*

(i) if X in (5.10.8) belongs to $L^\infty(H_S)$, then the Hamiltonian part of $\theta_0(X)$ vanishes and only the dissipative part remains.

(ii) if H_S has a nondegenerate spectrum then θ_0 maps $L^\infty(H_S)$ into itself and has the form, for any $X \in L^\infty(H_S)$,

$$\theta_0(X) = \sum_{ij}\sum_{\omega\in F}\Big(2\mathrm{Re}\,(g_i|g_j)^-_\omega\left[E^*_\omega(D_i)\,XE_\omega(D_j) - XE^*_\omega(D_i)\,E_\omega(D_j)\right]$$

$$+2\mathrm{Re}\,(g_i|g_j)^+_\omega\left[E_\omega(D_i)\,XE^*_\omega(D_j) - XE_\omega(D_i)\,E^*_\omega(D_j)\right]\Big) \in L^\infty(H_S)\,.$$

Proof 5.11.2. If X is an arbitrary bounded operator on $\mathcal{H}_S$ belonging to the commutant $L^\infty(H_S)'$ then (5.8.9) implies that, for any $\omega \in F$ and for any i,j, the operators are

$$E_\omega(D_i)\,XE^*_\omega(D_j) = \sum_{\varepsilon_r,\varepsilon_{r'}\in F_\omega} P_{\varepsilon_r-\omega}D_iP_{\varepsilon_r}XP_{\varepsilon_{r'}}D^*_jP_{\varepsilon_{r'}-\omega}$$

$$= \sum_{\varepsilon_r\in F_\omega} P_{\varepsilon_r-\omega}D_iP_{\varepsilon_r}XP_{\varepsilon_r}D^*_jP_{\varepsilon_r-\omega} \tag{5.11.1}$$

$$E^*_\omega(D_i)\,XE_\omega(D_j) = \sum_{\varepsilon_r,\varepsilon_{r'}\in F_\omega} P_{\varepsilon_r}D^*_iP_{\varepsilon_r-\omega}XP_{\varepsilon_{r'}-\omega}D_jP_{\varepsilon_{r'}}$$

$$= \sum_{\varepsilon_r\in F_\omega} P_{\varepsilon_r}D^*_iP_{\varepsilon_r-\omega}XP_{\varepsilon_r-\omega}D_jP_{\varepsilon_r}\,; \tag{5.11.2}$$

so the condition of Lemma 5.11.1 is satisfied and both operators (5.11.1–2) belong to the commutant of $L^\infty(H_S)$.

5.12 The Langevin Equation: Generic Systems

In this section we introduce an interesting class of Hamiltonians, H_S, which we call *generic*. The interest in them lies in the fact that they are canonically associated to a lattice of spin 1/2 systems. Here the lattice is the set F of all the Bohr frequencies of H_S.

Definition 5.12.1. *A quantum system with Hamiltonian H_S will be called generic, if:*

(i) The spectrum Spec H_S *of the Hamiltonian is nondegenerate.*
(ii) For any positive Bohr frequency $\omega > 0$: $|F_\omega| = 1$, i.e. there exists a unique pair of energy levels $\varepsilon_{1_\omega}, \varepsilon_{2_\omega} \in$ Spec H_S *such that $\omega = \varepsilon_{2_\omega} - \varepsilon_{1_\omega}$.*

The term *generic* means that *the eigenvalues of H_S are irregularly displaced.* For example, the spectrum of the 1-dimensional harmonic oscillator satisfies (i) but not (ii). Thus it is not generic in the above sense. On the other hand, if among the eigenvalues of H_S there are no rational relations (no linear combination of eigenvalues with rational coefficients nonidentically equal to zero can be equal to zero), then H_S is generic.

Given a generic quantum system, and a Bohr frequency ω, denote $|1_\omega\rangle$ and $|2_\omega\rangle$ the two eigenstates corresponding to the two energy levels, $\varepsilon_{1_\omega}, \varepsilon_{2_\omega}$, so that

$$H_S|1_\omega\rangle = \varepsilon_{1_\omega}|1_\omega\rangle \quad , \quad H_S|2_\omega\rangle = \varepsilon_{2_\omega}|2_\omega\rangle \quad , \quad \omega = \varepsilon_{2_\omega} - \varepsilon_{1_\omega} > 0\,. \tag{5.12.1}$$

In these notations,

$$E_\omega(D) = |1_\omega\rangle\langle 1_\omega|D|2_\omega\rangle\langle 2_\omega| = \langle 1_\omega|D|2_\omega\rangle|1_\omega\rangle\langle 2_\omega| =: \kappa_\omega(D)|1_\omega\rangle\langle 2_\omega| \tag{5.12.2}$$

or equivalently one can say that in the basis $(|1_\omega\rangle, |2_\omega\rangle)$ the operator $E_\omega(D)$ is proportional to the Pauli matrix that flips the spin down:

$$|1_\omega\rangle\langle 2_\omega| =: \sigma_\omega^- = \begin{pmatrix} 0 & 0 \\ 1 & 0 \end{pmatrix}\,. \tag{5.12.3}$$

With these notations we see that the restriction of H_S on the space generated by $|1_\omega\rangle$ and $|2_\omega\rangle$ is

$$\sigma_\omega^z = \frac{\omega}{2}\sigma^z + \left(\frac{\varepsilon_1+\varepsilon_2}{2}\right)1_\omega = \begin{pmatrix} \omega/2 & 0 \\ 0 & -\omega/2 \end{pmatrix} + \begin{pmatrix} (\varepsilon_1+\varepsilon_2)/2 & 0 \\ 0 & (\varepsilon_1+\varepsilon_2)/2 \end{pmatrix}$$
$$= \begin{pmatrix} \varepsilon_2 & 0 \\ 0 & \varepsilon_1 \end{pmatrix}\,.$$

Using this we define

$$c_\omega(t,k) := \sigma_\omega^+ \otimes b_\omega(t,k)\,, \tag{5.12.4}$$

$$g_\omega(k) := \sum_j \langle 1_\omega, D_j 2_\omega\rangle g_j(k)\,, \tag{5.12.5}$$

$$b_\omega(t,g) = \int dk\overline{g_\omega}(k) b_\omega(t,k)\,; \tag{5.12.6}$$

the white noise Hamiltonian (5.9.6) takes the form

$$h(t) = \sum_{\omega\in F}\int dk\overline{g_\omega}(k)\sigma_\omega^+ b_\omega(t,k) + \text{h.c.} \tag{5.12.7}$$

Let us denote $\mathcal{H}_\omega$ the subspace of the system space $\mathcal{H}_S$ generated by 1_ω, 2_ω. If $\omega \neq \omega'$, then $\mathcal{H}_\omega$ is orthogonal to $\mathcal{H}_{\omega'}$ because these subspaces are generated by eigenvectors of H_S corresponding to different eigenvalues. Moreover, since H_S has a discrete spectrum, one has $\mathcal{H}_S = \bigoplus_\omega \mathcal{H}_\omega$. In other words, *if H_S is generic, any generic quantum system S, interacting with a boson field with a generalized dipole interaction Hamiltonian, is equivalent, in the stochastic limit, to a (possibly infinite) orthogonal sum of independent 2-level systems.* The simplest case corresponds to a single 2-level system. This gives the spin–boson Hamiltonian to be discussed in Sect. 5.14.

Let us now investigate the Langevin equation in the case of an Hamiltonian H_S, generic in the sense of Definition 5.12.1. In this case, in the notations (5.12.2–3), we have:

$$E_\omega(X) = \kappa_\omega(X)\sigma_\omega^- \quad , \quad E_\omega^*(X) = \overline{\kappa_\omega(X)}\sigma_\omega^+ ; \tag{5.12.8}$$

introducing the notations

$$\gamma_{\text{re}}^-(\omega) = \sum_{ij} 2\text{Re}\,(g_i|g_j)_\omega^- \overline{\kappa_\omega(D_i)}\kappa_\omega(D_j) , \tag{5.12.9}$$

$$\gamma_{\text{im}}^-(\omega) = \sum_{ij} \text{Im}(g_i|g_j)_\omega^- \overline{\kappa_\omega(D_i)}\kappa_\omega(D_j) , \tag{5.12.10}$$

$$\gamma_{\text{re}}^+(\omega) = \sum_{ij} 2\text{Re}\,(g_i|g_j)_\omega^+ \kappa_\omega(D_i)\overline{\kappa_\omega(D_j)} , \tag{5.12.11}$$

$$\gamma_{\text{im}}^+(\omega) = \sum_{ij} \text{Im}(g_i|g_j)_\omega^+ \kappa_\omega(D_i)\overline{\kappa_\omega(D_j)} , \tag{5.12.12}$$

the causal Ito table (5.10.4) for the stochastic differentials becomes

$$dB_\omega(t)dB_\omega^*(t) = \gamma_{\text{re}}^-(\omega)dt ,$$

$$dB_\omega^*(t)dB_\omega(t) = \gamma_{\text{re}}^+(\omega)dt ,$$

$$dB_\omega(t)dB_\omega(t) = dB_\omega^*(t)dB_\omega^*(t) = 0 .$$

Using this the Langevin equation for the algebra of observables takes the form

$$\begin{aligned} dj_t(X) &= \sum_\alpha j_t \circ \theta_\alpha(X)dM^\alpha(t) \\ &= \sum_{n=-1,1;\omega} j_t \circ \theta_{n\omega}(X)dM^{n\omega}(t) + j_t \circ \theta_0(X)dt , \end{aligned} \tag{5.12.13}$$

where the stochastic differentials and the structure maps θ_α are given by

$$dM^{-1,\omega}(t) = dB_\omega(t) = \sum_j \overline{\kappa_\omega(D_j)}dB_{j\omega}(t), \quad \theta_{-1,\omega}(X) = -i[X, \sigma_\omega^+] \tag{5.12.14}$$

$$dM^{1,\omega}(t) = dB_\omega^*(t) = \sum_j \kappa_\omega(D_j)dB_{j\omega}^*(t), \quad \theta_{1,j\omega}(X) = -i[X, \sigma_\omega^-] \tag{5.12.15}$$

$$\theta_0(X) = \sum_{\omega \in F} \Big[-i\gamma_{\text{im}}^{-}(\omega)[X, |2_\omega\rangle\langle 2_\omega|] + i\gamma_{\text{im}}^{+}(\omega)[X, |1_\omega\rangle\langle 1_\omega|]$$
$$+ \gamma_{\text{re}}^{-}(\omega)\left(\langle 1_\omega|X|1_\omega\rangle |2_\omega\rangle\langle 2_\omega| - \frac{1}{2}\{X, |2_\omega\rangle\langle 2_\omega|\}\right)$$
$$+ \gamma_{\text{re}}^{+}(\omega)\left(\langle 2_\omega|X|2_\omega\rangle |1_\omega\rangle\langle 1_\omega| - \frac{1}{2}\{X, |1_\omega\rangle\langle 1_\omega|\}\right)\Big] . \qquad (5.12.16)$$

In summary, the stochastic flow associated with a generic system can be described in terms of the following ingredients:

- A *lattice* F (the set of the Bohr frequencies of H_S).
- An Hilbert space $\mathcal{H}_S$ (the *system state space*).
- A system Hamiltonian H_S acting on $\mathcal{H}_S$ and with a discrete spectrum, i.e. H_S has the following spectral decomposition:

$$H_S = \sum_r \varepsilon_r P_{\varepsilon_r} ,$$

where the index r labels the spectral projections of H_S. For example, for a nondegenerate eigenvalue ε_r of H_S the corresponding spectral projection is

$$P_{\varepsilon_r} = |\varepsilon_r\rangle\langle\varepsilon_r| , \qquad (5.12.17)$$

where $|\varepsilon_r\rangle$ is the corresponding eigenvector.

5.13 The Stochastic Golden Rule Versus Standard Perturbation Theory

Consider an Hamiltonian of the form[(1)]

$$H_\lambda = H_{\text{A}} \otimes 1 + 1 \otimes H + \lambda Q \otimes \varphi(g) ,$$

where the atomic Hamiltonian H_{A} is a finite rank operator

$$H_{\text{A}} = \sum E_j |E_j\rangle\langle E_j| ,$$

Q denotes the position of the atom and

$$\varphi(g) = A(g) + A^+(g)$$

is a Boson field.[(2)] Using the expansion

$$Q = \sum \langle E_j, QE_k\rangle |E_j\rangle\langle E_k| = \sum c_{jk} |E_j\rangle\langle E_k|$$

(which also has meaning without the finite rank assumption if we suppose the sum to be weakly convergent on the dense subspace of $\mathcal{H}_{\text{A}}$ given by the finite linear combinations of $|E_j\rangle$), one can write the interacting Hamiltonian in the form

$$\begin{aligned} H_{\mathrm{I}} &= \sum_{jk} |E_j\rangle\langle E_k| \otimes [A^+(c_{jk}g) + A(c_{jk}g)] \\ &= \sum_{j} |E_j\rangle\langle E_j| \otimes [A^+(c_{jj}g) + A(c_{jj}g)] \\ &\quad + \sum_{j<k} \Big\{ |E_j\rangle\langle E_k| \otimes [A^+(c_{jk}g) + A(\bar{c}_{jk}g)] \\ &\quad + |E_k\rangle\langle E_j| \otimes [A^+(\bar{c}_{jk}g) + A(c_{jk}g)] \Big\} \end{aligned}$$

$$\begin{aligned} &= \sum_{j} |E_j\rangle\langle E_j| \otimes [A^+(c_{jj}g) + A(c_{jj}g)] \\ &\quad + \sum_{j<k} \Big[|E_j\rangle\langle E_k| \otimes A^+(c_{jk}g) + |E_k\rangle\langle E_j| \otimes A(c_{jk}g) \\ &\quad + |E_j\rangle\langle E_k| \otimes A(\bar{c}_{jk}g) + |E_k\rangle\langle E_j| \otimes A^+(\bar{c}_{jk}g) \Big] . \end{aligned}$$

Introducing the notations

$$g_{jk} = \begin{cases} c_{jk}g & \text{if } j \le k\,, \\ \bar{c}_{jk}g & \text{if } j > k\,, \end{cases}$$

we write H_{I} in the form

$$H_{\mathrm{I}} = \sum_{(j,k)\in F} \left[|E_j\rangle\langle E_k| \otimes A^+(g_{jk}) + \text{h.c.} \right] ,$$

and, with the further notations

$$F := \{(j,k) : j,k = 1 \dots N\} \tag{5.13.1}$$

$$D_{(j,k)} := |E_j\rangle\langle E_k| \,; \qquad (j,k) \in F\,, \tag{5.13.2}$$

we finally obtain the canonical form (4.8.7):

$$H_{\mathrm{I}} = \sum_{d\in F} \left[D_d \otimes A^+(g_d) + \text{h.c.} \right] \tag{5.13.3}$$

under the assumption

$$E_j - E_j \ne F_{j'} - E_{k'} \quad \text{if} \quad (j,k) \ne (j',k')\,, \quad j<k, j'<k'\,. \tag{5.13.4}$$

Written in the form of (5.13.3), the interaction Hamiltonian H_{I} is precisely a multiplicative interaction without the rotating-wave approximation.

An application of the stochastic golden rule, described in Sect. 4.14, immediately gives the stochastic equation

$$dU_t = \sum_{d\in F} \left(iD_d \otimes dA_t^+(g_d) + iD_d^+ \otimes dA_t(g_d) - D_d^+ (g|g)_d^- D_d dt\right) U_t \,,$$

where $(g|g)_d^-$ is defined as in (4.5.5) with $d = \omega_q$. The vacuum-to-vacuum (partial) amplitude associated with U_t is then the solution of the ordinary differential equation

$$\frac{d}{dt} \langle \Phi, U_t \Phi \rangle = -\sum_{-d} D_d^+ D_d (g_d|g_d)_d^- \langle \Phi, U_t \Phi \rangle \;, \quad \langle \Phi, U_0 \Phi \rangle = \langle \Phi, \Phi \rangle = 1 \,,$$

the solution of which is given by

$$\langle \Phi, U_t \Phi \rangle = e^{-t \sum_d D_d^+ D_d (g_d|g_d)_d^-} \,.$$

Recalling the specific form of (5.13.1) of the index set F and the form of (5.13.2) of the operators D_d, one finds that if $d = (j,k)$ then

$$D_d = D_{\omega_d} = D_{E_j - E_k} = |E_j\rangle\langle E_k| \,;$$

therefore

$$D_d^+ D_d = |E_k\rangle\langle E_k| \,.$$

Consequently, by the spectral theorem

$$e^{-t \sum_d D_d^+ D_d (g_d|g_d)_d^-} = \sum_k e^{-t \sum_j (g_j|g_j)_{jk}^-} |E_k\rangle\langle E_k| \,, \tag{5.13.5}$$

with

$$(g_d|g_d)_d = |c_d|^2 (g|g)_d = |\langle E_j, Q E_k\rangle|^2 (g|g)_d \,.$$

Remark 5.13.1. If the assumption (5.13.4) is not verified, all the formulae remain the same with the one difference that the index d in (5.13.3), instead of the form (j,k), has the form (j,ν,k), where ν is an index counting the number of pairs for which (5.13.4) is not satisfied.

5.14 Spin–Boson Hamiltonian

In this section, as an illustration of the stochastic limit for Hamiltonian not satisfying the generalized rotating wave approximation, we investigate a two-state system coupled to an environment; this model is an example of the so-called *spin–boson Hamiltonian*, widely used in physics and chemistry. Examples include the motion of defects in some crystallizing solids, some chemical reactions, some approaches to the theory of quantum measurement and many other topics quoted in the survey paper [LCDFGZ87]. Here we shall be mainly interested in the context of the so-called *macroscopic quantum coherence phenomena*, related to the motion of the magnetic flux trapped in a ring of superconducting material (SQUID or *superconducting quantum interference device*) model. Extending a result first discovered in [LCDFGZ87], we prove the emergence of two interesting regimes for the system: the *purely oscillating* one and the *purely damping* one. The spin–boson Hamiltonian is

$$H_\lambda = -\frac{1}{2}\,\Delta\sigma_x + \frac{1}{2}\,\varepsilon\sigma_z + \int dk\omega(k)a^+(k)a(k) + \lambda\sigma_z[A(g^*) + A^+(g)]\,, \tag{5.14.1}$$

where σ_x and σ_z are Pauli matrixes, ε and Δ are real parameters interpreted respectively as the energy difference of the states localized in the two wells in absence of tunneling and as the matrix element for tunneling between the wells. We set $\Delta > 0$ and denote

$$A^+(g) = \int a^+(k)g(k)dk\,, \qquad A(g^*) = \int a(k)g^*(k)dk\,, \tag{5.14.2}$$

where $a(k)$ and $a^+(k)$ are bosonic annihilation and creation operators

$$[a(k), a^+(k')] = \delta(k - k')$$

which describe the environment. The free reservoir Hamiltonian is, as usual, the second quantization of a function of momentum $\omega(p)$ with

$$\omega(k) \geq 0 \qquad \text{and} \qquad \omega(0) = 0\,. \tag{5.14.3}$$

From here and until the end of this chapter we shall write $\omega(k)$, instead of $\omega_1(k)$, for the 1-particle energy density of the reservoir. Therefore the time-evolved interaction Hamiltonian is, in the notation of (5.14.2),

$$H_{\mathrm{I}}(t) = \sigma_z(t)[A(e^{-it\omega(p)}g^*) + A^+(e^{it\omega(p)}g)] \tag{5.14.4}$$

$$\sigma_z(t) = e^{itH_S}\sigma_z e^{-itH_S}\,. \tag{5.14.5}$$

The free system Hamiltonian H_S is

$$H_S = -\frac{1}{2}\,\Delta\sigma_x + \frac{1}{2}\,\varepsilon\sigma_z\,, \tag{5.14.6}$$

and its eigenvalues and eigenvectors are

$$H_S|e_\pm\rangle = \lambda_\pm|e_\pm\rangle\,, \tag{5.14.7}$$

where

$$\lambda_\pm = \pm\frac{1}{2}\,\Delta\nu\,, \tag{5.14.8}$$

$$|e_\pm\rangle = \frac{1}{\sqrt{1+\mu_\mp^2}} \begin{pmatrix} 1 \\ \mu_\mp \end{pmatrix} \tag{5.14.9}$$

and

$$\mu_\pm = \frac{\varepsilon}{\Delta} \pm \nu\,, \quad \nu = \sqrt{1+\left(\frac{\varepsilon}{\Delta}\right)^2}\,; \tag{5.14.10}$$

thus the possible Bohr frequencies are

$$\omega_1 = \nu\Delta > 0 \quad , \qquad \omega_2 = 0 \quad , \qquad \omega_3 = -\nu\Delta < 0\,. \tag{5.14.11}$$

Note, for future use, that

$$\langle e_\pm|\sigma_z|e_\pm\rangle = \frac{1-\mu_\mp^2}{1+\mu_\mp^2} \quad , \qquad \langle e_+|\sigma_z|e_-\rangle = \langle e_-|\sigma_z|e_+\rangle = 1/\nu\,. \tag{5.14.12}$$

Therefore denoting

$$D := |e_+\rangle\langle e_-|\,, \tag{5.14.13}$$

we find

$$\sigma_z(t) = \frac{1-\mu_-^2}{1+\mu_-^2}\,DD^+ + \frac{1-\mu_+^2}{1+\mu_+^2}\,D^+D + \nu^{-1}e^{it\nu\Delta}\,D + \nu^{-1}e^{-it\nu\Delta}D^+\,. \tag{5.14.14}$$

The interaction Hamiltonian (5.14.4) can now be written in the canonical form (4.8.7), i.e.

$$H_{\mathrm{I}}(t) = \sum_{\alpha=1}^{3} [D_\alpha^+ \otimes A(e^{-it\omega_\alpha} g^*) + \mathrm{h.c.}]\,, \tag{5.14.15}$$

where the three spectral frequencies correspond respectively to the down, zero, and up transitions of the 2-level system, i.e.

$$\omega_1(k) = \omega(k) - \nu\Delta\;, \quad \omega_2(k) = \omega(k)\;, \quad \omega_3(k) = \omega(k) + \nu\Delta\,, \tag{5.14.16}$$

$$D_1 = \nu^{-1}D^+\;, \quad D_2 = \frac{1-\mu_-^2}{1+\mu_-^2}\,DD^+ + \frac{1-\mu_+^2}{1+\mu_+^2}\,D^+D\;, \quad D_3 = \nu^{-1}D^+\,. \tag{5.14.17}$$

The Fock spectral function has the form

$$J(\omega) = \int dk|g(k)|^2\delta(\omega(k)-\omega)\,; \tag{5.14.18}$$

therefore, since $\omega(k)$ is positive and $\omega(0) = 0$, it will vanish if ω is negative or zero.

The spectral function $G_{-+}^{(\omega)}$, here denoted simply J, evaluated at the Bohr frequencies ω_α gives the covariance of the master fields associated with these frequencies, and because of (5.14.16) and (5.14.18), they vanish for $\alpha = 2, 3$. This implies that, in the stochastic limit, there are no white noises b_2 and b_3 corresponding to the Bohr frequencies ω_2 and ω_3 in (5.14.11).

5.15 The Damping and Oscillating Regimes: Fock Case

In the same notations as in the previous section, let us introduce the notations

$$\gamma = \nu^{-2}\pi J(\nu\Delta) \quad , \qquad I(\omega) = \text{P.P.} \int_0^\infty \frac{d\omega' J(\omega')}{\omega' - \omega} \, , \tag{5.15.1}$$

$$\sigma = \nu^{-2}[I(-\nu\Delta) - I(\nu\Delta)] + \left[\left(\frac{1-\mu_-^2}{1+\mu_-^2} \right)^2 - \left(\frac{1-\mu_+^2}{1+\mu_+^2} \right)^2 \right] I(0) \, , \tag{5.15.2}$$

$$\varphi = \nu^{-2} I(-\nu\Delta) + \left(\frac{1-\mu_-^2}{1+\mu_-^2} \right)^2 I(0) \, , \tag{5.15.3}$$

where P.P. means the principal part of the integral. In this section we shall consider the stochastic limit of the Hamiltonian (5.14.4) in the Fock case. In this case the stochastic golden rule gives, for the normally ordered form, the limiting evolution equation:

$$\partial_t U(t) = Db^+(t)U(t) - D^+U(t)b(t) - (\gamma + i\sigma)D^+DU(t) - i\varphi U(t) \, , \tag{5.15.4}$$

where $b(t)$, $b^+(t)$ denotes the Fock white noise corresponding to the Bohr frequency ω_1. It satisfies

$$[b(t), b^+(t')] = \gamma\delta(t - t') \, ,$$

with γ given by (5.15.1) and ν (in γ) given by (5.14.10). Thus we see that the negative or null Bohr frequencies contribute to the drift term through the constants σ and φ. This is what we call the *Cheshire cat effect.*

All the information about the model is encoded into the constants γ, σ and φ, and these constants are expressed in terms of the spectral density $J(\omega)$ depending on the parameters of the original Hamiltonian (ε and Δ) and on the temperature (which in this section does not play a role since we are in the Fock case). On these bases let us discuss the pure damping or pure oscillating behaviour of the vacuum transition amplitude:

$$\langle U(t) \rangle = e^{-i\varphi t} + e^{-i\varphi t}(e^{-(\gamma + i\sigma)t} - 1)D^+D \, . \tag{5.15.5}$$

Taking a trace over the spin variables, one obtains (since $\operatorname{Tr} D^+D = 1$):

$$\langle \operatorname{Tr} U(t) \rangle = e^{-[\gamma + i(\sigma + \varphi)]t} \, . \tag{5.15.6}$$

Since γ, σ and φ are real [see (5.15.1–3)], the quantity (5.15.2) has a purely oscillating behaviour if and only if there is no damping, i.e.

$$\gamma = 0 \, . \tag{5.15.7}$$

However the oscillations cannot vanish, because the quantity

$$\sigma + \varphi = \nu^{-2} I(-\nu\Delta) + \left(\frac{1-\mu^2}{1+\mu^2}\right)^2 I(0) > 0 \tag{5.15.8}$$

is strictly positive for positive $J(\omega)$ [because ν, $\Delta > 0$, see (5.14.10)] and $I(\omega)$ is given by (5.15.1). The stochastic approximation to the vacuum expectation value of the Heisenberg evolution of σ_z is given by

$$P(t) = \langle U^*(t)\sigma_z(t)U(t)\rangle \,. \tag{5.15.9}$$

From (5.15.4) one obtains an equation for $P(t)$ whose solution is

$$\begin{aligned} P(t) =& \nu^{-1} e^{-\gamma t}\left(D^+ e^{i(\sigma-\nu\Delta)t} + D e^{-i(\sigma-\nu\Delta)t}\right) \\ &+ D^+ D\left(\frac{1-\mu_+^2}{1+\mu_+^2} - \frac{1-\mu_-^2}{1+\mu_-^2}\right) e^{-2\gamma t} + \frac{1-\mu_-^2}{1+\mu_-^2} \,. \end{aligned} \tag{5.15.10}$$

Let us distinguish two cases: Let $\varepsilon = 0$ (or, equivalently, $\nu = 1$). One has

$$P(t) = e^{-\gamma t}(D^+ e^{i(\sigma-\Delta)t} + D e^{-i(\sigma-\Delta)t}) \,, \tag{5.15.11}$$

where γ, σ and $I(\omega)$ are now

$$\gamma = \pi J(\Delta) \,, \quad \sigma = I(-\Delta) - I(\Delta) \,, \quad I(\omega) = \text{P.P.} \int \frac{d\omega' J(\omega')}{\omega' - \omega} \,. \tag{5.15.12}$$

Two interesting regimes can now appear:

(i) *No oscillations*, i.e. pure damping regime. This happens when

$$\sigma - \Delta = 0 \,, \tag{5.15.13}$$

i.e. J and Δ satisfy the integral equation

$$\int dx \frac{J(x)}{x+\Delta} - \text{P.P.} \int dx \frac{J(x)}{x-\Delta} = \Delta \,. \tag{5.15.14}$$

In this case, since $\mu_\pm = \pm 1$ one has

$$P(t) = e^{-\gamma t}(D^+ + D) \,. \tag{5.15.15}$$

(ii) *Pure oscillations*. This regime is defined by the condition

$$\gamma = \pi J(\Delta) = \int dk |g(k)|^2 \delta(\omega(k) - \Delta) = 0 \,. \tag{5.15.16}$$

It is clear that condition (5.15.12) defines an *off-resonance condition*. If (5.15.12) is satisfied, then

$$P(t) = D^+ e^{i(\sigma-\Delta)t} + D e^{-i(\sigma-\Delta)t} \,,$$

where

$$\sigma - \Delta = \int \frac{dx J(x)}{x+\Delta} - \text{P.P.} \int \frac{dx J(x)}{x-\Delta} - \Delta .$$

Finally let us compute the correlator

$$C(t) = \frac{1}{2} \langle \{U_t^* \sigma_z U_t, \sigma_z\} \rangle = \frac{1}{2} \{P(t), P(0)\} ,$$

where $\{ \cdot \}$ denotes anticommutator. We have

$$\begin{aligned} C(t) =& \frac{1}{2} e^{-(\gamma+i\sigma+i\nu\Delta)t} \Bigg[\nu^{-2} + \nu^{-1} D \left(\frac{1-\mu_-^2}{1+\mu_-^2} + \frac{1-\mu_+^2}{1+\mu_+^2} \right) \\ & + \text{h.c.} + \frac{1-\mu_-^2}{1+\mu_-^2} \nu^{-1}(D+D^+) + \frac{1-\mu_-^2}{1+\mu_-^2} DD^+ + \frac{1-\mu_+^2}{1+\mu_+^2} D^+ D \\ & + e^{-2\gamma} \left(\frac{1-\mu_+^2}{1+\mu_+^2} - \frac{1-\mu_-^2}{1+\mu_-^2} \right) \nu^{-1}(D+D^+) + 2D^+ D \frac{1-\mu_+^2}{1+\mu_+^2} \Bigg] . \end{aligned}$$

The trace of $C(t)$ is

$$\operatorname{Tr} C(t) = 2\nu^{-2} e^{-\gamma t} \cos(\sigma+\nu\Delta)t + 2e^{-2\gamma t} \left(\frac{1-\mu_+^2}{1+\mu_+^2} - \frac{1-\mu_-^2}{1+\mu_-^2} \right) \left(\frac{1-\mu_+^2}{1+\mu_+^2} \right) .$$

Therefore the qualitative behaviour of $C(t)$ is the same as for $P(t)$.

5.16 The Damping and the Oscillating Regimes: Nonzero Temperature

Due to the very special form of (5.14.13) of the coefficient D, which implies in particular that $DD^+ + D^+D = 1$, for a nonzero temperature, we obtain a stochastic evolution equation of the same form as (5.15.4) only with new constants γ, σ and φ. More precisely,

$$\gamma = \nu^{-2} \pi [J_+(\nu\Delta) + J_-(\nu\Delta)] , \tag{5.16.1}$$

$$\begin{aligned} \sigma =& \left[\left(\frac{1-\mu_+^2}{1+\mu_+^2} \right)^2 - \left(\frac{1-\mu_-^2}{1+\mu_-^2} \right)^2 \right] [I_+(0) + I_-(0)] \\ & + \nu^{-2} [I_+(-\nu\Delta) - I_+(\nu\Delta) + I_-(-\nu\Delta) - I_-(\nu\Delta)] , \end{aligned} \tag{5.16.2}$$

where the spectral densities $J_\pm$ are expressed in terms of the Fock spectral density (5.14.18) and the inverse temperature β as follows:

$$J_+(\omega) = \frac{J(\omega)}{1-e^{-\beta\omega}} , \quad J_-(\omega) = \frac{J(\omega) e^{-\beta\omega}}{1-e^{-\beta\omega}} . \tag{5.16.3}$$

The functions $I_\pm(\omega)$ are defined by

$$I_\pm(\omega) = \text{P.P.} \int d\omega' \frac{J_\pm(\omega')}{\omega' - \omega} . \tag{5.16.4}$$

For $P(t)$ and $C(t)$ and their expectations one has the same expressions as for zero temperature but now with new constants γ and σ depending on temperature:

$$\gamma = \nu^{-2} \pi J(\nu\Delta) \coth \frac{\beta\nu\Delta q}{2} .$$

5.17 No Rotating-Wave Approximation Without Cutoff

In the present section we begin to realize the third step of the approach described in Sect. 4.8, namely we shall consider the general multiplicative interaction Hamiltonian (4.8.1), i.e.

$$H_{\text{I}} = \int d^3k \, \{D(k) \otimes a^+(k) + D^+(k) \otimes a(k)\} . \tag{5.17.1}$$

In Sect. 4.9.3 it was explained in what sense this interaction can be considered as *cutoff-free*. We suppose that the system Hamiltonian H_S has a discrete spectrum and, as in (4.8.4), we introduce the Bohr frequencies ω_q of this Hamiltonian so that, by construction

$$\omega_q \neq \omega_q \quad \text{if} \quad q \neq p . \tag{5.17.2}$$

With these notations, the evolved interaction Hamiltonian assumes the canonical form (5.1.2), namely

$$H_{\text{I}}(t) = \int dk \sum_q D_q(k) \otimes e^{it(\omega_k - \omega_q)} a_k^+ + \text{ h.c.} , \tag{5.17.3}$$

where the system operators $D_q(k)$ are given by (4.8.6) and satisfy

$$e^{itH_S} D_q(k) e^{-itH_S} = e^{it\omega_q} D_q(k) . \tag{5.17.4}$$

The rescaled Schrödinger equation, in the interaction representation, associated to the interaction Hamiltonian (5.17.3) is

$$\partial_t U_{t/\lambda^2}^{(\lambda)} = -\frac{i}{\lambda} H_{\text{I}}(t/\lambda^2) U_{t/\lambda^2}^{(\lambda)} = -i \sum_q \left(\frac{1}{\lambda} a_q^{(\lambda)+}(t/\lambda^2) + \text{h.c.} \right) U_{t/\lambda^2} , \tag{5.17.5}$$

where $a_q^{(\lambda)}(t), a_q^{(\lambda)+}(t)$ are the rescaled operators

$$a_q^{(\lambda)}(t) := \int dk D_q(k) \otimes \frac{1}{\lambda} e^{i(t/\lambda^2)(\omega_k - \omega_q)} a_k^+ . \tag{5.17.6}$$

From Theorem 5.2.1 we know that the rescaled fields $e^{i(t/\lambda^2)(\omega_k-\omega_q)}a_k^+/\lambda$ converge to independent white noises $b_q(t,k)$ whose covariance depend on the state of the a field. Therefore the rescaled fields $a_q^+(t/\lambda^2)/\lambda$ will converge to the white noise $b_q(t)$ given by

$$b_q(t) = \int dk D_q^+(k) \otimes b_q(t,k)\,, \tag{5.17.7}$$

and (5.17.5) will converge to the limit white noise equation

$$\partial_t U_t = -i \sum_q [b_q^+(t) + b_q(t)] U_t = -i \sum_q \left(\int dk D_q(k) \otimes b_q^+(t,k) + \text{h.c.} \right) U_t\,. \tag{5.17.8}$$

The normal form of equation (5.17.10), according to the stochastic golden rule is

$$\partial_t U_t = -i \sum_q \int dk : [D_q(k) \otimes b_q^+(t,k) + \text{h.c.}]\, U_t : -YU_t\,, \tag{5.17.9}$$

where the normal order $:\ :$ is meant in the sense of (4.14.4) and Y is the operator transport coefficient,

$$Y = \int_{-\infty}^{0} dt \langle H_{\mathrm{I}}(0) H_{\mathrm{I}}(t) \rangle\,, \tag{5.17.10}$$

where $\langle \cdot \rangle$ denotes expectation with respect to the reference state of the a field and the explicit expression of this coefficient for an arbitrary mean zero Gaussian state is given in Proposition 5.4.1.

5.18 The Drift Term for Gauge-Invariant States

If the reference state of the a field is a mean zero Gaussian gauge-invariant state with covariance

$$\begin{pmatrix} \langle a_k^+ a_{k'} \rangle & 0 \\ 0 & \langle a_k a_{k'}^+ \rangle \end{pmatrix} = \begin{pmatrix} n(k) & 0 \\ 0 & m(k) \end{pmatrix} \delta(k-k')\,, \tag{5.18.1}$$

then the integral (5.17.10) is easily calculated and gives

$$\begin{aligned} Y = & \sum_q \int_{-\infty}^{0} dt \int dk |D_q(k)|^2 m(k) e^{it(\omega_k - \omega_q)} \\ & + \sum_q \int_{-\infty}^{0} dt \int dk |D_q^+(k)|^2 n(k) e^{it(\omega_k - \omega_q)}\,. \end{aligned} \tag{5.18.2}$$

Assuming that the functions $|D_q(k)|^2 = D_q^+(k)D_q(k)$ and $|D_q^+(k)|^2 = D_q(k)D_q^+(k)$ decay sufficiently rapidly for large k (in the sense of its matrix elements in some o.n. basis), we can exchange the two integrals in (5.18.2) and apply formula (4.21.1). This gives

$$Y = \pi \sum_q \int dk|D_q(k)|^2 m(k)\delta(\omega_k - \omega_q) - i\sum_q \text{P.P.} \int dk \frac{|D_q(k)|^2 m(k)}{\omega_k - \omega_q}$$
$$+ \pi \sum_q \int dk|D_q^+(k)|^2 n(k)\delta(\omega_k - \omega_q) - i\sum_q \text{P.P.} \int dk \frac{|D_q^+(k)|^2 n(k)}{\omega_k - \omega_q} . \tag{5.18.3}$$

From a mathematical point of view there is not much difference between this case and the one dealt with in Sect. 5.7. In fact the only difference is that the integral is taken over an operator function, and therefore it should be interpreted as a weak integral, i.e. as an integral over the matrix elements of the operators $|D_q(k)|^2$. However from a physical point of view the difference is considerable because it gives a microscopic interpretation of the form factor and suggests that *in the interaction of two fields one can act as the cutoff function of the other*. This fact may have important consequences in the general of the removal of cutoffs in quantum field theory.

Another important consequence of (5.18.2) is that if the reference state of the a field is a mean zero Gaussian gauge-invariant state then the Ito correction term commutes with the system Hamiltonian:

$$[Y, H_S] = 0 . \tag{5.18.4}$$

This follows immediately from (5.17.4) and (5.18.2). We shall see that as a consequence of this fact the Markov semigroups induced by the stochastic flow on the observables of the system give rise to a classical Markov (jump) process among the energy levels of the system. In fact we have a stronger property: denoting for each Bohr frequency ω_q

$$Y_q = \sum_q \int_{-\infty}^0 dt \int dk|D_q(k)|^2 m(k)e^{it(\omega_k - \omega_q)}$$
$$+ \sum_q \int_{-\infty}^0 dt \int dk|D_q^+(k)|^2 n(k)e^{it(\omega_k - \omega_q)} , \tag{5.18.5}$$

then one has

$$[Y_q, H_S] = 0 , \tag{5.18.6}$$

which means that each independent white noise $b_q^+(t, k)$ gives rise to a classical Markov (jump) process among the energy levels of the system.

5.19 The Free Evolution of the Master Field

In the present section we will extend to the case of multiple (possibly an infinite number) resonances the construction made in Sect. 4.26. By analogy with (4.26.5) we start from the normally ordered form of the white noise Hamiltonian equation satisfied by V_t^o, which is

$$\partial_t V_t^o = -i \sum_q \left[D_q b_q^+(g_q, t) V_t^o - D_q^+ V_t b_q(g_q, t) \right] - \left\{ Y + \frac{i}{\hbar} K_o \right\} V_t^o , \tag{5.19.1}$$

where K_o denotes the generator of $\tilde{S}_t$. Then we define the free evolution of the master field as not $\tilde{S}_t \otimes 1$, but

$$R_t := \tilde{S}_t \otimes e^{-it\tilde{\omega}} , \tag{5.19.2}$$

where $e^{-it\tilde{\omega}}$ is the free evolution characterized by the property

$$e^{-it\tilde{\omega}} b_q(k) e^{-it\tilde{\omega}} = e^{-it\omega_q} b_q(k) .$$

If we now define

$$V_t = (e^{itH_S} \otimes R_t) U_t , \tag{5.19.3}$$

then it is easy to verify we have a bona fide quantum dynamics.

5.20 The Stochastic Limit of the Generalized Spin–Boson Hamiltonian

In the present section we apply the stochastic golden rule to deduce the Langevin and master equations for the general model discussed in Sects. 4.2–4. Recall that the total Hamiltonian is

$$H = H_0 + \lambda H_{\rm I} = H_S + H_R + \lambda H_{\rm I} , \tag{5.20.1}$$

where H_R is the free Hamiltonian of a bosonic reservoir R:

$$H_R = \int \omega(k) a^+(k) a(k) dk ,$$

acting in the representation space $\mathcal{F}$ corresponding to the mean zero gauge-invariant Gaussian state $\langle \cdot \rangle$ with correlation function

$$\langle a(k) a^+(k') \rangle = [n(k) + 1] \delta(k - k') \tag{5.20.2}$$

$$\langle a^+(k) a(k') \rangle = n(k) \delta(k - k') , \tag{5.20.3}$$

where the function $n(k)$ is equal to the number density of bosons with the frequency k. One of the examples is the (Gaussian) bosonic equilibrium state at temperature β^{-1}, or more generally, any of the states of the form $\phi_R(X) = \operatorname{Tr}\rho X$, where ρ is a function of the number operators $a^+(k)a(k)$.

We will consider a factorizable initial state on the algebra of observables. This means that for the observable $X = X_S \otimes X_R$, where X_R is the reservoir observable and X_S is an observable of the system, the expectation $\langle \cdot \rangle$ satisfies the property $\langle X_S \otimes X_R\rangle = \langle X_S\rangle_S \langle X_R\rangle_R$. In the following if no confusion is possible we will not indicate the indices S, R of the states of the system and reservoir. The system Hamiltonian has the following spectral decomposition:

$$H_S = \sum_\sigma E(\sigma)|\sigma\rangle\langle\sigma| ,$$

where the index σ labels the eigenvectors of H_S. The interaction Hamiltonian $H_{\rm I}$ (acting in $\mathcal{H}_S \otimes \mathcal{F}$) has the form

$$H_{\rm I} = D^* \otimes A(g) + D \otimes A^+(g) , \quad A(g) = \int dk \overline{g}(k)a(k) ,$$

where $A(g)$ is a smeared quantum field with cutoff function (form factor) $g(k)$. Using the formula for the free evolution of bosonic fields

$$e^{itH_R}a(k)e^{-itH_R} = e^{-it\omega(k)}a(k) ,$$

we obtain for the free evolution of the interaction Hamiltonian,

$$\begin{aligned} H_{\rm I}(t) &= e^{itH_0} H_{\rm I} e^{-itH_0} = H_{\rm I}(t) \qquad (5.20.4)\\ &= \sum_{\sigma\sigma'} |\sigma\rangle\langle\sigma'|\langle\sigma|D^*|\sigma'\rangle \int dk\overline{g}(k)e^{-it[\omega(k)+E(\sigma')-E(\sigma)]}a(k) + \text{h.c.} \end{aligned}$$

We will suppose that the conditions of Definition 5.8.1 are satisfied.

The rescaled interaction Hamiltonian is expressed in terms of the rescaled creation and annihilation operators:

$$a_{\lambda,\sigma\sigma'}(t,k) = \frac{1}{\lambda}\, e^{-i\frac{t}{\lambda^2}\,[\omega(k)+E(\sigma')-E(\sigma)]} a(k) .$$

After the stochastic limit the sum of all rescaled annihilation operators corresponding to the same transition energy from the configuration σ' to σ generates one nontrivial quantum white noise,

$$b_{\sigma\sigma'}(t,k) = \lim_{\lambda\to 0} a_{\lambda,\sigma\sigma'}(t,k) = \lim_{\lambda\to 0} \frac{1}{\lambda}\, e^{-i\frac{t}{\lambda^2}\,[\omega(k)+E(\sigma')-E(\sigma)]} a(k) , \qquad (5.20.5)$$

with the relations: if $E(\sigma') - E(\sigma) = E(\sigma_0') - E(\sigma_0)$, then

$$\begin{aligned} [b_{\sigma\sigma'}(t,k), b^+_{\sigma_0\sigma_0'}(t',k')] &= \lim_{\lambda\to 0} [a_{\lambda,\sigma_0\sigma_0'}(t,k), a^+_{\lambda,\sigma\sigma'}(t',k')]\\ &= \lim_{\lambda\to 0} \frac{1}{\lambda^2}\, e^{-i\frac{t-t'}{\lambda^2}\,[\omega(k)+E(\sigma')-E(\sigma)]}\delta(k-k')\\ &= 2\pi\delta(t-t')\delta\big(\omega(k)+E(\sigma')-E(\sigma)\big)\delta(k-k') , \end{aligned} \qquad (5.20.6)$$

while if $E(\sigma') - E(\sigma) \neq E(\sigma_0') - E(\sigma_0)$, one has

$$[b_{\sigma\sigma'}(t,k), b^+_{\sigma_0\sigma_0'}(t',k')] = 0 .$$

Note that (5.20.5) implies that $b_{\sigma,\sigma'}(t,k)$ depends on the pair σ, σ' only through the difference $E(\sigma') - E(\sigma)$. In other words,

$$b_{\sigma,\sigma'}(t,k) = b_{\sigma_0,\sigma_0'}(t,k) \quad \text{if} \quad E(\sigma') - E(\sigma) = E(\sigma_0') - E(\sigma_0) . \tag{5.20.7}$$

The stochastic limit of the interaction Hamiltonian is therefore equal to

$$h(t) = \sum_{\sigma\sigma'} |\sigma\rangle\langle\sigma'|\langle\sigma|D^*|\sigma'\rangle \int dk\overline{g}(k) b_{\sigma\sigma'}(k,t) + \text{h.c.} \tag{5.20.8}$$

The state on the master field (white noise) $b_{\sigma\sigma'}(t,k)$, corresponding to the state (5.20.2–3) of the field, is the mean zero gauge-invariant Gaussian state with correlations

$$\langle b^+_{\sigma\sigma'}(t,k) b_{\sigma\sigma'}(t',k')\rangle = 2\pi\delta(t-t')\delta\big(\omega(k) + E(\sigma') - E(\sigma)\big)\delta(k-k')n(k)$$

$$\langle b_{\sigma\sigma'}(t,k) b^+_{\sigma\sigma'}(t',k')\rangle = 2\pi\delta(t-t')\delta\big(\omega(k) + E(\sigma') - E(\sigma)\big)\delta(k-k')(n(k)+1)$$

and vanishes for noises corresponding to transitions with different $\Delta(E) = E(\sigma) - E(\sigma')$.

Now let us investigate the evolution equation in the interaction picture for our model. According to the general scheme of the stochastic limit, we obtain the (singular) white noise equation

$$\frac{d}{dt} U_t = -ih(t) U_t . \tag{5.20.9}$$

According to the stochastic golden rule of Sect. 5.9, the normally ordered form of (5.20.9) is the quantum stochastic differential equation

$$dU_t = [-idH(t) - Gdt] U_t , \tag{5.20.10}$$

where $h(t)$ is the white noise Hamiltonian [AcKo00b], given by the stochastic limit of the interaction Hamiltonian, and

$$dH(t) = \sum_{\sigma\sigma'} \big(|\sigma\rangle\langle\sigma'|\langle\sigma'|D|\sigma\rangle^* dB_{\sigma\sigma'}(t) + |\sigma'\rangle\langle\sigma|\langle\sigma'|D|\sigma\rangle dB^+_{\sigma\sigma'}(t)\big) ; \tag{5.20.11}$$

the stochastic differentials $dB_{\sigma\sigma'}(t)$ are defined by the prescription

$$dB_{\sigma\sigma'}(t) = \int dk\overline{g}(k) \int_t^{t+dt} b_{\sigma\sigma'}(\tau,k) d\tau = \lim_{\varepsilon\to 0+} \int dk , \tag{5.20.12}$$

which implies in particular that $[F_t, dB_{\sigma\sigma'}(t)] = 0$ for any F_t which depends on the noise before t (adapted). According to the stochastic golden rule,

(5.14.4) is obtained as follows: the first term in (5.14.4) is just the limit of the first term of the iterated series solution of (5.20.1),

$$\lim_{\lambda\to 0}\frac{1}{\lambda}\int_t^{t+dt} H_{\mathrm{I}}\left(\frac{\tau}{\lambda^2}\right) d\tau \,.$$

The second term Gdt, the drift, is equal to the limit of the expectation value in the reservoir state of the second term in the iterated series solution for (5.20.1)

$$\lim_{\lambda\to 0}\frac{1}{\lambda^2}\int_t^{t+dt} dt_1 \int_t^{t_1} dt_2 \left\langle H_{\mathrm{I}}\left(\frac{t_1}{\lambda^2}\right) H_{\mathrm{I}}\left(\frac{t_2}{\lambda^2}\right)\right\rangle .$$

Making the change of variables $\tau = t_2 - t_1$, we obtain

$$\lim_{\lambda\to 0}\frac{1}{\lambda^2}\int_t^{t+dt} dt_1 \int_{t-t_1}^{0} d\tau \left\langle H_{\mathrm{I}}\left(\frac{t_1}{\lambda^2}\right) H_{\mathrm{I}}\left(\frac{t_1}{\lambda^2}+\frac{\tau}{\lambda^2}\right)\right\rangle . \tag{5.20.13}$$

Computing the expectation value and using the fact that the limits of oscillating factors of the form $\lim_{\lambda\to 0} e^{(ict_1/\lambda^2)}$ vanish unless the constant c is equal to zero, we see that we can have a nonzero limit only when all oscillating factors of the type $e^{(ict_1/\lambda^2)}$ (with t_1) in (5.20.13) cancel. The third step of the stochastic golden rule in Sect. 5.9 gives

$$G = \sum_{\sigma\sigma'} |\langle\sigma'|D|\sigma\rangle|^2 \int dk |g(k)|^2 \int_{-\infty}^{0} d\tau \Big(|\sigma\rangle\langle\sigma| e^{i\tau[\omega(k)+E(\sigma')-E(\sigma)]}[n(k)+1] \\ + |\sigma'\rangle\langle\sigma'| e^{-i\tau[\omega(k)+E(\sigma')-E(\sigma)]} n(k)\Big) .$$

Therefore, from

$$\int_{-\infty}^{t} e^{it\omega} = \frac{-i}{\omega - i0} = \pi\delta(\omega) - i\mathrm{P.P.}\frac{1}{\omega} , \tag{5.20.14}$$

we obtain the following expression for the drift G:

$$\begin{aligned} &\sum_{\sigma\sigma'} |\langle\sigma'|D|\sigma\rangle|^2 \int dk |g(k)|^2 \left(\frac{-i|\sigma\rangle\langle\sigma|(n(k)+1)}{\omega(k)+E(\sigma')-E(\sigma)-i0} \right. \\ &\qquad\qquad \left. + \frac{i|\sigma'\rangle\langle\sigma'|n(k)}{\omega(k)+E(\sigma')-E(\sigma)+i0} \right) \\ &= \sum_{\sigma\sigma'} |\langle\sigma'|D|\sigma\rangle^2 \left(|\sigma\rangle\langle\sigma| (g|g)^-_{\sigma\sigma'} + |\sigma'\rangle\langle\sigma'| \overline{(g|g)^+_{\sigma\sigma'}} \right) . \end{aligned} \tag{5.20.15}$$

5.21 The Langevin Equation

Now we will find the Langevin equation following the general scheme of Sect. 5.10. Let X be an observable of the system S and let the stochastic flow j_t be defined by

$$j_t(X) = U_t^* X U_t \,, \tag{5.21.1}$$

where U_t satisfies (5.20.10–12).

Applying the general argument of Sect. 5.10 with all g_j equal to a single function g and $\omega = (\sigma, \sigma')$, we obtain the Langevin equation

$$\begin{aligned} dj_t(X) &= \sum_\alpha j_t \circ \theta_\alpha(X) dM^\alpha(t) \\ &= \sum_{n=-1,1;\sigma\sigma'} j_t \circ \theta_{n\sigma\sigma'}(X) dM^{n\sigma\sigma'}(t) + j_t \circ \theta_0(X) dt \,, \end{aligned} \tag{5.21.2}$$

where

$$dM^{-1,\sigma\sigma'}(t) = dB_{\sigma\sigma'}(t), \quad \theta_{-1,\sigma\sigma'}(X) = i\langle\sigma'|D|\sigma\rangle^*[X, |\sigma\rangle\langle\sigma'|] \tag{5.21.3}$$

$$dM^{1,\sigma\sigma'}(t) = dB^+_{\sigma\sigma'}(t), \quad \theta_{1,\sigma\sigma'}(X) = i\langle\sigma'|D|\sigma\rangle[X, |\sigma'\rangle\langle\sigma|] \tag{5.21.4}$$

$$\begin{aligned} \theta_0(X) &= \sum_{\sigma\sigma'} \theta^{(0,-1)}_{0,\sigma\sigma'} + \theta^{(0,1)}_{0,\sigma\sigma'} + \theta^{(-1)}_{0,\sigma\sigma'} + \theta^{(1)}_{0,\sigma\sigma'}(X) \\ &= \sum_{\sigma\sigma'} |\langle\sigma'|D|\sigma\rangle|^2 \Big[-i\mathrm{Im}\,(g|g)^-_{\sigma\sigma'}[X, |\sigma\rangle\langle\sigma|] + i\mathrm{Im}\,(g|g)^+_{\sigma\sigma'}[X, |\sigma'\rangle\langle\sigma'|] \\ &\quad + 2\mathrm{Re}\,(g|g)^-_{\sigma\sigma'} \left(|\sigma\rangle\langle\sigma|\langle\sigma'|X|\sigma'\rangle - \frac{1}{2}\{X, |\sigma\rangle\langle\sigma|\} \right) \\ &\quad + 2\mathrm{Re}\,(g|g)^+_{\sigma\sigma'} \left(|\sigma'\rangle\langle\sigma'|\langle\sigma|X|\sigma\rangle - \frac{1}{2}\{X, |\sigma'\rangle\langle\sigma'|\} \right) \Big] \end{aligned} \tag{5.21.5}$$

is a quantum Markovian generator as described in Sect. 1.13. Note that the factors $\mathrm{Re}\,(g|g)^\pm_{\sigma\sigma'}$ are > 0 only for $E(\sigma) > E(\sigma')$ and vanish otherwise.

5.22 Convergence to Equilibrium: Connections with Quantum Measurement

In this section we will investigate the master equation for the density matrix ρ. We will show that if the reservoir is in the equilibrium state at temperature β^{-1} then for the system with decoherence the solution of the master equation $\rho(t)$ with $t \to \infty$ tends to the classical Gibbs state with the same temperature β^{-1}. This phenomenon realizes the quantum measurement procedure in the sense that the quantum state (density matrix) collapses into a classical state. To show this we use the control of quantum decoherence that is allowed by the stochastic approximation of quantum theory.

Let us consider the evolution for a system state given by a density matrix ρ, which is a positive normalized linear functional $\rho(X) = \mathrm{Tr}\,\rho X$ on observables (we will identify the state and the corresponding density matrix). The evolution of the state is defined by the dual of the flow j_t given by (5.21.1) (see also Sect. 1.19), i.e.

$$\rho_t = j_t^*(\rho) = \rho \circ j_t \,. \tag{5.22.1}$$

Therefore from (5.21.2) we get the dual Langevin equation

$$d\rho_t(X) = \rho \circ dj_t(X) = \rho \circ \sum_\alpha j_t \circ \theta_\alpha(X) dM^\alpha(t) = \sum_\alpha \rho_t \left[\theta_\alpha(X) dM^\alpha(t)\right] .$$

Taking partial expectation with respect to the reference state of the reservoir, only the stochastic differential dt in this formula will survive, and we obtain the master equation

$$\frac{d}{dt}\rho_t(X) = \rho_t \circ \theta_0(X) \,. \tag{5.22.2}$$

Let us consider the density matrix $\rho = \rho_S \otimes \rho_R$ and investigate the master equation (5.22.2). We begin with the nondiagonal element

$$\rho_S = |\mu\rangle\langle\nu|, \qquad \mu \neq \nu \,.$$

Using (5.22.2) and calculating $\rho_t \circ \theta_0$ for $t = 0$ we obtain

$$A|\mu\rangle\langle\nu| \,,$$

where the number A is given by the following:

$$\begin{aligned} A = \sum_\sigma \Big(& - i\mathrm{Im}\,(g|g)^-_{\mu\sigma} |\langle\sigma|D|\mu\rangle|^2 + i\mathrm{Im}\,(g|g)^-_{\nu\sigma} |\langle\sigma|D|\nu\rangle|^2 \\ & + i\mathrm{Im}\,(g|g)^+_{\sigma\mu} |\langle\mu|D|\sigma\rangle|^2 - i\mathrm{Im}\,(g|g)^+_{\sigma\nu} |\langle\nu|D|\sigma\rangle|^2 \\ & - \mathrm{Re}\,(g|g)^-_{\mu\sigma} |\langle\sigma|D|\mu\rangle|^2 - \mathrm{Re}\,(g|g)^-_{\nu\sigma} |\langle\sigma|D|\nu\rangle|^2 \\ & - \mathrm{Re}\,(g|g)^+_{\sigma\mu} |\langle\mu|D|\sigma\rangle|^2 - \mathrm{Re}\,(g|g)^+_{\sigma\nu} |\langle\nu|D|\sigma\rangle|^2 \Big) \end{aligned} \tag{5.22.3}$$

and therefore

$$\rho_t = \exp(At)|\mu\rangle\langle\nu| \,.$$

We see that if any of $\mathrm{Re}\,(g|g)^\pm_{\alpha\beta} |\langle\beta|D|\alpha\rangle|^2$ in (5.22.3) is nonzero then the corresponding nondiagonal matrix element of the density matrix decays exponentially. In this case we say that the system *has decoherence*. This produces a diagonalization of the density matrix. We emphasize that our criteria for the quantum decoherence use in a crucial way the stochastic limit of quantum theory: the system has the decoherence property if corresponding *generalized susceptibility coefficients* $\mathrm{Re}\,(g|g)^\pm$ have a nonzero real part.

Consider now $\rho = \rho_S \otimes \rho_R$, the system density matrix $\rho_S \in \mathcal{C}$, where $\mathcal{C}$ is the algebra of spectral projections of the system Hamiltonian H_S, and consider (5.22.2). The evolution defined by this master equation will map the algebra $\mathcal{C}$ into itself and therefore will be a classical evolution. We will show that this classical evolution in fact describes quantum phenomena. To this goal we define the evolved density matrix of the system:

$$\rho_{S,t} = \sum_{\sigma} \rho(\sigma,t)|\sigma\rangle\langle\sigma| ,$$

where $\rho_{S,t} \in \mathcal{C}$. Let X be a system observable. Then (5.22.2) takes the form

$$\sum_{\sigma} \frac{d}{dt}\rho(\sigma,t)\langle\sigma|X|\sigma\rangle$$

$$= \sum_{\sigma\sigma'} |\langle\sigma'|D|\sigma\rangle|^2 \Big[\big(\rho(\sigma,t)2\mathrm{Re}\,(g|g)^-_{\sigma\sigma'} - \rho(\sigma',t)2\mathrm{Re}\,(g|g)^+_{\sigma\sigma'}\big) \cdot \big(\langle\sigma'|X|\sigma'\rangle - \langle\sigma|X|\sigma\rangle\big)\Big]$$

$$= \sum_{\sigma\sigma'} \langle\sigma|X|\sigma\rangle \Big(\rho(\sigma',t)\left(2\mathrm{Re}\,(g|g)^-_{\sigma'\sigma}|\langle\sigma|D|\sigma'\rangle|^2 + 2\mathrm{Re}\,(g|g)^+_{\sigma\sigma'}|\langle\sigma'|D|\sigma\rangle|^2\right) - \rho(\sigma,t)\left(2\mathrm{Re}\,(g|g)^+_{\sigma'\sigma}|\langle\sigma|D|\sigma'\rangle|^2 + 2\mathrm{Re}\,(g|g)^-_{\sigma\sigma'}|\langle\sigma'|D|\sigma\rangle|^2\right)\Big) .$$

Because the observable X is arbitrary, we obtain finally for the master equation for ρ

$$\frac{d}{dt}\rho(\sigma,t) = \sum_{\sigma'} \Big(\rho(\sigma',t)\left(2\mathrm{Re}\,(g|g)^-_{\sigma'\sigma}|\langle\sigma|D|\sigma'\rangle|^2 + 2\mathrm{Re}\,(g|g)^+_{\sigma\sigma'}|\langle\sigma'|D|\sigma\rangle|^2\right) - \rho(\sigma,t)\left(2\mathrm{Re}\,(g|g)^+_{\sigma'\sigma}|\langle\sigma|D|\sigma'\rangle|^2 + 2\mathrm{Re}\,(g|g)^-_{\sigma\sigma'}|\langle\sigma'|D|\sigma\rangle|^2\right)\Big) . \tag{5.22.4}$$

Let us note that if $\rho(\sigma,t)$ satisfies the *detailed balance* condition

$$\rho(\sigma,t)2\mathrm{Re}\,(g|g)^-_{\sigma\sigma'} = \rho(\sigma',t)2\mathrm{Re}\,(g|g)^+_{\sigma\sigma'} , \tag{5.22.5}$$

which implies that $\rho(\sigma,t)$ is a stationary solution for (5.22.4) (in fact one can prove the uniqueness of this solution).

Let us investigate (5.22.5) for the equlibrium state of the field. In this case

$$\begin{aligned} 2\mathrm{Re}\,(g|g)^-_{\sigma\sigma'} &= 2\pi \int dk\, |g(k)|^2 \delta\big(\omega(k) + E(\sigma') - E(\sigma)\big) \frac{1}{1 - e^{-\beta\omega(k)}} \\ &= 2\pi \int dk\, |g(k)|^2 \delta\big(\omega(k) + E(\sigma') - E(\sigma)\big) \frac{1}{1 - e^{-\beta[E(\sigma)-E(\sigma')]}} \\ &= \frac{C}{1 - e^{-\beta[E(\sigma)-E(\sigma')]}} , \end{aligned}$$

$$2\mathrm{Re}\,(g|g)^{+}_{\sigma_r^x\sigma,\sigma} = \frac{C}{e^{\beta[E(\sigma)-E(\sigma')]}-1}\,.$$

From (5.22.5) we obtain

$$\frac{\rho(\sigma,t)}{1-e^{-\beta[E(\sigma)-E(\sigma')]}} = \frac{\rho(\sigma',t)}{e^{\beta[E(\sigma)-E(\sigma')]}-1}$$

or

$$\rho(\sigma,t)e^{\beta E(\sigma)} = \rho(\sigma',t)e^{\beta E(\sigma')}\,.$$

This means that the stationary solution (5.22.5) of (5.22.4) describes the equlibrium state of the system

$$\rho(\sigma,t) = \frac{e^{-\beta E(\sigma)}}{\sum_{\sigma'} e^{-\beta E(\sigma')}}\,.$$

In summary, for a system with decoherence, the density matrix will tend for $t\to\infty$ to the stationary solution (5.22.5) of (5.22.4), which is the Gibbs distribution. It is natural to identify this phenomenon with the quantum measurement procedure. Several phenomenological models have been proposed for this phenomenon, but in the stochastic limit everything is deduced from the dynamics.

5.23 Control of Coherence

In this section we discuss different regimes of qualitative behaviour for the considered model generalizing the considerations in Sects. 5.15 and 5.16.

The master equation (5.22.4) at first sight looks completely classical. In Sect. 5.17 we derived this equation using quantum arguments. Now we will show that (5.22.4) in fact describes a quantum behaviour. To show this we consider the following example: Let us rewrite (5.22.4) using the particular form (5.14.8) of $(g|g)^{\pm}$. Using (5.20.14) and (5.14.8) we obtain

$$\begin{aligned}\frac{d}{dt}\rho(\sigma,t) =& \sum_{\sigma'} 2\pi \int dk\,|g(k)|^2 \Big((\rho(\sigma',t)-\rho(\sigma,t)) \\ & n(k)\left[\delta\big(\omega(k)+E(\sigma)-E(\sigma')\big)|\langle\sigma|D|\sigma'\rangle| \right.\\ & \left. +\,\delta\big(\omega(k)+E(\sigma')-E(\sigma)\big)|\langle\sigma'|D|\sigma\rangle|^2\right] \\ & +\rho(\sigma',t)\delta\big(\omega(k)+E(\sigma)-E(\sigma')\big)|\langle\sigma|D|\sigma'\rangle|^2 \\ & -\rho(\sigma,t)\delta\big(\omega(k)+E(\sigma')-E(\sigma)\big)|\langle\sigma'|D|\sigma\rangle|^2\Big)\,.\end{aligned} \tag{5.23.1}$$

Thus we see that (5.14.7) describes a macroscopic quantum effect. To show this let us take the spectrum of the system Hamiltonian (the set of system states $\Sigma=\{\sigma\}$) as follows: let Σ contain two groups Σ_1 and Σ_2 of states with an energy gap between these groups [or, for simplicity, two states σ_1 and σ_2 with $E(\sigma_2) > E(\sigma_1)$]. Let the state $\langle \cdot \rangle$ of the bosonic field be taken in such a way that the density $n(k)$ of number of quanta of bosonic field is supported in those k satisfying

$$0 < \omega(k) < \omega_0 < |E(\sigma_1) - E(\sigma_2)|, \qquad k \in \operatorname{Supp} n(k) . \tag{5.23.2}$$

This means that highly energetic bosons are absent. It is natural to consider the state $\langle \cdot \rangle$ as a sum of an equilibrium state of the temperature β^{-1} and a nonequilibrium part. Therefore the density $n(k)$ will be nonzero for small k because the equilibrium state satisfies this property.

Under the considered assumption (5.23.2), the integral of the δ-function $\delta\big(\omega(k) + E(\sigma_1) - E(\sigma_2)\big)$ with $n(k)$ in (5.14.7) identically equal to zero. Therefore the right-hand side of (5.14.7) will be equal to

$$\sum_{\sigma'} 2\pi \int dk\, |g(k)|^2 \Big(\rho(\sigma', t)\delta\big(\omega(k) + E(\sigma) - E(\sigma')\big) |\langle\sigma|D|\sigma'\rangle|^2$$
$$- \rho(\sigma, t)\delta\big(\omega(k) + E(\sigma') - E(\sigma)\big) |\langle\sigma'|D|\sigma\rangle|^2 \Big) .$$

It is natural to consider this value as small with respect to the corresponding integral with $n(k)$ and to identify it with the spontaneous emission of bosons by the system. In this case the density matrix $\rho(\sigma, t)$ will be almost constant in time. We have found an effect of conservation of quantum coherence: in the absence of bosons with the energy $\omega(k)$ equal to $E(\sigma_1) - E(\sigma_2)$ the system cannot jump between the states σ_1 and σ_2 (or, at least, this transition is very slow), because in the stochastic limit such a jump corresponds to a quantum white noise that must be on a mass shell.

At the same time, the transitions between states inside the groups Σ_1 and Σ_2 are not forbidden by (5.23.2), because these transitions are connected with the soft bosons (with small k) that are present in the equilibrium part of $\langle \cdot \rangle$. We find that under the above assumptions (5.14.7) describes the transition of a system to intermediate equilibrium, where the transitions between groups of states Σ_1 and Σ_2 are forbidden.

If the state $\langle \cdot \rangle$ does not satisfy the property (5.23.2), then the system undergoes fast transitions between states σ_1 and σ_2. We can switch on such transitions by switching on the bosons with frequency $\omega(k) = E(\sigma_2) - E(\sigma_1)$. Therefore it is natural to identify the integrals with $n(k)$ on the right-hand side of (5.14.7) with the induced radiation and absorption of bosons.

We see that (5.14.7) describes a macroscopic quantum effect controlled by the distribution of bosons $n(k)$ which is physically controlled, for example, by filtering.

5.24 Dynamics of Spin Systems

Starting with the work of Glauber [Glau63] the dynamics of infinite classical lattice systems has been considered by many authors and has led to the study of the ergodic and equlibrium properties of a new class of classical Markov semigroups (see [Lig85] for a general survey and for further references). Quantum analogues of these semigroups have also been considered by several authors (e.g. [MaBu], [Mats93], [MajZeg96a], [Spitz71], [Sull75]). However the problem of deriving these Markovian semigroups, and more generally the stochastic flows, as limits of Hamiltonian systems, was open both in the classical and in the quantum case. On the other hand the stochastic golden rules, provide a natural tool with which to associate a stochastic flow, driven by a white noise equation, with any discrete system interacting with a quantum field. Moreover another quite general result of the stochastic limit is that the Markov semigroup, canonically associated with the flow, leaves invariant the abelian subalgebra generated by the spectral projections of the Hamiltonian of the discrete system, so that the associated Markov process describes the jumps among these energy levels.

Since a quantum spin system in a finite volume is obviously a discrete system, the above results naturally suggest the conjecture that by coupling such a system to a quantum field via a suitable interaction, applying the stochastic golden rule and taking the thermodynamic limit one might obtain a class of quantum flows which, when restricted to an appropriate abelian subalgebra, gives rise to the classical interacting particle systems studied in classical statistical mechanics.

In the present section we consider the open Ising model, describing the interaction of a system S of spins (or, more generally, 2-level systems) with a reservoir, represented by a bosonic quantum field. The total Hamiltonian is

$$H = H_0 + \lambda H_{\mathrm{I}} = H_S + H_R + \lambda H_{\mathrm{I}} . \tag{5.24.1}$$

The spin variables are labelled by the lattice Z^d, and for each finite subset $\Lambda \subseteq Z^d$, the system Hilbert space is

$$\mathcal{H}_S = \mathcal{H}_\Lambda = \otimes_{r\in\Lambda} C^2 \tag{5.24.2}$$

and the system Hamiltonian has the form

$$H_S = H_\Lambda = -\frac{1}{2} \sum_{r,s\in\Lambda} J_{rs}\sigma_r^z\sigma_s^z , \tag{5.24.3}$$

where σ_r^x, σ_r^y, σ_r^z are Pauli matrices ($r \in \Lambda$) at the rth site in the tensor product. For any r, $s \in \Lambda$,

$$J_{rs} = J_{sr} \in R, \qquad J_{rr} = 0 . \tag{5.24.4}$$

In the present chapter we consider a system Hamiltonian that describes the interaction of spin with a finite number of other spins (finite range potential). The simplest example is the nearest-neighbour interaction (Ising model), discussed below. The interaction Hamiltonian H_{I} (acting in $\mathcal{H}_S \otimes \mathcal{F}$) has the form

$$H_{\mathrm{I}} = \sum_{r\in\Lambda} \sigma_r^x \otimes \psi(g_r), \quad \psi(g) = A(g) + A^*(g), \quad A(g) = \int dk\, \overline{g}(k) a(k), \tag{5.24.5}$$

where ψ is a field operator, $A(g)$ is a smeared quantum field with cutoff function (form factor) $g(k)$. We denote $\Sigma(r)$ the (finite) set of spins interacting with the spin r [so $r \neq \Sigma(r)$]. When no confusion is possible, we use the same symbol $\Sigma(r)$ to also denote the set of configurations of corresponding spins: $\sigma(r) = \{\varepsilon_s\}\big(s \in \Sigma(r)\big)$.

We denote

$$E\big(r, \sigma(r)\big) = \sum_{s\in\Sigma(r)} J_{rs}\varepsilon_s = H_S\big(\sigma_r^z = -1, \sigma(r)\big) - H_S\big(\sigma_r^z = -1, \sigma(r)\big) \tag{5.24.6}$$

the energy difference between two configurations with all spins fixed, and given by $\sigma(r)$, with the exception of ε_r, which changes value from -1 to 1. With these notations the free evolution of the interaction Hamiltonian takes the form

$$H_{\mathrm{I}}(t) = \sum_{r\in\Lambda} \sum_{\sigma(r)\in\Sigma(r)} \int dk\, \overline{g}(k) F^*_{r,\sigma(r)} e^{-it\left[\omega(k) - E\left(r,\sigma(r)\right)\right]} a(k) + \text{h.c.}\,, \tag{5.24.7}$$

where we denote

$$G_{r,\sigma(r)} = D_r \prod_{s\in\Sigma(r)} |\varepsilon_s\rangle\langle\varepsilon_s|\,, \quad F_{r,\sigma(r)} = G^*_{r,-\sigma(r)} + G_{r,\sigma(r)}\,. \tag{5.24.8}$$

To prove (5.24.7) we used a suitable rearrangement of the indices $\sigma(r)$ and the property

$$E\big(r, -\sigma(r)\big) = -E\big(r, \sigma(r)\big)\,,$$

where $-\sigma(r)$ is the configuration of spins with all spins opposite to the spins in $\sigma(r)$, i.e.

$$\varepsilon_s\big(-\sigma(r)\big) = -\varepsilon_s\big(\sigma(r)\big)\,.$$

The operator $F_{r,\sigma(r)}$ flips the spin at site r (or kills a spin configuration).

Let us denote $R \in \mathcal{S}$ the pair of indices $\big(r, \sigma(r)\big)$, where r takes values in Λ. In these notations the free evolution of the interaction Hamiltonian (5.24.7) takes the form

$$H_{\mathrm{I}}(t)=\sum_{R\in\mathcal{S}}\int dk\,\overline{g}(k)F_R^*e^{-it[\omega(k)-E(R)]}a(k)+\text{h.c.} \tag{5.24.9}$$

The drift term is equal to

$$G=\sum_{R\in\mathcal{S}}\int dk\,|g(k)|^2\int_{-\infty}^{0}d\tau\Big(F_R^*F_Re^{i\tau[\omega(k)-E(R)]}\frac{1}{1-e^{-\beta(\omega(k)-\mu)}}$$
$$+F_RF_R^*e^{-i\tau[\omega(k)-E(R)]}\frac{1}{e^{\beta(\omega(k)-\mu)}-1}\Big),$$

and therefore, from

$$\int_{-\infty}^{0}e^{it\omega}=\frac{-i}{\omega-i0}=\pi\delta(\omega)-i\mathrm{P.P.}\frac{1}{\omega}, \tag{5.24.10}$$

we obtain the following expression for the drift G:

$$\sum_{R\in\mathcal{S}}\int dk\,|g(k)|^2\Big(\frac{-iF_R^*F_R}{\omega(k)-E(R)-i0}\frac{1}{1-e^{-\beta[\omega(k)-\mu]}}$$
$$+\frac{iF_RF_R^*}{\omega(k)-E(R)+i0}\frac{1}{e^{\beta[\omega(k)-\mu]}-1}\Big)$$
$$=\sum_{R\in\mathcal{S}}\Big(F_R^*F_R(g|g)_R^-+F_RF_R^*\overline{(g|g)_R^+}\Big). \tag{5.24.11}$$

The drift term contains sums over R which are divergent for large Λ. Therefore in the Schrodinger picture there will be a divergence in the thermodynamic limit. In the present section we will consider the evolution in the Heisenberg picture (the Langevin equation), and we will see that in this context no divergence arises. The only nonvanishing products for quantum stochastic differentials are

$$dB_R(t)dB_R^*(t)=2\mathrm{Re}\,(g|g)_R^-dt\,,\quad dB_R^*(t)dB_R(t)=2\mathrm{Re}\,(g|g)_R^+dt. \tag{5.24.12}$$

The Langevin equation has the form

$$dj_t(X)=\sum_\alpha j_t\circ\theta_\alpha(X)dM^\alpha(t)$$
$$=\sum_{n=-1,1;R\in\mathcal{S}_+}j_t\circ\theta_{nR}(X)dM^{nR}(t)+j_t\circ\theta_0(X)dt\,, \tag{5.24.13}$$

where

$$dM^{-1,R}(t)=dB_R(t),\quad \theta_{-1,R}(X)=-i[X,F_R^*],\quad R\in\mathcal{S}_+ \tag{5.24.14}$$

$$dM^{1,R}(t)=dB_R^*(t),\quad \theta_{1,R}(X)=-i[X,F_R],\quad R\in\mathcal{S}_+ \tag{5.24.15}$$

$$\begin{aligned}\theta_0(X) &= \sum_{R\in\mathcal{S}} \left(\theta_{0,R}^{(0,-1)} + \theta_{0,R}^{(0,1)} + \theta_{0,R}^{(-1)} + \theta_{0,R}^{(1)}\right)(X) \\ &= \sum_{R\in\mathcal{S}} \Big(-i\mathrm{Im}\,(g|g)_R^-[X, F_R^* F_R] + i\mathrm{Im}\,(g|g)_R^+[X, F_R F_R^*] \\ &\quad + \mathrm{Re}\,(g|g)_R^-(2F_R^* X F_R - \{X, F_R^* F_R\}) \\ &\quad + \mathrm{Re}\,(g|g)_R^+(2F_R X F_R^* - \{X, F_R F_R^*\})\Big)\end{aligned} \tag{5.24.16}$$

is a quantum Markovian generator. Note that the factors $\mathrm{Re}\,(g|g)_R^{\pm}$ are >0 only for $R \in \mathcal{S}_+$ and vanish for $R \in \mathcal{S}_-$.

In the present section we apply the master equation (5.22.4) to derivation of the (classical) Glauber dynamics. Glauber dynamics is the dynamics of a one-dimensional spin lattice with nearest-neighbour interaction (see [Glau63]). We will prove that the Glauber dynamics can be considered to be a dynamics generated by the master equation of the type (5.22.4) derived from a stochastic limit for a quantum spin system interacting with the bosonic quantum field.

We take the bosonic reservoir space $\mathcal{F}$ corresponding to the bosonic equilibrium state at temperature β^{-1}. Thus the reservoir state is Gaussian with mean zero and correlations given by

$$\langle a^*(k)a(k')\rangle = \frac{1}{e^{\beta\omega(k)} - 1}\delta(k - k').$$

We consider the system to be one-dimensional spin lattice (for the general case see [AcKo00b]). We consider a system Hamiltonian that describes the interaction of spin with the nearest neighbours (Ising model).

The eigenvectors $|\sigma\rangle$ of the system Hamiltonian H_S can be identified with the spin configurations (sequences of ± 1), which label the natural basis in $\mathcal{H}_S$ consisting of tensor products of eigenvectors of σ_r^z (spin-up and spin-down vectors $|\varepsilon_r\rangle$, corresponding to eigenvalues $\varepsilon_r = \pm 1$). We define the action of operator σ_r^x on spin configuration σ using the action of σ_r^x on corresponding eigenvector $|\sigma\rangle$: the operator σ_r^x flips the spin at the rth site in the sequence σ (maps $|\varepsilon_r\rangle$ to $|-\varepsilon_r\rangle$). From the form of H_S and H_I, it follows that for each two eigenvectors corresponding to spin configurations σ, σ' the matrix element $\langle\sigma|D|\sigma'\rangle$ from (5.22.4) will be nonzero only if the configurations σ, σ' differ exactly at one site. If configurations σ, σ' differ exactly at one site then $\langle\sigma|D|\sigma'\rangle = 1$. Therefore from the master equation (5.17.3) we obtain

$$\begin{aligned}\frac{d}{dt}\rho(\sigma,t) = \sum_{r\in\Lambda} \Big[&\rho(\sigma_r^x\sigma, t)\left(2\mathrm{Re}\,(g|g)^-_{\sigma_r^x\sigma,\sigma} + 2\mathrm{Re}\,(g|g)^+_{\sigma,\sigma_r^x\sigma}\right) \\ &- \rho(\sigma,t)\left(2\mathrm{Re}\,(g|g)^+_{\sigma_r^x\sigma,\sigma} + 2\mathrm{Re}\,(g|g)^-_{\sigma,\sigma_r^x\sigma}\right)\Big],\end{aligned} \tag{5.24.17}$$

which gives the Glauber dynamics of the system of spins (see [Glau63]). Here

$$2\mathrm{Re}\,(g|g)^-_{\sigma,\sigma^x_r\sigma} = 2\pi\int dk\,|g(k)|^2\delta\big(\omega(k)-J_{r-1,r}\varepsilon_{r-1}-J_{r,r+1}\varepsilon_{r+1}\big)\frac{1}{1-e^{-\beta\omega(k)}}\,; \tag{5.24.18}$$

analogous representations are valid for all the other $(g|g)^\pm$. Let us consider for simplicity the case of a one-dimensional system with a translationally invariant Hamiltonian:

$$J_{rs} = J_{r,r+1} = J > 0\,.$$

Then $(g|g)^\pm$ given by (5.24.18) are nonzero only if $\varepsilon_{r-1} = \varepsilon_{r+1} = 1$; we obtain for (5.24.18)

$$2\mathrm{Re}\,(g|g)^-_{\sigma,\sigma^x_r\sigma} = 2\pi\int dk\,|g(k)|^2\delta(\omega(k)-2J)\frac{1}{1-e^{-2\beta J}} = \frac{C}{1-e^{-2\beta J}}\,.$$

Therefore for (5.24.17) we obtain the following equation for the Glauber dynamics for a translationally invariant Hamiltonian (compare with [Glau63]):

$$\frac{d}{dt}\rho(\sigma,t) = \frac{C}{1-e^{-2\beta J}}\Bigg(\sum_{r\in\Lambda;E(\sigma)>E(\sigma^x_r\sigma)}\left[e^{-2\beta J}\rho(\sigma^x_r\sigma,t)-\rho(\sigma,t)\right] \tag{5.24.19}$$
$$+\sum_{r\in\Lambda;E(\sigma)<E(\sigma^x_r\sigma)}\left[\rho(\sigma^x_r\sigma,t)-e^{-2\beta J}\rho(\sigma,t)\right]\Bigg)\,.$$

The detailed balance stationary solution of (5.24.19) satisfy the following: for two spin configurations σ, $\sigma^x_r\sigma$ that differ by the flip of spin at site r, the energy of the corresponding configurations differ by $2J$. The expectation $\rho(\mu)$, $\mu = \sigma, \sigma^x_r\sigma$, of the configuration with the higher energy will be $e^{-2\beta J}$ times less.

Complementary Material

5.25 Nonstationary White Noises

Theorem 5.2.1 is applicable to states which are invariant under the free evolution; this includes many examples of physical interest (vacuum, temperature, ground states), but not some non-gauge-invariant states, for which the field $a^{\pm}_{\omega}(t,k) = e^{\pm it(\omega(k)-\omega)}a_k$ is not necessarily stationary. We want to show that this class of states leads to a class of processes which could be called *nonstationary white noises* whose diagonal covariance is like a usual white noise, but whose off-diagonal covariance terms have a correlation of the form $\delta(t+t')$, and which maybe can be used in the description of some nonequilibrium phenomena. The following result indicates that it should be possible to develop a theory of the stochastic limit, in which the reference reservoir state is not invariant under free evolution.

Theorem 5.25.1. *Let $a(k)$, $a^+(k)$ ($k \in \mathbb{R}^d$) be a mean zero squeezing Gaussian quantum field in the sense of Definition 2.3.1 with respect to a given expectation value $\langle \cdot \rangle$. For a fixed real valued function $\omega(k)$ and real number ω, define the new field*

$$a_\omega(t,k) := e^{-it(\omega(k)-\omega)}a(k)$$

and suppose that the covariance matrix, with respect to the time variable, of the process $a^{\pm}(t,k)$ satisfies condition (5.2.11) and that

$$\langle a^{\varepsilon}(k)a^{\varepsilon'}(k')\rangle = G_{\varepsilon\varepsilon'}(k')\delta(k-k') \, .$$

Then the limit (5.9.2) exists and is the Gaussian noise $b_\omega(t,k)$ [of the same Gaussian type as $a(k)$] whose covariance has diagonal elements

$$\langle b^{\varepsilon}_{\omega}(t,k)b^{\varepsilon'}_{\omega'}(t',k')\rangle = \delta_{\omega,\omega'}\delta(t-t')\delta\big(\omega(k)-\omega\big)G_{\varepsilon\varepsilon'}(k')\delta(k-k') \qquad (5.25.1)$$

($\varepsilon \neq \varepsilon'$) and-off diagonal elements equal to

$$\langle b_{\omega}(t,k)b_{\omega'}(t',k')\rangle = \delta_{\omega,\omega'}\delta(t+t')\delta\big(\omega(k)+\omega\big)G_{\varepsilon\varepsilon'}(k')\delta(k-k') \, . \qquad (5.25.2)$$

Proof 5.25.1. The diagonal terms of the covariance are gauge invariant, so they are dealt with by Theorem 5.9.1. Therefore we only have to deal with the squeezing terms $(1/\lambda^2)\langle a_\omega(t/\lambda^2,k)a_{\omega'}(t'/\lambda^2,k')\rangle$. Introducing test functions with compact support φ_1 and φ_2, and a momentum test function g, this becomes

$$\frac{1}{\lambda^2}\int dt\varphi_1(t)\int d\tau\varphi_2(\tau)e^{-i\omega\tau/\lambda^2-i\omega' t/\lambda^2}e^{-i\omega(k)(\tau+t)/\lambda^2}G_{--}(k)\delta(k-k') \, .$$

After the change of variables $\tau \to \sigma = (\tau + t)/\lambda^2$, this becomes

$$\int dt\varphi_1(t) \int d\sigma\varphi_2(\sigma\lambda^2 - t)e^{-i\omega\sigma + i\omega t/\lambda^2 - i\omega' t/\lambda^2} e^{-i\sigma\omega(k)} G_{--}(k)\delta(k-k') .$$

For $\omega \neq \omega'$ this converges to zero by the Riemann–Lebesgue lemma. For $\omega = \omega'$ it converges to

$$\int dt\varphi_1(t)\varphi_2(-t) \int d\sigma e^{-i\omega\sigma} e^{-i\sigma\omega(k)} G_{--}(k)\delta(k-k') = \hat{G}_{--}(k) ,$$

and this proves the statement.

Remark 5.25.1. Note however that the terms in (5.25.1) will be nonzero only for $\omega > 0$, and this implies that the terms in (5.25.2) will be zero. Conversely, if the terms in (5.25.2) are nonzero, $\omega < 0$ and the terms in (5.25.1) are zero.

5.26 Notes

Introduction to Chapter 5

(1) The stochastic resonance is a phenomenon whereby an added noise helps to amplify a signal. If the noise strength is chosen properly then the power spectrum of the response of a system has a sharp peak at some input frequence [Gam98, Gri98]. An interesting study of stochastic resonance in a driven spin–boson system by using the stochastic limit has been performed in [ImYuOh99]. The stochastic limit for rapidly decaying systems has been considered in [KiYuIm00].

Section 5.5

(1) First shown in the paper [AcGoLu95].
(2) Introduced in the paper [AcKoVo97].

Section 5.13

(1) The Hamiltonian in this section was considered by Jakšic and Pillet [JaPi94] using time-dependent perturbation theory and (5.13.5) gives precisely the decay rates found by them, which are the same as those predicted by the Fermi golden rule. These authors also give an estimate of the error for finite λ, valid if the system space is finite dimensional. For a general discussion of the estimate of the error in the stochastic limit, see [ArVo00]. The first rigorous investigation of the large time behaviour for a three-dimensional model of scalar field in the Euclidean formulation was given in [GlJa73].
(2) In the Fock case, if $\omega^{-1/2}g \in L^2(\mathbb{R}^3)$ then the operator H_λ is dominated by $H_A \otimes 1 + 1 \otimes H$, and therefore it is essentially self-adjoint on $\mathcal{H}_A \otimes D(H_b)$. In the finite temperature case, it is proved in [JaPi94] that if

$\omega^{-1}\alpha$ and $\omega\alpha$ both belong to $L^2(\mathbb{R}^3)$ then even if H_{I} is not a relatively bounded perturbation of H_0 the operator H_λ is essentially self-adjoint on $\mathcal{H}_{\mathrm{A}} \otimes D(H_b) \otimes D(H_b)$. Note that, in the stochastic limit, we get a bona fide unitary evolution even without this condition, i.e. even if the original Hamiltonian is not essentially self-adjoint.

Sections 5.20–5.24

These sections include the results obtained in the papers [AcKo99a, AcKo99b, AcKo00a, AcKo00b].

6. Measurements and Filtering Theory

6.1 Input–Output Channels

We consider the following problem: a field interacts with a system S and after this interaction no direct measurement is made on S, but one extracts information on S by measuring the field before (*input*) and after (*output*) the interaction. For example, in the case of an atom which decays emitting radiation, if one knows its initial state and detects the radiated photons, then one can deduce some information on its new state. The idea is to deduce information on the system emitting the radiation from the measured radiation. The emitted radiation is a typical example of an *output process.* Other typical examples of output fields are the field operators (*quadratures*) evolved at time t. More generally, one is interested in the statistics of the output field with respect to a given initial state. Often, e.g. in quantum optics, by *signal* one means the mean value of the output process and by *noise* its variance. Typical choices of input fields are $A_j^\epsilon(t)$, $N_{jk}(t)$ or linear combinations thereof. The corresponding output fields are

$$A^\epsilon_{j,\text{out}}(t) = U_t^+ A_j^\epsilon(t) U_t \quad , \quad N_{jk,\text{out}}(t) = U_t^+ N_{jk}(t) U_t \, ,$$

where U_t is an evolution operator involving the interaction between the system and the field.

Example 6.1.1. The system is an atom, and the input field is a laser beam; the output field is the emitted fluorescence light.

Example 6.1.2. The system is an optical cavity; the input and output fields are light beams.

We know that, after the stochastic limit, the field is approximated by a quantum white noise and the interaction is described by a stochastic equation. We want to show how the white noise description allows one to simplify the *input–output scheme.* Before discussing the problem in full generality, we want to illustrate its statement and solution in a simple example. This is done in the following section.

6.2 The Filtering Problem in Classical Probability

The problem of filtering in classical probability theory is the following: Given the two random processes $X \equiv (X_t) \equiv$ process describing the state of the system (input process) and $Y \equiv (Y_t) \equiv$ observed (output) process (e.g. radiation), *which information on X is obtained by measurement of Y?* Mathematically this is translated into the following problem: *evaluate the conditional distribution of the X-process given knowledge of the Y-process up to time t.*

One would like, moreover, to express the solution in a recursive way, i.e. given any new information on the Y-process, one has a rule to update the corresponding prediction on the X-process.

In classical probability, the conditional expectation of a functional of the X-process given the past history of Y is the optimal estimate, with respect to the minimum mean quadratic deviation criterium of that functional given the past history of Y.

If instead of a general functional of the X-process, one only considers functionals at a given time t [i.e. $f(X_t)$], then one speaks of the *prediction problem*). Finally the *interpolation problem* is the estimate of $f(X_t)$ given $Y_{[t,u]}$; $u > t$.

The mathematical formulation of the problem is as follows: let $(\Omega, \mathcal{F}, P)$ be the probability space of the (X,Y)-process and $(\Omega, \mathcal{F}_0, P^0)$ the space of the Y-process. Denoting

$$\mathcal{F}^Y_{t]} := \sigma - \text{algebra generated by } \{Y_s : s \leq t\},$$

$$\mathcal{F}^X := \sigma - \text{algebra generated by } \{X_s : s \in \mathbb{R}\},$$

the filtering problem is to calculate the conditional expectation value

$$E_{t]} := E\left(\cdot | \mathcal{F}^Y_{t]}\right).$$

One starts with an $\mathcal{H}_S$-valued functional of the Y-process, i.e. $\varphi_t(\omega) \in \mathcal{H}_S$ with

$$\varphi_t \in L^2(\Omega, P^0, \mathcal{H}_S).$$

Suppose that this process is given by a functional of the Y-process with values in the operators on $\mathcal{H}_S$,

$$\varphi_t = T_\omega(t)\psi \in \mathcal{H}_S,$$

and consider the expectation value

$$\begin{aligned}\langle \varphi_t, F_t(Y^0)(X \otimes 1)\varphi_t \rangle &= \int \langle \varphi_t(\omega), X\varphi_t(\omega) \rangle F_t(Y^0)(\omega) dP^0(\omega) \\ &= \int \langle \psi, T^+_\omega(t) X T_\omega(t)\psi \rangle F_t(Y^0)(\omega) dP^0(\omega) \\ &= \int dP^0(\omega) F_t(Y^0)(\omega) \langle \psi, P^t_\omega(X)\psi \rangle .\end{aligned}$$

To postulate a model for X_t and Y_t is the same as postulating that X_t and Y_t obey a differential equation (stochastic or not). Equivalently, this means that

$$U_t(\psi \otimes \Phi)(\omega) = \varphi_t = T_\omega(t)\psi \in \mathcal{H}_S\,,$$

so that

$$\begin{aligned}\langle \varphi_t, F_t(Y^0)(X \otimes 1)\varphi_t\rangle &= \langle \psi \otimes \Phi, U_t^+ F_t(Y^0)(X \otimes 1)U_t \psi \otimes \Phi\rangle \\ &= \langle \psi \otimes \Phi, F_t(Y) X_t \psi \otimes \Phi\rangle\,.\end{aligned}$$

6.3 Field Measurements

In this section we illustrate the basic ideas of quantum filtering by considering a particular example: the measurement of the field process. To fix the ideas we consider the scalar boson case, i.e. our state space has the form $\mathcal{H}_S \otimes \Gamma\big(L^2(\mathbb{R})\big)$, where $\Gamma\big(L^2(\mathbb{R})\big)$ is the boson Fock space over $L^2(\mathbb{R})$ and $\mathcal{H}_S$ is the system space. The vacuum vector is denoted Φ_0, and for the creation and annihilation processes we use the notations

$$A_t = A(\chi_{[0,t]}) \quad , \qquad A_t^+ = A(\chi_{[0,t]})\,;$$

the *field operator* (process) is defined by

$$Q_t = A_t^+ + A_t\,, \tag{6.3.1}$$

and we know that it is a classical process:

$$[Q_t, Q_s] = 0 \quad , \qquad \forall\, s,t \in \mathbb{R}\,. \tag{6.3.2}$$

More precisely, if we define (for any t in $\mathbb{R}$) $\mathcal{A}_{t]}(Q)$ to be the von Neumann algebra generated by Q_s with $s \le t$ and P to be the restriction of the vacuum expectation value on $\mathcal{A}_{t]}(Q)$, then one has the identification

$$\mathcal{A}_{t]}(Q) \cong L^\infty(\Omega, \mathcal{F}_{t]}, P)\,, \tag{6.3.3}$$

where $(\Omega, \mathcal{F}, P)$ is the Wiener probability space and $(\mathcal{F}_{t]})$ the associated past filtration. Let us fix some system operators K, L (i.e. acting on the space $\mathcal{H}_S$), and let us consider a unitary operator $U_{0t} = U_t$, which satisfies the stochastic differential equation

$$dU_t = (-Kdt + LdA_t^+ - L^+ dA_t)U_t\,. \tag{6.3.4}$$

It is known that under general conditions on L and K (say L, K are bounded or polynomials in some creation and annihilation operators) (6.3.4) gives a unique unitary solution.

Now fix a vector ψ in the system space $\mathcal{H}_S$ and consider the vector

$$\Phi = \psi \otimes \Phi_0 .$$

Then, using the fact that dA commutes with U_t and $dA\Phi_0 = 0$, we see that the vector

$$\chi_t := U_t\psi \otimes \Phi_0 \tag{6.3.5}$$

satisfies

$$\begin{aligned} d\chi_t &= (-Kdt + LdA^+)U_t\psi \otimes \Phi_0 = \big(-Kdt + L(dA^+ + dA)\big)\,\chi_t \\ &= (-Kdt + LdQ_t)\chi_t\,, \end{aligned} \tag{6.3.6}$$

which shows that χ_t is a functional of the Q-process. The initial condition of (6.3.6) is $\chi_0 = \Phi = \psi \otimes \Phi_0$, and its solution can be obtained in two steps:

(i) First consider the *operator equation*

$$d\hat{\chi}_t = (-Kdt + LdQ_t)\hat{\chi}_t \tag{6.3.7}$$

with initial condition

$$\hat{\chi}_0 = 1 \quad \text{(the identity operator)}. \tag{6.3.8}$$

(ii) Then apply the operator $\hat{\chi}_t$ to the initial state $\psi \otimes \Phi_0$ to obtain χ_t:

$$\chi_t = \hat{\chi}_t\psi \otimes \Phi_0 . \tag{6.3.9}$$

Note the difference between the two expressions (6.3.5) and (6.3.9) of the same vector χ_t: in (6.3.9) the operator $\hat{\chi}_t$ is a functional of the Q-process; therefore it can be identified to a multiplication operator using the identification (6.3.3). More precisely the space $\mathcal{H}_S \otimes \Gamma\big(L^2(\mathbb{R})\big)$ can be identified to the square-integrable, $\mathcal{H}_S$-valued functionals on the Wiener space,

$$\mathcal{H}_S \otimes \Gamma\big(L^2(\mathbb{R})\big) \cong L^2(\Omega, \mathcal{F}, P) \otimes \mathcal{H}_S \cong L^2(\Omega, \mathcal{F}, P; \mathcal{H}_S)\,, \tag{6.3.10}$$

and in this identification, χ_t can be identified to a vector in the space $L^2(\Omega, \mathcal{F}_{t]}, P) \otimes \mathcal{H}_S = L^2(\Omega, \mathcal{F}_{t]}, P; \mathcal{H}_S)$, i.e. to a random vector, meaning by this an $\mathcal{F}_{t]}$-measurable function,

$$\chi_t : \omega \in \Omega \to \chi_t(\omega) \in \mathcal{H}_S .$$

As a consequence of this, if F^Q is a functional of the Q-process and R is an arbitrary system operator, then one has the identity

$$\begin{aligned} \langle \chi_t, R \otimes F^Q \chi_t \rangle &= \int_\Omega F^Q(\omega)\langle \chi_t(\omega), R\chi_t(\omega)\rangle P(d\omega) \\ &= E^P\left(F^Q\langle \chi_t(\cdot), R\chi_t(\cdot)\rangle\right)\,, \end{aligned} \tag{6.3.11}$$

where now P is the usual Wiener measure. Using the identity (6.3.9) we can be even more explicit, in fact:

$$\langle \chi_t, R \otimes F^Q \chi_t \rangle = \langle \psi \otimes \Phi_0, \hat{\chi}_t^* \left(R \otimes F^Q \right) \hat{\chi}_t \psi \otimes \Phi_0 \rangle \; ;$$

since F^Q commutes with $\hat{\chi}_t$, this is equal to

$$\langle \psi \otimes \Phi_0, \left(1 \otimes F^Q\right) \hat{\chi}_t^* (R \otimes 1) \hat{\chi}_t \psi \otimes \Phi_0 \rangle = \langle \Phi_0, F^Q \langle \psi, \hat{\chi}_t^* (R \otimes 1) \hat{\chi}_t \psi \rangle \Phi_0 \rangle \,, \tag{6.3.12}$$

where $\langle \psi, (\cdot) \psi \rangle$ denotes the partial expectation value with respect to the system state ψ. Recalling that the restriction of the Φ_0 expectation on the functionals of Q gives the Wiener measure P and that in the identification (6.3.10) the operator $\hat{\chi}_t$ is identified to (the multiplication by) a function of ω with values in the operators on $\mathcal{H}_S$, the right-hand side of (6.3.12) becomes

$$\int_\Omega F^Q(\omega) \Big\langle \psi, \hat{\chi}_t^*(\omega) R \hat{\chi}_t(\omega) \psi \Big\rangle P(d\omega) \,. \tag{6.3.13}$$

Now, in the above notations, let us define the *output field* process

$$X_t := U_t^* (1_S \otimes Q_t) U_t \tag{6.3.14}$$

and the *output system* process

$$Y_t = U_t^* (R \otimes 1) U_t \,. \tag{6.3.15}$$

It will be shown in the following section that, for any $t' \geq t$,

$$X_t = U_{t'}^* (1_S \otimes Q_t) U_{t'} \,, \tag{6.3.16}$$

which implies that also the output field process X is commutative, and therefore it makes sense to speak of a functional of the process X. Let $F_{t]}$ be such a functional, adapted up to time t (i.e. depending only on X_s with $s \leq t$). Then (6.3.16) implies (see the following section for a proof) that there exists a functional $F_{t]}^Q$, of the Q-process, adapted up to time t, such that, for any $t' \geq t$,

$$F_{t]} = U_t^* F_{t]}^Q U_t = U_{t'}^* F_{t]}^Q U_{t'} \,. \tag{6.3.17}$$

Therefore, recalling (6.3.5) and (6.3.9) any expectation value of the form

$$\langle \psi \otimes \Phi_0, F_{t]} Y_{t'} \psi \otimes \Phi_0 \rangle \tag{6.3.18}$$

with $t' \geq t$ and $F_{t]}$, $Y_{t'}$ given respectively by (6.3.17) and (6.3.15) can be written in the form

$$\left\langle \psi \otimes \Phi_0, U_{t'}^* F_{t]}^Q (R_{t'} \otimes 1) U_{t'} \psi \otimes \Phi_0 \right\rangle = \left\langle \chi_{t'}, \left(R_{t'} \otimes F_{t]}^Q \right) \chi_{t'} \right\rangle , \tag{6.3.19}$$

where we identify, here and in the following, $F_{t]}^Q$ with $1_S \otimes F_{t]}^Q$.

Now, because of (6.3.16), the random variables X_s, with $s \leq t$, commute with the random variables Y_t, with $t' \geq t$. Therefore the algebra generated by these random variables is commutative and we denote $E_{(t)}^{X,Y}$ the probability measure obtained by restriction of the state $\langle \psi \otimes \Phi_0, \cdot \psi \otimes \Phi_0 \rangle$ on this algebra. With these notations one has

$$\begin{aligned}
E_{(t)}^{X,Y}(F_{t]} Y_{t'}) &= \langle \psi \otimes \Phi_0, F_{t]} Y_t, \psi \otimes \Phi_0 \rangle = \left\langle \psi \otimes \Phi_0, U_{t'}^* F_{t]}^Q R, U_{t'} \psi \otimes \Phi_0 \right\rangle \\
&= \langle \chi_{t'}, F^Q R \chi_{t'} \rangle = E^P \left(F_{t]}^Q \langle \chi_{t'}, R \chi_{t'} \rangle \right) \\
&= E^P \left(F_{t]}^Q E_{t]}^P (\langle \chi_{t'}, R \chi_{t'} \rangle) \right) \\
&= E^P \left(F_{t]}^Q E_{t]}^Q \left(\left\langle \frac{\chi_{t'}}{\|\chi_t\|}, R \frac{\chi_{t'}}{\|\chi_t\|} \right\rangle \right) \langle \chi_t, \chi_t \rangle \right) \\
&= E^P \left(\left\langle \chi_t, F_{t]}^Q E_{t]}^Q \left(\left\langle \frac{\chi_{t'}}{\|\chi_t\|}, R_t \frac{\chi_{t'}}{\|\chi_t\|} \right\rangle \chi_t \right) \right) \right) \\
&= E^X \left(F_{t]} U_t^* E_{t]}^Q \left(\left\langle \frac{\chi_{t'}}{\|\chi_t\|}, \mathbb{R} \frac{\chi_{t'}}{\|\chi_t\|} \right\rangle \right) U_t \right) .
\end{aligned} \tag{6.3.20}$$

In summary, if $t \leq t'$

$$E_{t]}^{U_{0t'}^* Q U_{0t'}} (U_{t'}^* R U_{t'}) = U_t^* E_{t]}^Q \left(\left\langle \frac{\chi_{t'}}{\|\chi_t\|}, R \frac{\chi_{t'}}{\|\chi_t\|} \right\rangle \right) U_t . \tag{6.3.21}$$

Remark 6.3.1. If we compute $E_{t]}^Q$ using classical probability, then we find some functional G of the Q-process:

$$E_{t]}^Q \left(\langle \chi_{t'/\|\chi_t\|}, R \chi_{t'}/\|\chi_t\| \rangle \right) = G(\{Q_s\}_{s \leq t}) .$$

Therefore, because of (6.3.14) and (6.3.16), the right-hand side of (6.3.21) is simply $G(\{X_s\}_{s \leq t})$, a functional of the output field process X.

6.4 Properties of the Input and Output Processes

To fix the ideas in this section, the field shall be supposed to be the standard boson Fock white noise, and we denote $\mathcal{H}_R$ its state space, $\mathcal{H}_S$ the state space of the system, and $\mathcal{H} = \mathcal{H}_S \otimes \mathcal{H}_R$ the state space of the composite system. On $\mathcal{H}_R$ there is a time filtration of Hilbert spaces $\{\mathcal{H}_{t]}\}$ which generates the whole space $\mathcal{H}_R$:

$$\mathcal{H}_{s]} \subset \mathcal{H}_{t]} \uparrow \mathcal{H}_R , \qquad s \leq t , \tag{6.4.1}$$

and we identify an operator $A_{t]}$, acting on $\mathcal{H}_{t]}$, to the operator $A_{t]} \otimes 1_{[t}$, acting on $\mathcal{H}_R$, i.e. we consider $\mathcal{B}(\mathcal{H}_{t]})$ as a subalgebra of $\mathcal{B}(\mathcal{H}_R)$ with the same identity. If U_t is the unitary solution of a stochastic differential equation with constant coefficients [e.g. (6.3.4)] and u_t^0 is the *time shift*, then the family

$$U_{s,t} := u_s^0(U_{t-s}) \tag{6.4.2}$$

satisfies the identity $U_{s,t}U_{r,s} = U_{r,t}$; it is a unitary evolution on the space. The evolution U_{st} is adapted in the sense that

$$U_{s,t}\Big(\mathcal{H}_S \otimes \mathcal{H}_{s]}\Big) \subseteq \mathcal{H}_S \otimes \mathcal{H}_{t]}\,. \tag{6.4.3}$$

Thus, if we define the new filtration

$$\mathcal{B}(\mathcal{H}_S \otimes \mathcal{H}_{t]}) =: \mathcal{A}_{t]} \quad , \qquad \mathcal{B}(\mathcal{H}_S \otimes \mathcal{H}_R) =: \mathcal{A}\,, \tag{6.4.4}$$

our models can be described by the general situation described in the following.

Definition 6.4.1. *Let $\mathcal{A}$ be a $*$-algebra and $(\mathcal{A}_{s]})$ a past filtration in $\mathcal{A}$ indexed by $s \in \mathbb{R}$. A* localized left unitary evolution *(also called a* localized left unitary multiplicative functional*) is a 2-parameter family $U_{s,t}$ $(s \le t)$ of unitary operators in $\mathcal{A}$ such that*

$$r \le s \le t \Rightarrow U_{st}U_{rs} = U_{r,t}\,, \tag{6.4.5}$$

$$U_{t,t} = 1\,, \quad \forall t\,, \tag{6.4.6}$$

$$U_{st} \in \mathcal{A}_{t]}\,, \quad \forall s \le t\,. \tag{6.4.7}$$

Given such an evolution, we define, for any fixed $s \in \mathbb{R}$, the input algebra *$\mathcal{B}_{s]}$ to be the commutant of the family $\{U_{tt'} \ :\ s \le t \le t'\}$, i.e. the family of all operators $F_{s]}$ such that*

$$F_{s]}U_{tt'} = U_{tt'}F_{s]}\,, \quad t' \ge t \ge s\,. \tag{6.4.8}$$

The family $(\mathcal{B}_{s]})$ is increasing $(s \le s' \to \mathcal{B}_{s]} \subseteq \mathcal{B}_{s']})$ and its set-theory union (i.e. without closure) shall be called the input algebra *(for a given U_{st}).*

Remark 6.4.1. In the space $\mathcal{H} = \mathcal{H}_S \otimes \mathcal{H}_R$, the typical elements of the input algebra are the adapted *pure noise processes*, i.e. those of the form: $1_S \otimes X_s$.

Lemma 6.4.1. *Let $(U_{s,t})$ be a localized left unitary evolution and let it be given that*

(i) An adapted process $Q = (Q_t)$: $(Q_t \in \mathcal{B}_{t]})$ in the input algebra $\mathcal{B}$ (input process).

(ii) An adapted process $R = (R_t)$ commuting with Q (input system process), i.e.

$$[Q_s, R_t] = 0\,, \quad \forall s, t\,. \tag{6.4.9}$$

Define the output process $X = (X_t)$ *by*

$$X_t := U_{0t}^* Q_t U_{0t} \ , \quad \forall t \, , \tag{6.4.10}$$

and the output system process $Y = (Y_t)$ *by*

$$Y_t := U_{0t}^* R_t U_{0t} \ , \quad \forall t \, . \tag{6.4.11}$$

Then

$$U_{0T}^+ Q_t U_{0T} = X_t \ , \quad \forall T \geq t \, , \tag{6.4.12}$$

$$X_s Y_t = Y_t X_s \ , \quad \forall t \geq s \, . \tag{6.4.13}$$

If Q *is commutative, so is* X*. Finally denoting*

$$\mathcal{A}_{t]}(X) \qquad [\text{or } \mathcal{A}_{t]}(Q)] \tag{6.4.14}$$

the closure of the polynomial algebra generated by the X_s *(or* Q_s*) with* $s \leq t$*, in any topology in which the maps* $z \mapsto U_{0t}^+ z U_{0t}$ *are continuous for any* t*, then*

$$\mathcal{A}_{t]}(X) = \mathcal{A}_{t]}(U_{0T}^+ Q U_{0T}) = U_{0T}^+ \mathcal{A}_{t]}(Q) U_{0T} \ , \quad \forall T \geq t \, . \tag{6.4.15}$$

Proof 6.4.1. Let $T \geq t$. Since Q is an input process, then using (6.4.5), (6.4.8) and the unitarity of $U_{s,t}$, we find

$$U_{0T}^* Q_t U_{0T} = U_{0t}^* U_{tT}^* Q_t U_{tT} U_{0T} = U_{0t} Q_t U_{tT}^* U_{tT} U_{0T} = X_t \, ,$$

and this proves (6.4.12). Using (6.4.12) we find, for $t \geq s$,

$$[X_s, Y_t] = [V_{0t}^* Q_s U_{0t}, U_{0t}^* R_t U_{0t}] = U_{0t}^* [Q_s, R_t] U_{0t} = 0,$$

which is (6.4.13). Equation (6.4.15) follows from (6.4.10) plus continuity since $z \mapsto U_{st}^* z U_{st}$ is a $*$-isomorphism. If Q is commutative, we can choose $R = Q$, i.e. $X = Y$, so X is commutative because of (6.4.13).

Remark 6.4.2. Typical examples of families Q and R, satisfying the conditions of Lemma 6.4.1 are realized on an Hilbert space of the form $\mathcal{H}_S \otimes \mathcal{H}_R$ by taking the output and the system process respectively as

$$Q_t = 1_S \otimes \tilde{Q}_t \qquad \left(\tilde{Q}_t \text{ acting on } \mathcal{H}_R\right) \tag{6.4.16}$$

and

$$R_t = R \otimes 1_R \qquad (R \text{ acting on } \mathcal{H}_S) \, . \tag{6.4.17}$$

Corollary 6.4.1. *On the algebra of the input processes* ($Q_t \in \mathcal{B}_{t]}$)*, the wave automorphism*

$$\omega_\infty(Q_t) := \lim_{T \to +\infty} U_T^* Q_t U_T =: X_t^{\text{out}} =: X_t \tag{6.4.18}$$

exists and globally defines the output field $X_{\text{out}}(t)$ *in the sense that the limit exists and the equality holds.*

Proof 6.4.2. This is clear from (6.4.12).

Remark 6.4.3. Formally $X_{\text{out}}(t) = U_\infty^* X_t U_\infty$, but this is only a formal notation, since, while the limit (6.4.15) exists trivially, usually there is no limit such as $\lim_{T \to +\infty} U_T = U_\infty$.

6.5 The Filtering Problem in Quantum Theory

Let φ be a state on $\mathcal{A}$ and denote $\{\mathcal{H}, \pi, \Phi\}$ the GNS triple of $\{\mathcal{A}, \varphi\}$. In the following for simplicity we omit π from the notations.

In the notations of the previous section, let be given a unitary evolution (U_{st}), an input process Q and a system process R. Define the output process X_t and the output system process Y_t and the associated algebras $\mathcal{A}(Q)$, $\mathcal{A}(X)$ as in Lemma 6.4.1. Then $Y_{t'}$ commutes with $\mathcal{A}_{t]}(X)$ for $t' \geq t$. Therefore [Sak71] there is an element, denoted $E^X_{t]}(Y_{t'})$, in the commutant of $\mathcal{A}_{t]}(X)$ characterized by

$$E^X\left(F_{t]}Y_{t'}\right) = E^X\left(F_{t]}E^X_{t]}(Y_{t'})\right) , \quad \forall\, F_{t]} \in \mathcal{A}_{t]}(X) , \tag{6.5.1}$$

where E^X denotes the restriction of the state $\langle \Phi, (\cdot)\Phi \rangle$ on the algebra

$$\mathcal{A}(X) = \text{ strong closure of } \cup_{t\geq 0} \mathcal{A}_{t]}(X) .$$

Defining in a similar way E^Q, $E^Q_{t]}(R_{t'})$, we are interested in expressing $E^X_{t]}(Y_{t'})$ in terms of $E^Q_{t]}$. This is *the filtering problem.*

Theorem 6.5.1. *In the above notations suppose that, for each $t \geq 0$ there exist an adapted closed invertible operator $\hat{\chi}_t$ affiliated to $\mathcal{A}_{t]}(Q)' \cap \mathcal{A}_{t]}$ such that*

$$U_{0t}\Phi = \hat{\chi}_t \Phi . \tag{6.5.2}$$

Then, denoting $P^Q_{t]}$ the orthogonal projection onto the cyclic space of $\mathcal{A}_{t]}(Q)$, one has

$$E^X_{t]}(Y_{t'}) = U^*_{0t}\hat{\chi}^{*-1}_t P^Q_{t]}\hat{\chi}^*_{t'}R_{t'}\hat{\chi}_{t'}P^Q_{t]}\hat{\chi}^{-1}_t U_{0t} . \tag{6.5.3}$$

Remark 6.5.1. If the vector Φ is cyclic for $\mathcal{A}_{t]}(Q)'$ in the space $\mathcal{H}_{t]}$, then an operator $\hat{\chi}_t$ as above always exists [Sak71].

Proof 6.5.1. Given $F_{t]} \in \mathcal{A}_{t]}(X)$, since $t' \geq t$, using (6.4.10) we deduce the existence of $F^Q_{t]} \in \mathcal{A}_{t]}(Q)$ such that

$$F_{t]} = U^+_{0t'}F^Q_{t]}U_{0t'} ; \tag{6.5.4}$$

therefore, using (6.4.12) with $T = t'$

$$E^X(F_{t]}Y_{t'}) = \left\langle U_{0t'}\Phi, F^Q_{t]}R_{t'}, U_{0t}\Phi \right\rangle = \left\langle \chi_{t'}, F^Q_{t]}R_{t'}, \chi_{t'} \right\rangle , \tag{6.5.5}$$

where the vector $\chi_{t'}$ is defined by

$$\chi_{t'} := U_{0t'}\Phi , \quad \forall\, t' . \tag{6.5.6}$$

By assumption there exists a closed operator $\hat{\chi}_t$ affiliated to $\mathcal{A}_{t']}(Q)'$ such that

$$\chi_{t'} = U_{0t'}\Phi = \hat{\chi}_{t'}\Phi\,. \tag{6.5.7}$$

Using (6.5.7) in (6.5.5) we find

$$E^X(F_{t]}Y_{t'}) = \left\langle \Phi, F^Q_{t]}\hat{\chi}^*_{t'}R_{t'}\hat{\chi}_{t'}\Phi \right\rangle\,. \tag{6.5.8}$$

Since by (6.4.13) $R_{t'}$ is in the commutant of $\mathcal{A}_{t]}(Q)$, it follows that $\hat{\chi}^*_{t'}R_{t'}\hat{\chi}_{t'}$ is also in $\mathcal{A}_{t]}(Q)'$. Denoting $P^Q_{t]}$ the orthogonal projection onto the cyclic space of $\mathcal{A}_{t]}(Q)$, (6.5.6) implies that

$$\begin{aligned} E^X(F_{t]}Y_{t'}) &= \left\langle \Phi, F^Q_{t]}P^Q_{t]}\hat{\chi}^*_{t'}R_{t'}\hat{\chi}_{t'}P^Q_{t]}\Phi \right\rangle \\ &= \left\langle \chi_t, F^Q_{t]}\hat{\chi}^{*-1}_t P^Q_{t]}\hat{\chi}^*_{t'}R_{t'}\hat{\chi}_{t'}P^Q_{t]}\hat{\chi}^{-1}_t\chi_t \right\rangle \\ &= \left\langle U_{0t}\Phi, F^Q_{t]}\hat{\chi}^{*-1}_t P^Q_{t]}\hat{\chi}^*_{t'}R_{t'}\hat{\chi}_{t'}P^Q_{t]}\hat{\chi}^{-1}_t U_{0t}\Phi \right\rangle \\ &= \left\langle \Phi, F_t\left[U^*_{0t}\hat{\chi}^{*-1}_t P^Q_{t]}\hat{\chi}^*_{t'}R_{t'}\hat{\chi}^*_{t'}P^Q_{t]}\hat{\chi}^{-1}_t U_{0t}\right]\Phi \right\rangle\,. \end{aligned}$$

Since $t \le t'$, it follows that $\mathcal{A}_{t]}(Q)' \supseteq \mathcal{A}_{t']}(Q)'$ and, because of (6.4.13), the term in square brackets in (6.5.7) is in $\mathcal{A}_{t]}(X)'$. It follows that

$$E^X_{t]}(Y_{t'}) = U^*_{0t}\hat{\chi}^{*-1}_t P^Q_{t]}\hat{\chi}^*_{t'}R_{t'}\hat{\chi}_{t'}P^Q_{t]}\hat{\chi}^{-1}_t U_{0t}\,.$$

6.6 Filtering of a Quantum System Over a Classical Process

The analysis in the present section will help us to acquire a better insight on the meaning of (6.5.3), which is very general, but rather abstract.

Any experimentally realizable measurement must be performed on a set of mutually compatible observables. If this set is maximal, its commutant will coincide with the algebra generated by it, which is abelian and therefore corresponds to a classical stochastic process. This leads to the problem of analyzing the filtering process of a quantum system over a classical process. This means that the algebras $\mathcal{A}_{t]}(Q)$ [and therefore $\mathcal{A}_{t]}(X)$, by (6.4.15)] are abelian. In the notations of Sect. 6.5, let us moreover suppose that $\hat{\chi}_t$ is bounded, i.e.

$$\hat{\chi}_t \in \mathcal{A}_{t]}(Q)' \cap \mathcal{A}_{t]}(Q)\,. \tag{6.6.1}$$

Because of (6.4.13), this implies that $\hat{\chi}_t$ commutes with all the $R_{t'}$. Suppose that R_t are normal operators. Then the von Neumann algebra generated by $\mathcal{A}_{t]}(X)$ and Y_t is abelian, and for any $t' \ge t$, the conditional expectation

$$E^X_{t]}(Y_{t'}) = E^X\left(Y_{t'}|\mathcal{A}_{t]}(X)\right) = E^X\left(Y_{t'}|U^*_{0t'}\mathcal{A}_{t]}(Q)U_{0t'}\right)$$

is meant in the usual sense of classical probability. Denote $\mathcal{B}^Q_{t]\vee t'}$ ($\mathcal{B}^X_{t]\vee t'}$) the abelian von Neumann algebra generated by $\mathcal{A}_{t]}(Q)$ and $R_{t'}$ [$\mathcal{A}_{t]}(X)$ and $Y_{t'}$] and denote E^Q the restriction of $\langle\Phi,(\cdot)\Phi\rangle$ to $\mathcal{B}^Q_{t]\vee t'}$. Then, under the assumption (6.6.1) and recalling (6.4.9) and (6.5.5),

$$\begin{aligned} E^X(F_{t]}Y_{t'}) &= E^X\left(U^*_{0t'}F^Q_{t]}R_{t'}U_{0t'}\right) = \langle U_{0t'},\Phi, F_{t]}QR_{t'}U_{0t'},\Phi\rangle \\ &= \left\langle \Phi, \hat{\chi}^*_{t'}F^Q_{t]}R_{t'}\hat{\chi}_{t'}\Phi\right\rangle = E^Q(F^Q_{t]}|\hat{\chi}_{t'}|^2R_{t'})\,, \end{aligned} \tag{6.6.2}$$

which shows that the output measure is absolutely continuous with respect to the input measure and moreover

$$\frac{dE^X\left(U^*_{0t'}(\cdot)U_{0t'}\right)}{dE^Q(\cdot)}|_{\mathcal{F}_{t']}} = |\hat{\chi}_{t'}|^2\,, \quad \forall t' \geq t\,. \tag{6.6.3}$$

In particular $|\hat{\chi}_t|^2$ is a martingale. From this we deduce

$$\begin{aligned} E^X(F_{t]}Y_{t'}) &= E^Q\left(F^Q_{t]}E^Q_{t]}(|\hat{\chi}_{t'}|^2R_{t'})\right) = E^Q\left(|\chi_t|^2F^Q_{t]}E^Q_{t]}\left(|\hat{\chi}_{t'}|^2R_{t'}\right)\right) \\ &= E^X\left(F_{t]}U^*_{0t}E^Q_{t]}\left(\frac{|\hat{\chi}_{t'}|^2}{E^Q_{t]}(|\hat{\chi}_{t'}|^2)}R_{t'}\right)U_{0t}\right)\,. \end{aligned}$$

Therefore, recalling that $X = U^*QU$, we find the basic filtering formula:

$$E^X_{t]}(U^*_{0t'}R_{t'}U_{0t}) = U^*_{0t}E^Q_{t]}\left(\frac{|\hat{\chi}_{t'}|^2}{E^Q_{t]}(|\hat{\chi}_{t'}|^2)}R_{t'}\right)U_{0t}\,. \tag{6.6.4}$$

Recalling that $X_t = U^*_{0t}Q_tU_{0t}$, we see that the basic filtering formula expresses a dynamical covariance property.

Remark 6.6.1. Formula (6.6.4) above is a generalization of (6.3.21) to a state space not necessarily of the form $\mathcal{H}_S\otimes\mathcal{H}_R$.

6.7 Nondemolition Processes

Definition 6.7.1. *A process (Y_t) is called a* nondemolition process *with respect to the process X_t if for any fixed time t the t future of the X-process commutes with the t past of the Y-process. In symbols:*

$$[X_s, Y_t] = 0 \quad , \forall s \geq t\,. \tag{6.7.1}$$

Two processes X, Y are called mutually nondemolition processes *if they commute, i.e. if*

$$[X_s, Y_t] = 0 , \quad \forall s, t . \tag{6.7.2}$$

A process (Y_t) is called classical *(sometimes also* self-nondemolition*) if each Y_t is a normal operator,*

$$Y_t^+ Y_t = Y_t Y_t^+ , \ \forall t , \tag{6.7.3}$$

and operators at different times commute,

$$[Y_s, Y_t^+] = [Y_s, Y_t] = 0 , \ \forall s, t . \tag{6.7.4}$$

Definition 6.7.2. *A classical process Y which is nondemolition with respect to a process X is called a* measurement *of X.*

Remark 6.7.1. In classical filtering theory, the output processes are also called *observable processes.* In the quantum case it might happen that even if each variable of a stochastic process (Y_t) is an observable (i.e. a self-adjoint operator) the process as a whole is not observable, since observables at different times do not commute. For this reason, in the quantum case, an observable process is required to be a classical process by definition.

Lemma 6.7.1. *Let $X \equiv (X_t)$ and $Y \equiv (Y_t)$ be commuting processes such that Y is nondemolition for X and (Y_t) is an output process in the sense of Lemma 6.4.1, i.e.*

$$[Y_t, U_{t',v}] = 0 , \ \forall v \geq t' \geq t . \tag{6.7.5}$$

Then the output processes

$$\tilde{X}_t := U_t^* X_t U_t \quad , \quad \tilde{Y}_t := U_t^* Y_t U_t$$

are such that $\tilde{Y}$ is nondemolition for $\tilde{X}$.

Proof 6.7.1. Fix $t < v$. Then, by the same argument as in Lemma 6.4.1,

$$\tilde{Y}_t = \quad U_t^* Y_t U_t = U_t^* U_{t,v}^* Y_t U_{t,v} U_t = U_v^* Y_t U_v .$$

Hence

$$[\tilde{Y}_t, \tilde{X}_v] = U_v^* [Y_t, X_v] U_v = 0 .$$

The above Lemma is frequently used in the situation described in the following.

Corollary 6.7.1.

(i) *If $Y \equiv (Y_t)$ is a classical process satisfying (6.7.5), then the output process $(\tilde{Y}_t)$ is also classical.*

(ii) *If, for each t, $X_t = X$, X is an operator in the system space, and $Y = (Y_t)$ is a process satisfying (6.7.5) and commuting with any operator in the system space, then $(\tilde{Y}_t)$ is nondemolition for $(\tilde{X}_s)$.*

(iii) *In the assumptions of (ii), if $Y \equiv (Y_t)$ is a classical process then it is a measurement of X, in the sense of Definition 6.7.2.*

Proof 6.7.2. (i) Put $X = Y$ in Lemma 6.7.1. (ii) Choose (X_t) in Lemma 6.7.1 to be the constant process equal to X.

Note that in general (X_t) is a noncommutative process even if X is normal.

6.8 Standard Scheme to Construct Examples of Nondemolition Measurements

From (6.4.16–17) we know that if $Y \equiv (Y_t)$ and $X \equiv (X_t)$ are stochastic processes of the form

$$\begin{aligned} X_t &:= U^*_{o,t}(X \otimes 1_A)U_{o,t} , \quad X \hat{\in} \mathcal{B}(\mathcal{H}_S) , \\ Y_t &:= 1_S \otimes Y^{\circ}_t , \quad Y \hat{\in} \mathcal{B}(\mathcal{H}_{t]}) , \end{aligned} \tag{6.8.1}$$

and if the process Y_t satisfies

$$U^*_{s,t}(1_S \otimes Y_s)U_{s,t} = 1_S \otimes Y_s \tag{6.8.2}$$

for any $s < t$, then the output process

$$\tilde{Y}_s := U^*_{o,s} Y_s U_{o,s} , \quad s > 0 ,$$

is nondemolition for (X_t); if (Y_t) is a commutative process, then the same is true for $(\tilde{Y}_s)$.

The crucial condition (6.8.2) is always satsified when Y_t has the form (6.8.1) and the unitary evolution $U_{s,t}$ is the solution of a stochastic differential equation with coefficients living on the (system) space $\mathcal{H}_S$ and noises living on $\mathcal{H}_R$ with increments commuting with the past.

6.9 Discrete Time Nondemolition Processes

The following definition of *simultaneous measurement* of two operators was taken from [HaYa86].

Definition 6.9.1. *A* simultaneous measurement *of two operators a_1, a_2, acting on an Hilbert space H is defined by a map T from operators on H to operators on another Hilbert space $\mathcal{H}$ such that*

$$[T(a_1), T(a_2)] = 0 . \tag{6.9.1}$$

Typically a simultaneous measurement of a_1, a_2 *is realized by choosing another Hilbert space* H_N*, called the* noise space *and fixing two operators* A, B *acting on* H_N *such that*

$$[a_1, a_2] = [B, A] \,. \tag{6.9.2}$$

One then defines $\mathcal{H} = H \otimes H_N$ *and*

$$T(a_1) = a_1 \otimes 1 + 1 \otimes A \ , \ T(a_2) = a_2 \otimes 1 + 1 \otimes B \,. \tag{6.9.3}$$

Because of (6.9.2), (6.9.1) is satisfied.

Since $T(a_1)$ and $T(a_2)$ commute, they can be simultaneously measured with arbitrary precision. A simultaneous measurment of the two quantities $T(a_1), T(a_2)$ is called a simultaneous measurement of a_1 and a_2.

6.10 Notes

The discussion of the filtering problem in this chapter is inspired by the approach of [Bel80, Bel85]. This approach has been used to construct phenomenological models of quantum measurement processes in the spirit of [Zur82] (see for example [BeMe95], [Dio88], [Ex85], [Gis82], [GPR90], [Zur82]).

The main point of the present chapter is to underline the possibility of a potentially fruitful application, to the quantum measurement problem, of a combination of the stochastic limit and the filtering technique. In fact, using the stochastic limit, the above-mentioned phenomenological models can be deduced from the first-principle Hamiltonians, thus giving a physical meaning (and structure) to the "noises" introduced by hand in these models. In this direction also the dependence of the rate of decoherence on the "macroscopicity" of the system (i.e. the number of particles), discussed in Sect. 5.22, seems to confirm the accepted physical intuition of this phenomenon.

7. Idea of the Proof and Causal Normal Order

In this chapter we begin the proof of the main results discussed in Chap. 4 by illustrating the first two new features which appear as a consequence of the interaction, with respect to the convergence of free fields, discussed in Chap. 3. These are:

(i) the causal δ-function;
(ii) the time consecutive principle.

These two notions are used to bring to normal order the white noise Hamiltonian equation and to prove that *this normally ordered form is equivalent to a quantum stochastic differential equation.*

This result constitutes the starting point of the *white noise approach to (classical and quantum) stochastic calculus* (see [AcLuVo99]).

7.1 Term-by-Term Convergence of the Series

Theorem 7.1.1. *Under the analytical assumptions (4.5.5) and (4.9.3), the iterated series for the rescaled evolution operator* $U^{(\lambda)}_{t/\lambda^2}$,

$$U^{(\lambda)}_{t/\lambda^2} = 1 + \sum_{n=1}^{\infty}(-i)^n \int_0^t dt_1 \int_0^{t_1} dt_2 \ldots \int_0^{t_{n-1}} dt_n \\ \frac{1}{\lambda} H_{\mathrm{I}}(t_1/\lambda^2) \cdot \frac{1}{\lambda} H_{\mathrm{I}}(t_2/\lambda^2) \cdot \ldots \cdot \frac{1}{\lambda} H_{\mathrm{I}}(t_n/\lambda^2), \tag{7.1.1}$$

converges weakly on the standard domain (number plus coherent vectors), uniformly for bounded t. Moreover the term-by-term limit, in the sense of matrix elements (see Sect. 1.15), of (7.1.1) exists and is equal to

$$\lim_{\lambda\to 0} U^{(\lambda)}_{t/\lambda^2} = U_t := 1 + \sum_{n=1}^{\infty}(-i)^n \int_0^t dt_1 \int_0^{t_1} dt_2 \ldots \int_0^{t_{n-1}} dt_n H_{t_1} H_{t_2} \ldots H_{t_n}, \tag{7.1.2}$$

where the series converges uniformly for bounded t on the standard domain. The operator U_t is unitary and is the iterated series solution of the white noise Hamiltonian equation

$$\partial_t U_t = -iH_t U_t = -i(D^+ b_t + D b_t^+)U_t, \quad U_0 = 1. \tag{7.1.3}$$

Proof 7.1.1. From Theorem 3.2.1 we know that $(1/\lambda)H_{\mathrm{I}}(t/\lambda^2)$ converges to $H_t = (D^+ b_t + D b_t^+)$. Therefore the series on the right-hand side of (7.1.2) is obtained as the term-by-term limit of the right-hand side of (7.1.1). By taking the term-by-term derivative of this series we find (7.1.3).

7.2 Vacuum Transition Amplitude: The Fourth-Order Term

In this section we give an idea of the method of proof of Theorem 7.1.1 by checking directly (7.1.2) up to the fourth-order term of the series expansion. Even in the fourth-order approximation we shall not prove the convergence of all the matrix elements of the rescaled evolution operator $U^{(\lambda)}_{t/\lambda^2}$ but only of the vacuum matrix elements, i.e. the vacuum transition amplitude $\left\langle U^{(\lambda)}_{t/\lambda^2}\right\rangle$.

From Theorem 7.1.1 it follows in particular that

$$\lim_{\lambda\to 0}\left\langle U^{(\lambda)}_{t/\lambda^2}\right\rangle = \langle U_t\rangle\,,$$

where we use the same symbol $\langle\,\cdot\,\rangle$ for the vacuum expectation of both the original a field and the master field. On the other hand, from (4.17.4) we know that

$$\lim_{\lambda\to 0}\left\langle U^{(\lambda)}_{t/\lambda^2}\right\rangle = \langle U_t\rangle = \exp\left(-tD^+D\int_{-\infty}^{0} d\sigma G_{-+}(\sigma)\right) = e^{-t\gamma_- D^+ D}, \tag{7.2.1}$$

where G_{-+} (for simplicity in the following we shall only write G) is the 2-point *effective* vacuum correlation of the limit white noise,

$$G_{-+}(t) := \langle \Phi, b_{t+\tau} b_\tau^+ \Phi\rangle = \int_{\mathbb{R}^d} e^{-i\omega(k)t}|g(k)|^2 dk = \langle g, S_t g\rangle = G(t), \tag{7.2.2}$$

and

$$\gamma_- := \int_{-\infty}^{0} dt G_{-+}(t) = \int_{-\infty}^{0} dt\langle g, S_t g\rangle = \int_{-\infty}^{0} dt \int_{\mathbb{R}^d} e^{-it(\omega_1(k)-\omega_o)} \mid g(k)\mid^2 . \tag{7.2.3}$$

We shall proceed as follows: we truncate up to the fourth order the series expansions (7.1.1) and find the limit of these terms; then we explain the rules to compute the iterated integrals in the series (7.1.2) (Sects. 7.4 and 7.5); finally we apply these rules to calculate the iterated integrals in (7.1.2) (Sects. 7.4 and 7.5) and we find that they coincide with the limits of the corresponding terms in (7.1.2). This computation also illustrates another important general feature of the stochastic limit: *only the time consecutive diagrams survive* (see below for the definition). The proof of the analogue result for the general, nth term of the series, as well as the proof of the fact that the term-by-term convergence implies the overall convergence of the series, is contained in Chap. 16.

By truncating (7.1.1) up to the fourth order one finds:

$$
\begin{aligned}
U^{(\lambda)}_{t/\lambda^2} &= 1 - i\lambda \int_0^{t/\lambda^2} dt_1 \left(D^+ a^{(\lambda)}_{t_1} + D a^{(\lambda)+}_{t_1}\right) \\
&+ (-i)^2\lambda^2 \int_0^{t/\lambda^2} dt_1 \int_0^{t_1} dt_2 \left(D^+ a^{(\lambda)}_{t_1} + D a^{(\lambda)+}_{t_1}\right)\left(D^+ a^{(\lambda)}_{t_2} + D a^{(\lambda)+}_{t_2}\right) \\
&+ (-i)^3\lambda^3 \ldots \\
&+ (-i)^4\lambda^4 \int_0^{t/\lambda^2} dt_1 \int_0^{t_1} dt_2 \int_0^{t_2} dt_3 \int_0^{t_3} dt_4 \\
&\qquad \left(D^+ a^{(\lambda)}_{t_1} + D a^{(\lambda)+}_{t_1}\right)\left(D^+ a^{(\lambda)}_{t_2} + D a^{(\lambda)+}_{t_2}\right) \\
&\qquad \cdot \left(D^+ a^{(\lambda)}_{t_3} + D a^{(\lambda)+}_{t_3}\right)\left(D^+ a^{(\lambda)}_{t_4} + D a^{(\lambda)+}_{t_4}\right).
\end{aligned}
$$

Since the vacuum expectation values of an odd number of operators $a^{(\lambda)}$ and $a^{(\lambda)+}$ vanish, the vacuum expectation of the above expression is equal to

$$
\begin{aligned}
\left\langle U^{(\lambda)}_{t/\lambda^2}\right\rangle &= 1 + (-i)^2\lambda^2 \int_0^{t/\lambda^2} dt_1 \int_0^{t_1} dt_2 D^+ D \left\langle a^{(\lambda)}_{t_1} a^{(\lambda)+}_{t_2}\right\rangle \\
&+ (-i)^4\lambda^4 \int_0^{t/\lambda^2} dt_1 \int_0^{t_1} dt_2 \int_0^{t_2} dt_3 \int_0^{t_3} dt_4 \left\{D^+D^+DD \left\langle a^{(\lambda)}_{t_1} a^{(\lambda)}_{t_2} a^{(\lambda)+}_{t_3} a^{(\lambda)+}_{t_4}\right\rangle\right. \\
&+ \left. D^+DD^+D \left\langle a^{(\lambda)}_{t_1} a^{(\lambda)+}_{t_2} a^{(\lambda)}_{t_3} a^{(\lambda)+}_{t_4}\right\rangle\right\}.
\end{aligned}
$$

Using the Gaussian property of the expectation value this can be written in the form

$$
\begin{aligned}
\left\langle U^{(\lambda)}_{t/\lambda^2}\right\rangle &= 1 + (-i)^2 D^+ D\lambda^2 \int_0^{t/\lambda^2} dt_1 \int_0^{t_1} dt_2 G(t_2 - t_1) \\
&+ (-i)^4\lambda^4 \int_0^{t/\lambda^2} dt_1 \int_0^{t_1} dt_2 \int_0^{t_2} dt_3 \int_0^{t_3} dt_4 \\
&\{D^+D^+DD \cdot [G(t_3 - t_1)G(t_4 - t_2) + G(t_4 - t_1)G(t_3 - t_2)] \\
&+ D^+DD^+D G(t_2 - t_1)G(t_4 - t_3)\}.
\end{aligned}
\tag{7.2.4}
$$

By Lemma 1.8.1 the limit, as $\lambda \to 0$, of the second-order term in (7.2.3) exists and is equal to the operator transport coefficient Sect. 1.8:

$$-tD^+D\int_{-\infty}^{0} dsG(s) = -tD^+D\gamma_- = -tY. \qquad (7.2.5)$$

Now we consider the fourth-order terms in (7.2.4). This illustrates an important difference between vanishing and nonvanishing terms in the limit $\lambda \to 0$. Namely that when $\lambda \to 0$ only the integral of $G(t_2-t_1)G(t_4-t_3)$, corresponding to the *time-consecutive pairings* $(t_1,t_2),(t_3,t_4)$ tends to a nonvanishing limit, while all the terms corresponding to the *non-time-consecutive pairings* $(t_1,t_3),(t_2,t_4)$ and $(t_1,t_4),(t_2,t_3)$ vanish in the limit; this gives the main idea of the *time-consecutive principle* discussed in detail in Sects. 11.11, 15.5–7. First let us consider the time consecutive term in (7.2.4), i.e. $(t_1,t_2),(t_3,t_4)$.

Lemma 7.2.1. *The following limiting relation holds*

$$\lim_{\lambda\to 0}\lambda^4\int_0^{t/\lambda^2} dt_1\int_0^{t_1} dt_2\int_0^{t_2} dt_3\int_0^{t_3} dt_4 G(t_2-t_1)G(t_4-t_3)$$
$$= \frac{t^2}{2}\left[\int_{-\infty}^{0} d\sigma G(\sigma)\right]^2 . \qquad (7.2.6)$$

Proof 7.2.1. In order to bring the integral

$$\lambda^4\int_0^{t/\lambda^2} dt_1\int_0^{t_1} dt_2\int_0^{t_2} dt_3\int_0^{t_3} dt_4 G(t_2-t_1)G(t_4-t_3) \qquad (7.2.7)$$

to a convenient form, we introduce new variables: $t_4 \to \sigma_4 := t_4 - t_3$. Then (7.2.7) becomes

$$\lambda^4\int_0^{t/\lambda^2} dt_1\int_0^{t_1} dt_2\int_0^{t_2} dt_3\int_{-t_3}^{0} d\sigma_4 G(t_2-t_1)G(\sigma_4). \qquad (7.2.8)$$

Now change variable from t_2 to a variable σ_2: $t_2 \to \sigma_2 := t_2 - t_1$. Then (7.2.8) is equal to

$$\lambda^4\int_0^{t/\lambda^2} dt_1\int_{-t_1}^{0} d\sigma_2\int_0^{\sigma_2+t_1} dt_3\int_{-t_3}^{0} d\sigma_4 G(\sigma_2)G(\sigma_4). \qquad (7.2.9)$$

After the rescalings

$$t_1 \to \sigma_1 = \lambda^2 t_1, \quad t_3 \to \sigma_3 = \lambda^2 t_3,$$

(7.2.9) takes the form

$$\int_0^t d\sigma_1\int_{-\sigma_1/\lambda^2}^{0} G(\sigma_2)d\sigma_2\int_0^{\sigma_1+\lambda^2\sigma_2} d\sigma_3\int_{-\sigma_3/\lambda^2}^{0} d\sigma_4 G(\sigma_4). \qquad (7.2.10)$$

When $\lambda \to 0$, by dominated convergence, (7.2.10) tends to

$$\int_0^t d\sigma_1 \int_{-\infty}^0 G(\sigma_2) d\sigma_2 \int_0^{\sigma_1} d\sigma_3 \int_{-\infty}^0 d\sigma_4 G(\sigma_4)$$

$$= \int_0^t d\sigma_1 \int_0^{\sigma_1} d\sigma_3 \left[\int_{-\infty}^0 d\sigma G(\sigma)\right]^2 = \frac{t^2}{2} \left[\int_{-\infty}^0 d\sigma G(\sigma)\right]^2 , \quad (7.2.11)$$

and this proves (7.2.6).

7.3 Vacuum Transition Amplitude: Non-Time-Consecutive Diagrams

Now we prove that the non-time-consecutive terms (in square brackets) in (7.2.4), i.e.

$$(t_1, t_3), (t_2, t_4), \quad (t_1, t_4), (t_2, t_3), \quad (7.3.1)$$

vanish in the stochastic limit, i.e. in contrast to the previous case we will now obtain a vanishing limit when $\lambda \to 0$.

Lemma 7.3.1. *One has*

$$\lim_{\lambda \to 0} \lambda^4 \int_0^{t/\lambda^2} dt_1 \int_0^{t_1} dt_2 \int_0^{t_2} dt_3 \int_0^{t_3} dt_4 \; G(t_3 - t_1) G(t_4 - t_2) = 0. \quad (7.3.2)$$

Proof 7.3.1. Let us change the variable t_4 to a variable $\sigma_4 := t_4 - t_2$ in (7.3.2). Then (7.3.2) becomes

$$\lambda^4 \int_0^{t/\lambda^2} dt_1 \int_0^{t_1} dt_2 \int_0^{t_2} dt_3 \int_{-t_2}^{t_3 - t_2} d\sigma_4 G(t_3 - t_1) G(\sigma_4). \quad (7.3.3)$$

Now change from a variable t_3 to a variable $\sigma_3 = t_3 - t_1$. Then (7.3.3) becomes

$$\lambda^4 \int_0^{t/\lambda^2} dt_1 \int_0^{t_1} dt_2 \int_{-t_1}^{t_2 - t_1} d\sigma_3 \int_{-t_2}^{\sigma_3 + t_1 - t_2} d\sigma_4 G(\sigma_3) G(\sigma_4). \quad (7.3.4)$$

After the rescalings

$$t_2 \to \sigma_2 := \lambda^2 t_2, \quad t_1 \to \sigma_1 := \lambda^2 t,$$

(7.3.4) takes the form

$$\int_0^t d\sigma_1 \int_0^{\sigma_1} d\sigma_2 \int_{\frac{-\sigma_1}{\lambda^2}}^{\frac{\sigma_2 - \sigma_1}{\lambda^2}} d\sigma_3 \int_{-\sigma_2/\lambda^2}^{\sigma_3 + \frac{\sigma_1 - \sigma_2}{\lambda^2}} d\sigma_4 G(\sigma_3) G(\sigma_4). \quad (7.3.5)$$

The integral (7.3.5) is dominated in modulus by

$$\int_0^t d\sigma_1 \int_0^{\sigma_1} d\sigma_2 \int_{-\sigma_1/\lambda^2}^{(\sigma_2-\sigma_1)/\lambda^2} d\sigma_3 \int_{-\infty}^{\infty} d\sigma_4 |G(\sigma_3)|\,|G(\sigma_4)|$$
$$= \int_0^t d\sigma_1 \int_0^{\sigma_1} d\sigma_2 \int_{-\sigma_1/\lambda^2}^{(\sigma_2-\sigma_1)/\lambda^2} d\sigma_3 |G(\sigma_3)| \int_{-\infty}^{\infty} d\sigma_4 |G(\sigma_4)|. \quad (7.3.6)$$

Since $G(\sigma_3)$ is an integrable function and since $\sigma_2 - \sigma_1 \leq 0$, when $\lambda \to 0$ one has, for almost all σ_1 and σ_2,

$$\lim_{\lambda\to 0} \int_{-\sigma_1/\lambda^2}^{(\sigma_2-\sigma_1)/\lambda^2} d\sigma_3 G(\sigma_3) = 0. \quad (7.3.7)$$

Therefore the whole expression (7.3.6) vanishes as $\lambda \to 0$. The relation (7.3.2) is proved.

Lemma 7.3.2. *One has*

$$\lim_{\lambda\to 0} \lambda^4 \int_0^{t/\lambda^2} dt_1 \int_0^{t_1} dt_2 \int_0^{t_2} dt_3 \int_0^{t_3} dt_4 G(t_4 - t_1) G(t_3 - t_2) = 0. \quad (7.3.8)$$

Proof 7.3.2. In (7.3.8) change the variable t_4 to a variable $\sigma_4 := t_4 - t_1$. Then it becomes

$$\lambda^4 \int_0^{t/\lambda^2} dt_1 \int_0^{t_1} dt_2 \int_0^{t_2} dt_3 \int_{-t_1}^{t_3-t_1} d\sigma_4 G(\sigma_4) G(t_3 - t_2). \quad (7.3.9)$$

Now change from a variable t_3 to a variable $\sigma_3 := t_3 - t_2$. Then (7.3.9) is equal to

$$\lambda^4 \int_0^{t/\lambda^2} dt_1 \int_0^{t_1} dt_2 \int_{-t_2}^{0} d\sigma_3 \int_{-t_1}^{\sigma_3+t_2-t_1} d\sigma_4 G(\sigma_4) G(\sigma_3). \quad (7.3.10)$$

After the rescalings

$$t_2 \to \sigma_2 = \lambda^2 t_2, \quad t_1 \to \sigma_1 = \lambda^2 t_1,$$

(7.3.10) takes the form

$$\int_0^t d\sigma_1 \int_0^{\sigma_1} d\sigma_2 \int_{-\sigma_2/\lambda^2}^{0} d\sigma_3 \int_{-\sigma_1/\lambda^2}^{\sigma_3+\frac{\sigma_2-\sigma_1}{\lambda^2}} d\sigma_4 G(\sigma_4) G(\sigma_3). \quad (7.3.11)$$

When $\lambda \to 0$ (7.3.11) is dominated in modulus by

$$\int_0^t d\sigma_1 \int_0^{\sigma_1} d\sigma_2 \int_{-\sigma_1/\lambda^2}^{\frac{\sigma_2-\sigma_1}{\lambda^2}} d\sigma_4 |G(\sigma_4)| \int_{-\infty}^{\infty} d\sigma_3 |G(\sigma_3)|. \quad (7.3.12)$$

Since $G(\sigma_4)$ is an integrable function, one has, for almost every σ_1 and σ_2,

$$\lim_{\lambda\to 0}\int_{-\sigma_1/\lambda^2}^{\frac{\sigma_2-\sigma_1}{\lambda^2}} d\sigma|G(\sigma_4)| = 0.$$

Therefore the whole expression (7.3.12) vanishes as $\lambda \to 0$. The relation (7.3.8) is proved.

7.4 The Causal δ-Function and the Time-Consecutive Principle

The simplest model in which the stochastic limit can be performed, the spin–boson Hamiltonian discussed in Sect. 5.8, has led to the emergence of quantum white noise b_t, b_t^+ and of the *the white noise Hamiltonian equation*

$$\partial_t U_t = -i(D^+ b_t + D b_t^+)U_t, \quad U_o = 1. \tag{7.4.1}$$

In the present chapter we begin to study this equation.

From the right-hand side of (7.4.1), the formal unitarity of U_t is apparent because it has an Hamiltonian form with the formally Hermitean Hamiltonian $H_t := (D^+b_t + Db_t^+)$. However, since b_t, b_t^+ are operator-valued distributions, one has to specify how to interpret equation (7.4.1) or how to give a precise meaning to the iterated series (7.4.1), involving products of such operators. This is done using the fact that the linear space generated by the number vectors (smeared with some test function) is invariant under the action of both b_t and b_t^+. Therefore one can interpret (7.4.1) in a weak sense with respect to them, i.e. considering the ordinary differential equation obtained by taking matrix elements of both sides of (7.4.1) with respect to these vectors and then exchanging the time derivative with the scalar product.

Having interpreted in a rigorous way the distribution (7.4.1), the next step is *to work with it*, in particular to prove that it has a solution, that the solution is unitary and that it coincides with the limit (7.1.2). To this goal it is convenient to write the right-hand side of (7.4.1) in normal form i.e. instead of the term $b_t U_t$, we want to have the term $U_t b_t$. This shall give rise to a commutator that we are going to compute. In doing so an additional term shall arise in the equation: this is what, in the stochastic calculus terminology, is called the *Ito correction term* (see Sect. 4.13) and it has a deep physical meaning being related to the fluctuation–dissipation theorem (see Sect. 4.16) and the *dispersion relations* (see Sect. 4.21).

To carry on this calculation we develop a calculus for the operator-valued distributions b_t, b_t^+, i.e. for quantum white noise. The details of this calculus are in [AcLuVo99]. In the present chapter we state without proofs its two main rules, namely:

(i) *the causal commutator rule;*
(ii) *the time–consecutive principle.*

Recall that the white noise operators b_t, b_t^+ satisfy the commutation relations (2.15.1a,b,c); however crucial rule in dealing correctly with the normal form of U_t is the *causal commutator rule*: When bringing to normal order the operators b_t, b_t^+ in the iterated series for U_t, the need for commutation between b_σ and b_τ^+, arises only when $\sigma \geq \tau$, and in such a case we replace the usual commutation rule $[b_\sigma, b_\tau^+] = \gamma\delta(\sigma - \tau)$ (see (5.2.13)) by the causal commutation rule, defined by

$$[b_\sigma, b_\tau^+] = \gamma_- \delta_+(\sigma - \tau), \quad \sigma \geq \tau, \tag{7.4.2}$$

where δ_+ denotes the δ-functions on the standard simplex, described in Sect. 7.5, and γ_- is given by (4.20.1).

Remark 7.4.1. Note that the causal commutator is defined only for σ (the time index of the annihilator) greater than τ (the time index of the creator); for $\sigma < \tau$, the causal commutator (7.4.2) is not defined. So the causal commutator rule should not be interpreted as a new commutation relation, but only as a rule to bring to normal order the creation and annihilation operators which appear in the iterated series for U_t.

Remark 7.4.2. The scalar products arising from commutators of creation and annihilation operators *inside the iterated series*, i.e. those arising when putting in normally ordered form the products $H_{t_1} H_{t_2} \ldots H_{t_n}$, are called *internal lines* and those arising from the action of some annihilator on some vector are called *external lines*. Therefore the causal commutator rule can be rephrased as follows: all the internal lines correspond to causal commutators of the form (7.4.2); all the external lines correspond to standard white noise commutators with the usual δ-function.

Remark 7.4.3. This rule has its roots in the peculiar features of the theory of δ-functions (and in general of distributions) on the simplex

$$\Delta_t^{(n)} := \{t \geq t_1 \geq \ldots \geq t_n \geq 0\}, \tag{7.4.3}$$

which is discussed in Sect. 7.5 and in [AcLuVo99]. In fact the original Hamiltonian theory can be considered as a regularized form of the limit white noise theory, and since the nth term of the iterated series is expressed as an integral over the nth simplex, the commutation rules performed under this integral give rise in the limit not to the standard commutator computed below but to the causal commutator (7.4.2).

To get a feel of the origin of the difference between the ordinary commutator and the causal commutator it is useful to compare the non-time-ordered limit:

$$\lim_{\lambda\to 0}\int_0^t d\tau\int_0^t d\tau\left[\frac{1}{\lambda}a_{\tau/\lambda^2},\frac{1}{\lambda}a^+_{\sigma/\lambda^2}\right]\varphi(\tau)\psi(\sigma)=\gamma\int_0^t\varphi(\tau)\psi(\tau)d\tau \quad (7.4.4)$$

with the time-ordered limit

$$\lim_{\lambda\to 0}\int_0^t d\tau\int_0^\tau d\sigma\left[\frac{1}{\lambda}a_{\tau/\lambda^2},\frac{1}{\lambda}a^+_{\sigma/\lambda^2}\right]\varphi(\tau)\psi(\sigma)=\gamma_-\int_0^t\varphi(\tau)\psi(\tau)d\tau, \quad (7.4.5)$$

where γ is defined by

$$\gamma:=\int_{-\infty}^{+\infty}dtG(t)=\int_{-\infty}^{+\infty}dt\langle g,S_tg\rangle=\int_{-\infty}^{+\infty}dt\int_{\mathbb{R}^d}e^{-it(\omega_1(k)-\omega_o)}\mid g(k)\mid^2 \quad (7.4.6)$$

and γ_- by (4.20.1). Both limits are meant in the sense of correlators, and their validity follows immediately from Lemma 1.8.1.

Finally remember that the nth-order term in the iterated series is obtained by integrating over the simplex $\Delta_t^{(n)}$, given by (7.4.3), the product of operators $H_{\mathrm{I}}(t_1)\dots H_{\mathrm{I}}(t_n)$ indexed by the ordered n-tuple $t\geq t_1\geq\dots\geq t_n\geq 0$. We say that two times t_j, t_k in this n-tuple are *consecutive* if $j=k+1$ or $k=j+1$. By bringing the product $H_{\mathrm{I}}(t_1)\dots H_{\mathrm{I}}(t_n)$ to normal order, several scalar products of the form $\langle g,S_{t_q-t_p}g\rangle$ will appear and in fact they are in one-to-one correspondence with the above-defined internal lines.

The *time-consecutive principle*, which is probably the most fundamental principle of the stochastic limit theory, states that: if a term, obtained by bringing the product $H_{\mathrm{I}}(t_1)\dots H_{\mathrm{I}}(t_n)$ to normal order, contains a scalar products of the form $\langle g,S_{t_q-t_p}g\rangle$ with $q-1\neq p$, then in the stochastic limit this term vanishes. Stated otherwise and slightly more generally: in the stochastic limit only those terms whose only internal lines correspond to pairings between time-consecutive operators $H_{\mathrm{I}}(t_q)$ and $H_{\mathrm{I}}(t_p)$, survive.

The proof of this principle depends on some estimates on Feynman diagrams which are described in Sects. 15.1–4 and will be given in detail in Sects. 15.5–7. In the following section we show how this principle is reflected in the structure of the master field, but to this goal the usual theory of distributions has to be extended to the standard simplex $\Delta_1^{(n)}$: this extension is nontrivial because the boundary terms here play a crucial role.

7.5 Theory of Distributions on the Standard Simplex

Here we state some simple rules to work with δ-functions on the standard simplex

$$\Delta_t^{(n)}:=t\geq t_1\geq\dots\geq t_n\geq 0. \quad (7.5.1)$$

Such a δ-function is called *causal* because, as it will be clear from the following developments, its emergence is the expression of the *causality condition* which is coded into the decreasing order $(t \geq t_1 \geq \cdots \geq t_n)$ of the time variables in the iterated series (7.4.3). The distribution theory developed to handle this situation differs from the standard one [Vl83, ChBr73] because the space of the test functions considered here is more irregular. By taking matrix elements with respect to an appropriate set of vectors, all the identities in this section can be applied to the case when $\varphi(t)$ is an operator-valued function. We shall give proofs only in the simplest cases, and we refer to [AcLuVo99] for a full mathematical treatment. If f is an integrable function on $\mathbb{R}_+$ such that $\int_0^\infty f(\sigma)d\sigma = 1$, then one has

$$\lim_{\lambda\to 0}\int_0^t \frac{1}{\lambda^2} f\left(\frac{t-t_1}{\lambda^2}\right) dt_1 = \lim_{\lambda\to 0}\int_0^{t/\lambda^2} f(\sigma)d\sigma = 1.$$

This relation we write symbolically using δ_+-functions as

$$\int_0^t \delta_+(t-t_1)dt_1 = 1. \tag{7.5.2}$$

For example, using (7.5.2) one has

$$\int_0^t dt_1 \int_0^{t_1} dt_2 \delta_+(t_1-t_2) = \int_0^t dt_1 = t. \tag{7.5.3}$$

Similarly, using repeatedly (7.5.2) and an appropriate generalization of the standard rules for tensor products of distributions, one can write the expression

$$\int_0^t dt_1 \int_0^{t_1} dt_2 \int_0^{t_2} dt_3 \int_0^{t_3} dt_4 \delta_+(t_1-t_2)\delta_+(t_3-t_4)$$

in the form

$$\int_0^t dt_1 \int_0^{t_1} dt_2 \delta_+(t_1-t_2) \int_0^{t_2} dt_3 \int_0^{t_3} dt_4 \delta_+(t_3-t_4).$$

Then one finds

$$\int_0^t dt_1 \int_0^{t_1} dt_2 \int_0^{t_2} dt_3 \int_0^{t_3} dt_4 \delta_+(t_1-t_2)\delta_+(t_3-t_4) = \frac{t^2}{2}. \tag{7.5.4}$$

On the other hand, it is intuitively clear that

$$\int_0^t dt_1 \int_0^{t_1} dt_2 \int_0^{t_2} dt_3 \delta_+(t_1-t_3) = 0, \tag{7.5.5}$$

because the dt_3 integral (considered as a function of t_2) is zero for all values of t_2 with the exception of $t_2 = t_1$.

It is instructive to prove the identity (7.5.5) by using a regularized version of δ_+. In this case the identity (7.4.2) reads

$$\lim_{\lambda\to 0}\int_0^t dt_1 \int_0^{t_1} dt_2 \int_0^{t_2} dt_3 \frac{1}{\lambda^2} f\left(\frac{t_1-t_3}{\lambda^2}\right)$$

$$= \lim_{\lambda\to 0}\int_0^t dt_1 \int_0^{t_1} dt_2 \int_{(t_1-t_2)/\lambda^2}^{t_1/\lambda^2} d\sigma f(\sigma), \qquad (7.5.6)$$

and it is clear that the right-hand side of (7.5.6) is zero for any integrable function f on $\mathbb{R}$ because $t_1 \geq t_2$. More generally, let us introduce the following definition:

Definition 7.5.1. *Let m be a natural integer and let the indices $\{p_k\}$, $\{q_k\}$ ($k = 1 \dots m$) define a partition of the set $\{1, 2 \dots 2m\}$ such that for each k*

$$q_k + 1 \leq p_k. \qquad (7.5.7)$$

An integral of the form

$$\int_0^t dt_1 \int_0^{t_1} dt_2 \dots \int_0^{t_{2m-1}} dt_{2m} \prod_{k=1}^{m} \delta_+(t_{q_k} - t_{p_k}) \qquad (7.5.8)$$

shall be called time-consecutive (type I) if (7.5.8) is equal to

$$\int_0^t dt_1 \int_0^{t_1} dt_2 \dots \int_0^{t_{2m-1}} dt_{2m} \delta_+(t_1-t_2)\delta_+(t_3-t_4)\dots\delta_+(t_{2m-1}-t_{2m}), \qquad (7.5.9)$$

i.e. if in (7.5.7) the equality holds for each $k = 1 \dots m$.

Otherwise, i.e. if there exists an index $\alpha \in \{1 \dots m\}$ for which

$$q_\alpha + 1 < p_\alpha, \qquad (7.5.10)$$

it shall be called non-time-consecutive (type II).

In [AcLuVo99] we proved the following:

Lemma 7.5.1. *For any $m \in \mathbb{N}$ and $t > 0$ one has, for type I integrals,*

$$\int_0^t dt_1 \int_0^{t_1} dt_2 \dots \int_0^{t_{2m-1}} dt_{2m} \delta_+(t_1-t_2)\delta_+(t_3-t_4) \dots \delta_+(t_{2m-1}-t_{2m}) = \frac{t^m}{m!} \qquad (7.5.11)$$

while for type II integrals, i.e. if the indices $\{p_k\}$ and $\{q_k\}$ satisfy condition (7.5.10), it follows that

$$\int_0^t dt_1 \int_0^{t_1} dt_2 \dots \int_0^{t_{2m-1}} dt_{2m} \prod_{k=1}^{m} \delta_+(t_{q_k} - t_{p_k}) = 0. \qquad (7.5.12)$$

7.6 The Second-Order Term of the Limit Vacuum Amplitude

In Sects. 7.2 and 7.3 we computed the limit of the second-order term of the iterated series for $U_t^{(\lambda)}$ from the Hamiltonian form (7.1.1). It is instructive to see how one can obtain the same expression directly from the white noise equation (7.1.3). More precisely, in this and in Sect. 7.7 we show that the second-order term in the iterated series expansion of U_t in $\langle U_t \rangle$ gives the correct (operator) decay rate, i.e. the exponent of the right-hand side of (7.2.1). The second-order term of the vacuum expectation value of the iterated series (7.1.2) is

$$\begin{aligned}(-i)^2 \int_0^t dt_1 \int_0^{t_1} dt_2 \left\langle D^+ b_{t_1} D b_{t_2}^+ \right\rangle &= -D^+ D \int_0^t dt_1 \int_0^{t_1} dt_2 \langle b_{t_1} b_{t_2}^+ \rangle \\ &= -D^+ D \gamma_- \int_0^t dt_1 \int_0^{t_1} dt_2 \delta_+(t_1 - t_2) = -\gamma_- t D^+ D, \quad (7.6.1)\end{aligned}$$

where we have used (7.4.2). Note that the second-order term in the iterated series of the limit equation is the limit as $\lambda \to 0$ of the second-order term in the expansion of $U_{t/\lambda^2}^{(\lambda)}$.

7.7 The Fourth-Order Term of the Limit Vacuum Amplitude

Now let us perform for the fourth-order term the calculation made in the previous section for the second-order one. In this case one has

$$\begin{aligned}(-i)^4 \int_0^t dt_1 \int_0^{t_1} dt_2 \int_0^{t_2} dt_3 \int_0^{t_3} dt_4 \left\{ D^+ D^+ D D \left\langle b_{t_1} b_{t_2} b_{t_3}^+ b_{t_4}^+ \right\rangle \right. \\ \left. + D^+ D D^+ D \left\langle b_{t_1} b_{t_2}^+ b_{t_3} b_{t_4}^+ \right\rangle \right\} . \quad (7.7.1)\end{aligned}$$

Using the rules for Gaussian expectation values, this is equal to

$$\begin{aligned}(-i)^4 \int_0^t dt_1 \int_0^{t_1} dt_2 \int_0^{t_2} dt_3 \int_0^{t_3} dt_4 \gamma_-^2 \{ D^+ D^+ D D \cdot [\delta_+(t_4 - t_2) \delta_+(t_3 - t_1) \\ + \delta_+(t_4 - t_1) \delta_+(t_3 - t_2)] + D^+ D D^+ D \delta_+(t_2 - t_1) \delta_+(t_4 - t_3) \}. \\ (7.7.2)\end{aligned}$$

We know from Lemma 7.5.1 that the terms in square brackets containing non-time-consecutive δ_+-functions vanish. Using (7.5.11) to compute the integral of the product of time-consecutive δ_+-functions $\delta_+(t_1 - t_2)\delta_+(t_3 - t_4)$, we find that (7.7.1) is equal to $\frac{1}{2}\gamma_-(D^+ D)^2 t^2$. This is equal to the limit computed in Lemma 7.2.1.

7.8 Higher-Order Terms of the Vacuum–Vacuum Amplitude

One can proceed in this way, computing all the terms in the iterated expansion (7.1.2) and the check that they coincide with the corresponding terms in the limit (7.1.1).

Using the results of Sect. 7.5 above, it is easy to verify that in the normally ordered form of the iterated series (7.1.2) for U_t all the non-time-consecutive (type II) integrals vanish identically and only the time-consecutive (type I) integrals survive.

This suggests the natural conjecture that in the limit $\lambda \to 0$ of the iterated series (7.1.1) for $U^{(\lambda)}_{t/\lambda^2}$ all the non-time-consecutive (type II) integrals vanish identically and only the time-consecutive (type I) integrals survive. In Chap. 15 we shall prove that this is in fact the case.

7.9 Proof of the Normal Form of the White Noise Hamiltonian Equation

Now we apply the rules introduced in the previous sections to the computation of the commutator $[b_t, U_t]$. All the operator identities in this section are meant weakly in the number vectors.

Lemma 7.9.1. *Consider the boson Fock white noise Hamiltonian equation (7.1.3) i.e.*

$$\partial_t U_t = -iH_t U_t = -i(D^+ b_t + D b_t^+) U_t \tag{7.9.1}$$

and suppose that the iterated series (7.1.1) converges to U_t weakly on the n-particle vectors and that the map $t \mapsto U_t$ is weakly continuous on these vectors. Then, under the causal commutator condition (7.4.2), i.e.

$$[b_t, b_\tau^+] = \gamma_- \delta_+(t-\tau), \quad t \geq \tau,$$

the following identities hold:

$$[b_t, U_t] = b_t U_t - U_t b_t = -i\gamma_- D U_t, \tag{7.9.2}$$

$$[U_t^*, b_t^+] = U_t^* b_t^+ - b_t^+ U_t^* = i\overline{\gamma}_- U_t^* D^+, \tag{7.9.3}$$

$$[b_t, U_t^*] = i\gamma_- U_t^* D. \tag{7.9.4}$$

Proof 7.9.1. Denote $U_t^{(n)}$ the nth iterate of the integral form of (7.9.1), i.e.

$$iU_t^{(n+1)} = i + \int_0^t (D^+ b_\sigma U_\sigma^{(n)} + D b_\sigma^+ U_\sigma^{(n)}) d\sigma, \quad U_t^{(0)} = 1. \tag{7.9.5}$$

For $n = 0$, $U_t^{(n)} = 1$ and the identities

$$\int_0^\tau d\sigma b_\sigma [b_t, U_\sigma^{(n)}] = \int_0^\tau d\sigma b_\sigma^+ [b_t, U_\sigma^{(n)}] = 0 \tag{7.9.6}$$

clearly hold for any τ. Assuming by induction that (7.9.6) holds for n and using (7.9.5), one finds, for any $t \geq \tau$,

$$[b_t, iU_\tau^{(n+1)}] = \int_0^\tau D\gamma_- \delta_+(t-\sigma)U_\sigma^{(n)} + Db_\sigma^+[b_t, U_\sigma^{(n)}] + D^+ b_\sigma [b_t, U_\sigma^{(n)}]\, d\sigma. \tag{7.9.7}$$

Using the definition of δ_+ we know that the right-hand side of (7.9.7) is equal to

$$\chi_{[0,\tau]}(t)\gamma_- DU_\tau^{(n)} + \int_0^\tau d\sigma \{Db_\sigma^+[b_t, U_\sigma^{(n)}] + D^+ b_\sigma [b_t, U_\sigma^{(n)}]\} \tag{7.9.8}$$

By the induction assumption the integral term in (7.9.8) is zero. Therefore,

$$[b_t, iU_\tau^{(n+1)}] = \chi_{[0,\tau]}(t)\gamma_- DU_t^{(n)}, \quad n \geq 1. \tag{7.9.9}$$

Taking matrix elements in the number vectors of both sides of (7.9.9) and letting $n \to \infty$ with $\tau = t$, we obtain

$$[b_t, iU_t] = \gamma_- DU_t,$$

which is equivalent to (7.9.2). By taking the adjoints of both sides of (7.9.2), one finds (7.9.3). Finally, from

$$\partial_t U^* = iU_t^*(b_t^+ D + b_t D^+)$$

or

$$U_t^* - 1 = i\int_0^t U_s^*(b_s^+ D + b_t D^+)ds,$$

we deduce

$$\begin{aligned}[b_t, U_t^*] &= i\int_0^t [b_t, U_s^*](b_s^* D + b_t D^+)ds + i\int_0^t U_s^* D\gamma_- \delta_+(t-s)ds \\ &= i\int_0^t [b_t, U_s](b_s^* D + b_t D^+)ds + i\gamma_- U_t^* D.\end{aligned}$$

By the same argument used for $[b_t, U_t]$, we conclude that

$$[b_t, U_t^*] = i\gamma_- U_t^* D,$$

which is (7.9.4).

7.10 The Unitarity Condition for the Limit Equation

Note that, while the Hamiltonian in (7.9.1) is formally self-adjoint, it has meaning only as a sesquilinear form, since it is not an operator but only an operator-valued distribution. The normally ordered form of this equation is not in Hamiltonian form; therefore the unitarity of its solution has to be proved. This is done in the present section. In fact here we prove only the isometricity condition ($U_t^* U_t = 1$), which is easy. The co-isometricity condition ($U_t U_t^* = 1$) is more delicate but can be proved on similar lines.

Theorem 7.10.1. *Let D be a bounded operator. Then the solution of*

$$i\partial_t U_t = D^+ U_t b_t + D b_t^+ U_t + \beta D^+ D U_t, \tag{7.10.1}$$

with

$$[b_t, b_\tau^+] = \gamma_- \delta_+(\tau - t), \tag{7.10.2}$$

is isometric, i.e. $U_t^* U_t = 1$, if and only if

$$\beta = -\gamma_- + i\lambda, \tag{7.10.3}$$

where λ is an arbitrary real number.

Remark 7.10.1. The equation found in Sect. 4.12 corresponds to the case $\lambda = 0$.

Proof 7.10.1. Taking the adjoint of both sides of (7.10.1), we find

$$i\partial U_t^* = -b_t^+ U_t^* D - U_t^* b_t D^+ - \overline{\beta} U_t^* D^+ D. \tag{7.10.4}$$

Therefore, differentiating the product $U_t^* U_t$ by the usual Leibniz rule,

$$\begin{aligned} i\partial(U_t^* U_t) &= i\partial U_t^* \, U_t + iU_t^* \partial U_t \\ &= (-b_t^+ U_t^* D - U_t^* b_t D^+ - \overline{\beta} U_t^* D^+ D) U_t \\ &\quad + U_t^* (D^+ U_t b_t + D b_t^+ U_t + \beta D^+ D U_t) \\ &= - b_t^+ U_t^* D U_t - U_t^* b_t D^+ U_t - \overline{\beta} U_t^* D^+ D U_t \\ &\quad + U_t^* D^+ U_t b_t + U_t^* D b_t^+ U_t + \beta U_t^* D^+ D U_t. \end{aligned}$$

Since D commutes with the b_t^ε, this is equal to

$$[U_t^*, b_t^+] D U_t - U_t^* D^+ [b_t, U_t] + (\beta - \overline{\beta}) U_t^* D^+ D U_t. \tag{7.10.5}$$

We know from Lemma 7.9.1 that the commutator $[b_t, U_t]$ is $-i\gamma_- D U_t$. Using this we find

$$\begin{aligned} i\partial(U_t^* U_t) &= i\gamma_- U_t^* D^+ D U_t + i\overline{\gamma}_- U_t^* D^+ D U_t + (\beta - \overline{\beta}) U_t^* D^+ D U_t \\ &= (i\overline{\gamma}_- + i\gamma_- + \beta - \overline{\beta}) U_t^* D^+ D U_t. \end{aligned}$$

This shows that the isometricity condition

$$\partial_t(U_t^* U_t) = 0, \quad U_o^* U_o = 1,$$

is equivalent to

$$i\gamma_- + i\overline{\gamma}_- + \beta - \overline{\beta} = 0 \Leftrightarrow \text{ Re } \gamma_- = - \text{ Im } \beta \tag{7.10.6}$$

and this is equivalent to (7.10.3).

7.11 Normal Form of the Thermal White Noise Equation: Boson Case

In this section we outline the proof of the normally ordered white noise equation obtained in Sect. 5.9 by application of the stochastic golden rule. This normally ordered form in the nonFock case is a more delicate problem and we are going to discuss it in this section in the case of a gauge-invariant state. The white noise calculus for nonFock noises is reduced to the Fock case through the following two remarks.

Remark 7.11.1. Since a finite temperature white noise is a sum of two independent Fock white noises (see Sect. 2.18), the calculus for it follows the general rule described in Sect. 7.9.

Remark 7.11.2. If an equation is driven by a family of independent thermal white noises, its normal form is the sum of the normal forms corresponding to each of the independent white noises.

By analogy with (4.12.5) and in the notation of (2.18.3), let us introduce the semiscalar products

$$\gamma_{1-} := \int_{-\infty}^{0} d\sigma \int dk e^{i\sigma\omega(k)} |g(k)|^2 (n(k) + 1) = \gamma_{2-} + \gamma_-, \tag{7.11.1}$$

$$\gamma_{2-} := \int_{-\infty}^{0} d\sigma \int dk e^{-i\sigma\omega(k)} |g(k)|^2 n(k) = \gamma_{1-} - \gamma_-. \tag{7.11.2}$$

Proposition 7.11.1. *Let U_t be the iterated series solution of (7.9.1) where $b(t)$ is a white noise of the form (5.9.4a,b) and $b_t = b_1(t) + b_2(t)$ is its representation as described in Sect. 2.18. Then the following identities hold:*

$$[b_1(t), U_t] = -i\gamma_{1-} D U_t, \tag{7.11.3}$$

$$[b_2(t), U_t] = -i\gamma_{2-} D^+ U_t. \tag{7.11.4}$$

Proof 7.11.1. Consider the integral form of (7.9.1), i.e.

$$U_t = 1 - i \int_0^t ds (D b_s^+ + D^+ b_s) U_s. \tag{7.11.5}$$

Using the canonical representation (2.18.3) of b_t in terms of Fock white noises, one writes (7.11.5) in the form

$$U_t = 1 - i\int_0^t ds\left[Db_1^+(s) + Db_2(s) + D^+b_1(s) + D^+b_2^+(s)\right]U_s.$$

Therefore, using Lemma 7.9.1, one has

$$\begin{aligned}
[b_1(t), U_t] &= -i\int_0^t D[b_1(t), b_1^+(s)]U_s \\
&= -i\int_0^t D\gamma_1\delta_+(t-s)U_s ds = -i\gamma_{1-}DU_s, \\
[b_2(t), U_t] &= -i\int_0^t dsD^+[b_2(t), b_2^+(s)]U_s \\
&= -i\int_0^t dsD^+\gamma_{2-}\delta_+(t-s)U_s = -i\gamma_{2-}D^+U_t,
\end{aligned}$$

and this proves (7.11.3) and (7.11.4).

Theorem 7.11.1. *The normal form of (7.11.5) is*

$$\partial_t U_t = -i(Db^+(t)U_t + D^+U_t b(t)) - \gamma_{1-}D^+DU_t - \gamma_{2-}DD^+U_t. \tag{7.11.6}$$

Proof 7.11.2. Using the canonical representation (2.18.3) of b_t we write (7.11.5) in the form

$$\begin{aligned}
\partial_t U_t = {} & Db_1^+(t)U_t + DU_t b_2(t) + D[b_2(t), U_t] + D^+U_t b_1(t) \\
& + D^+[b_1(t), U_t] + D^+b_2^+(t)U_t,
\end{aligned}$$

and (7.11.6) follows from the identities (7.11.3) and (7.11.4).

7.12 From White Noise Calculus to Stochastic Calculus

A normally ordered white noise Hamiltonian equation can also be interpreted as a stochastic differential equation. The following is a simple formal rule which allows one to pass from one type of equation to the other:

(i) Start from the normally ordered form of the white noise Hamiltonian equation:

$$\partial_t U_t = -i(Db_t^+U_t + D^+U_t b_t) - YU_t; \tag{7.12.1}$$

(ii) replace ∂_t by $\frac{d}{dt}$ and multiply both sides of (7.12.1) by dt:

$$dU_t = -i(Db_t^+ dtU_t + D^+U_t b_t dt) - YU_t dt; \tag{7.12.2}$$

(iii) replace $b_t^{\pm} dt$ by $dB_t^{\pm}$:

$$dU_t = -i(DdB_t^+ U_t + D^+ U_t dB_t) - YU_t dt. \tag{7.12.3}$$

(iv) In the Boson case one can, in addition to the above, use the commutation rule,

$$dB_t^+ U_t = U_t dB_t^+, \tag{7.12.4}$$

to obtain the final form of the stochastic differential equation

$$dU_t = (-iDdB_t^+ - iD^+ dB_t - Ydt)U_t. \tag{7.12.5}$$

8. Chronological Product Approach to the Stochastic Limit

In this chapter we give a proof of the convergence of the rescaled evolution operator $U^{(\lambda)}_{t/\lambda^2}$ corresponding to the interaction Hamiltonian considered in Sect. 4.10. For these simple Hamiltonians (in particular with both dipole and rotating-wave approximation), the chronological product technique works well, but for more complicated interactions the combinatorics becomes intricate and the more powerful analytical technique developed in Part III is more effective.

8.1 Chronological Products

Definition 8.1.1. *The* chronological *(or* time-ordered*)* product *of n operators $X_{t_1}, \dots, X_{t_n}$ depending on a parameter $t_j \in \mathbb{R}$ is defined by (see, for example, [BoSch87])*

$$T[X_1(t_1)\dots X_n(t_n)] = X(t_{\pi_1})\dots X(t_{\pi_n}), \tag{8.1.1}$$

where π is the unique permutation of $\{1\dots n\}$ such that

$$t_{\pi_1} \le t_{\pi_2} \le \dots \le t_{\pi_n}. \tag{8.1.2}$$

The uniqueness of π holds if in the n-tuple $(t_1, \dots, t_n)$ there are no equal coordinates, but the set of n-tuples for which this condition is not satisfied has zero measure and products of the form (8.1.1) *shall appear only under iterated integrals. Therefore in the following we shall consider π to be uniquely determined.*

Note that (8.1.1) *is equivalent to*

$$T(A_{t_1}\dots A_{t_k}) = \sum_{\pi\in\mathcal{P}_k} \theta(t_{\pi_1} > \dots > t_{\pi_k}) \cdot A_{t_{\pi_1}} \dots A_{t_{\pi_k}}, \tag{8.1.3}$$

where $\mathcal{P}_k$ is the permutation group and

$$\theta(t_{\pi_1} > t_{\pi_2} > \dots > t_{\pi_k}) = \begin{cases} 1 & \text{if } t_{\pi_1} > t_{\pi_2} > \dots > t_{\pi_k} \\ 0 & \text{otherwise} \end{cases} .$$

The chronological exponent is defined as the solution of

$$i\partial_t U(t_0,t) = H_{\rm I}(t)U(t_0,t), \quad U(t_0,t_0) = 1, \tag{8.1.4}$$

$$H_{\rm I}(t) := e^{itH_0} \cdot H_{\rm I} \cdot e^{-itH_0}. \tag{8.1.5}$$

Its connection with the chronological product is obtained by writing the iterated series for the solution of (8.1.4), i.e.

$$\begin{aligned} U(t_0,t) &= 1 + \sum_{n\geq 1}(-i)^n \int_{t_0}^{t} dt_1 \int_{t_0}^{t_1} dt_2 \ldots \int_{t_0}^{t_n} dt_n H_1(t_1)H_1(t_2)\ldots H_1(t_n) \\ &= 1 + \sum_{n\geq 1} \frac{(-i)^n}{n!} \int_{t_0}^{t} dt_1 \ldots \int_{t_0}^{t} T[H_1(t_1)\ldots H_1(t_n)] \\ &= T\exp\left(-i\int_{t_0}^{t} H_1(s)ds\right). \end{aligned} \tag{8.1.6}$$

So, at least formally, the exponential chronological product is defined by

$$U(t_0,t) = T\exp\left(-i\int_{t_0}^{t} H_1(s)ds\right).$$

We want to apply the chronological product technique to find the limit of the rescaled evolution operator

$$\begin{aligned} U^{(\lambda)}_{t/\lambda^2} &= T\exp\left(-i\int_0^{t/\lambda^2} \lambda V_\tau d\tau\right) \\ &= \sum_{h=0}^{\infty} \frac{(-i)}{h!} \lambda^n \int_0^{t/\lambda^2} dt_1 \ldots \int_0^{t/\lambda^2} dt_n T[H_{\rm I}(t_1)\ldots H_{\rm I}(t_n)], \end{aligned} \tag{8.1.7}$$

where in the notations of Sect. 4.10

$$H_{\rm I}(t) = \{D^+ \otimes A(S_t g) + D \otimes A^+(S_t g)\} = D^+ A_t + D A_t^+.$$

8.2 Chronological Product Approach to the Stochastic Limit

Let H be an Hilbert space and $\mathcal{H}_R = \Gamma(H)$ be a corresponding bosonic Fock space. Let for any $f \in H$ be given a family of creation and annihilation operators $A_t(f)$, $A_t^+(f)$ depending on real parameter t with the commutator

$$[A_t(f), A_\tau^+(g)] = G(\tau - t), \tag{8.2.1}$$

where $G(t) = G(t|f,g)$ is a continuous integrable function

$$\int |G(t|f,g)|dt < \infty. \tag{8.2.2}$$

Let D and D^+ be bounded operators in an Hilbert space $\mathcal{H}_s$ and let $U_t^{(\lambda)}$ be the evolution operator in the Hilbert space $\mathcal{H} = \mathcal{H}_R \otimes \mathcal{H}_s$ depending on a real parameter λ and satisfying

$$\frac{dU_t^{(\lambda)}}{dt} = -i\lambda\{D^+ \otimes A_t(g) + D \otimes A_t^+(g)\}U_t^{(\lambda)}, \tag{8.2.3}$$
$$U_t^{(\lambda)} = I.$$

We will prove the existence and compute the limit

$$\lim_{\lambda\to 0} U_{t/\lambda^2}^{(\lambda)} = U_t. \tag{8.2.4}$$

We understand the limit in the sense of matrix elements between collective vectors. In fact we first compute the expectation value of $U_{t/\lambda^2}^{(\lambda)}$ in the form of a T product and then compute the limit. We will show that the limit satisfies a quantum stochastic differential equation. To this goal we introduce the notation

$$H_{\rm I}(t) = D_t^+ A_t + D_t A_t^+, \tag{8.2.5}$$
$$D_t = D, \quad \forall t \geq 0, \tag{8.2.6}$$

i.e. we introduce a fictitious time dependence also in the system operator D to keep into account the order of D and D^+ in the iterated series expansion (8.2.7). Taking the vacuum expectation value we obtain

$$\left\langle U_{t/\lambda^2}^{(\lambda)} \right\rangle = \sum_{m=0}^{\infty} \frac{(-i)^{2m}}{(2m)!} \lambda^{2m} \int_0^{t/\lambda^2} dt_1 \ldots \int_0^{t/\lambda^2} dt_{2m} \left\langle T\big(H_{\rm I}(t_1) \ldots H_{\rm I}(t_{2m})\big) \right\rangle.$$

Note that, from the identity

$$H_{\rm I}(t_1) \ldots H_{\rm I}(t_n) = D_{t_1}^{1-\epsilon_1} A_{t_1}^{\epsilon_1} \ldots D_{t_n}^{1-\epsilon_n} A_{t_n}^{\epsilon_n} = D_{t_1}^{1-\epsilon_1} \ldots D_{t_n}^{1-\epsilon_n} A_{t_1}^{\epsilon_1} \ldots A_{t_n}^{\epsilon_n},$$

we deduce

$$\begin{aligned} T\big(H_{\rm I}(t_1) \ldots H_{\rm I}(t_n)\big) &= H_{\rm I}(t_{\pi 1}) \ldots H_{\rm I}(t_{\pi n}) = D_{t_{\pi 1}}^{1-\epsilon_{\pi 1}} A_{t_{\pi 1}}^{\epsilon_{\pi 1}} \ldots D_{t_{\pi n}}^{1-\epsilon_{\pi n}} A_{t_{\pi n}}^{\epsilon_{\pi n}} \\ &= D_{t_{\pi 1}}^{1-\epsilon_{\pi 1}} \ldots D_{t_{\pi n}}^{1-\epsilon_{\pi n}} A_{t_{\pi 1}}^{\epsilon_{\pi 1}} \ldots A_{t_{\pi n}}^{\epsilon_{\pi n}} \\ &= T\big(D_{t_1}^{1-\epsilon_1} \ldots D_{t_n}^{1-\epsilon_n}\big) T\big(A_{t_1}^{\epsilon_1} \ldots A_{t_n}^{\epsilon_n}\big). \end{aligned} \tag{8.2.7}$$

Because of the vacuum expectation value, in the expression $\langle T(H_{\rm I}(t_1) \ldots H_{\rm I}(t_{2m}))\rangle$, only the factors with an equal number of creators and annihilators survive. Since, given $k = 1 \ldots 2m$, there are exactly $\binom{2m}{k}$ terms with k annihilators and $2m - k$ creators and since, because of the time-ordered product, all the terms with $m = k$ have the same expectation, it follows that

$$
\begin{aligned}
I_{2m} := {} & \lambda^{2m} \int_0^{t/\lambda^2} dt_1 \ldots \int_0^{t/\lambda^2} dt_{2m} \langle T(H_I(t_1) \ldots H_I(t_{2m}))\rangle \\
= {} & \frac{(2m)!}{m!m!} \lambda^{2m} \int_0^{t/\lambda^2} dt_1 \ldots \int_0^{t/\lambda^2} dt_m \int_0^{t/\lambda^2} d\tau_1 \ldots \int_0^{t/\lambda^2} d\tau_m \\
& T(D^+_{t_1} \ldots D^+_{t_m} D_{\tau_1} \ldots D_{\tau_m}) \cdot \langle T(A_{t_1} \ldots A_{t_m} A^+_{\tau_1} \ldots A^+_{\tau_m})\rangle
\end{aligned}
$$

By performing the vacuum expectation and keeping into account the identity

$$
\langle T(A_{t_1} \ldots A_{t_m} A^+_{\tau_1} \ldots A^+_{\tau_m})\rangle = m! \prod_{k=1}^{m} G_+(t_k - \tau_k),
$$

where we have introduced of the Feynman propagator

$$
G_+(t-\tau) = \langle T(A_t A^+_\tau)\rangle = \theta(t-\tau)\langle A_t A^+_\tau\rangle, \tag{8.2.8}
$$

which by definition satisfies the causality condition

$$
G_+(\sigma) = 0 \quad \text{if } \sigma < 0 \tag{8.2.9}
$$

and the $m!$ arises because, for any permutation π of the set $\{1 \ldots m\}$, the permutation $(t_k, \tau_k) \mapsto (t_{\pi_k}, \tau_{\pi_k})$ leaves both the scalar and the time-ordered product invariant, we find

$$
\begin{aligned}
I_{2m} = {} & \frac{(2m)!}{m!} \lambda^{2m} \int_0^{t/\lambda^2} dt_1 \ldots \int_0^{t/\lambda^2} dt_m \int_0^{t/\lambda^2} d\tau_1 \ldots \int_0^{t/\lambda^2} d\tau_m \\
& \cdot T(D^+_{t_1} \ldots D^+_{t_m} D_{\tau_1} \ldots D_{\tau_m}) \cdot \prod_{k=1}^{m} G_+(t_k - \tau_k)
\end{aligned}
$$

Using the fact that we can arbitrarily commute the D_{t_α} operators inside the chronological product, we find for the rescaled vacuum-to-vacuum amplitude:

$$
\begin{aligned}
\left\langle U^{(\lambda)}_{t/\lambda^2} \right\rangle &= \sum_{m=0}^{\infty} \frac{(-1)^m}{(2m)!} I_{2m} \\
&= \sum_{m=0}^{\infty} \frac{(-1)^m \lambda^{2m}}{(2m)!} T \prod_{k=1}^{m} \int_0^{t/\lambda^2} dt_k \int_0^{t/\lambda^2} d\tau_k G_+(t_k - \tau_k) D^+_{t_k} D_{\tau_k} \\
&= T \exp\left(-\lambda^2 \int_0^{t/\lambda^2} d\sigma \int_0^{t/\lambda^2} d\tau G_+(\sigma - \tau) D^+_\sigma D_\tau \right).
\end{aligned}
$$

Lemma 8.2.1. *In the notation of (8.2.8) one has*

$$
\left\langle U^{(\lambda)}_{t/\lambda^2} \right\rangle = T \exp\left(- \int_0^t d\sigma \int_0^{\frac{t-\sigma}{\lambda^2}} d\tau G_+(\tau) D^+_{\frac{\sigma}{\lambda^2}+\tau} D_{\frac{\sigma}{\lambda^2}} \right). \tag{8.2.10}
$$

Proof 8.2.1. With the change of variables $t_1 - t_2 = \sigma_1$, we obtain from (8.2.13)

$$\left\langle U^{(\lambda)}_{t/\lambda^2} \right\rangle = T\exp\left(-\lambda^2 \int_0^{t/\lambda^2} dt_2 \int_{t_2}^{\frac{t}{\lambda^2} - t_2} d\sigma_4 D^+_{t_2+\sigma_1} G_+(\sigma_1) D_{t_2} \right)$$
$$= T\exp\left(- \int_0^t d\sigma_2 \int_{\frac{\sigma_2}{\lambda^2}}^{\frac{t-\sigma_2}{2}} d\sigma_1 D^+_{\frac{\sigma_2}{\lambda^2}+\sigma_1} G_+(\sigma_1) D_{\frac{\sigma_2}{\lambda^2}} \right).$$

By the causality condition (8.2.9) this is equal to

$$T\exp\left(- \int_0^t d\sigma_2 \int_0^{\frac{t-\sigma_2}{\lambda^2}} d\sigma_1 G_+(\sigma_1) D^+_{\frac{\sigma_2}{\lambda^2}+\sigma_1} D_{\frac{\sigma_2}{\lambda^2}} \right),$$

and this is (8.2.10).

Remark 8.2.1. Lemma 8.2.1 above suggests a natural guess for the limit

$$\lim_{\lambda\to 0} \left\langle U^{(\lambda)}_{t/\lambda^2} \right\rangle; \tag{8.2.11}$$

in fact, recalling from (8.2.6) that $D_t = D$ is independent of t, one might be tempted to rewrite the right-hand side of (8.2.10) as

$$T\exp\left(- \int_0^t d\sigma \int_0^{(t-\sigma)/\lambda^2} d\tau G_+(\tau) D^+ D \right), \tag{8.2.12}$$

which, as $\lambda \to 0$, converges to

$$\exp\left(-tD^+D \int_0^{+\infty} G_+(\tau) d\tau \right). \tag{8.2.13}$$

Of course the naive replacement $D_t \mapsto D$ is not justified in the right-hand side of (8.2.10) (i.e. for finite λ), which is a time-ordered product. But in the following section, we show that this replacement becomes rigorous in the limit $\lambda \to 0$ at any order of perturbation theory [the estimates of Sect. 15.1 will then imply that the limit (8.2.11) is effectively given by (8.2.13)]. This can be expected on the basis of the following heuristic arguments: Let us try to understand which terms in the expression of I_{2m} give a nonvanishing contribution in the limit $\lambda \to 0$. Using (8.2.7) we see that the first operator on the left must be an annihilator ($\varepsilon_1 = 1$). By the time-consecutive principle, $A^{\varepsilon_2}_{t_2}$ must be a creator ($\varepsilon_2 = 0$). Iterating this argument and using (8.2.6), we see that the time-ordered system part (corresponding to nonvanishing terms) is equal to $(D^+D)^m$. So the replacement of (8.2.10) by (8.2.12) is justified in the limit $\lambda \to 0$. Now let us prove that these heuristic considerations lead to the correct result. In the following section we shall give a full proof.

8.3 The Limit of the nth Term, Time-Ordered Product Approach: Vacuum Expectation

Lemma 8.3.1. *In the notations of (8.2.1), (8.2.8) and denoting* $G(t) := \theta(t)G_+(t)$, *one has*

$$\lim_{\lambda\to 0}\frac{(2m)!}{m!}\lambda^{2m}\int_0^{t/\lambda^2}dt_1\dots\int_0^{t/\lambda^2}dt_m\int_0^{t/\lambda^2}d\tau_1\dots\int_0^{t/\lambda^2}d\tau_m$$
$$\cdot T(D^+_{t_1}\dots D^+_{t_m}D_{\tau_1}\dots D_{\tau_m})\prod_{k=1}^m G(t_k-\tau_k) = (D^+D)^m\frac{t^m}{m!}\left[\int_{-\infty}^0 G(\sigma)d\sigma\right]^m .$$

Proof 8.3.1. By changing variables

$$t_k\to\sigma_k = t_k/\lambda^2,\quad k=1\dots m,$$

one obtains

$$I_{2m} = \frac{(2m)!}{m!}\int_0^t d\sigma_1\dots\int_0^t d\sigma_m\int_0^{t/\lambda^2}d\tau_1\dots\int_0^{t/\lambda^2}d\tau_m$$
$$\cdot T\left(D^+_{\frac{\sigma_1}{\lambda^2}}\dots D^+_{\frac{\sigma_m}{\lambda^2}}D_{\tau_1}\dots D_{\tau_m}\right)\cdot G\left(\frac{\sigma_1}{\lambda^2}-\tau_1\right)\dots G\left(\frac{\sigma_m}{\lambda^2}-\tau_m\right).$$

Now, changing

$$\tau_k\to s_k = \frac{\sigma_k}{\lambda^2}-\tau_k\ ,\quad k=1\dots m,$$

I_{2m} becomes

$$\frac{(2m)!}{m!}\int_0^t d\sigma_1\dots\int_0^t d\sigma_m\int_0^{\frac{\sigma_1}{\lambda^2}}ds_1\dots\int_0^{\frac{\sigma_m}{\lambda^2}}ds_m$$
$$\cdot T\left(D^+_{\frac{\sigma_1}{\lambda^2}}\dots D^+_{\frac{\sigma_m}{\lambda^2}}D_{\frac{\sigma_1}{\lambda^2}-s_1}\dots D_{\frac{\sigma_m}{\lambda^2}-s_m}\right)\cdot G(s_1)\dots G(s_m), \tag{8.3.1}$$

where in the last m integrals in (8.3.1) we have used the causality condition $G(\sigma)=0$ for $\sigma<0$ to replace by zero the (negative) lower bounds $(\sigma_k-t)/\lambda^2$ $k=1\dots m$ of the integrals.

We shall now prove that possibly only the integrand over the simplex

$$\frac{\sigma_1}{\lambda^2}\geq\frac{\sigma_1}{\lambda^2}-s_1\geq\frac{\sigma_2}{\lambda^2}\geq\frac{\sigma_2}{\lambda^2}-s_2\geq\dots$$

corresponding to the product

$$T\left(D^+_{\frac{\sigma_1}{\lambda^2}}D_{\frac{\sigma_1}{\lambda^2}-s_1}D^+_{\frac{\sigma_2}{\lambda^2}}D_{\frac{\sigma_2}{\lambda^2}-s_2}\dots D^+_{\frac{\sigma_n}{\lambda^2}}D_{\frac{\sigma_m}{\lambda^2}-s_m}\right)\cdot G(s_1)\dots G(s_m)$$

survives as $\lambda\to 0$. To this goal, first recall that by definition of the chronological product

$$T\left(D^+_{\frac{\sigma_1}{\lambda^2}} D_{\frac{\sigma_1}{\lambda^2}-s_1}\cdots\right) = \sum_{\pi\in\mathcal{P}_{2m}} \theta\left(\sigma_{\frac{\pi_1}{\lambda^2}} \geq \cdots\right) D^+_{\frac{\sigma\pi_1}{\lambda^2}} \cdots D_{\frac{\sigma\pi_2}{\lambda^2}-s_{\pi_2}} \cdots \tag{8.3.2}$$

Consider a term of the sum (8.3.2)

$$D^+_{\frac{\sigma\pi_1}{\lambda^2}} \cdots D_{\frac{\sigma\pi_1}{\lambda^2}-s_{\pi_1}} \cdots$$

and the corresponding integration domain

$$\frac{\sigma_{\pi_1}}{\lambda^2} \geq \cdots \geq \frac{\sigma_{\pi_1}}{\lambda^2} - s_{\pi_1} \geq \cdots$$

We shall use notations $\sigma_i = \sigma_{\pi_i}$, $s_i = s_{\pi_i}$ and we shall say that for an index k there is a *broken pair* if between $D^+_{\sigma_k/\lambda^2}$ and $D_{\sigma_k/\lambda^2-s_k}$ there is some other operator, i.e. if one has one of the following situations:

$$D^+_{\sigma_k/\lambda^2} \cdots D^+_{\sigma_\alpha/\lambda^2} \cdots D_{\sigma_k/\lambda^2-s_k}, \tag{8.3.3}$$

$$D^+_{\sigma_k/\lambda^2} \cdots D_{\sigma_{\alpha/\lambda^2-s_\alpha}} \cdots D_{\sigma_k/\lambda^2-s_k}. \tag{8.3.4}$$

In fact one needs only to consider case (8.3.3), because in case (8.3.4) if $D^+_{\sigma_\alpha/\lambda^2}$ is on the right of $D^+_{\sigma_k/\lambda^2}$ then we are back to case (8.3.3); if it is on the left then we are again back to case (8.3.3) but with the roles of k and α exchanged. In case (8.3.3) one has $\frac{\sigma_k}{\lambda^2} \geq \frac{\sigma_\alpha}{\lambda^2} \geq \frac{\sigma_k}{\lambda^2} - s_k$, a.e. Therefore

$$s_\alpha \geq \frac{\sigma_\alpha - \sigma_k}{\lambda^2} \geq 0 \quad \text{a.e.}$$

and the integral in (8.3.1) in case (8.3.3) becomes

$$\int_0^t d\sigma_1 \cdots \int_{(\sigma_k-\sigma_\alpha)/\lambda^2}^{\sigma_k/\lambda^2} ds_k G(s_k) \cdots \to 0,$$

where the dots denote other integrals from (8.3.1) which are uniformly bounded as $\lambda \to 0$. From (8.3.2) and the integrability of $G(s_1)$ it follows that (8.3.3) vanishes when $\lambda \to 0$.

In summary, as $\lambda \to 0$, the only nonvanishing terms in the integral of the sum (8.3.2) correspond to the integrals over the domains

$$\frac{\sigma_1}{\lambda^2} \geq \frac{\sigma_1}{\lambda^2} - s_1 \geq \frac{\sigma_2}{\lambda^2} \geq \frac{\sigma_2}{\lambda^2} - s_2 \geq \frac{\sigma_3}{\lambda^2} \cdots \frac{\sigma_m}{\lambda^2} \geq \frac{\sigma_m}{\lambda^2} - s_m. \tag{8.3.5}$$

In the domains (8.3.5) one has

$$T\left(D^+_{\frac{\sigma_1}{\lambda^2}} D_{\frac{\sigma_1}{\lambda^2}-s_1} D^+_{\frac{\sigma_2}{\lambda^2}} \cdots\right) = D^+D\ldots D^+D = (D^+D)^m.$$

Each domain in (8.3.5) is equivalent to $0 \leq \sigma_1 \leq t$, $\sigma_2 \leq \sigma_1 - \lambda^2 s_1$,

$$\sigma_3 \leq \sigma_2 - \lambda^2 s_2, \ldots, \sigma_m \leq \sigma_{m-1} - \lambda^2 s_m,$$

and the integral in (8.3.1) in this domain of integration is

$$\int_0^t d\sigma_1 \int_0^{\sigma_1/\lambda^2} ds_1 G_+(s_1) \int_0^{\sigma_1-\lambda^2 s_1} d\sigma_2 \int_0^{\sigma_2/\lambda^2} ds_2 G_+(s_2)$$
$$\dots \int_0^{\sigma_{m-1}-\lambda^2 s_{m-1}} d\sigma_m \int_0^{\sigma_m/\lambda^2} ds_m G_+(s_m),$$

which tends to

$$\begin{aligned}\int_0^t d\sigma_1 \int_0^\infty ds_1 G_+(s_1) \int_0^{\sigma_1} d\sigma_2 \int_0^\infty ds_2 G_+(s_2)&\\ \dots \int_0^{\sigma_{m-1}} d\sigma_m \int_0^\infty d\sigma_m \int_0^\infty ds_m G_+(s_m)&\\ = \int_0^t d\sigma_1 \int_0^{\sigma_1} d\sigma_2 \dots \int_0^{\sigma_{m-1}} d\sigma_m \left[\int_0^\infty ds G_+(s)\right]^m&\\ = \frac{t^m}{m!}\left[\int_0^\infty ds G_+(s)\right]^m .& \qquad (8.3.6)\end{aligned}$$

Note that if one permutation in (8.3.2) gives a nonzero contribution in the limit then all the permutations with this property are obtained from this one by permuting the pairs D^+D considered as a single object. Since these pairs are m, this gives exactly $m!$ terms of the form (8.3.6). The $(2m)!$ in (8.3.1) is cancelled by the corresponding factor in the expansion of the time-ordered exponential. Therefore Lemma 8.3.1 is proved.

8.4 The Stochastic Limit, Time-Ordered Product Approach: General Case

In this section we shall compute the limit as $\lambda \to 0$ of the following matrix elements

$$\left\langle \psi_\lambda(\varphi_1 \otimes f_1), U^{(\lambda)}_{t/\lambda^2} \psi_\lambda(\varphi_2 \otimes f_2)\right\rangle \qquad (8.4.1)$$

where $\psi_\lambda(\varphi \otimes f)$ is a collective coherent vector, i.e. a coherent vector in the Fock space $\mathcal{H}_R$ of the form

$$\psi_\lambda(\varphi \otimes f) = \exp\left(\int_R \varphi(\sigma)\frac{1}{\lambda}\left(A_{\sigma/\lambda^2}(f) + A^+_{\sigma/\lambda^2}(f)\right) d\sigma\right)\Phi, \qquad (8.4.2)$$

where Φ is the vacuum vector in $\mathcal{H}_R$, f is an element in the space $\mathcal{K}$ (see Sect. 4.5) and φ is a test function. To write $U^{(\lambda)}_t$ in the T-product form, we introduce, as in (8.2.9), auxiliary functions $D_t = D$ and $D^+_t = D$. These functions are constant, but it is convenient to use such notations when presenting $U^{(\lambda)}_t$ in the T-product form.

Lemma 8.4.1. *One has the following equality*

$$
\begin{aligned}
U_t^{(\lambda)} &= T\exp\left(-i\lambda\int_0^t \left(D_\tau^+\otimes A_\tau(g)+D_\tau\otimes A_\tau^+(g)\right)d\tau\right)\\
&= T\exp\left(-i\lambda\int_0^t D_\tau A_\tau(g)d\tau\right)\exp\left(-i\lambda\int_0^t D_\tau^+ A_\tau(g)\right)\\
&\cdot T\exp\left(-\lambda^2\int_0^t du\int_0^t dv D_u^+ G_-(v-u|g,g)D_v\right). \qquad (8.4.3)
\end{aligned}
$$

Proof 8.4.1. In (8.4.3) we bring the T-ordered product of operators A_τ and A_τ^+ to the normal form. The propagator $G_-(v-u|g,f)$ is defined by means of the usual commutation relations

$$[A_v(g),A_u^+(f)] = G_-(v-u|g,f) \qquad (8.4.4)$$

and the causality condition

$$G_-(t|g,f)=\begin{cases} G(t|g,f) & \text{if } t\le 0\\ 0 & \text{if } t\ge 0\end{cases}\ .$$

Theorem 8.4.1. *If the function $G(\sigma)$ satisfies the causality condition, one has the following relation:*

$$
\begin{aligned}
&\lim_{\lambda\to 0}\left\langle \psi_\lambda(\varphi_1\otimes f_1),U^{(\lambda)}_{t/\lambda^2}\psi_\lambda(\varphi_2\otimes f_2)\right\rangle\\
&= T\exp\Bigg[-i\int_0^t d\tau\big((f_1|g)\varphi_1(\tau)D_\tau+(g|f_2)\varphi_2(\tau)D_\tau^+\\
&\qquad -(g|g)_-D^+_{\tau+o}D_\tau+(f_1|f_2)\varphi_1(\tau)\varphi_2(\tau)\big)\Bigg], \qquad (8.4.5)
\end{aligned}
$$

where

$$(f|g)=\int_{-\infty}^{\infty}G_-(t|f,g)dt,\quad (f|g)_-=\int_{-\infty}^{0}G_-(t|f,g)dt. \qquad (8.4.6)$$

Proof 8.4.2. By using (8.4.3) one has

$$
\begin{aligned}
&\left\langle \psi_\lambda(\varphi_1\otimes f_1),U^{(\lambda)}_{t/\lambda^2}\psi_\lambda(\varphi_2\otimes f_2)\right\rangle\\
&= T\exp\Bigg\{-i\int_0^{t/\lambda^2}d\tau\int d\sigma\bigg[\varphi_1(\sigma)G\Big(\tau-\frac{\sigma}{\lambda^2}|f_1,g\Big)D_\tau\\
&\qquad -\varphi_2(\sigma)G\Big(\frac{\sigma}{\lambda^2}-\tau|g,f_2\Big)D_\tau^+\bigg]\\
&\qquad -\int_0^{t/\lambda^2}dU\int_0^{t/\lambda^2}dvD_u^+G_-(v_-u|g,g)D_v\\
&\qquad +\int\overline{\varphi_1}(\sigma)\varphi_2(\sigma_2)\frac{1}{\lambda^2}G\Big(\frac{\sigma_2-\sigma_1}{\lambda^2}|f_1,f_2\Big)d\sigma_1 d\sigma_2\Bigg\}. \qquad (8.4.7)
\end{aligned}
$$

By rescaling time according to $\tau \to \tau/\lambda^2$, one obtains

$$T\exp\Big[-i\int_0^t d\tau\int d\sigma\overline{\varphi_1}(\sigma)\frac{1}{\lambda^2}G\left(\frac{\tau-\sigma}{\lambda^2}\,|f_1,g\right)D_{\tau/\lambda^2}$$
$$+\varphi_2(\sigma)\frac{1}{\lambda^2}G\left(\frac{\sigma-\tau}{\lambda^2}\,|g,f_2\right)D^+_{\tau/\lambda^2}$$
$$-\int_0^t du\int_0^t dvD^+_{u/\lambda^2}\frac{1}{\lambda^2}G_-\left(\frac{v-u}{\lambda^2}\,|g,g\right)D_{v/\lambda^2}$$
$$+\int\overline{\varphi_1}(\sigma_1)\varphi_2(\sigma_2)\frac{1}{\lambda^2}G\left(\frac{\sigma_2-\sigma_1}{\lambda^2}\,|f_1,f_2\right)d\sigma_1 d\sigma_2\Big]. \tag{8.4.8}$$

Since D_{τ/λ^2}, D^+_{τ/λ^2}, D^+_{u/λ^2}, D_{v/λ^2} are constant functions and the time label is only used to keep track of the order in the operator products, one can make

$$D_{\tau/\lambda^2} = D_\tau\ ,\quad D^+_{\tau/\lambda^2} = D^+_\tau\ ,\quad D^+_{u/\lambda^2} = D_u\ ,\quad D_{v/\lambda^2} = D_v$$

under the sign of the T product because this does not change the order. One therefore obtains

$$T\exp\left[-i\int_0^t d\tau\int d\sigma\overline{\varphi_1}(\sigma)\frac{1}{\lambda^2}G\left(\frac{\tau-\sigma}{\lambda^2}|f_1,g\right)D_\tau\right.$$
$$+\varphi_2(\sigma)\frac{1}{\lambda^2}G\left(\frac{\sigma-\tau}{\lambda^2}|g,f_2\right)D^+_\tau$$
$$-\int_0^t du\int_0^t dvD^+_u\frac{1}{\lambda^2}G_-\left(\frac{v-u}{\lambda^2}|g,g\right)D_v$$
$$\left.+\int\int\overline{\varphi_1}(\sigma_2)\varphi_2(\sigma_2)\frac{1}{\lambda^2}G\left(\frac{\sigma_2-\sigma_1}{\lambda^2}|f_1,f_2\right)d\sigma_1 d\sigma_2\right]. \tag{8.4.9}$$

To evaluate the limit of (8.4.9) when $\lambda \to 0$, one uses the following identity (see Lemma 1.9.1):

$$\lim_{\lambda\to 0}\int\frac{1}{\lambda^2}G\left(\frac{\sigma}{\lambda^2}\right)F(\sigma)d\sigma = F(0)\int G(\sigma)d\sigma, \tag{8.4.10}$$

which holds if $G(\sigma)$ is a continuous integrable function on the real line and $F(\sigma)$ is any continuous function.

9. Functional Integral Approach to the Stochastic Limit

This chapter is added for completeness; to read it is not essential for an understanding the remainder of the book. We shall illustrate here how the functional integral approach can be used to derive the stochastic limit. In other words, we consider the basic formula for the stochastic limit of a scalar field,

$$\lim_{\lambda\to 0}\frac{1}{\lambda}\varphi\left(\frac{t}{\lambda^2},x\right)=\Phi(t,x)\,,$$

in the functional integral approach. We shall use freely the functional integral formalism, assuming that the reader is familiar with it. For some useful references on functional integrals see [AHK76], [HKPS93], [Hi01], [SmSh91], [AcVo98].

9.1 Statement of the Problem

The free fields are related to the creation and annihilation operators through the identity

$$\varphi\left(t,x\right)=\frac{1}{\left(\sqrt{2\pi}\right)^{3/2}}\int\frac{d^3k}{\sqrt{2\omega\left(k\right)}}\left(e^{it\omega(k)-ikx}a^+\left(k\right)+\text{h.c.}\right)\,.$$

In the stochastic limit one studies the asymptotic behaviour, as $\lambda\to 0$, of the rescaled quantum field $\varphi(t,x)$,

$$\varphi(t,x)\to\frac{1}{\lambda}\varphi\left(\frac{t}{\lambda^2},x\right)\,, \tag{9.1.1}$$

[(t,x) being the time–space variables]. The basic result of the stochastic limit is

$$\lim_{\lambda\to 0}\frac{1}{\lambda}\varphi\left(\frac{t}{\lambda^2},x\right)=\Phi(t,x)\,, \tag{9.1.2}$$

where $\Phi(t,x)$ is a new quantum field. The crucial new property of the field Φ which drastically simplifies the analysis is that Φ is δ-correlated with respect to time, i.e. it is a quantum white noise. We will use (9.1.2) not only for free fields but also for interacting fields. On the other hand the functional integral is a powerful tool in quantum theory, and in particular, it is very convenient for making changes of variables such as the time-field rescaling (9.1.1). Therefore the way to apply this technique to receive insight into the form of the stochastic limit of quantum field theory is a natural one.

In the rest of this chapter we will illustrate how this limit can be performed in the functional integral formalism by considering several examples. However, the reader should be warned that, in the presence of nonlinearities, a more detailed analysis should be performed. The functional integral approach gives a very quick answer: one makes a change of variables in the action functional, and after it a power of λ multiplies the time derivative in the action, which therefore disappears in the limit. The white noise character of the limit is indicated precisely by the absence of the time derivative in the limiting action functional (which gives the δ-function in the calculation of the propagator). In this chapter we work with functional integrals at a formal level, without giving rigorous proofs. We use the following notations: the coordinates in the Minkowski space are (t, x^i), $i = 1, 2, 3$.

9.2 The Stochastic Limit of the Free Massive Scalar Field

The scalar field φ is a real-valued function (operator-valued distribution) with the Lagrangian

$$\mathcal{L} = \frac{1}{2}\,(\partial_\mu \varphi)^2 - \frac{m^2}{2}\,\varphi^2 \tag{9.2.1}$$

and associated action

$$S = \frac{1}{2}\int_{\mathbb{R}\times\mathbb{R}^3} \{[\partial_t \varphi(t,x)]^2 - [\partial_i \varphi(t,x)]^2 - m^2 \varphi(t,x)\} dt dx\,. \tag{9.2.2}$$

In the stochastic limit we want to find the asymptotic behaviour of the rescaled field,

$$\varphi(t,x) \to \frac{1}{\lambda}\,\varphi\left(\frac{t}{\lambda^2}\,, x\right)\,. \tag{9.2.3}$$

In the functional integral approach the correlation functions have the representation

$$\langle \varphi(t_1,x_1)\ldots\varphi(t_k,x_k)\rangle = \frac{1}{Z}\int \varphi(t_1,x_1)\ldots\varphi(t_k,x_k)e^{iS}\mathcal{D}\varphi\,. \tag{9.2.4}$$

Theorem 9.2.1. *For the free massive scalar field the limit of the rescaled correlators*

$$G = \frac{1}{Z}\int \frac{1}{\lambda}\phi\left(\frac{t_1}{\lambda^2}, x_1\right)\cdots\frac{1}{\lambda}\phi\left(\frac{t_n}{\lambda^2}, x_n\right) e^{iS}\mathcal{D}\phi \tag{9.2.5}$$

exists and is

$$\frac{1}{Z}\int \psi(x_1)\cdots\psi(x_n)e^{iS_{\text{eff}}}\,\mathcal{D}\psi\,, \tag{9.2.6}$$

where

$$S_{\text{eff}} = -\frac{1}{2}\int\left(\sum_{i=1}^{3}(\partial_i\psi)^2 + m^2\psi^2\right)d^4x \tag{9.2.7}$$

is the limiting (effective) action, corresponding to the propagator ($k^2 = k_1^2 + k_1^2 + k_3^2$):

$$i\int\frac{e^{ik_0t-ikx}}{k^2+m^2}\,dk_0d^3k \;= i2\pi\delta(t)\int\frac{e^{-ikx}}{k^2+m^2}d^3k = i2\pi^2\delta(t)\frac{e^{-m|x|}}{|x|}\,.$$

Proof 9.2.1. With the change of variables

$$\frac{1}{\lambda}\phi\left(\frac{t}{\lambda^2}, x\right) = \psi(t,x)\,,$$

one has

$$\phi(t) = \lambda\psi\left(\lambda^2t\right)\,,$$
$$\partial_t\phi(t) = \lambda^3\psi'\left(\lambda^2t\right)$$

and

$$S = \int dt\int d^3x\left(\frac{1}{2}\left[\partial_t\varphi(t)\right]^2 - \frac{1}{2}\left[\partial_i\varphi(t)\right]^2 - \frac{m^2}{2}\varphi^2\right)$$
$$= \int dt\int d^3x\left(\frac{1}{2}\lambda^6\left[\psi'\left(\lambda^2t\right)\right]^2 - \frac{1}{2}\lambda^2\left[\partial_i\psi(\lambda^2t)\right]^2 - \frac{m^2}{2}\lambda^2\psi(\lambda^2t)^2\right)\,.$$

Then, with the further change of variables $\lambda^2t = \tilde{t} \to t$, this becomes

$$S = \int dtd^3x\left(\frac{1}{2}\lambda^4[\partial_t\psi(t)]^2 - \frac{1}{2}\left[\partial_i\psi(t)\right]^2 - \frac{m^2}{2}\psi(t)^2\right)\,.$$

Therefore the correlation function G from (9.2.5) becomes

$$G = \frac{1}{Z}\int\psi(x_1)\cdots\psi(x_n)\exp\left[i\frac{1}{2}\int\left[\lambda^4\left(\partial_0\psi\right)^2 - \left(\partial_i\psi\right)^2 - m^2\psi^2\right]d^4x\right]\,\mathcal{D}\psi\,,$$

and its limit as $\lambda \to 0$ exists and is equal to (9.2.6). Note that the functional Jacobian, arising under the above change of variables, is a constant and therefore it cancels with the normalization factor Z in (9.2.5).

Remark 9.2.1. One can compute the correlators (9.2.6) by using the Gaussian generating functional

$$\langle\phi(t_1,x_1)\dots\phi(t_k,x_k)\rangle = (-i)^k \frac{\delta}{\delta J(t_1,x_1)} \cdots \frac{\delta}{\delta J(t_k,x_k)} Z(J)\,, \tag{9.2.8}$$

where

$$\begin{aligned} Z(J) &= \int \exp\Big[-\int \frac{1}{2}\Big([\partial_i\phi(t,x)]^2 + \frac{m^2}{2}\phi(t,x)^2 + J(t,x)\phi(t,x)\Big)dtdx\Big]\mathcal{D}\phi \\ &= \exp\Big(\frac{i}{2}\int J(t_1,x_1)\mathcal{D}(t_1-t_2,x_1-x_2)J(t_2,x_2)dt_1dt_2dx_1dx_2\Big) \end{aligned} \tag{9.2.9}$$

and

$$\begin{aligned} \mathcal{D}(t,x) &= \frac{1}{(2\pi)^4}\int_{\mathbb{R}^1\times\mathbb{R}^3} \frac{e^{ip_0t-ipx}}{p^2+m^2}dp_0dp \\ &= \frac{1}{(2\pi)^3}\delta(t)\int_{\mathbb{R}^3}\frac{e^{-ipx}dp}{p^2+m^2} = \frac{1}{2\pi}\delta(t)\frac{e^{-m|x|}}{|x|}\,, \end{aligned} \tag{9.2.10}$$

which is the correlator for a white noise (in the t variable).

9.3 The Stochastic Limit of the Free Massless Scalar Field

Let us consider the correlations of the free massless scalar field:

$$I = \int \frac{1}{\sqrt{\lambda}}\varphi\left(\frac{x_0}{\lambda},\bar{x}\right)\frac{1}{\sqrt{\lambda}}\varphi\left(\frac{y_0}{\lambda},\bar{y}\right)\dots\exp\left(i\int\frac{1}{2}(\partial_\mu\varphi)^2dx^4\right)\mathcal{D}\varphi\,. \tag{9.3.1}$$

The usual change of variables

$$\frac{1}{\sqrt{\lambda}}\varphi\left(\frac{x_0}{\lambda},\bar{x}\right) = \Phi(x_0,\bar{x}) \tag{9.3.2}$$

gives

$$\varphi(x_0,\bar{x}) = \sqrt{\lambda}\Phi(\lambda x_0,\bar{x}) \tag{9.3.3}$$

and

$$\frac{\partial}{\partial x_0}\varphi(x_0,\bar{x}) = \sqrt{\lambda}\lambda\Phi'(\lambda x_0,\bar{x})\,. \tag{9.3.4}$$

Making also the change of variables in the action

$$\lambda x_0 = \tilde{x}_0 \tag{9.3.5}$$

$$\int \lambda^3\Phi'^2(\tilde{x}_0,\bar{x})\frac{d\tilde{x}_0}{\lambda} = \lambda^2\int[\partial_0\Phi(x_0,\bar{x})]^2dx_0\,, \tag{9.3.6}$$

one obtains

$$I = \int \Phi(x)\Phi(y) \exp\left(\frac{i}{2}\int [\lambda^2(\partial_0\Phi)^2 - (\partial_k\Phi)^2] dx^4\right) \mathcal{D}\Phi . \tag{9.3.7}$$

Therefore the effective action for the theory in the limit $\lambda \to 0$ will be

$$S_{\text{eff}} = -\frac{1}{2}\int (\partial_k\Phi)^2 dx^4 \tag{9.3.8}$$

$(k = 1, 2, 3)$, corresponding to the white noise propogator

$$G_{\text{eff}}(x) = i2\pi\delta(x_0)\frac{1}{|\,\bar{x}\,|} . \tag{9.3.9}$$

9.4 Polynomial Interactions

In this section we shall show how the functional integral approach can be used to derive the stochastic limit for a model with interaction. In the limit we will obtain an interacting white noise. We consider the basic formula (9.1.2) for the stochastic limit of the scalar field,

$$\lim_{\lambda\to 0}\frac{1}{\lambda}\varphi\left(\frac{t}{\lambda^2}, x\right) = \phi(t, x),$$

in the functional integral approach. Let us take the following classical action:

$$S_\lambda(\varphi) = \int dtdx \left(\frac{1}{2}[\partial_t\varphi(t,x)]^2 - \frac{1}{2}[\partial_i\varphi(t,x)]^2 - \frac{1}{2}m^2\varphi(t,x)^2 - g\lambda^k\varphi(t,x)^n\right), \tag{9.4.1}$$

where m, g, k, n are real parameters of the model, $x = (x_1 \dots x_d)$, and $\partial_i = \partial/\partial x_i$.

Theorem 9.4.1. *If $k = 2 - n$ in (9.4.1), then the following limiting relation holds for formal functional integrals:*

$$\lim_{\lambda\to 0}\int \frac{1}{\lambda}\varphi\left(\frac{t_1}{\lambda^2}, x_1\right)\cdots\frac{1}{\lambda}\varphi\left(\frac{t_p}{\lambda^2}, x_p\right) e^{iS_\lambda(\varphi)}\mathcal{D}\varphi$$
$$= \int \phi(t_1, x_1)\dots\phi(t_p, x_p)e^{iI(\phi)}\mathcal{D}\phi , \tag{9.4.2}$$

where the action

$$I(\phi) = \int dtdx \left(-\frac{1}{2}\left[\partial_i\phi(t,x)\right]^2 - \frac{1}{2}m^2\phi^2(t,x) - g\phi^n(t,x)\right) \tag{9.4.3}$$

defines an interacting white noise.

Proof 9.4.1. The change of variables

$$\frac{1}{\lambda}\varphi\left(\frac{t}{\lambda^2}, x\right) = \phi(t,x)$$

gives

$$\varphi(t,x) = \lambda\phi(\lambda^2 t, x)$$

and

$$\partial_t\varphi(t,x) = \lambda^3\phi'(\lambda^2 t, x)\,.$$

Making also the change of variables in the action

$$\lambda^2 t = \tau\,,$$

one obtains

$$S_\lambda(\varphi) = \int d\tau dx\left(\frac{1}{2}\lambda^2[\phi'(\tau,x)]^2 - \frac{1}{2}[\partial_i\phi(\tau,x)]^2 - \frac{1}{2}m^2\phi(\tau,x)^2 - \lambda^{k+n-2}g\phi^n(\tau,x)\right).$$

Therefore

$$\int \frac{1}{\lambda}\varphi\left(\frac{t_1}{\lambda^2}, x_1\right)\dots\frac{t}{\lambda}\varphi\left(\frac{t_1}{\lambda^2}, x_p\right)e^{iS_\lambda(\varphi)}\mathcal{D}\varphi = \int \phi(t_1,x_1)\dots\phi(t_p,x_p)e^{iI_\lambda(\phi)}\mathcal{D}\phi\,,$$

where

$$I_\lambda(\phi) = \int dtdx\left(\frac{1}{2}\lambda^2[\partial_t\phi(t,x)]^2 - \frac{1}{2}[\partial_i\phi(t,x)]^2 - \frac{1}{2}m^2\phi(t,x)^2 - \lambda^{k+n-2}g\phi^n(t,x)\right).$$

Now if $k = 2-n$, then

$$\lim_{\lambda\to 0} I_\lambda(\phi) = I(\phi)\,,$$

and we obtain the relation (9.4.2).

Example 9.4.1. As an example one has that interactions of the form $g\lambda\varphi$, $g\varphi^2$ and $g\lambda^{-2}\varphi^4$ lead in the stochastic limit to interactions of the form $g\phi$, $g\phi^2$ and $g\phi^4$, respectively.

Remark 9.4.1. From Proof 9.4.1 it follows that if in the action (9.4.1) instead of $g\lambda^k\varphi^n$ one has a more general interaction of the form

$$V(\varphi) = \sum_i g_i\lambda^{k_i}\varphi^{n_i}$$

then after the change of variables one obtains

$$V(\phi) = \sum_i g_i\lambda^{k_i+2-n_i}\phi^{n_i}\,.$$

Remark 9.4.2. The expressions on the right-hand sides of (9.4.1) and (9.4.3) are formal ones. In particular the second one defines an interacting white noise with the free covariance

$$\int \phi(t_1, x_1)\phi(t_2, x_2) \exp\left[i \int dtdx \left(-\frac{1}{2}\left[\partial_i \phi(t,x)\right] - \frac{1}{2} m^2 \phi^2(t,x)\right)\right] \mathcal{D}\phi$$

$$= iD(t_1 - t_2, x_1 - x_2) = i\frac{1}{(2\pi)^d}\,\delta(t_1 - t_2) \int_{R^d} \frac{e^{ik(x_1 - x_2)}dk}{k^2 + m^2}\,.$$

In order to give a precise meaning to the interacting white noise [and to (9.4.1)], one has to introduce a regularization.

9.5 The Stochastic Limit of the Electromagnetic Field

The same method can be used for the electromagnetic field whose action is

$$S = \frac{1}{4}\int F_{\mu\nu}^2 d^4x\,,$$

where summation over the μ, ν indices (which run from 0 to 3) is understood and

$$F_{\mu\nu} = \partial_\mu A_\nu - \partial_\nu A_\mu$$

are the field operators; A_μ is the vector potential of the electromagnetic field, and we have to add to the action a gauge-fixing term that we shall discuss later. Now let us consider the following correlation functions:

$$G = \int \frac{1}{\lambda} A_{\mu_1}\left(\frac{x_1^0}{\lambda^2}, \bar{x}_1\right) \frac{1}{\lambda} \cdots A_{\mu_n}\left(\frac{x_n^0}{\lambda^2}, \bar{x}_n\right) e^{iS}\mathcal{D}A_\mu\,.$$

With the change of variables

$$\frac{1}{\lambda} A_\mu\left(\frac{x_0}{\lambda^2}, \bar{x}\right) = B_\mu(x_0, \bar{x}) \quad , \qquad A_\mu(x_0, \bar{x}) = \lambda B_\mu(\lambda^2 x_0, \bar{x})\,,$$

one obtains

$$\partial_0 A_i(x_0, \bar{x}) = \lambda^3 B_i'(\lambda^2 x_0, \bar{x})\,.$$

Hence by using

$$F_{\mu\nu}^2 = F_{oi}^2 + F_{ij}^2\,,$$

where $i, j = 1, 2, 3$, one obtains

$$G = \int B_{\mu_1}(x_1) \cdots B_{\mu_n}(x_n) e^{i\tilde{S}}\mathcal{D}A_\mu\,,$$

where

$$\tilde{S} = \frac{1}{4}\int dx_0 d^3x[\lambda^2 B_i'(\lambda^2 x_0, \bar{x}) - \lambda\partial_i B_0(\lambda^2 x_0, \bar{x})]^2$$
$$+ \lambda^2[\partial_i B_j(\lambda^2 x_0, \bar{x}) - \partial_j B_i(\lambda^2 x_0, \bar{x})]^2 + \text{gauge fixing}\,.$$

So one has

$$G = \int B_{\mu_1}(x_1)\cdots B_{\mu_n}(x_n)$$

$$\cdot \exp\frac{i}{4}\int \left[(\lambda^2\partial_0 B_i - \partial_i B_0)^2 + (\partial_i B_j - \partial_j B_i)^2 + \text{gauge fixing}\right]\mathcal{D}B_\mu\,.$$

The gauge-fixing term can be taken to have the form

$$\frac{1}{2}\int (\partial_\mu A_\mu)^2 dx\,.$$

After the change of variables it is equal to

$$\frac{1}{2}\int \left(\lambda^4(\partial_0 B_0)^2 - \frac{1}{\lambda}(\partial_i B_i)^2\right) d^4x\,. \tag{9.5.1}$$

Therefore, if we want the limit $\lambda \to 0$ to exist, we must choose

$$\partial_i B_i = 0\,, \tag{9.5.2}$$

which corresponds to the effective action

$$S_{\text{eff}} = \frac{1}{4}\int \left[(\partial_i B_0)^2 + (\partial_i B_j - \partial_j B_i)^2\right] dx^4$$

with the gauge condition (9.5.2). This S_{eff} leads to the following white noise type propagators:

$$\langle B_0 B_i\rangle = 0$$

$$\langle B_0(x)B_0(y)\rangle = i\delta(x_0 - y_0)\frac{1}{\mid \bar{x} - \bar{y}\mid}$$

$$\langle B_i(x)B_j(y)\rangle = i\delta(x_0 - y_0)\left(\delta_{ij} - \frac{\partial_i\partial_j}{\Delta}\right)\frac{1}{\mid \bar{x} - \bar{y}\mid}\,,$$

where $\Delta = \partial_i\partial_i$ and δ_{ij} is the Kronecker symbol.

10. Low-Density Limit: The Basic Idea

In this chapter we consider the low-density limit and the associated equations, including the Boltzmann equation. Here we limit ourselves to a quick outline of the main ideas underlying the emergence of the number process in the stochastic limit and its connection with the 2-particle scattering operator. A full discussion of the low-density limit and of the associated transport equations (see [AcLu91a-b-c], [AcLu92]) will be given in a separate book, entirely devoted to this problem.

10.1 The Low-Density Limit: Fock Case, No System

We will work in the same context and notations as in Sect. 4.11, in particular the $a_k^{\pm}$ field is boson and referred to the Fock representation. For simplicity we begin by considering the case when the system operators are the identity ($D = 1$), so that the equation for the evolution operator is

$$\partial_t U_{t/\lambda^2}^{(\lambda)} = -i\left(\frac{1}{\lambda}\, a_{t/\lambda^2}^{+}(g_0)\frac{1}{\lambda}\, a_{t/\lambda^2}(g_1) + \frac{1}{\lambda}\, a_{t/\lambda^2}^{+}(g_1)\frac{1}{\lambda}\, a_{t/\lambda^2}(g_0)\right) U_{t/\lambda^2}^{(\lambda)} . \tag{10.1.1a}$$

By applying formally the first two steps of the stochastic golden rule of Sect. 4.14, we would be led to conclude that the equation for the limit evolution operator is

$$\partial_t U_t = -i[b_t^{+}(g_0)b_t(g_1) + b_t^{+}(g_1)b_t(g_0)]U_t = -iH_tU_t, \tag{10.1.1b}$$

where $b_t(g_0)$, $b_t^{+}(g_1)$ are the boson white noises obtained, through Theorem 3.2.1, as limits of the rescaled fields $(1/\lambda)a_{t/\lambda^2}$. We obtain a quadratic form in the white noise operators which is the density of the number process.

Note the important difference between the Hamiltonian considered in this section, in which the quadratic term has a factor $(1/\lambda)^2$, and the polynomial Hamiltonians considered in Chap. 11, in which the quadratic term has a factor $(1/\lambda)$. This difference infers that after writing (10.1.1a) in integral form and changing variables the (small) factor λ *no longer appears in front of the integral, but only in the time rescaling*:

$$U_{t/\lambda^2}^{(\lambda)} = 1 - i \int_0^{t/\lambda^2} ds \left[a_s^+(g_0)\, a_s(g_1) + a_s^+(g_1) a_s(g_0)\right] U_s^{(\lambda)}. \quad (10.1.1c)$$

Therefore this type of quadratic expression is quite different from the *type-$\mathcal{N}$* blocks discussed in Chap. 11. On the other hand, by inspection of (10.1.1c), one would expect a connection with some kind of *wave operator*. Another difference is that now we have a quadratic analogue of the (first-order) white noise Hamiltonian equations of the type (4.11.11), already discussed in the weak-coupling limit case. We will give a rigorous meaning to (10.1.1b), which involves an Hamiltonian that is a formal quadratic expression in the operator-valued distributions $b_t^{\#}(g_j)$ $(j = 0, 1)$, by bringing it to its normally ordered form using the causal commutator of Sect. 7.4:

$$[b_t(g_i), b_\tau^+(g_j)] = \delta_+(t - \tau)(g_i|g_j)_-. \quad (10.1.1d)$$

Lemma 10.1.1. *Suppose that* $(g_0|g_1) = 0$, *then:*

$$[b_t(g_1), U_t] = -i(g_1|g_1)_- b_t(g_0) U_t, \quad (10.1.2)$$
$$[b_t(g_0), U_t] = -i(g_0|g_0)_- b_t(g_1) U_t, \quad (10.1.3)$$

in the sense that these identities hold identically for each term of the iterated series solution of (10.1.1b).

Proof 10.1.1. The integral form of (10.1.1b) is

$$U_t = 1 - i \int_0^t H_{t_1} U_{t_1} dt_1. \quad (10.1.1')$$

Replacing the right-hand side of (10.1.1') in the commutator, one finds

$$[b_t(g_1), U_t] = -i \int_0^t [b_t(g_1), H_{t_1}] U_{t_1} dt_1 + (-i) \int_0^t H_{t_1} [b_t(g_1), U_{t_1}] dt_1.$$

The commutator $[b_t(g_1), H_{t_1}]$, easily computed using (10.1.1d), is equal to $\delta_+(t - t_0)(g_1|g_1)_- b_t(g_0)$. The commutator $[b_t(g_1), U_{t_1}]$ is a non-time-consecutive term (as one immediately sees by writing U_{t_1} in its integral form); therefore it vanishes identically because of the time-consecutive principle. Now, for simplicity, let us introduce the notations

$$b_t(g_1) \equiv b_1, \quad b_t(g_0) \equiv b_0, \quad U_t \equiv U, \quad (g_0|g_0)_- \equiv \gamma_0, \quad (g_1|g_1)_- \equiv \gamma_1. \quad (10.1.4)$$

Thus (10.1.1b, 2, 3) become respectively

$$\partial_t U = -i(b_0^+ b_1 + b_1^+ b_0) U, \quad (10.1.1b')$$
$$[b_1, U] = -i\gamma_1 b_0 U, \quad (10.1.2')$$
$$[b_0, U] = -i\gamma_0 b_1 U. \quad (10.1.3')$$

Lemma 10.1.2. *In the assumption of Lemma 10.1.1 and in the notations of (10.1.4), the normally ordered form of the product* $b_1 U$ *is*

$$b_1 U = \frac{1}{1+\gamma_0\gamma_1}(Ub_1 - i\gamma_1 U b_0). \tag{10.1.5}$$

Similarly, the normally ordered form of the product $b_0 U$ *is*

$$b_0 U = \frac{1}{1+\gamma_0\gamma_1}(Ub_0 - i\gamma_0 U b_1). \tag{10.1.6}$$

Proof 10.1.2. Using (10.1.2') we find

$$b_1 U = [b_1, U] + Ub_1 = -i\gamma_1 b_0 U + Ub_1,$$

and using (10.1.3') we find

$$b_1 U = -i\gamma_1([b_0, U] + Ub_0) + Ub_1 = -i\gamma_1(-i\gamma_0 b_1 U + Ub_0) + Ub_1.$$

Therefore

$$b_1 U(1+\gamma_0\gamma_1) = -i\gamma_1 U b_0 + Ub_1,$$

which is equivalent to (10.1.5). Exchanging the roles of 0 and 1 in (10.1.3), one obtains (10.1.6).

Theorem 10.1.1. *The normal form of (10.1.1b) is*

$$i\partial_t U = \frac{1}{(1+\gamma_0\gamma_1)}[b_0^+ U(b_1 - i\gamma_1 b_0) + b_1^+ U(b_0 - i\gamma_0 b_1)],$$

or coming back to the original notations:

$$\begin{aligned}\partial_t U_t = &\frac{(-i)}{1+(g_0|g_0)_-(g_1|g_1)_-}\Big[b_t^+(g_0)U_t\big(b_t(g_1) - i(g_1|g_1)_- b_t(g_0)\big)\\ &+ b_t^+(g_1)U_t(b_t(g_0) - i(g_0|g_0)_- b_t(g_1))\Big].\end{aligned}$$

Proof 10.1.3. Clear from Lemma 10.1.1.

10.2 The Low-Density Limit: Fock Case, Arbitrary System Operator

Keeping the assumptions of Sect. 10.1, we consider the case of an arbitrary system operator D.

Theorem 10.2.1. *The normal form of the stochastic evolution equation*

$$\partial_t U = \sum_{\varepsilon=0} D_\varepsilon b_\varepsilon^+ b_{1-\varepsilon} U = (D_0 b_0^+ b_1 + D_1 b_1^+ b_0)U \tag{10.2.1}$$

is

$$\begin{aligned}\partial_t U = &D_0 b_0^+(1-\gamma_0\gamma_1 D_1 D_0)^{-1}(\gamma_1 D_1 U b_0 + Ub_1)\\ &+ D_1 b_1^+(1-\gamma_0\gamma_1 D_0 D_1)^{-1}(\gamma_0 D_0 U b_1 + Ub_0).\end{aligned} \tag{10.2.2}$$

Proof 10.2.1. As in the proof of Theorem 10.1.1

$$[b_1, U] = \gamma_1 D_1 b_0 U$$
$$[b_0, U] = \gamma_0 D_0 b_1 U.$$

From this we deduce

$$\begin{aligned} b_1 U &= [b_1, U] + U b_1 = \gamma_1 D_1 b_0 U + U b_1 \\ &= \gamma_1 D_1([b_0, U] + U b_0) + U b_1 \\ &= \gamma_1 D_1(\gamma_0 D_0 b_1 U + U b_0) + U b_1 \\ &= \gamma_0 \gamma_1 D_1 D_0 b_1 U + \gamma_1 D_1 U b_0 + U b_1. \end{aligned}$$

Therefore

$$(1 - \gamma_0 \gamma_1 D_1 D_0) b_1 U = \gamma_1 D_1 U b_0 + U b_1$$

or

$$b_1 U = (1 - \gamma_0 \gamma_1 D_1 D_0)^{-1} (\gamma_1 D_1 U b_0 + U b_1). \tag{10.2.3}$$

Similarly one can show that

$$b_0 U = (1 - \gamma_0 \gamma_1 D_0 D_1)^{-1} (\gamma_0 D_0 U b_1 + U b_0). \tag{10.2.4}$$

From (10.2.3) and (10.2.4), (10.2.2) immediately follows.

10.3 Comparison of the Distribution and the Stochastic Approach

From the earlier results in the low-density limit (10.1.1b) [AcLu91a] we know that the stochastic equation associated with the interaction Hamiltonian (10.1.1a) is

$$dU_t = \left(\sum_{\varepsilon} R_{\varepsilon,1-\varepsilon}(0) dN_t(g_\varepsilon g_{1-\varepsilon}) + R_{\varepsilon,\varepsilon}(0) dN_t(g_\varepsilon, g_\varepsilon) \right) U_t, \tag{10.3.1}$$

where

$$R_{\varepsilon,1-\varepsilon}(0) := D_g(\varepsilon) D_\varepsilon, \quad R_{\varepsilon,\varepsilon}(0) := \gamma_{1-\varepsilon} D_\varepsilon D_{1-\varepsilon} D_g(\varepsilon),$$
$$D_g(\varepsilon) := (1 - \gamma_\varepsilon \gamma_{1-\varepsilon} D_\varepsilon D_{1-\varepsilon})^{-1}.$$

The practical rule to pass from the stochastic to the distribution equation, and vice versa, is to perform the following replacements:

$$\frac{dN_t}{dt}(g_\varepsilon, g_{\varepsilon'}) U \Leftrightarrow b_\varepsilon^+ U b_{\varepsilon'}.$$

These replacements bring the stochastic equation (10.3.1) to the form

$$\begin{aligned}\frac{dU_t}{dt} &= \sum_\varepsilon R_{\varepsilon,1-\varepsilon}(0) b_\varepsilon^+ U b_{1-\varepsilon} + R_{\varepsilon,\varepsilon}(0) b_\varepsilon^+ U b_\varepsilon \\ &= (1-\gamma_0\gamma_1 D_0 D_1)^{-1} D_0 b_0^+ U b_1 + \gamma_1 D_0 D_1 (1-\gamma_0\gamma_1 D_0 D_1)^{-1} b_0^+ U b_0 \\ &\quad + (1-\gamma_0\gamma_1 D_1 D_0)^{-1} D_1 b_1^+ U b_0 + \gamma_0 D_0 D_1 (1-\gamma_0\gamma_1 D_0 D_1)^{-1} b_1^+ U b_1.\end{aligned}$$

Using the identity

$$D_1(1-cD_0D_1)^{-1} = (1-cD_1D_0)^{-1}D_1,$$

which is proved by expanding $(1-x)^{-1} = \sum x^n$, one can verify that the right-hand side of (10.3.1) is equal to the right-hand side of (10.2.2).

The connection with the scattering operator is given by the identities

$$S_{\varepsilon,\varepsilon}(0) - 1 = -2\pi i R_{\varepsilon,\varepsilon'}(0) \mid g_\varepsilon\rangle\langle g_\varepsilon \mid,$$

so that (10.3.1) becomes

$$\begin{aligned}dU_t &= \sum_\varepsilon R_{\varepsilon,1-\varepsilon}(0) dN_t\big(\mid g_\varepsilon\rangle\langle g_{1-\varepsilon} \mid\big) + \sum_\varepsilon R_{\varepsilon,\varepsilon}(0) dN_t\big(\mid g_\varepsilon\rangle\langle g_\varepsilon \mid\big) U_t \\ &= \sum_{\varepsilon,\varepsilon'} R_{\varepsilon,\varepsilon'}(0) dN_t\big(\mid g_\varepsilon\rangle\langle g_{\varepsilon'} \mid\big) U_t = -\frac{1}{2\pi i} dN_t\big[S(0)-1\big] U_t.\end{aligned}$$

To make a connection with the notations used in the papers [AcLu90], [AAFL91], [AcLu91a, AcLu91b, AcLu91c], [AcLu92], recall that

$$-\frac{1}{2\pi i}[S(0)-1] = T(0).$$

10.4 LDL General, Fock Case, No System Operator, $\omega = 0$

The extension of the above discussion to the case in which the assumption of Lemma 10.1.1 is not satisfied leads to the consideration of an interaction Hamiltonian of the form (here too we assume $D = 1$)

$$H_{\mathrm{I}} = \sum_{mn} \omega_{mn} A^+(g_m) A(g_n) + \overline{\omega}_{mn} A^+(g_n) A(g_m), \tag{10.4.1}$$

where (g_n) is an orthogonal basis of the 1-particle space and the free evolution is

$$e^{iH_0t} A^+(g_m) e^{-iH_0t} = A^+(S_t g_m).$$

The white noise Hamiltonian equation associated with this interaction is

$$\begin{aligned}\frac{dU_t}{dt} &= -i\left(\sum_{mn}\omega_{mn}b_t^+(g_m)b_t(g_n) + \text{h.c.}\right)U_t \\ &= -i\left(\sum_{mn}\omega_{mn}b_m^+b_n + \overline{\omega}_{mn}b_n^+b_m\right)U_t, \qquad (10.4.2)\end{aligned}$$

and the commutator, computed with the same arguments as in Lemma 10.1.1, is

$$[b_k, U_t] = -i\gamma_k\sum_n(\omega_{kn} + \overline{\omega}_{nk})b_nU_t;$$

therefore

$$b_kU_t = [b_k, U] + Ub_k = -i\gamma_k\sum_n(\omega_{kn} + \overline{\omega}_{nk})b_nU + Ub_k$$

or equivalently

$$\sum_n[\delta_{kn} + i\gamma_k(\omega_{kn} + \overline{\omega}_{nk})]b_nU_t = U_tb_k. \qquad (10.4.3)$$

Introducing the matrix

$$t_{kn} = \delta_{kn} + i\gamma_k(\omega_{kn} + \overline{\omega}_{nk})$$

(10.4.3) becomes

$$\sum_n t_{kn}b_nU = Ub_k. \qquad (10.4.4)$$

Assuming that the matrix $t = (t_{kn})$ is invertible, from (10.4.4) one has

$$b_nU = \sum_k t_{nk}^{-1}Ub_k. \qquad (10.4.5)$$

From (10.4.2) and (10.4.5) we find the normal form of white noise Hamiltonian equation

$$\partial_tU = -i\sum_{mk}b_m^+T_{mk}Ub_k, \qquad (10.4.6)$$

where

$$T_{mk} = \sum_n(\omega_{mn} + \overline{\omega}_{nm})t_{nk}^{-1}. \qquad (10.4.7)$$

In terms of stochastic differentials, (10.4.6) can be written in the form

$$dU(t) = dN_t(T)U(t), \qquad (10.4.8)$$

where

$$N_t(T) = N(\chi_{[0,t]} \otimes T).$$

11. Six Basic Principles of the Stochastic Limit

All the relevant results concerning the stochastic limit in the weak coupling case can be obtained by applying the following principles:

- the stochastic universality class principle,
- the block principle,
- the stochastic bosonization principle,
- the orthogonalization principle,
- the stochastic resonance principle,
- the time-consecutive principle.

The proofs of these principles are based on a few basic results: the *connected component theorem* of Sect. 15.1, the *factorization theorem* of Sect. 15.2, the *multiple integral lemma* of Sect. 15.10 and the *multiple simplex theorems* of Sects. 15.9 and 15.11, which have been abstracted from a multiplicity of results concerning specific models. In this chapter, starting from Sect. 11.3 we shall state these six principles.

We discuss their main implications starting from the stochastic limit in the general framework of polynomial interactions. By this we mean that the interaction Hamiltonians considered are polynomials in the smoothed *fields*, e.g. $A(g)^4 + A^+(g)^4$. The more interesting case of polynomials in the field densities will be deal with in Part II. For (boson) polynomials of degree greater than 2 the convergence results have to be interpreted in the sense of term-by-term convergence of the iterated series for the propagator. However it should be noted that, after the stochastic limit, the iterated series will be, in general, convergent.

In particular we shall discuss:

(i) The canonical form of the interaction Hamiltonians to which our technique can be applied and the notion of *stochastic universality class* (i.e. the class of all the interaction Hamiltonians which give rise to the same stochastic limit). This motivates the introduction of the notion of an *asymptotically effective interacting Hamiltonian*, which roughly speaking corresponds to the simplest choice in the universality class. From it we deduce the form of the rescaled fields (collective vectors), and from them the structure of the master space is obtained by applying the block, orthogonalization and stochastic resonance principles.

(ii) The extension of the stochastic golden rule of Sect. 4.14 to general polynomial interactions. In this chapter we shall only formulate the results, giving no proofs.

11.1 Polynomial Interactions with Cutoff

We shall work, as in Chap. 4, in the system–reservoir (S, R) framework, described by the following:

(i) The state space $\mathcal{H}_S$ of the system S, and the 1-particle state space $\mathcal{H}_1$ of the reservoir R which are both pre–Hilbert spaces.

(ii) The state space of the composite system (S, R):

$$\mathcal{H} \otimes \mathcal{H}_S = \Gamma(\mathcal{H}_1) \otimes \mathcal{H}_S ,$$

where $\mathcal{H}$ is a Gaussian second quantized space (Fock, finite temperature, squeezed, etc.) based on $\mathcal{H}_1$. An operator L, acting on the Hilbert space $\mathcal{H}_S$ ($\mathcal{H}$) will be often identified with $1 \otimes L$ ($L \otimes 1$).

(iii) The initial state of the reservoir: $\varphi(\cdot) := \langle \Phi, \cdot \Phi \rangle$.

(iv) The system Hamiltonian H_S and the field (or reservoir) Hamiltonian H_R, so that the total free Hamiltonian is

$$H_o = H_S + H_R .$$

(v) The interaction Hamiltonian H_{I} is a polynomial Hamiltonian, i.e. a finite sum of terms of the following type:

$$\begin{aligned} &[\text{a polynomial of (smeared) creators and annihilators}] \\ &\otimes [\text{an operator on } \mathcal{H}_S] . \end{aligned}$$

The role of the cutoff is played here by the test functions, "smearing" the creators and annihilators, whose physical meaning has been discussed in Sect. 4.9.3. For a given coupling constant λ, between the system and the reservoir, the total Hamiltonian $H^{(\lambda)}$ is defined as usual, i.e.

$$H_S + H_R + \lambda H_{\mathrm{I}} = H_o + \lambda H_{\mathrm{I}} =: H^{(\lambda)}$$

The object to be investigated is the evolution operator in the interaction picture, i.e.

$$U_t := e^{itH_o} e^{-itH^{(\lambda)}} , \tag{11.1.1}$$

which satisfies the Schrödinger equation in the interaction picture,

$$\partial_t U_t^{(\lambda)} := -i\lambda H_{\mathrm{I}}(t)\, U_t , \qquad U_0^{(\lambda)} = \mathbf{1} , \tag{11.1.2}$$

where $H_{\mathrm{I}}(t)$ denotes the time-evolved (with the free evolution) interaction Hamiltonian,

$$H_{\mathrm{I}}(t) := e^{itH_o} H_{\mathrm{I}} e^{-itH_o} \, . \tag{11.1.3}$$

The iterated series solution of the integral form of the differential equation (11.1.2) is

$$U_t^{(\lambda)} = 1 + (-i\lambda) \int_0^t H_{\mathrm{I}}(t_1) dt_1 + (-i\lambda)^2 \int_0^t \int_0^{t_1} H_{\mathrm{I}}(t_1) H_{\mathrm{I}}(t_2) dt_1 dt_2 + \cdots \tag{11.1.4}$$

Similarly, if $X \in \mathcal{B}(\mathcal{H}) \otimes \mathcal{H}_S$ is any operator of the composite system, then

$$X_-(t) := U_t^{(\lambda)*} X U_t^{(\lambda)} \quad , \quad X_+(t) := U_t^{(\lambda)} X U_t^{(\lambda)*}$$

satisfy respectively the backward and forward Heisenberg equations in the interaction representation whose iterated series solution are respectively

$$\begin{aligned} X_-(t) = {} & 1 + (-i\lambda) \int_0^t [H_{\mathrm{I}}(t_1), X] dt_1 \\ & +(-i\lambda)^2 \int_0^t \int_0^{t_1} \big[H_{\mathrm{I}}(t_1), [H_{\mathrm{I}}(t_2), X]\big] dt_1 dt_2 + \cdots \end{aligned} \tag{11.1.5a}$$

$$\begin{aligned} X_+(t) = {} & 1 + i\lambda \int_0^t [X, H_{\mathrm{I}}(t_1)] dt_1 \\ & +(i\lambda)^2 \int_0^t \int_0^{t_1} \big[[X, H_{\mathrm{I}}(t_2)], H_{\mathrm{I}}(t_1)\big] dt_1 dt_2 + \cdots \end{aligned} \tag{11.1.5b}$$

Remark 11.1.1. In (11.1.4) and (11.1.5a,b), the equalities are formal. We do not investigate here the convergence of the series appearing on the right-hand sides of these equations. In many interesting cases (e.g. for Fermi fields), once the precise form of the interaction H_{I} is given, this convergence can be proved. When the series does not converge, our results will be interpreted in the sense of term-by-term convergence.

11.2 Assumptions on the Dynamics: Standard Models

Following the general scheme of Chap. 4 the description of the system–reservoir model is completed by assigning the specific form of the free system, free reservoir and interaction Hamiltonian. In this chapter we shall make three types of assumptions:

(I) The *free reservoir Hamiltonian* H_R is the second quantization of a 1-particle evolution (see Definition 2.4.2) denoted by

$$S_t^1 := e^{-itH_1} = e^{-it\omega_1(p)} \, ,$$

where ω_1 is a positive function and p is a momentum operator. For any contraction A, in the 1-particle reservoir space, we shall denote by $\Gamma(A)$ the second quantization of A so that the free reservoir Hamiltonian H_R is

$$H_R := \left.\frac{d}{dt}\right|_{t=0} \Gamma\left(e^{-itH_1}\right) = \left.\frac{d}{dt}\right|_{t=0} \Gamma\left(e^{-it\omega_1(p)}\right) .$$

Moreover all the test functions that appear in the interaction Hamiltonian belong to a subspace $\mathcal{K}$ of the 1-particle space $L^2(\mathbf{R}^3)$ [more generally $L^2(\mathbf{R}^d)$] such that any pair of functions $f, g \in \mathcal{K}$ satisfies the basic analytical condition (4.5.4), i.e.

$$\int_{\mathbf{R}} |\langle f, S_t^1 g\rangle| \, dt < +\infty .$$

(II) The *interaction Hamiltonian* has the following general structure:

$$H_{\mathrm{I}} := \left(\sum_j P_j(A^+, A) \otimes D_j + \text{ h.c.}\right) , \tag{11.2.1}$$

where

(i) $P_j(A^+, A)$ is a polynomial, without a constant term, in the non-commutative indeterminates given by creation and annihilation operators. Moreover the polynomials $P_j(A, A^+)$ will always be assumed to be *normally ordered.* If this is not the case, one can always put them in normal order, still obtaining a polynomial in the creation or annihilation operators and absorb the constant term, coming from the normal ordering procedure, into the free Hamiltonian.

(ii) $\{D_j\}$ is either a family of bounded system operators or satisfies the following two conditions:
- there is a dense subspace $\mathcal{D}_S$ of the system space $\mathcal{H}_S$ such that for any j, $\mathcal{D}_S \subseteq \mathrm{Dom}(D_j)$; $D_j\mathcal{D}_S \subseteq \mathcal{D}_S$;
- there are two constant $\alpha > 0$, $\beta > 0$ such that for any j and for any $\xi, \eta \in \mathcal{D}_S$,

$$\sum_{n=1}^{\infty} \alpha^n \sum_{\varepsilon\in\{0,1\}^n} \frac{\left\langle \xi, D_j^{\varepsilon(1)} \cdots D_j^{\varepsilon(n)} \eta \right\rangle}{\left[\binom{n}{\beta}\right]!} < \infty . \tag{11.2.2}$$

These conditions are automatically satisfied if D_j are bounded operators or if the system space is also a second quantized space, the D_j are creation or annihilation operators (or polynomials in these variables) and $\mathcal{D}_S$ is the algebraic span of the number vectors.

(iii) The free system Hamiltonian H_S is related to the D_j operators as follows: for any j, there exist a set of system self-adjoint operators $\{\omega_{j,k}\}$ and a set of system operators $\{D_{j,k}\}$ such that

- $\mathcal{D} \subset \mathrm{Dom}(D_{j,k}) \cap \mathrm{Dom}(\omega_{j,k})$;
-

$$e^{itH_S} D_j e^{-itH_S} = \sum_k D_{j,k} e^{it\omega_{j,k}} ; \tag{11.2.3}$$

- $\left[\omega_{j,k}, D_{j',k'}\right]$ is a scalar.

(III) These (S, R) models are called *quasi-standard.* If all the $\omega_{j,k}$ are real numbers, they are called *standard.* In Sect. 5.1 we showed that whenever the system Hamiltonian H_S has discrete spectrum the model is standard. The same is true if D_j commute with the system Hamiltonian H_S. So standard models include many interesting examples: N-level systems; H_S is the Laplacian on a compact Riemannian manifold, etc. The stochastic limit can also be performed for non-quasi-standard model (e.g. quantum electrodynamics [QED] without the dipole approximation); however, in this case additional technical difficulties and totally new features arise: the breaking of the usual Bose–Einstein (or Fermi–Dirac) statistics and the emergence of new statistics, the emergence of *Hilbert modules*, rather than Hilbert spaces, as natural state spaces of composite systems and the associated new (entangled) commutation relations, etc. These features will be discussed in the Chaps. 12–14. From now to the end of this chapter we shall assume, unless otherwise specified, that the model considered is *standard.* Given the above conditions, the time-evolved interaction Hamiltonian is

$$\begin{aligned} H_{\mathrm{I}}(t) &= e^{itH_o} H_{\mathrm{I}} e^{-itH_o} = \sum_j e^{itH_o} \left\{ P_j(A^+, A) \otimes D_j \right\} e^{-itH_o} + \mathrm{h.c.} \\ &= \sum_j \left[e^{itH_R} P_j(A^+, A) e^{-itH_R} \right] \otimes e^{itH_S} D_j e^{-itH_S} + \mathrm{h.c.} \\ &= \sum_j \sum_k P_j(A_t, A_t^+) \otimes D_{j,k} e^{it\omega_{j,k}} + \mathrm{h.c.} \end{aligned} \tag{11.2.4}$$

For a standard model $\omega_{j,k} = \varepsilon_j - \varepsilon_k$, where the system energy levels ε_j are as in (4.6.1) and we have introduced the notation

$$A_t(f) := A(S_t^1 f) . \tag{11.2.5}$$

Moreover, introducing the set of all resonance frequencies

$$\Omega := \{\omega_{j,k} \ : \ j, k \in \mathbb{N}\} = (\omega_q)_{q \in \mathbb{N}} , \tag{11.2.6}$$

so that for any pair of indices $q \neq q'$ one has

$$\omega_q \neq \omega_{q'} , \tag{11.2.7}$$

$H_{\rm I}(t)$ becomes

$$H_{\rm I}(t) = \sum_{\omega_q} e^{it\omega_q} \sum_{\{j,k\,:\,\omega_{j,k}=\omega_q\}} P_j(A_t, A_t^+) \otimes D_{j,k} + \text{h.c.}$$

Thus, defining the new operator

$$D_{j,q} := \sum_{\{k\,:\,\omega_q=\omega_{j,k}\}} D_{j,k}\,,$$

we write the interaction Hamiltonian in standard form

$$H_{\rm I}(t) = \sum_{\omega_q} e^{it\omega_q} \sum_j P_j(A_t, A_t^+) \otimes D_{j,q} + \text{h.c.}\,, \tag{11.2.8}$$

where the sum over the index j is extended to the set

$$I_q := \{j : \text{there exists a } k \text{ with } \omega_{j,k} = \omega_q\}\,.$$

In summary, for a standard model one has

$$H_{\rm I}(t) = \sum_{j,k} \left\{ e^{it\omega_{j,k}} \Gamma\left(S_t^1\right) P_j(A, A^+) \Gamma\left(S_{-t}^1\right) \otimes D_{j,k} \right\} + \text{h.c.}$$

So, also in the case of general polynomial interactions, the time variable, which appears originally both in the field and system parts, can be moved to the field part and the system part becomes time independent.

11.3 Polynomial Interactions: Canonical Forms, Fock Case

For a quasi-standard model, without loss of generality, we can assume more about polynomials P_j. Each P_j can be assumed to have the form

$$\sum_{\substack{n,m\\ n+m\geq 1}} c_{j,n,m} \cdot [\text{a product of } n \text{ creators}] \cdot [\text{a product of } m \text{ annihilators}]\,,$$

where, $c_{j,n,m} \in \mathbf{C}$ and the condition $n+m \geq 1$ is a consequence of $P_j(0,0) = 0$. Moreover, by moving the constant $c_{j,n,m}$ to the system part, we can suppose that

$$H_{\rm I} = \sum_{\substack{n,m\\ n+m\geq 1}} \left(\sum_j P_{j,n,m}(A, A^+) \otimes D_{j,n,m} + \text{h.c.} \right)\,,$$

where, any $P_{j,n,m}(A,A^+)$ is an operator of the form

$$[\text{a product of } n \text{ creators}] \cdot [\text{a product of } m \text{ annihilators}] .$$

Furthermore we can always suppose that $n \geq m$ (the annihilators are not more numerous than the creators). In fact if this is not the case we simply exchange the roles of the j-summation in (2) with its hermitean conjugate. More precisely, if $n < m$, we replace $P_{j,n,m}(A,A^+)$, $D_{j,n,m}$ by $\left(P_{j,n,m}(A,A^+)\right)^*, (-D_{j,n,m})^*$.

By denoting

$$P_{j,n,m}(A,A^+) = A^+(f_{j,n,1})\cdots A^+(f_{j,n,n})\,A(g_{j,m,1})\cdots A(g_{j,m,m}) \, ,$$

where $f_{j,n,1},\cdots,f_{j,n,n},g_{j,m,1},\cdots,g_{j,m,m}$ are test functions in $\mathcal{H}_1$, we can separate, in the interaction Hamiltonian, the *pure creation* terms and the *pure annihilation* terms ($\mathcal{A}$- and $\mathcal{A}^+$-type blocks) from the *mixed* terms ($\mathcal{N}$-type blocks) and rewrite it in the form

$$\begin{aligned} H_{\mathrm{I}} = &\sum_{n\geq 1}\left(\sum_j A^+(f_{j,n,1})\cdots A^+(f_{j,n,n})\otimes D_{j,n,0} + \text{h.c.}\right) \\ &+ \sum_{\substack{n,m\\ n\geq m\geq 1}}\left(\sum_j P_{j,n,m}(A,A^+)\otimes D_{j,n,m} + \text{h.c.}\right) . \end{aligned} \tag{11.3.1}$$

Therefore, the time-evolved Hamiltonian has the following form:

$$\begin{aligned} H_{\mathrm{I}}(t) = &\sum_{n\geq 1}\Bigg(\sum_{j,k}\Gamma\left(S_t^1\right)A^+(f_{j,n,1})\cdots A^+(f_{j,n,n})\,\Gamma\left(S_{-t}^1\right) \\ &\qquad\otimes D_{j,n,0,k}e^{it\omega_{j,n,k}} + \text{h.c.}\Bigg) \\ &+ \sum_{\substack{n,m\\ n\geq m\geq 1}}\Bigg(\sum_{j,k}\Gamma\left(S_t^1\right)P_{j,n,m}(A,A^+)\Gamma\left(S_{-t}^1\right) \\ &\qquad\otimes D_{j,n,m,k}e^{it\omega_{j,n,m,k}} + \text{h.c.}\Bigg) \\ = &\sum_{n\geq 1}\left(\sum_{j,k}A^+\left(S_t^1 f_{j,n,1}\right)\cdots A^+\left(S_t^1 f_{j,n,n}\right)\otimes D_{j,n,0,k}e^{it\omega_{j,n,k}} + \text{h.c.}\right) \\ &+ \sum_{\substack{n,m\\ n\geq m\geq 1}}\Bigg(\sum_{j,k}\Gamma\left(S_t^1\right)P_{j,n,m}(A,A^+)\Gamma\left(S_{-t}^1\right) \\ &\qquad\otimes D_{j,n,m,k}e^{it\omega_{j,n,m,k}} + \text{h.c.}\Bigg) . \end{aligned} \tag{11.3.2a}$$

Moreover, for each $n \geq 1$, we can relabel the (countable) set $\{\omega_{j,n,k}\}_{j,k}$, replacing the pair of indices (j,k) by a single index h so that this set is denoted $\left\{\omega_h^{(n)}\right\}_h$ where, for any n, if $h \neq r$, one has

$$\omega_h^{(n)} \neq \omega_r^{(n)} .$$

With these notations,

$$\begin{aligned} H_{\mathrm{I}}(t) = \sum_{n\geq 1} \Bigg[\sum_{h,j} & A^+\left(S_t^1 f_{j,n,1}\right) \cdots A^+\left(S_t^1 f_{j,n,n}\right) \\ & \otimes \left(\sum_k D_{j,n,0,k} \chi_{\left\{\omega_h^{(n)}\right\}}(\omega_{j,n,k}) \right) e^{it\omega_h^{(n)}} + \mathrm{h.c.} \Bigg] \\ + \sum_{\substack{n,m \\ n\geq m\geq 1}} & \left(\sum_{j,k} \Gamma\left(S_t^1\right) P_{j,n,m}(A,A^+) \Gamma\left(S_{-t}^1\right) \right. \\ & \left. \otimes D_{j,n,m,k} e^{it\omega_{j,n,m,k}} + \mathrm{h.c.} \right) \end{aligned}$$

and, adding together all the coefficients belonging to the same Bohr frequency

$$D_{j,h}^{(n)} := \sum_k \chi_{\left\{\omega_h^{(n)}\right\}}(\omega_{j,n,k}) D_{j,n,0,k}$$

$\left[\chi_{\left\{\omega_h^{(n)}\right\}}(x) = 0 \text{ if } x \neq \omega_h^{(n)} \text{ and } = 1 \text{ if } x = \omega_h^{(n)}\right]$, the interaction Hamiltonian becomes

$$\begin{aligned} H_{\mathrm{I}}(t) = \sum_{n\geq 1} & \left(\sum_{h,j} A^+\left(S_t^1 f_{j,n,1}\right) \cdots A^+\left(S_t^1 f_{j,n,n}\right) \otimes D_{j,h}^{(n)} e^{it\omega_h^{(n)}} + \mathrm{h.c.} \right) \\ + \sum_{\substack{n,m \\ n\geq m\geq 1}} & \left(\sum_h \sum_{j,k} e^{it\omega_{j,n,m,k}} \Gamma\left(S_t^1\right) P_{j,n,m}(A,A^+) \Gamma\left(S_{-t}^1\right) \right. \\ & \left. \otimes D_{j,n,m,h} + \mathrm{h.c.} \right) . \end{aligned} \tag{11.3.2b}$$

When it will not be relevant for our discussion, we will specify in the time-evolved Hamiltonian (11.3.2b) neither the number of operators in each block, nor the precise forms of the test functions, nor the operators in the system part, nor the frequency terms. In these cases, for (11.3.2b), we will use the symbolic notation

$$H_{\mathrm{I}}(t) = i\left(\mathcal{A}^+(t) \otimes D + \mathrm{h.c.}\right) + i\mathcal{N}(t) \otimes D' , \tag{11.3.3}$$

which underlines the role of the $\mathcal{A}$-type and $\mathcal{N}$-type blocks. Since the adjoint of an $\mathcal{N}$-type block is still an $\mathcal{N}$-type block, we denote all of them with the single symbol $\mathcal{N}(t)$.

11.4 Polynomial Interactions: Canonical Forms, Gauge-Invariant Case

Formula (11.3.2b) is useful if the reference state is the Fock state, but if the reference state is a general gauge-invariant state, e.g. a Gibbs state at a certain temperature, a more refined analysis is required which keeps into account the possibility to describe, by means of a Bogoliubov transformation, the representation associated with such a state as a product of a Fock representation and an anti-Fock representation. To obtain, for these states, the analogues of the canonical forms, discussed in Sect. 11.3 above, we proceed as follows:

First of all, using a Bogoliubov transformation, we write the reservoir reference state as a vector state in its cyclic (GNS) representation:

$$\langle \Phi \otimes \Phi_\iota, \cdot \Phi \otimes \Phi_\iota \rangle \ ;$$

introducing the parity operator θ on the Fermi Fock space, i.e.

$$\theta = \oplus_{n=0}^{\infty} (-1)^n \ ,$$

we write the creator with test function g as

$$\begin{aligned} &A^+ (Q_+ g) \otimes 1 + 1 \otimes A(\iota Q_- g) \quad \text{for the boson case and} \\ &A^+ (Q_+ g) \otimes 1 + \theta \otimes A(\iota Q_- g) \quad \text{for the Fermi case,} \end{aligned}$$

where $Q_\pm$ are operators on the 1-particle space satisfying

$$\begin{aligned} |Q_+|^2 - |Q_-|^2 &= 1 \quad \text{for the boson case, and} \\ |Q_+|^2 + |Q_-|^2 &= 1 \quad \text{for the Fermi case.} \end{aligned}$$

Therefore the annihilator with test function g is

$$\begin{aligned} &A(Q_+ g) \otimes 1 + 1 \otimes A^+ (\iota Q_- g) \quad \text{for the boson case and} \\ &A(Q_+ g) \otimes 1 + \theta \otimes A^+ (\iota Q_- g) \quad \text{for the Fermi case,} \end{aligned}$$

where ι is the conjugation operator ($\iota f := \overline{f}$) which embeds a test function f of the 1-particle space $\mathcal{H}_1$ into its dual space $\overline{\mathcal{H}}_1$ (if the elements of $\mathcal{H}_1$ are thought of as *ket* vectors, then the elements of $\overline{\mathcal{H}}_1$ are *bra* vectors). In this way any polynomial $P_j(A^+, A)$ is a sum of polynomials of the form $P_{j,1}(A^+, A) \otimes P_{j,2}(A^+, A)$ in the boson case and $P_{j,1}(A^+, A, \theta) \otimes P_{j,2}(A^+, A)$ in the Fermi case; $P_{j,1}(A^+, A, \theta)$ is a polynomial in the variables A, A^+ and θ. Moreover because of the relations (where $\{\cdot, \cdot\}$ denotes the anticommutator)

$$\theta^2 = 1, \ \theta^* = \theta, \ \{A(f), \theta\} = \{A^+(f), \theta\} = 0, \quad \text{for any test function } f \, , \tag{11.4.1a}$$

we can write $P_{j,1}(A^+, A, \theta)$ either as $\theta P_{j,1}(A^+, A)$ or $P_{j,1}(A^+, A)$. In this case all polynomials $P_{j,h}(A^+, A)$ can be assumed to be normally ordered.

By the same arguments used to deduce the canonical form (11.3.2b), we can assume that the interaction H_{I} in the boson case has the form

$$\begin{aligned} H_{\mathrm{I}} = & \sum_{n\geq 1}\left(\sum_j A^+(f_{j,n,1})\cdots A^+(f_{j,n,n})\otimes \mathbf{1}\otimes D_{j,n,+} + \text{h.c.}\right) \\ & + \sum_{n\geq 1}\left(\sum_j \mathbf{1}\otimes A^+(\iota g_{j,n,1})\cdots A^+(\iota g_{j,n,n})\otimes D_{j,n,-} + \text{h.c.}\right) \\ & + \sum_{n,m\geq 1}\left(\sum_j A^+(f'_{j,n,1})\cdots A^+(f'_{j,n,n})\right. \\ & \left.\otimes A^+(\iota g'_{j,m,1})\cdots A^+(\iota g'_{j,m,m})\otimes D_{j,n,m} + \text{h.c.}\right) \\ & + \sum_{\substack{n,m \\ n,m\geq 1}}\left(\sum_j \mathcal{P}_{j,n,m}(A,A^+)\otimes L_{j,n,m} + \text{h.c.}\right), \end{aligned} \tag{11.4.1b}$$

where $\mathcal{P}_{j,n,m}(A,A^+)$ is a normally ordered product of n creators and m annihilators. In the Fermi case,

$$\begin{aligned} H_{\mathrm{I}} = & \sum_{n\geq 1}\left(\sum_j F_{j,n,+}(\theta)\, A^+(f_{j,n,1})\cdots A^+(f_{j,n,n})\otimes \mathbf{1}\otimes D_{j,n,+} + \text{h.c.}\right) \\ & + \sum_{n\geq 1}\left(\sum_j F_{j,n,-}(\theta)\, A^+(\iota g_{j,n,1})\cdots A^+(\iota g_{j,n,n})\otimes D_{j,n,-} + \text{h.c.}\right) \\ & + \sum_{n,m\geq 1}\left(\sum_j F_{j,n,m}(\theta)\, A^+(f'_{j,n,1})\cdots A^+(f'_{j,n,n})\right. \\ & \left.\otimes A^+(\iota g'_{j,m,1})\cdots A^+(\iota g'_{j,m,m})\otimes D_{j,n,m} + \text{h.c.}\right) \\ & + \sum_{\substack{n,m \\ n,m\geq 1}}\left(\sum_j \mathcal{Q}_{j,n,m}(A,A^+,\theta)\otimes L_{j,n,m} + \text{h.c.}\right), \end{aligned} \tag{11.4.2a}$$

where $\mathcal{Q}_{j,n,m}(A,A^+,\theta)$ is a normally ordered product of n creators, m annihilators and the functions $F_{j,n,m}(\theta)$, $F_{j,n,\pm}(\theta)$, of the parity operator, are either the identity or the parity operator itself. More explicitly

$$F_{j,n,m}(\theta) = \theta^m(-1)^{nm} \quad , \quad F_{j,n,+}(\theta) = 1 \quad , \quad F_{j,n,-}(\theta) = \theta^n . \tag{11.4.2b}$$

At the end of this section we shall prove these formulae.

The evolved interaction Hamiltonian then has the following form in the boson case:

$$\begin{aligned}H_{\mathrm{I}}(t) = &\sum_{n\geq 1}\left(\sum_{j,k} A^{+}\left(S_t^1 f_{j,n,1}\right)\cdots A^{+}\left(S_t^1 f_{j,n,n}\right)\otimes \mathbf{1}\right.\\ &\left.\qquad \otimes D_{j,n,+,k}e^{it\omega_{j,n,+,k}} + \text{h.c.}\right)\\ &+\sum_{n\geq 1}\left(\sum_{j,k}\mathbf{1}\otimes A^{+}\left(\iota S_t^1 g_{j,n,1}\right)\cdots A^{+}\left(\iota S_t^1 g_{j,n,n}\right)\right.\\ &\left.\qquad \otimes D_{j,n,-,k}e^{it\omega_{j,n,-,k}} + \text{h.c.}\right)\\ &+\sum_{n,m\geq 1}\left(\sum_{j,k} A^{+}\left(S_t^1 f'_{j,n,1}\right)\cdots A^{+}\left(S_t^1 f'_{j,n,n}\right)\right.\\ &\left.\otimes A^{+}\left(\iota S_t^1 g'_{j,m,1}\right)\cdots A^{+}\left(\iota S_t^1 g'_{j,m,m}\right)\otimes D_{j,n,m,k}e^{it\omega_{j,n,m,k}} + \text{h.c.}\right)\\ &+\sum_{\substack{n,m\\ n,m\geq 1}}\left(\sum_{j,k}\Gamma\left(S_t^1\right)\mathcal{P}_{j,n,m}(A,A^{+})\Gamma\left(S_{-t}^1\right)\right.\\ &\left.\qquad \otimes L_{j,n,m,k}e^{it\omega'_{j,n,m,k}} + \text{h.c.}\right). \end{aligned} \tag{11.4.3a}$$

Using the explicit form of the action of the free evolution on $A(g), A^*(g)$, we can write

$$\begin{aligned}H_{\mathrm{I}}(t) = &\sum_{n\geq 1}\left(\sum_{h,j} A^{+}\left(S_t^1 f_{j,n,1}\right)\cdots A^{+}\left(S_t^1 f_{j,n,n}\right)\otimes \mathbf{1}\otimes D_{j,h,+}^{(n)}e^{it\omega_{h,+}^{(n)}} + \text{h.c.}\right)\\ &+\sum_{n\geq 1}\left(\sum_{h,j}\mathbf{1}\otimes A^{+}\left(\iota S_t^1 g_{j,n,1}\right)\cdots A^{+}\left(\iota S_t^1 g_{j,n,n}\right)\otimes D_{j,h,-}^{(n)}e^{it\omega_{h,-}^{(n)}} + \text{h.c.}\right)\\ &+\sum_{n,m\geq 1}\left(\sum_{h,j} A^{+}\left(S_t^1 f'_{j,n,1}\right)\cdots A^{+}\left(S_t^1 f'_{j,n,n}\right)\right.\\ &\left.\qquad\otimes A^{+}\left(\iota S_t^1 g'_{j,m,1}\right)\cdots A^{+}\left(\iota S_t^1 g'_{j,m,m}\right)\otimes D_{j,h}^{(n,m)}e^{it\omega_h^{(n,m)}} + \text{h.c.}\right)\\ &+\sum_{\substack{n,m\\ n,m\geq 1}}\left(\sum_{j,k}\Gamma\left(S_t^1\right)\mathcal{P}_{j,n,m}(A,A^{+})\Gamma\left(S_{-t}^1\right)\right.\\ &\left.\qquad \otimes L_{j,n,m,k}e^{it\omega'_{j,n,m,k}} + \text{h.c.}\right), \end{aligned} \tag{11.4.3b}$$

where for any n, m and $h \neq r$

$$\omega_{h,+}^{(n)} \neq \omega_{r,+}^{(n)} \quad , \quad \omega_{h,-}^{(n)} \neq \omega_{r,-}^{(n)} \quad , \quad \omega_h^{(n,m)} \neq \omega_r^{(n,m)} .$$

In the Fermi case,

$$\begin{aligned} H_{\mathrm{I}}(t) = \sum_{n\geq 1} \Bigg(\sum_{j,k} & A^+\left(S_t^1 f_{j,n,1}\right) \cdots A^+\left(S_t^1 f_{j,n,n}\right) \otimes \mathbf{1} \\ & \otimes D_{j,n,+,k} e^{it\omega_{j,n,k}} + \text{h.c.} \Bigg) \\ + \sum_{n\geq 1} \Bigg(\sum_{j,k} & \theta^n \otimes A^+\left(\iota S_t^1 g_{j,n,1}\right) \cdots A^+\left(\iota S_t^1 g_{j,n,n}\right) \\ & \otimes D_{j,n,-,k} e^{it\omega_{j,n,k}} + \text{h.c.} \Bigg) \\ + \sum_{n,m\geq 1} \Bigg(\sum_{j,k} & A^+\left(S_t^1 f'_{j,n,1}\right) \cdots A^+\left(S_t^1 f'_{j,n,n}\right) \theta^m \\ & \otimes A^+\left(\iota S_t^1 g'_{j,m,1}\right) \cdots A^+\left(\iota S_t^1 g'_{j,m,m}\right) \otimes D_{j,n,m,k} e^{it\omega_{j,n,m,k}} + \text{h.c.} \Bigg) \\ + \sum_{\substack{n,m \\ n,m\geq 1}} \Bigg(\sum_{j,k} & \Gamma\left(S_t^1\right) \mathcal{Q}_{j,n,m}(A, A^+) \Gamma\left(S_{-t}^1\right) \\ & \otimes L_{j,n,m,k} e^{it\omega'_{j,n,m,k}} + \text{h.c.} \Bigg) ; \end{aligned} \tag{11.4.4a}$$

with the same argument as above, this is equal to

$$\begin{aligned} H_{\mathrm{I}}(t) = \sum_{n} \Bigg(\sum_{h,j} & A^+\left(S_t^1 f_{j,n,1}\right) \cdots A^+\left(S_t^1 f_{j,n,n}\right) \otimes \mathbf{1} \\ & \otimes D_{j,h,+}^{(n)} e^{it\omega_{h,+}^{(n)}} + \text{h.c.} \Bigg) \\ + \sum_{n} \Bigg(\sum_{h,j} & \theta^n \otimes A^+\left(\iota S_t^1 g_{j,n,1}\right) \cdots A^+\left(\iota S_t^1 g_{j,n,n}\right) \\ & \otimes D_{j,h,-}^{(n)} e^{it\omega_{h,-}^{(n)}} + \text{h.c.} \Bigg) \\ + \sum_{n,m} \Bigg(\sum_{h,j} & A^+\left(S_t^1 f'_{j,n,1}\right) \cdots A^+\left(S_t^1 f'_{j,n,n}\right) \theta^m \end{aligned}$$

$$\otimes A^{+}\left(\iota S_t^1 g'_{j,m,1}\right)\cdots A^{+}\left(\iota S_t^1 g'_{j,m,m}\right)\otimes \mathbf{1}\otimes D_{j,h}^{(n,m)} e^{it\omega_h^{(n,m)}} + \text{h.c.}\Bigg)$$

$$+\sum_{\substack{n,m\\ n,m\geq 1}}\left(\sum_{j,k}\Gamma\left(S_t^1\right)\mathcal{Q}_{j,n,m}(A,A^{+},\theta)\Gamma\left(S_{-t}^1\right)\right.$$

$$\left.\otimes L_{j,n,m,k}e^{it\omega'_{j,n,m,k}}+\text{h.c.}\right), \tag{11.4.4b}$$

where $\omega_{h,\pm}^{(n)}$ satisfy the same relations as in the boson case.

Remark 11.4.1. In some cases we will write $H_{\mathrm{I}}(t)$ in a finer form, separating each term on the right-hand side of (11.4.4b) according to the number of operators in any block, i.e.

$$H_{\mathrm{I}}(t)=\sum_n\left(\sum_{h,j}A^{+}\left(S_t^1 f_{j,2n,1}\right)\cdots A^{+}\left(S_t^1 f_{j,2n,2n}\right)\otimes\mathbf{1}\right.$$

$$\left.\otimes\, D_{j,h,+}^{(2n)}e^{it\omega_{h,+}^{(2n)}}+\text{h.c.}\right)$$

$$+\sum_n\left(\sum_{h,j}\mathbf{1}\otimes A^{+}\left(\iota S_t^1 g_{j,2n,1}\right)\cdots A^{+}\left(\iota S_t^1 g_{j,2n,2n}\right)\right.$$

$$\left.\otimes\, D_{j,h,-}^{(2n)}e^{it\omega_{h,-}^{(2n)}}+\text{h.c.}\right)$$

$$+\sum_{n\text{ even},\ m\text{ even}}\left(\sum_{h,j}A^{+}\left(S_t^1 f'_{j,n,1}\right)\cdots A^{+}\left(S_t^1 f'_{j,n,n}\right)\right.$$

$$\left.\otimes\, A^{+}\left(\iota S_t^1 g'_{j,m,1}\right)\cdots A^{+}\left(\iota S_t^1 g'_{j,m,m}\right)\otimes\mathbf{1}\otimes D_{j,h}^{(n,m)}e^{it\omega_h^{(n,m)}}+\text{h.c.}\right)$$

$$+\sum_{n\text{ odd},\ m\text{ odd}}\left(\sum_{h,j}A^{+}\left(S_t^1 f'_{j,n,1}\right)\cdots A^{+}\left(S_t^1 f'_{j,n,n}\right)\theta\right.$$

$$\left.\otimes\, A^{+}\left(\iota S_t^1 g'_{j,m,1}\right)\cdots A^{+}\left(\iota S_t^1 g'_{j,m,m}\right)\otimes\mathbf{1}\otimes D_{j,h}^{(n,m)}e^{it\omega_h^{(n,m)}}+\text{h.c.}\right)$$

$$+\sum_n\left(\sum_{h,j}A^{+}\left(S_t^1 f_{j,2n-1,1}\right)\cdots A^{+}\left(S_t^1 f_{j,2n-1,2n-1}\right)\otimes\mathbf{1}\right.$$

$$\left.\otimes\, D_{j,h,+}^{(2n-1)}e^{it\omega_{h,+}^{(2n-1)}}+\text{h.c.}\right)$$

$$+\sum_n \left(\sum_{h,j} \theta \otimes A^+\left(\iota S_t^1 g_{j,2n-1,1}\right) \cdots A^+\left(\iota S_t^1 g_{j,2n-1,2n-1}\right) \right.$$

$$\left. \otimes D_{j,h,-}^{(2n-1)} e^{it\omega_{h,-}^{(2n-1)}} + \text{h.c.} \right)$$

$$+ \sum_{n \text{ odd}\ ,\ m \text{ even}} \left(\sum_{h,j} A^+\left(S_t^1 f'_{j,n,1}\right) \cdots A^+\left(S_t^1 f'_{j,n,n}\right) \right.$$

$$\left. \otimes A^+\left(\iota S_t^1 g'_{j,m,1}\right) \cdots A^+\left(\iota S_t^1 g'_{j,m,m}\right) \otimes \mathbf{1} \otimes D_{j,h}^{(n,m)} e^{it\omega_h^{(n,m)}} + \text{h.c.} \right)$$

$$+ \sum_{n \text{ even}\ ,\ m \text{ odd}} \left(\sum_{h,j} A^+\left(S_t^1 f'_{j,n,1}\right) \cdots A^+\left(S_t^1 f'_{j,n,n}\right) \theta \right.$$

$$\left. \otimes A^+\left(\iota S_t^1 g'_{j,m,1}\right) \cdots A^+\left(\iota S_t^1 g'_{j,m,m}\right) \otimes \mathbf{1} \otimes D_{j,h}^{(n,m)} e^{it\omega_h^{(n,m)}} + \text{h.c.} \right)$$

$$+ \sum_{\substack{n,m \\ n,m \geq 1}} \left(\sum_{j,k} \Gamma\left(S_t^1\right) \mathcal{Q}_{j,n,m}(A, A^+, \theta) \Gamma\left(S_{-t}^1\right) \right.$$

$$\left. \otimes L_{j,n,m,k} e^{it\omega'_{j,n,m,k}} + \text{h.c.} \right) .$$

Similar to what was done at the end of Sect. 11.3, if in the time-evolved Hamiltonian $H_{\mathrm{I}}(t)$ the number of operators in each block, the precise form of the test functions, the operators in the system part, the frequencies and the space structure are not relevant, we write both in the boson and Fermi cases

$$H_{\mathrm{I}}(t) = i\left[F(\theta)\, \mathcal{A}^+(t) \otimes D + \text{h.c.}\right] + i\mathcal{N}(t) \otimes L\,, \tag{11.4.5}$$

where $F(\theta)$ is simply the identity if the boson field is considered.

In summary, in the stochastic limit problem, for a standard model, the most general time-evolved interaction Hamiltonians are given by (11.4.3a,b) and (11.4.4a,b). In particular, if the reference state is the Fock vacuum, the most general form of the time-evolved interaction Hamiltonian is given in (11.3.2a,b).

Our last goal in this section is to derive the explicit form (11.4.2b) of the functions $F_{j,n,m}(\theta)$, $F_{j,n,\pm}(\theta)$. We are interested only in the coefficient $F(\theta)$ of the $\mathcal{A}$-type blocks as in (11.4.5). By taking conjugates this gives the coefficient of the $\mathcal{A}$-type blocks. Of the other ones we say that they "do not contribute". To this goal first note that the term $A^+(f_{j,n,1}) \cdots A^+(f_{j,n,n}) \otimes \mathbf{1}$ comes from the expansion of

$$
\begin{aligned}
A^+(x_1)\cdots A^+(x_n) &= \prod_{k=1}^{n}\left[A^+\left(Q_+x_k\right)\otimes 1+\theta\otimes A\left(\iota Q_-x_k\right)\right] \\
&= \prod_{k=1}^{n} A^+\left(Q_+x_k\right)\otimes 1 + \text{other terms}\,;
\end{aligned}
\tag{11.4.6}
$$

the other terms in (11.4.6) do not contribute to the product $A^+\left(f_{j,n,1}\right)\cdots A^+\left(f_{j,n,n}\right)\otimes\mathbf{1}$. Therefore $F_{j,n,+}\left(\theta\right)=1$. Similarly, the term $1\otimes A^+\left(\iota g_{j,n,1}\right)\cdots A^+\left(\iota g_{j,n,n}\right)$ comes from the expansion of

$$
\begin{aligned}
A(x_1)\cdots A(x_n) &= \prod_{k=1}^{n}\left[A\left(Q_+x_k\right)\otimes 1+\theta\otimes A^+\left(\iota Q_-x_k\right)\right] \\
&= \prod_{k=1}^{n}\theta\otimes A^+\left(\iota Q_-x_k\right) + \text{other terms} \\
&= \theta^n\otimes\prod_{k=1}^{n} A^+\left(\iota Q_-x_k\right) + \text{other terms}\,;
\end{aligned}
\tag{11.4.7}
$$

the other terms do not contribute. Therefore $F_{j,n,-}\left(\theta\right)=\theta^n$. Finally, since we have assumed that any term of the polynomial $P_j(A,A^+)$ is normally ordered, one knows that the term

$$
A^+\left(f'_{j,n,1}\right)\cdots A^+\left(f'_{j,n,n}\right)\otimes A^+\left(\iota g'_{j,m,1}\right)\cdots A^+\left(\iota g'_{j,m,m}\right)
\tag{11.4.8}
$$

comes from the expansion of

$$
\begin{aligned}
&A^+(x_1)\cdots A^+(x_n)A(x_{n+1})\cdots A(x_{n+m}) \\
&= \prod_{k=1}^{n} A^+\left(Q_+x_k\right)\theta^m\otimes\prod_{h=1}^{m} A^+\left(\iota Q_-x_{n+h}\right) + \text{ other terms}\,;
\end{aligned}
\tag{11.4.9}
$$

the other terms do not contribute. Applying the commutation relations (11.4.1a) one obtains

$$
\prod_{k=1}^{n} A^+\left(Q_+x_k\right)\theta^m = \theta^m\left(-1\right)^{nm}\prod_{k=1}^{n} A^+\left(Q_+x_k\right)\,,
$$

i.e. $F_{j,n,m}\left(\theta\right)$ in (11.4.2a) is given by

$$
\theta^m\left(-1\right)^{nm}\,.
$$

11.5 The Stochastic Universality Class Principle

In the general form (11.2.1) of the interaction Hamiltonians, the polynomials P_j can be arbitrary; however it may be that different interaction Hamiltonians give rise to the same stochastic limit. Therefore we can divide the possible polynomial interactions into equivalence classes in which all the elements inside a single class have the same stochastic limit. It is therefore natural to ask oneself if one can make the classification by choosing within each class the simplest possible representative. The answer to this question is positive, and it allows one to associate with each interaction Hamiltonian of the form (11.2.1) an *asymptotically effective interaction Hamiltonian*, which is simpler and gives rise to the same stochastic limit (i.e. both the same master fields and the same evolution equation). Note that *these effective Hamiltonians depend on the state* not only because the rigorous definition of the Hamiltonian depends on the state but also because the limit class depends on the initial state of the field.

To achieve this goal we have to single out, in the given interaction Hamiltonian, those terms which are essential and those which are negligible. Of course the universality class, and therefore the choice of the asymptotically effective interaction Hamiltonian, also depends on the type of stochastic limit one is considering (weak coupling, low density, anisotropic, etc.). In the weak coupling case this choice is made by means of a rule which is a corollary of the following:

Theorem 11.5.1. *Given a normally ordered polynomial interaction Hamiltonian H_{I} of the form (11.2.1), the interaction Hamiltonian obtained from H_{I} by keeping only the pure creation and pure annihilation monomials (the $\mathcal{A}$- and $\mathcal{A}^+$-type blocks in H_{I}) is in the same (weak coupling) universality class as H_{I}. If there are more independent fields, the blocks should be meant globally, i.e. a block containing an annihilator of one field and a creator of another is of $\mathcal{N}$-type (hence irrelevant).*

Proof 11.5.1. See Sect. 16.1.

Example 11.5.1. The interaction Hamiltonians

$$H_{\mathrm{I}} = i\left[A^+(f_1)\,A^+(f_2) - A(f_2)\,A(f_1) + iA^+(f)\,A(f)\right] \otimes \mathbf{1} \qquad (11.5.1)$$

and

$$H_{\mathrm{I}}' = i\left[A^+(f_1)\,A^+(f_2) - A(f_2)\,A(f_1)\right] \otimes \mathbf{1}\,, \qquad (11.5.2)$$

where $A^+(f)$ is a Bose field, give rise to the same stochastic limit. The same is true if the field is Fermi.

Remark 11.5.1. The case for more independent fields includes the case in which a single field in a nonFock representation once one goes to the canonical representation in terms of tensor products of Fock and anti-Fock fields.

The Hamiltonian obtained by applying the rule of Theorem 11.5.1 above is called the *asymptotically effective interaction Hamiltonian* associated to the given interaction Hamiltonian H_{I}. An interaction Hamiltonian $H_{\mathrm{I}}(t)$ which is a sum of purely creation and purely annihilation monomials will be said to be *in standard form*. The meaning of Theorem 11.5.1 is that the asymptotically effective interaction Hamiltonians which parametrize the universality classes of polynomial interactions are sums of terms of the form:

$$\int dk_1 \dots \int dk_n F_J(k_1 \dots k_n) a^+_{k_1} \dots a^+_{k_n}$$
$$\cdot \exp\left\{ it \left(\sum_{\alpha=1}^{n} \omega_1(k_\alpha) - \omega_q \right) \right\} D_{J,q} + \text{h.c.}\,, \tag{11.5.3}$$

where each sum can run over

- the natural integer n (degree of monomial);
- the frequency index $q = (q_1 \dots q_n)$ is a multi-index running into a finite or infinite set $\mathcal{I}_n$ and ω_q is different from ω_p if $q \neq p$;
- F_J are test functions;
- $D_{J,q}$ are system operators;
- the index J runs into a finite or infinite set.

The associated rescaled fields are then

$$a^{(\lambda)}(t, k_1 \dots k_n) =: \frac{1}{\lambda} a_{k_1} \dots a_{k_n} \exp\left\{ -i \left(\sum_{\alpha=1}^{n} \omega_1(k_\alpha) - \omega_{J_n} \right) t/\lambda^2 \right\}. \tag{11.5.4}$$

In terms of test functions, a typical asymptotically effective interaction Hamiltonian will have the form

$$H_{\mathrm{I}}(t) = \sum_n \sum_{j_1 \dots j_n, q} A^+\left(S^1_t f^{(n)}_{j_1}\right) \dots A^+\left(S^1_t f^{(n)}_{j_n}\right) e^{it\omega_q} D^{(n)}_{j_1 \dots j_n, q} + \text{h.c.}\,, \tag{11.5.5}$$

where $S^1_t = e^{it\omega_1(k)}$ $(j = 1 \dots n)$, $f^{(n)}_j$ are test functions, $D^{(n)}_{j_1 \dots j_n}$ are system operators and ω_q are real numbers. In Sect. 14.8 this general principle will be applied to the Yukawa interaction showing that it is in the same universality class of the tri-linear interaction considered in Sect. 14.2.

11.6 The Case of Many Independent Fields

If instead of considering a single field a_k, one considers many fields a^α_k, each with its own energy function $\omega^\alpha_1(k)$, then if the fields are independent all the results of the previous section remain valid without any change in their proofs. In fact, in all the proofs in the previous section, the only property of the expectation functional we have used is its Gaussianity, i.e. the fact that expectations of products of creation and annihilation operators are sums of products of 2-point functions. Therefore if we replace:

(i) the reference vector Φ by a tensor product of a certain number, say M, of reference vectors (one for each field): $\Phi_1 \otimes \Phi_2 \otimes \cdots \otimes \Phi_M$;
(ii) any product of annihilators by an M-tuple tensor product of (products of) annihilators according to the scheme:

$$B_1 \otimes \cdots \otimes B_M ,$$

where B_j are blocks of products of annihilators of the same field and different blocks may have a different number of factors;
(iii) any product of creators by an M-tuple tensor product of creator blocks according to the same scheme as in the previous item;
(iv) any $\mathcal{N}$-type block by an M-tuple tensor product of blocks in which either in a single block or in different blocks both creators and annihilators appear;

then all the above results, in particular Theorem 11.5.1, are still valid.

By a *block* we mean a product of the form $A_j\left(S_t^{(j)} f_1^{(j)}\right) \cdots A_j\left(S_t^{(j)} f_k^{(j)}\right)$, where j is the index of the field and *each creator has its own test function.* If this is not the case, as, for example, in the operator

$$\int_{\mathbb{R}^d} dk f(k) a_k \otimes b_k$$

corresponding to the product of the two independent fields a_k and b_k, then the block principle no longer holds and the picture changes drastically with the emergence of the Hilbert module and the interacting Fock spaces (see Chap. 14 and, in particular, the remark after (14.8.8), for further discussion of this point).

Once the interaction Hamiltonian has been put in standard form, the general principles of the stochastic limit, whose formulation will be completed in the remainder of this chapter, allow one to immediately write down the master field and the limit white noise equation.

11.7 The Block Principle: Fock Case

The block principle is a nonlinear generalization of Theorems 11.5.1 and 3.2.1. It states that any monomial block in a polynomial interaction such as (11.3.2b) or (11.4.4a), even if a product of many creators (annihilators) is before the limit, asymptotically, in the limit $\lambda \to 0$, acts on the master space like a single linear creator (annihilator), with a test function which depends on a number of variables equal to the degree of the block.

Because of the universality class principle, we can restrict ourselves to the case in which all the blocks consist of products of only creation or only annihilation operators [for more general Hamiltonians of this type see (11.4.5)]. We also restrict ourselves, to begin with, to the Fock case, so that the statistics of the field is uniquely determined by the commutator (or anticommutator, in the Fermi case). The Hamiltonian has the form (11.5.3), i.e.

$$\int dk_1 \ldots \int dk_n F_{J_n}(k_1, \ldots, k_n) a_{k_1} \ldots a_{k_n} \cdot \exp\left[-it\left(\sum_{\alpha=1}^{n} \omega_1(k_\alpha) - \omega_{J_n}\right)\right] D_{J_n}^{(n)} + \text{h.c.}, \tag{11.7.1}$$

where $J_n = (j_1 \ldots j_n)$ is a multi-index (the sum over n can run over a finite or infinite set of natural integers), F_{J_n} are test functions, and $D_{J_n}^{(n)}$ are system operators.

Lemma 11.7.1. *Defining*

$$a_{J_n}^{(\lambda)}(t, k_1 \ldots k_n) := \frac{1}{\lambda} a_{k_1} \ldots a_{k_n} \exp\left[-it\left(\sum_{\alpha=1}^{n} \omega_1(k_\alpha) - \omega_{J_n}\right)/\lambda^2\right], \tag{11.7.2}$$

one has, as $\lambda \to 0$,

$$a_{J_n}^{(\lambda)}(t, k_1 \ldots k_n) \to b_t^{J_n}(k_1 \ldots k_n), \tag{11.7.3}$$

where $b_t^{J_n}(k_1 \ldots k_n)$ *is the white noise on* $\mathbb{R}^{dn}$ *with a covariance depending on the reference state of the a field. Moreover the quantum white noises* $\{b_t^{J_n}(k_1 \ldots k_n) \,:\, J_n\}$ *are mutually independent in the sense of (2.1.7).*

Remark 11.7.1. For example, in the Fock case the master fields are uniquely determined by the conditions

$$b_{t,J_n}(k_1 \ldots k_n)\Phi = 0$$

and by the commutation relations

$$\begin{aligned}&[b_t^{J_n}(k_1 \ldots k_n) \,,\, b_{t'}^{J'_{n'}}(k'_1 \ldots k'_n)] \\ &= \delta_{n,n'}\delta_{\omega_{J_n},\omega_{J'_n}}\delta(t-t') \sum_{\sigma \in \mathcal{S}_n} \delta(k_1 - k'_{\sigma 1}) \ldots \delta(k_n - k'_{\sigma m}) \\ &\quad \cdot \delta\left(\sum_{\alpha=1}^{n} \omega_1(k_\alpha) - \omega_{J_n}\right).\end{aligned} \tag{11.7.4}$$

Proof 11.7.1. See Sect. 16.2.

11.8 The Stochastic Resonance Principle

We have already met in Sect. 5.2 the stochastic resonance principle in the case of interaction Hamiltonians linear in the a field. In the present section we extend the stochastic resonance principle to the nonlinear case as follows: Any two terms in the interaction Hamiltonian (11.7.1), corresponding to different resonance frequences ω_{J_n}, in the stochastic limit, become independent white noises. More precisely:

(i) blocks corresponding to different frequencies ω_{J_n} give rise to independent white noises;
(ii) blocks corresponding to the same frequency give rise to the same white noise, possibly integrated over different test functions.

11.9 The Orthogonalization Principle

Let $H_{\mathrm{I}}(t)$ be an interaction Hamiltonian in standard form (11.7.1). The orthogonalization principle states that for any two terms in $H_{\mathrm{I}}(t)$, if one is a product of m creators (annihilators) and the other is a product of n creators (annihilators), then in the limit they become independent master fields if $n \neq m$. The orthogonalization principle can be equivalently formulated as follows: *Blocks of different length give rise, in the stochastic limit, to independent white noises.*

Lemma 11.9.1. *Defining the rescaled fields $a^{(\lambda)}_{J_n}(t, k_1 \dots k_n)$ as in (11.7.2), then if either $m \neq n$ or $J_n \neq J_m$ one has*

$$a^{(\lambda)}_{J_n}(t, k_1 \dots k_n) \to b^{J_n}_t(k_1 \dots k_n)$$

$$a^{(\lambda)}_{J'_n}(t, k_1 \dots k_m) \to b^{J_m}_t(k_1 \dots k_m)\,,$$

where $b^{J_n}_t(k_1 \dots k_n)$ and $b^{J_m}_t(k_1 \dots k_m)$ are independent white noises.

Proof 11.9.1. See Sect. 16.2.

11.10 The Stochastic Bosonization Principle

The basic idea of *bosonization* is that a quadratic (or, more generally, even) polynomial in the fermion operators behaves like a boson operator. Up to now it has been difficult to substantiate this idea in more than one dimension, in the sense that several inequivalent proposals exist, but due to the absence of a definite proof in favour of one or the other, it is difficult to choose among them. The existing approaches to the problem are algebraic and exact but a satisfactory theory exists only in one dimension; the new approach suggested by the stochastic limit[3] is complementary to the existing ones, in the sense that it is an approximation (its basic idea being that bosonization does not take place exactly, but only after the scaling $t \to t/\lambda^2$ and the limit $\lambda \to 0$), but it has the advantage of working *independent of the dimension* provided this is greater than or equal to three.

Stochastic bosonization is a refinement of both the block and the orthogonalization principle for Fermi fields. Stochastic bosonization can be realized in any dimension greater than or equal to 3 (this is a common feature of the stochastic limit method). So we have here a situation complementary to *algebraic bosonization*, which is well understood only in one dimension.

Keeping in mind the universality class principle, which allows one to reduce the interaction Hamiltonian to a sum of pure creation blocks and their adjoints, the stochastic bosonization principle can be formulated as the following modification of the block principle: *if the interaction Hamiltonian is expressed in terms of Fermi fields then, in the stochastic limit,*

(i) any even monomial in the creation–annihilation operators in the original Hamiltonian gives a boson noise;

(ii) any odd monomial in the creation–annihilation operators in the original Hamiltonian gives a fermion noise;

(iii) any two monomials of different degrees in the creation–annihilation operators in the original Hamiltonian give two independent noises;

(iv) if, in the original model, the state of the field is nonFock but a general mean zero gauge-invariant Gaussian state (e.g. finite temperature), then in the master field superalgebra there exists a parity operator, called *superparity*, which acts nontrivially on the boson part.

Example 11.10.1. We shall see that for a general polynomial interaction the master field and the corresponding evolution equation can be rather complex. However the basic new feature of stochastic bosonization can be illustrated in the simplest nontrivial case, in which the interaction Hamitonian has only one quadratic annihilation block and a trivial system part, i.e. it has the form

$$H_{\mathrm{I}} = \left[A^{+}(f_1)\, A^{+}(f_2) + A(f_2)\, A(f_1)\right] \otimes \mathbf{1}\,. \tag{11.10.1}$$

We shall prove in Sect. 16.2 that, as $\lambda \to 0$, the time-rescaled evolution operator $U^{\lambda}_{t/\lambda^2}$ converges to a unitary operator $U(t)$ which is the (unique) solution of the white noise Hamiltonian equation

$$\frac{d}{dt}U(t) = -i\left\{b_t^{+}(f_1, f_2) \otimes 1 + b_t(f_1, f_2) \otimes 1\right\} U(t)\,, \quad U(0) = 1\,, \tag{11.10.2}$$

where $\left\{b_t^{+}(f_1, f_2)\,, b_t(f_1, f_2)\right\}_{t \geq 0}$ is the *boson* Fock white noise with covariance given by (11.10.5) below. Thus we can see that the original time-rescaled Schrödinger equation

$$\partial_t U_{t/\lambda^2} = -i\frac{1}{\lambda} H_{\mathrm{I}}\left(t/\lambda^2\right) U_{t/\lambda^2}\,, \quad U_0 = 1\,, \tag{11.10.3}$$

is approximated by the white noise equation (11.10.2) and that the time-rescaled evolved interaction *fermion* Hamiltonian $H_{\mathrm{I}}\left(t/\lambda^2\right)$ is approximated by the *boson* white noise Hamiltonian

$$\left\{b_t^+\left(f_1, f_2\right) \otimes 1 + b_t\left(f_1, f_2\right) \otimes 1\right\} .$$

Using the causal commutator rule of Sect. 7.4, we rewrite the white noise Hamiltonian equation in the form of a stochastic Schrödinger equation:

$$\begin{aligned} U(t) = \mathbf{1} + \int_0^t \Big\{ & dA_s^+\left(f_1, f_2\right) \otimes \mathbf{1} - dA_s^+\left(f_1, f_2\right) \\ & \otimes \mathbf{1} - \left(\left(f_1, f_2\right) \mid \left(f_1, f_2\right)\right)_- \otimes \mathbf{1} ds \Big\} U(s) , \end{aligned} \tag{11.10.4}$$

where

(i) $\{U(t)\}_{t\geq 0}$ lives on $\Gamma\left(\mathbf{L}^2\left(\mathbb{R}_+\right) \otimes \mathcal{K}_2\right) \otimes \mathcal{H}_S$ and $\Gamma\left(\mathbf{L}^2\left(\mathbb{R}_+\right) \otimes \mathcal{K}_2\right)$ is the *boson* Fock space over the Hilbert space $\mathbf{L}^2\left(\mathbb{R}_+\right) \otimes \mathcal{K}_2$;

(ii) for any $F \in \mathcal{K}_2, \{A_s^+(F)\}_{s\geq 0} := \left\{A^+(\chi_{[0,s)} \otimes F)\right\}_{s\geq 0}$ is the boson creation process and $dA_s^+(F), dA_s(F)$ are the associated stochastic differentials;

(iii) $\mathcal{K}_2$ is the Hilbert space obtained by completion of the space

$$\{(g_1, g_2) : g_j \in \mathcal{K},\ j = 1, 2\}$$

endowed with the scalar product

$$\begin{aligned} & \left(\left(g_1, g_2\right) \mid \left(g_1', g_2'\right)\right) \\ & := \int_{\mathbb{R}} \left(\left\langle g_1, S_t^1 g_1'\right\rangle \left\langle g_2, S_t^1 g_2'\right\rangle - \left\langle g_2, S_t^1 g_1'\right\rangle \left\langle g_1, S_t^1 g_2'\right\rangle\right) dt ; \end{aligned} \tag{11.10.5}$$

(iv) the coefficient of the drift (Ito correction term) in the stochastic Schrödinger equation (11.10.4) is given by

$$\begin{aligned} & \left(\left(g_1, g_2\right) \mid \left(g_1', g_2'\right)\right)_- \\ & \quad := \int_{-\infty}^1 \left(\left\langle g_1, S_t^1 g_1'\right\rangle \left\langle g_2, S_t^1 g_2'\right\rangle - \left\langle g_2, S_t^1 g_1'\right\rangle \left\langle g_1, S_t^1 g_2'\right\rangle\right) dt \\ & \quad = \int_{-\infty}^1 \det\left(\left\langle g_i, S_t^1 g_j'\right\rangle\right) dt . \end{aligned} \tag{11.10.6}$$

11.11 The Time-Consecutive Principle

The time-consecutive principle, already formulated in Sect. 7.4, states that the chronological order of the time variables in the n-simplex $\Delta_1^{(n)}$, which appears in the nth term of the iterated series [see (7.4.3)], allows one to separate those terms which contribute in a nontrivial way to the stochastic limit (the time-consecutive ones) from those irrelevant terms which in the limit give zero contribution. It holds also for polynomial interactions (see Sects. 15.5–7 for the proof).

Part II

Strongly Nonlinear Regimes

12. Particles Interacting with a Boson Field

In the general scheme of interaction, described in Sect. 4.8, the techniques described in Part I do not apply to the case when not only the reservoir R but also the system S has a continuous spectrum. The usual way to handle these more complex interactions consists in replacing the response term by the first few terms in its series expansion. For example, in the case of nonrelativistic quantum electrodynamics (QED) (e.g. in nonlinear optics, laser theory, etc.) this amounts to replacing the dipole approximation by a multipole expansion. These approximations however break momentum conservation, and we shall see that by introducing them one loses information on some subtle cancellations which occur due to fast oscillations.

In the present section these approximations are not introduced: we consider two interacting systems with a continuous spectrum (e.g. free particles coupled to a quantum field, two interacting fields, etc.) and we just keep the response terms. We work with translation-invariant Hamiltonians.

In this case the rescaled fields depend on the degrees of freedom of both interacting systems. For example, in the case of a particle coupled to a quantum field, the rescaled field depends not only on the quantum field but also on the coordinates and the momenta of the particle. As a consequence of this, after the stochastic limit, three new qualitative features arise:

(i) *Only the noncrossing diagrams survive* (see Sect. 12.4 for their definition and main property); in this case the master field, which approximates the quantum electromagnetic field in the stochastic limit, is no longer a boson or fermion or supersymmetric white noise, but a nonlinear generalization of the *Boltzmannian* (or *free*) *white noise* (see Sect. 12.9 for its definition). As explained in Sect. 2.2, the main difference between the usual (in particular, the boson) Gaussianity and the *Boltzmannian (or free) Gaussianity* is that in the Boltzmannian case the vacuum expectation values of products of creation and annihilation operators are obtained by summing the products of pair correlation functions not over all the pair partitions, as in the usual (boson or fermion) Gaussian case, but only over the *non crossing pair partitions.* In terms of graphs this means that the summation does not run over all graphs, but only over the so-called *rainbow graphs.*

(ii) *The state space of the master field is not a usual Fock space, but a new object, called an "interacting Fock space"* because the quanta in the n-particle spaces are not independent but interact nonlinearly.

(iii) *The canonical commutation relations (boson or fermion) are broken and replaced by the so-called "entangled commutation relations".* From a mathematical point of view this means that the state space of the master field has a nontrivial structure of *Hilbert module* (in the nonrelativistic QED case this is a module over the momentum algebra of the electron). The module structure reflects the fact that, after the stochastic limit, the two interacting systems become *entangled* into one. This phenomenon has some analogies with the *stochastic bosonization*, described in Sect. 11.10.

These new features are experimentally observable because they imply new statistics given by quite explicit formulae and accounting for new phenomena which are not accessible with the usual asymptotic techniques (e.g. perturbation theory). For example, in the nonrelativistic QED case, the vacuum (coherent) statistics of the field (number operator) is non-Gaussian (Poissonian), but a nonlinear modification of the Wigner semicircle law (of the Poisson law), to which it reduces exactly when the nonlinear factor arising in the interacting Fock space is put equal to a positive constant.

We shall prove, by considering an ample variety of different models, that these features are quite universal under the above assumptions (continuous spectrum and conservation of momentum), but we shall begin by illustrating their emergence in the case of a free particle, say a free atom (electron), minimally coupled to the nonrelativistic quantum electromagnetic (phonon) field.

This includes the cases of the QED Hamiltonian and the Frölich *polaron Hamiltonian*; the latter is obtained from the interaction Hamiltonian (12.1.2) below when one makes $p = 1$ and takes the dispersion of the field $\omega(k) = 1$. Therefore the results below are also applicable, with inessential modifications, to this case. Up to Sect. 12.21 the free Hamiltonian of the particle is just its kinetic energy. Starting from Sect. 12.21 we discuss how the picture changes if the particle Hamiltonian includes a potential depending only on the position of the particle. From the point of view of the new phenomena mentioned above, the transition from one to many particles makes no difference (provided that the relevant wave operators exist).

Since the dipole approximation is effective at low frequencies and small atomic dimensions, it is reasonable to expect that this equation will shed some light on a class of phenomena in which high frequencies and finite atomic dimensions play an essential role.

12.1 A Single Particle Interacting with a Boson Field

In this section we describe the model of a single particle interacting with a boson field. We consider the simplest case in which matter is represented by a single particle, say an electron or a free atom, whose position and momentum we denote respectively by $q = (q_1 \dots q_d)$ and $p = (p_1 \dots p_d)$ and satisfy the commutation relations

$$[q_h, p_k] = i\delta_{hk} \,.$$

The EM field is described by the usual boson operators (in fact, operator-valued distributions)

$$a(k) = \big(a_1(k) \dots a_d(k)\big) \quad , \quad a^+(k) = \big(a_1^+(k) \dots a_d^+(k)\big) \,,$$

satisfying the *canonical commutation relations*

$$[a_j(k), a_h^+(k')] = \delta_{jh}\delta(k - k') \,.$$

The Hamiltonian of a free atom interacting with a nonrelativistic quantum electromagnetic field is, neglecting polarization,

$$H = H_0 + \lambda H_{\mathrm{I}} = \int \omega(k) a^+(k) a(k)\, dk + \frac{1}{2} p^2 + \lambda H_{\mathrm{I}} \,, \tag{12.1.1}$$

where λ is a small constant, $\omega(k)$ is a dispersion law (a positive function on $\mathbb{R}^d$) and, for photons, $\omega(k) = |k|$.

The interaction between the particle and the EM field is expressed in terms of a potential $A(x)$, describing the field strength at the space point $x \in R^d$, more precisely $A(x)$ is the potential felt by the particle in position x as a consequence of its interaction with the field. The explicit form of the interaction Hamiltonian is, neglecting a quadratic term in λ,

$$H_{\mathrm{I}} = p \cdot A(q) + A(q) \cdot p := \int d^3k \left[g(k) p \cdot e^{ikq} a^+(k) + \overline{g}(k) p \cdot e^{-ikq} a(k)\right] + \text{h.c.} \,, \tag{12.1.2}$$

where p, q are as above and

$$A(q) = \int dk \quad \left[g(k) e^{ik \cdot q} \cdot a^+(k) + \overline{g}(k) e^{-ik \cdot q} \cdot a(k)\right] \,. \tag{12.1.3}$$

Denoting $U_t^{(\lambda)} = e^{itH_0} e^{-itH}$ the solution of the Schrödinger equation in the interaction picture associated with the Hamiltonian H, after the time rescaling $t \to t/\lambda^2$, we obtain the rescaled equation

$$\frac{\partial}{\partial t} U_{t/\lambda^2}^{(\lambda)} = -\frac{i}{\lambda} H_{\mathrm{I}}(t/\lambda^2)\, U_{t/\lambda^2}^{(\lambda)} \qquad , \qquad U_0^{(\lambda)} = 1 \,,$$

with $H_{\mathrm{I}}(t) = e^{itH_0} H_{\mathrm{I}} e^{-itH_0}$ (the *evolved interaction Hamiltonian*). The time dependence is defined by letting the original interaction H_{I}, given by

(12.1.2–3), evolve under the free Hamiltonian H_0 and then performing the time rescaling $t \to t/\lambda^2$. We want to study the limits

$$\lim_{\lambda\to 0} \frac{1}{\lambda} H_{\mathrm{I}}\left(\frac{t}{\lambda^2}\right) =: H_t = \int d^3k \left[g(k)p \cdot b^+(t,k) + \overline{g}(k)p \cdot b(t,k) + \text{h.c.}\right] , \tag{12.1.4}$$

$$\lim_{\lambda\to 0} U^{(\lambda)}_{t/\lambda^2} = U_t ,$$

and the corresponding white noise equation

$$\partial_t U_t = -iH_tU_t \quad , \qquad U_0 = 1 . \tag{12.1.5a}$$

A simple calculation shows that the rescaled Hamiltonian is equal to

$$\frac{1}{\lambda} H_{\mathrm{I}}\left(\frac{t}{\lambda^2}\right) = \int d^3k \left[g(k)p \cdot a^+_\lambda(t,k) + \overline{g}(k)p \cdot a_\lambda(t,k) + \text{h.c.}\right] ,$$

where the operators $a^\pm_\lambda(t,k)$ are the *rescaled fields*

$$a_\lambda(t,k) = \frac{1}{\lambda} e^{i[\omega(k)+kp]t/\lambda^2} e^{-ikq} a(k) . \tag{12.1.5b}$$

We will consider the limit of the correlation functions

$$\lim_{\lambda\to 0} \left\langle a^{\epsilon_N}_\lambda(t_N,k_N) a^{\epsilon_{N-1}}_\lambda(t_{N-1},k_{N-1}) \dots a^{\epsilon_1}_\lambda(t_1,k_1)\right\rangle , \tag{12.1.6}$$

where $\varepsilon = \{\epsilon_N \dots \epsilon_1\} \in \{1,0\}^N$, $\epsilon \in \{1,0\}$ ($\epsilon = 0$ for a, $\epsilon = 1$ for a^+) and $\langle \cdot \rangle$ denotes the Fock state of the reservoir, i.e. the mean zero boson Gaussian state with pair correlations vanishing on the off-diagonal terms and, on the diagonal ones, equal to

$$\langle a_k a^+_{k'}\rangle = \delta(k-k') ,$$

$$\langle a^+_{k'} a_k\rangle = 0 .$$

The other correlators can be calculated using Gaussianity. We will prove convergence of the correlators (12.1.6) and show that in the stochastic limit we have nontrivial cancellations, as a consequence of which in the limit the crossing diagrams vanish. More precisely we show that the above limit exists and has the form

$$\langle b^{\epsilon_N}(t_N,k_N) b^{\epsilon_{N-1}}(t_{N-1},k_{N-1}) \dots b^{\epsilon_1}(t_1,k_1)\rangle .$$

In Chapter 4 we saw that if one introduces the dipole approximation then the master field is the usual boson Fock white noise. Without the dipole approximation, the master field is a new type of white noise whose algebra is described by the relations

$$b(t,k)p = (p+k)b(t,k) , \tag{12.1.7}$$

$$b(t,k)b^+(t',k') = 2\pi\delta(t-t')\delta\big(\omega(k) + E(k+p) - E(p)\big)\delta(k-k') , \tag{12.1.8}$$

where $E(p) = \frac{1}{2}p^2$ is the energy of a free nonrelativistic particle and therefore the argument of the δ-function of energy is equal to

$$\tilde{\omega}(k) + kp\,,$$

where

$$\tilde{\omega}(k) := \omega(k) + \frac{1}{2}\,k^2\,. \tag{12.1.9}$$

Recalling that p is the atomic momentum, we see that the relations (12.1.7–8) show that the atom and the master field are not independent even at a kinematical level. This is what we call *entanglement.*[(1)] From (12.1.7), by considering polynomials in p and limits thereof, it is easy to see that, for any Schwartz function f one has

$$b(t,k)f(p) = f(p+k)b(t,k)\,. \tag{12.1.10}$$

The relation (12.1.8) is a generalization of the algebra of Boltzmannian or free creation–annihilation operators with commutation relations

$$a_k a_h^+ = \delta(k-h)\,, \tag{12.1.11}$$

and the corresponding statistics becomes a generalization of the Boltzmannian (or free) statistics. This generalization is due to the fact that the right-hand side of (12.1.7–8) is not a scalar but an operator (a function of the atomic momentum). This means that (12.1.7–8) are *module commutation relations.* For any fixed value $\bar{p}$ of the atomic momentum we obtain a copy of the free (or Boltzmannian) algebra. Therefore, the basis of the emergence of the new non-free statistics is the module commutation relation (12.1.10). Given (12.1.7–9), the statistics of the master field is uniquely determined by the condition

$$b(t,k)\Psi = 0\,, \tag{12.1.12}$$

where Ψ is the vacuum of the master field, via a module generalization of the free Wick theorem (Theorem 12.4.2). Therefore, just as the typical Gaussian statistics is coded into the typical bosonic (or fermionic) commutation relations through the Wick theorem, the new non-Gaussian (i.e. nonlinear) statistics is coded into the new (module) commutation relations through the module Wick theorem.

12.2 Dynamical q-Deformation: Emergence of the Entangled Commutation Relations

In order to determine the limit (12.1.4) one rewrites the rescaled interaction Hamiltonian

$$\frac{1}{\lambda} H_{\mathrm{I}}\left(\frac{t}{\lambda^2}\right) = \frac{1}{\lambda} A(t/\lambda^2) + \text{h.c.} = \int d^3kp \left(\overline{g}(k)a_\lambda(t,k) + g(k)a^+_\lambda(t,k)\right) + \text{h.c.} \tag{12.2.1}$$

in terms of the dynamically q-deformed (or *entangled rescaled*) *fields* $a_\lambda(t,k)$ defined by

$$\begin{aligned} a_\lambda(t,k) :&= \frac{1}{\lambda} e^{i\frac{t}{\lambda^2} H_0} e^{-ikq} a(k) e^{-i\frac{t}{\lambda^2} H_0} \\ &= \exp\left\{-i\frac{t}{\lambda^2}[\omega(k) + E(k+p) - E(p)]\right\} e^{-ikq} a(k)\,, \end{aligned} \tag{12.2.2}$$

where the term *entangled* reflects the fact that the new "fields" $a_\lambda(t,k)$ include both the particle and the field degrees of freedom. The algebra of the rescaled fields in the stochastic limit will give rise to the algebra of the master field. Using the standard commutation relation $[p,q] = -i$, it is now easy to prove that the operators $a_\lambda(t,k)$ satisfy the following q-deformed module relations:

$$\begin{aligned} a_\lambda(t,k)a^+_\lambda(t',k') =& a^+_\lambda(t',k')a_\lambda(t,k) \cdot q_\lambda(t-t',kk') \\ &+ \frac{1}{\lambda^2} q_\lambda\big(t-t',\omega(k) + E(k+p) - E(p)\big)\delta(k-k')\,, \end{aligned} \tag{12.2.3}$$

$$a_\lambda(t,k)f(p) = f(p+k)a_\lambda(t,k)\,, \tag{12.2.4a}$$

$$a_\lambda(t,k)\Phi = 0\,, \tag{12.2.4b}$$

where f is any Schwartz function and

$$q_\lambda(t-t',x) = \exp\left(-i\frac{t-t'}{\lambda^2}x\right) \tag{12.2.5}$$

is an oscillating factor. The above considerations show that the module q deformation of the commutation relations arise here as a result of the dynamics and are not put artificially ab initio. Now let us suppose that the master field

$$b(t,k) = \lim_{\lambda\to 0} a_\lambda(t,k) \tag{12.2.6}$$

exists. It is natural to conjecture that its algebra shall be obtained as the stochastic limit ($\lambda \to 0$) of the algebra (12.2.3–4). Since the factor $q_\lambda(t-t',x)$ is an oscillating exponent, one easily sees that in the sense of distributions

$$\lim_{\lambda\to 0} q_\lambda(t,x) = 0\,, \qquad \lim_{\lambda\to 0} \frac{1}{\lambda^2} q_\lambda(t,x) = 2\pi\delta(t)\delta(x)\,. \tag{12.2.7}$$

It is natural to expect that the limit of (12.2.4a,b) is

$$b(t,k)p = (p+k)b(t,k) \tag{12.2.8}$$

and the limit of (12.2.3) gives the module-free relation

$$b(t,k)b^+(t',k') = 2\pi\delta(t-t')\delta\big(\omega(k)+E(k+p)-E(p)\big)\delta(k-k')\,. \tag{12.2.9}$$

Remark 12.2.1. The operators $a_\lambda(t,k)$ also obey the relation

$$a_\lambda(t,k)a_\lambda(t',k') = a_\lambda(t',k')a_\lambda(t,k)q_\lambda^{-1}(t-t',kk')\,. \tag{12.2.10}$$

One can show that (12.2.10) disappears after the limit. In fact, if (12.2.9) would survive in the limit then, because of (12.2.5), it should give $b(t,k)\cdot b(t',k') = 0$, hence also $b^+(t,k)b^+(t',k') = 0$, so all the n-particle vectors with $n \geq 2$ would be zero. But we shall prove that this is not the case. Note that we can omit (12.2.10) even before the limit. For simplicity let us explain this fact for the example of a finite dimensional bosonic algebra. Consider the algebra $\mathcal{A}$ with generators a_i, a_j^+, relations

$$[a_i, a_j^+] = \delta_{ij} \tag{12.2.11}$$

and the state $\langle\,\cdot\,\rangle$ on this algebra, given by the vacuum expectation in the Fock representation. It is easy to prove the following lemma.

Lemma 12.2.1. *In the Fock representation of the algebra $\mathcal{A}$ with respect to the state $\langle\,\cdot\,\rangle$, we have the extra relation*

$$[a_i, a_j] = 0\,, \tag{12.2.12}$$

which is meant for the total set of number vectors.

Proof 12.2.1. The thesis follows from the identity

$$a_i a_j a_{\alpha_n}^+ \ldots a_{\alpha_1}^+ \Phi = \sum_{\substack{(a_{\alpha_h}^+, a_{\alpha_k}^+)\in\{1\ldots n\}^2 \\ \alpha_h \neq \alpha_k}} \delta_{j,\alpha_k}\delta_{i,\alpha_h} a_{\alpha_n}^+ \ldots \hat{a}_{\alpha_h}^+ \ldots \hat{a}_{\alpha_k}^+ \ldots a_{\alpha_1}^+ \Phi\,,$$

which is symmetric in i and j. Here the notation $\hat{a}$ means that this operator is omitted in the product.

We have to prove the existence of the stochastic limit of the n-point correlators. In the next section we begin by considering the 2-point function.

12.3 The Two-Point and Four-Point Correlators

In the present section we calculate the 2-point and 4-point correlation functions of the rescaled fields (12.2.2) in order to show how to make use of the dynamical q deformation and to illustrate the mechanism through which the crossing diagrams vanish in the limit. The 2-point correlator for the entangled quantum white noise $b(t,k)$ can be obtained by calculating the stochastic limit of the dynamically q-deformed relation (12.2.3):

$$\begin{aligned}\langle b(t,k)b^+(t',k')\rangle &= \lim_{\lambda\to 0}\frac{1}{\lambda^2}\,q_\lambda\big(t-t',\omega(k)+E(k-p)-E(p)\big)\delta(k-k')\\ &= 2\pi\delta(t-t')\delta\big(\omega(k)+E(k-p)-E(p)\big)\delta(k-k'),\end{aligned}$$

according to (12.2.7). Let us now calculate the 4-point correlation function for $b(t,k)$ and, using this example, show that (12.2.10), i.e.

$$a_\lambda(t,k)a_\lambda(t',k') = a_\lambda(t',k')a_\lambda(t,k)q_\lambda^{-1}(t-t',kk'),$$

disappears in the stochastic limit. Note that the limit, as $\lambda\to 0$, of the left-hand side of this equation is $b(t_1,k_1)b(t_2,k_2)$ and that, using (12.2.7) in a formal way, i.e. calculating the limit of the product on the right-hand side as the product of the limits, we would find zero [i.e. $b(t_2,k_2)b(t_1,k_1)$ times zero]. But we will prove that this is not the case and in fact the limit of the right-hand side of (12.2.10) is the same as that of the left-hand side. In other words, the limit of (12.2.10) is the trivial identity

$$b(t_1,k_1)b(t_2,k_2) = b(t_1,k_1)b(t_2,k_2). \tag{12.3.1}$$

To this goal, let us consider the 4-point correlator

$$\langle a_\lambda(t_1,k_1)a_\lambda(t_2,k_2)a_\lambda^+(t_2',k_2')a_\lambda^+(t_1',k_1')\rangle. \tag{12.3.2}$$

This corresponds to taking the matrix element of (12.2.10) between vacuum and two creation operators. According to (12.2.3) this correlator is equal to the sum of two terms:

$$\begin{aligned}&\frac{1}{\lambda^2}\,q_\lambda\big(t_2-t_2',\tilde\omega(k_2)+k_2(p+k_1)\big)\delta(k_2-k_2')\\ &\quad\cdot\frac{1}{\lambda^2}\,q_\lambda\big(t_1-t_1',\tilde\omega(k_1)+k_1p\big)\delta(k_1-k_1')q_\lambda(t_2-t_2',k_2k_2')\\ &+\frac{1}{\lambda^2}\,q_\lambda\big(t_1-t_2',\tilde\omega(k_1)+k_1p\big)\delta(k_1-k_2')\\ &\quad\cdot\frac{1}{\lambda^2}\,q_\lambda\big(t_2-t_1',\tilde\omega(k_2)+k_2p\big)\delta(k_2-k_1')q_\lambda(t_2-t_2',k_2k_2').\end{aligned}$$

In the stochastic limit $\lambda\to 0$; only the first term survives. The second term vanishes because $\lim_{\lambda\to 0} q_\lambda(t_2-t_2',k_2k_2')=0$. The first term survives because

$$\begin{aligned}&\frac{1}{\lambda^2}\,q_\lambda\big(t_2-t_2',\tilde\omega(k_2)+k_2(p+k_1)\big)q_\lambda(t_2-t_2',k_2k_2')\\ &=\frac{1}{\lambda^2}\,q_\lambda\big(t_2-t_2',\tilde\omega(k_2)+k_2(p+k_1)+k_2k_2'\big).\end{aligned}$$

Let us now consider the behaviour of (12.2.10) in the stochastic limit. According to (12.2.10) the correlator (12.3.2) is equal to

$$
\begin{aligned}
&\langle\, a_\lambda(t_2,k_2)a_\lambda(t_1,k_1)a_\lambda^+(t_2',k_2')a_\lambda^+(t_1',k_1')\rangle q_\lambda^{-1}(t_2-t_2',k_2k_2')\\
&=q_\lambda^{-1}(t_2-t_2',k_2k_2')\bigg(\frac{1}{\lambda^2}\,q_\lambda\big(t_1-t_2',\tilde\omega(k_1)+k_1(p+k_1)\big)\delta(k_1-k_2')\\
&\quad\cdot\frac{1}{\lambda^2}\,q_\lambda\big(t_2-t_1',\tilde\omega(k_2)+k_2p\big)\delta(k_2-k_1')q_\lambda(t_1-t_2',k_1k_2')\\
&\quad+\frac{1}{\lambda^2}\,q_\lambda\big(t_2-t_2',\tilde\omega(k_2)+k_2p\big)\delta(k_2-k_2')\\
&\quad\cdot\frac{1}{\lambda^2}\,q_\lambda\big(t_1-t_1',\tilde\omega(k_1)+k_1p\big)\delta(k_1-k_1')q_\lambda(t_1-t_2',k_1k_2')\bigg)\,.
\end{aligned}
$$

We see that due to the term $q_\lambda^{-1}(t_2-t_2',k_2k_2')$ only the second component of this correlator survives after the stochastic limit. Therefore the stochastic limit of

$$
\langle a_\lambda(t_2,k_2)a_\lambda(t_1,k_1)a_\lambda^+(t_2',k_2')a_\lambda^+(t_1',k_1')\rangle q_\lambda^{-1}(t_2-t_2',k_2k_2')
$$

is not equal to the product of the limit of the correlator and the limit of $q_\lambda^{-1}(t_2-t_2',k_2k_2')$ (i.e. to zero). We have proved that at least for the 4-point correlation function the identity (12.3.1) is the stochastic limit of the relation (12.2.10). To finish the proof we have to prove the existence of the stochastic limit of arbitrary n-point correlators. This is the subject of the next section.

12.4 The Stochastic Limit of the N-Point Correlator

In the present section we prove the existence of the limit of the q-deformed correlators

$$
\langle a_\lambda^{\varepsilon_1}(t_1,k_1)\dots a_\lambda^{\varepsilon_N}(t_N,k_N)\rangle\,, \tag{12.4.1}
$$

where a^ε means either a or a^+ ($\epsilon=0$ for a, $\epsilon=1$ for a^+) and $\langle\,\cdot\,\rangle$ denotes vacuum expectation. We also will prove that the limit of this correlator will be equal to the corresponding correlator of the master field:

$$
\langle b^{\varepsilon_1}(t_1,k_1)\dots b^{\varepsilon_N}(t_N,k_N)\rangle\,. \tag{12.4.2}
$$

In the product $a_\lambda^{\varepsilon_1}(t_1,k_1)\dots a_\lambda^{\varepsilon_N}(t_N,k_N)$ let us enumerate the annihilators as $a_\lambda(t_{m_j},k_{m_j})$, $j=1\dots J$, and the creators as $a_\lambda^+(t_{m_j'},k_{m_j'})$, $j=1\dots I$, $I+J=N$. This means that if $\varepsilon_m=0$ then $a_\lambda^{\varepsilon_m}(t_m,k_m)=a_\lambda(t_{m_j},k_{m_j})$ for $m=m_j$ (and the analogous condition for $\varepsilon_m=1$). Recall that a pair partition $\{(m_j<m_j'):j=1\dots n\}$, of the set $\{1\dots 2n\}$ is called a *non crossing pair partition* if, whenever $[m_j,m_j']\cap[m_k,m_k']\neq\emptyset$ ($[m_k,m_k']$ denotes the interval $\{x:m_k\le x\le m_k'\}$), then either $[m_j,m_j']\subseteq[m_k,m_k']$ or $[m_k,m_k']\subseteq[m_j,m_j']$.

Note that if $m_1<m_2<\dots<m_n$ is a subset of the set $\{1\dots 2n\}$ then either there is a unique noncrossing pair partition $\{(m_j<m_j'):j=1\dots n\}$ of

$\{1 \dots 2n\}$, or there is none. This is easily seen by induction on the number n. In fact, if such a partition exists, then $2n$ must be one of the m'_j, say m'_{j_n}. Denote (m_{j_n}, m'_{j_n}) the corresponding pair. Then the interval $[1, 2n]$ splits into disjoint intervals, one of which is $[m_{j_n}, m'_{j_n}]$, and to each of them one applies the induction hypothesis. For more information on noncrossing pair partitions, see Sect. 15.12.

Theorem 12.4.1. *The stochastic limit of the dynamically q-deformed correlator (12.4.1) exists. Moreover,*

(i) if the number of creators is not equal to the number of annihilators, then the correlator (12.4.1) is equal to zero (even before the limit);
(ii) if the number of creators is equal to the number of annihilators ($N = 2n$), then the limit is equal to the following:

$$\prod_{h=1}^{n} \delta\left(k_{m_h} - k_{m'_h}\right) 2\pi\delta\left(t_{m_h} - t_{m'_h}\right) \cdot \delta\left(\tilde{\omega}\left(k_{m_h}\right) + k_{m_h} p + \sum_{\alpha : m_\alpha < m_h < m'_\alpha} k_{m_\alpha} \cdot k_{m_h}\right), \tag{12.4.3}$$

where $\{(m_j, m'_j) : j = 1 \dots n\}$ is the unique noncrossing partition of $\{1 \dots 2n\}$ associated with $\varepsilon = \{\varepsilon_1 \dots \varepsilon_N\}$.

Proof 12.4.1. The proof of this theorem is by induction. We need to prove (12.4.3) for $N = 2n$. For $n = 1$ this relation is clear. Let us assume (12.4.3) for $N = 2n - 2$ and prove that the same is true for $N = 2n$; let m_n be the first annihilation index in (12.4.1) starting from the right. Then (12.4.1) is equal to

$$\langle a_\lambda(t_{m_1}, k_{m_1}) \cdots a_\lambda(t_{m_n}, k_{m_n}) a^+_\lambda(t_{m_n+1}, k_{m_n+1}) \dots a^+_\lambda(t_{2n}, k_{2n}) \rangle .$$

Using the commutation relation (12.2.3), this is equal to

$$\begin{aligned} &q_\lambda(t_{m_n} - t_{m_n+1}, k_{m_n} k_{m_n+1}) \\ &\cdot \langle a_\lambda(t_{m_1}, k_{m_1}) \cdots a^+_\lambda(t_{m_n+1}, k_{m_n+1}) a_\lambda(t_{m_n}, k_{m_n}) a^+_\lambda(t_{m_n+2}, k_{m_n+2}) \dots \rangle \\ &+ \frac{1}{\lambda^2} \delta(k_{m_n} - k_{m_n+1}) \langle a_\lambda(t_{m_1}, k_{m_1}) \cdots q_\lambda(t_{m_n} - t_{m_{n+1}}, \tilde{\omega}(k_{m_n}) + k_{m_n} p) \dots \rangle . \end{aligned} \tag{12.4.4}$$

Using the adjoint of the commutation relation (12.2.4a,b), we commute $q_\lambda\left(t_{m_n} - t_{m_n+1}, \tilde{\omega}(k) + k_{m_n} p\right)$ with all the annihilators on its right. Doing so the second term in (12.4.4) becomes

$$\begin{aligned} &\langle a_\lambda(t_{m_1}, k_{m_1}) \cdots \widehat{a}_\lambda(t_{m_n}, k_{m_n}) \widehat{a}^+_\lambda(t_{m_n+1}, k_{m_n+1}) \dots a^+_\lambda(t_{2n}, k_{2n}) \rangle \\ &\cdot \delta(k_{m_n} - k_{m_n+1}) \frac{1}{\lambda^2} q_\lambda \left(t_{m_n} - t_{m_{n+1}}, \tilde{\omega}(k_{m_n}) + k_{m_n} \left(p + \sum_{m'_j > m_n+1} k_{m'_j} \right) \right) . \end{aligned}$$

As usual $\widehat{a}$ means that we omit this operator in the product. Iterating this procedure [moving $a_\lambda(t_{m_n}, k_{m_n})$ further to the right] we find that

$$\begin{aligned}
&\langle a_\lambda(t_{m_1}, k_{m_1}) \cdots a_\lambda^+(t_{2n}, k_{2n})\rangle \\
&= \sum_{m'_\beta > m_n} \langle a_\lambda(t_{m_1}, k_{m_1}) \cdots \widehat{a}_\lambda(t_{m_n}, k_{m_n}) \cdots \widehat{a}_\lambda^+(t_{m'_\beta}, k_{m'_\beta}) \cdots a_\lambda^+(t_{2n}, k_{2n})\rangle \\
&\quad \cdot \prod_{m_n < m'_j < m'_\beta} q_\lambda\left(t_{m_n} - t_{m'_j}, k_{m_n} k_{m'_j}\right) \\
&\quad \cdot \delta(k_{m_n} - k_{m_n+1}) \frac{1}{\lambda^2} q_\lambda \left(t_{m_n} - t_{m'_\beta}, \tilde{\omega}(k_{m_n}) + k_{m_n}\left(p + \sum_{m'_\delta > m'_\beta} k_{m'_\delta}\right)\right),
\end{aligned}$$

where we make the convention that the product

$$\prod_{m_n < m'_j < m'_\beta} q_\lambda\left(t_{m_n} - t_{m'_j}, k_{m_n} k_{m'_j}\right) \tag{12.4.5}$$

is equal to 1 if $\beta = j$. Taking the stochastic limit of this recurrence relation we see that if the product (12.4.5) is nontrivial (not equal to 1) then the stochastic limit will be equal to zero. If the product (12.4.5) is equal to 1 then we obtain the recurrence relation for the stochastic limit of the correlator (12.4.1):

$$\begin{aligned}
&\lim_{\lambda\to 0} \langle a_\lambda(t_{m_1}, k_{m_1}) \cdots a_\lambda^+(t_{2n}, k_{2n})\rangle \\
&= \lim_{\lambda\to 0} \langle a_\lambda(t_{m_1}, k_{m_1}) \cdots \widehat{a}_\lambda(t_{m_n}, k_{m_n}) \widehat{a}_\lambda^+(t_{m_n+1}, k_{m_n+1}) \cdots a_\lambda^+(t_{2n}, k_{2n})\rangle \\
&\quad \cdot \delta(k_{m_n} - k_{m_n+1}) \frac{1}{\lambda^2} q_\lambda \left(t_{m_n} - t_{m'_\beta}, \tilde{\omega}(k_{m_n}) + k_{m_n}\left(p + \sum_{m'_\delta > m'_\beta} k_{m'_\delta}\right)\right).
\end{aligned} \tag{12.4.6}$$

Using this recurrence relation and the induction assumption, we obtain the statement of the theorem.

Theorem 12.4.2. *The correlators for the master field satisfying (12.2.8) with $\langle \cdot \rangle$ equal to the vacuum expectation in the free Fock space are equal to the stochastic limit of the corresponding correlators for the dynamically q-deformed field (calculated in Theorem 12.4.1).*

Proof 12.4.2. The proof is made by computing the correlation functions

$$\langle b^{\epsilon_1}(t_1, k_1) b^{\epsilon_2}(t_2, k_2) \ldots b^{\epsilon_N}(t_N, k_N)\rangle$$

using the commutation relations (12.2.9) and showing that they satisfy the same induction relation (12.4.6) of the stochastic limit of the rescaled correlators of the original system. As in the Proof 12.4.1 we begin from the first annihilator from the right. Using (12.2.9) the pairing $b(t_{m'_h}, k_{m'_h})b^+(t_{m_h}, k_{m_h})$ gives the factor

$$\delta(k_{m'_h} - k_{m_h})2\pi\delta(t_{m'_h} - t_{m_h})\delta\big(\tilde{\omega}(k_{m_h}) + k_{m_h}p\big),$$

which has to be brought onto the right of the remaining product of $b^+(t,k)$ using the commutation relation (12.2.8). This gives the factor

$$\sum_{\alpha:m_\alpha<m_h<m'_\alpha} k_{m_\alpha} \cdot k_{m_h} \quad \text{(phase shift)}$$

in the argument of the δ-function [see (12.4.3)]. The recurrence relation for the correlator thus obtained coincides with (12.4.6). This finishes the proof of the theorem.

Remark 12.4.1. Using the recurrence relation for the correlator we iterate this procedure until the product of creators and annihilators in the correlator is reduced to its normally ordered form. Since the expectation value $\langle \cdot \rangle$ is taken with respect to the Fock vacuum, only the "scalar" term (which in this case is a product of atomic operators) will survive.

12.5 The q-Deformed Module Wick Theorem

In the present section we find a q-deformed module extension of the Wick theorem. This will allow us to find the exact form of the q-deformed n-point correlators, i.e. the correlators before the stochastic limit:

$$\langle a_\lambda^{\varepsilon_1}(t_1, k_1) \dots a_\lambda^{\varepsilon_N}(t_N, k_N) \rangle . \tag{12.5.1}$$

Lemma 12.5.1. *In the notations of Sect. 12.2 one has*

$$\begin{aligned}
&a_\lambda(t,k)a_\lambda^{\varepsilon_1}(t_1,k_1)\dots a_\lambda^{\varepsilon_N}(t_N,k_N) \\
&\quad - \prod_{i=1}^{I} q_\lambda^{-1}\left(t - t_{m_i}, kk_{m_i}\right) \prod_{j=1}^{J} q_\lambda\left(t - t_{m'_j}, kk_{m'_j}\right) \\
&\qquad\qquad \cdot a_\lambda^{\varepsilon_1}(t_1,k_1)\dots a_\lambda^{\varepsilon_N}(t_N,k_N)a_\lambda(t,k) \\
&= \sum_{j=1}^{I} \delta\left(k - k_{m'_j}\right) \frac{1}{\lambda^2} q_\lambda\left(t - t_{m'_j}, \tilde{\omega}(k) + kp\right) \prod_{m_i<m'_j} q_\lambda\left(t - t_{m'_j}, kk_{m_i}\right) \\
&\quad \cdot \prod_{m'_i<m'_j} q_\lambda^{-1}\left(t - t_{m'_j}, kk_{m'_i}\right) \prod_{m_i<m'_j} q_\lambda^{-1}(t - t_{m_i}, kk_{m_i}) \\
&\quad \cdot \prod_{m'_i<m'_j} q_\lambda\left(t - t_{m'_i}, kk_{m'_i}\right) a_\lambda^{\varepsilon_1}(t_1,k_1)\dots \widehat{a}_\lambda^+(t_{m'_j}, k_{m'_j})\dots a_\lambda^{\varepsilon_N}(t_N,k_N) .
\end{aligned} \tag{12.5.2}$$

As usual the notation $\widehat{a}_\lambda^+$ means that we omit the operator a_λ^+ in this product.

Proof 12.5.1. We prove this lemma by induction over N. The first induction step is given by the relations (12.2.3) and (12.2.10). Now, assuming the validity of (12.5.2) for N, we will prove this formula for $N+1$ using the commutation relations (12.2.3), (12.2.4a,b) and (12.2.10).

(i) First case: $\varepsilon_{N+1} = 0$. In this case using (12.5.2) for N and (12.2.10) we obtain

$$\begin{aligned}
&a_\lambda(t,k)a_\lambda^{\varepsilon_1}(t_1,k_1)\dots a_\lambda^{\varepsilon_N}(t_{N+1},k_{N+1}) - q_\lambda^{-1}(t-t_{N+1},kk_{N+1}) \\
&\cdot \prod_{i=1}^{I} q_\lambda^{-1}(t-t_{m_i},kk_{m_i}) \prod_{j=1}^{J} q_\lambda\left(t-t_{m'_j},kk_{m'_j}\right) \\
&\qquad \cdot a_\lambda^{\varepsilon_1}(t_1,k_1)\dots a_\lambda(t_{N+1},k_{N+1})a_\lambda(t,k) \\
&= \sum_{j=1}^{I} \delta\left(k-k_{m'_j}\right)\frac{1}{\lambda^2} q_\lambda\left(t-t_{m'_j},\tilde\omega(k)+kp\right) \prod_{m_i<m'_j} q_\lambda\left(t-t_{m'_j},kk_{m_i}\right) \\
&\quad \cdot \prod_{m'_i<m'_j} q_\lambda^{-1}\left(t-t_{m'_j},kk_{m'_i}\right) \prod_{m_i<m'_j} q_\lambda^{-1}(t-t_{m_i},kk_{m_i}) \\
&\quad \cdot \prod_{m'_i<m'_j} q_\lambda\left(t-t_{m'_i},kk_{m'_i}\right) a_\lambda^{\varepsilon_1}(t_1,k_1)\dots \widehat{a}_\lambda^{+}\left(t_{m'_j},k_{m'_j}\right)\dots a_\lambda^{\varepsilon_N}(t_N,k_N),
\end{aligned}$$

which is exactly (12.5.2) for $N+1$.

(ii) Second case: $\varepsilon_{N+1} = 1$. In this case using (12.5.2) for N and (12.2.3) we obtain

$$\begin{aligned}
&a_\lambda(t,k)a_\lambda^{\varepsilon_1}(t_1,k_1)\dots a_\lambda^{\varepsilon_N}(t_{N+1},k_{N+1}) \\
&- \prod_{i=1}^{I} q_\lambda^{-1}(t-t_{m_i},kk_{m_i}) \prod_{j=1}^{J}\Bigg[q_\lambda\left(t-t_{m'_j},kk_{m'_j}\right) \\
&\qquad\qquad \cdot a_\lambda^{\varepsilon_1}(t_1,k_1)\dots a_\lambda^{\varepsilon_N}(t_N,k_N)\Bigg] \\
&\cdot \Big(a_\lambda^{+}(t_{N+1},k_{N+1})a_\lambda(t,k)q_\lambda(t-t_{N+1},kk_{N+1}) \\
&\quad + \delta(k-k_{N+1})\frac{1}{\lambda^2}q_\lambda(t-t_{N+1},\tilde\omega(k)+kp)\Big) \\
&= \sum_{j=1}^{I} \delta\left(k-k_{m'_j}\right)\frac{1}{\lambda^2} q_\lambda\left(t-t_{m'_j},\tilde\omega(k)+kp\right) \prod_{m_i<m'_j} q_\lambda\left(t-t_{m'_j},kk_{m_i}\right) \\
&\quad \cdot \prod_{m'_i<m'_j} q_\lambda^{-1}\left(t-t_{m'_j},kk_{m'_i}\right) \prod_{m_i<m'_j} q_\lambda^{-1}(t-t_{m_i},kk_{m_i})
\end{aligned}$$

$$\cdot \prod_{m'_i < m'_j} \left[q_\lambda \left(t - t_{m'_i}, k k_{m'_i} \right) \right.$$

$$\left. \cdot a_\lambda^{\varepsilon_1}(t_1, k_1) \ldots \widehat{a}_\lambda^+ \left(t_{m'_j}, k_{m'_j} \right) \ldots a_\lambda^{\varepsilon_N}(t_N, k_N) a_\lambda^+(t_{N+1}, k_{N+1}) \right] .$$

Now we use the commutation relation (12.2.4a,b) to move the factor

$$\delta(k - k_{N+1}) \frac{1}{\lambda^2} q_\lambda (t - t_{N+1}, \tilde{\omega}(k) + kp)$$

to the left of the remaining product of creators and annihilators. This gives (12.5.2) for $N + 1$ and thus finishes the proof of the lemma.

The following theorem uses the above result to describe the form of the N-point correlators.

Theorem 12.5.1.

(i) If the number of creators is not equal to the number of annihilators, then the correlator (12.5.1) is equal to zero;
(ii) if the number of creators is equal to the number of annihilators ($N = 2n$), then the correlator (12.5.1) is equal to the following sum over pair partitions:

$$\sum_{\sigma(\varepsilon)} \prod_{h=1}^{n} \delta \left(k_{m_h} - k_{m'_h} \right)$$

$$\cdot \frac{1}{\lambda^2} q_\lambda \left(\left(t_{m_h} - t_{m'_h} \right), \left(\tilde{\omega}(k_{m_h}) + k_{m_h} p + \sum_{m_\alpha < m_h < m'_\alpha} k_{m_\alpha} \cdot k_{m_h} \right) \right)$$

$$\cdot \prod_{\substack{(m_j, m'_j), (m_i, m'_i); i,j \\ =1 \ldots n : m_j < m_i < m'_j < m'_i}} q_\lambda \left(t_{m_i} - t_{m'_j}, k_{m_i} \cdot k_{m_j} \right) , \qquad (12.5.3)$$

where $\sigma(\varepsilon) = \{(m_j < m'_j) : j = 1 \ldots n\}$ is the set of all partitions of $\{1 \ldots 2n\}$ associated with $\varepsilon = (\varepsilon_1 \ldots \varepsilon_{2n})$.

Proof 12.5.2. The proof of this theorem is by induction over n. The first induction step is obvious. Let us assume that (12.5.1) is expressed by (12.5.3) for $N = 2n - 2$ and prove that the same is true for $N = 2n$. We consider the $2n$-point correlator

$$\langle a_\lambda^{\varepsilon_1}(t_1, k_1) \ldots a_\lambda^{\varepsilon_{2n}}(t_{2n}, k_{2n}) \rangle .$$

It is easy to see that if this correlator is not equal to zero, then the first operator is an annihilator and the last one a creator. Therefore, without loss of generality, we can consider the case when the correlator is as follows:

$$\langle a_\lambda(t_{m_1}, k_{m_1}) a_\lambda^{\varepsilon_2}(t_2, k_2) \ldots a_\lambda^{\varepsilon_{2n-1}}(t_{2n-1}, k_{2n-1}) a_\lambda^+(t_{m'_n}, k_{m'_n}) \rangle . \qquad (12.5.4)$$

From Lemma 12.5.1 it follows that (12.5.4) is equal to

$$\sum_{j=1}^{n} \delta\left(k_{m_1} - k_{m'_j}\right) \frac{1}{\lambda^2} q_\lambda\left(t_{m_1} - t_{m'_j}, \tilde{\omega}(k_{m_1}) + k_{m_1} p\right)$$
$$\cdot \prod_{m_i < m'_j < m'_i} q_\lambda\left(t_{m_1} - t_{m'_j}, k_{m_1} k_{m_i}\right) \prod_{m_i < m'_j} q_\lambda^{-1}(t_{m_1} - t_{m_i}, k_{m_1} k_{m_i})$$
$$\cdot \prod_{m'_i < m'_j} q_\lambda\left(t_{m_1} - t_{m'_i}, k_{m_1} k_{m'_i}\right) \left\langle \widehat{a}_\lambda\left(t_{m_1}, k_{m_1}\right) \ldots \widehat{a}_\lambda^{+}\left(t_{m'_j}, k_{m'_j}\right) \ldots \right\rangle, \tag{12.5.5}$$

where the expectation value in (12.5.5) contains a product of $2N - 2$ creators and annihilators. The product $\prod_{m_i < m'_j < m'_i} q_\lambda\left(t_{m_1} - t_{m'_j}, k_{m_1} k_{m_i}\right)$ in (12.5.5) arises from the products

$$\prod_{m_i < m'_j} q_\lambda\left(t - t_{m'_j}, k k_{m_i}\right) \prod_{m'_i < m'_j} q_\lambda^{-1}\left(t - t_{m'_j}, k k_{m'_i}\right)$$

in (12.5.2), because of cancellations of the corresponding terms due to the δ-functions $\delta\left(k_{m_i} - k_{m'_i}\right)$ in the correlator (12.5.3). By the induction assumption, for $N = 2n - 2$ we have

$$\prod_{m_i < m'_j} q_\lambda\left(t - t_{m'_j}, k k_{m_i}\right) \prod_{m'_i < m'_j} q_\lambda^{-1}\left(t - t_{m'_j}, k k_{m'_i}\right)$$
$$= \prod_{m'_i < m'_j} q_\lambda\left(t - t_{m'_j}, k k_{m_i}\right) \prod_{m_i < m'_j < m'_i} q_\lambda\left(t - t_{m'_j}, k k_{m_i}\right)$$
$$\cdot \prod_{m'_i < m'_j} q_\lambda^{-1}\left(t - t_{m'_j}, k k_{m_i}\right)$$
$$= \prod_{m_i < m'_j < m'_i} q_\lambda\left(t - t_{m'_j}, k k_{m_i}\right).$$

Using this let us now prove that (12.5.5) is equal to (12.5.3). This will prove the theorem. We have

$$\prod_{m_i < m'_j} q_\lambda^{-1}\left(t_{m_1} - t_{m_i}, k_{m_1} k_{m_i}\right) \prod_{m'_i < m'_j} q_\lambda\left(t_{m_1} - t_{m'_i}, k_{m_1} k_{m'_i}\right)$$
$$= \prod_{m'_i < m'_j} q_\lambda^{-1}\left(t_{m_1} - t_{m_i}, k_{m_1} k_{m_i}\right) \prod_{m_i < m'_j < m'_i} q_\lambda^{-1}\left(t_{m_1} - t_{m_i}, k_{m_1} k_{m_i}\right)$$
$$\cdot \prod_{m'_i < m'_j} q_\lambda\left(t_{m_1} - t_{m'_i}, k_{m_1} k_{m'_i}\right), \tag{12.5.6}$$

because $m_i < m_i'$. From (12.5.3) for $2n-2$ we have $k_{m_i} = k_{m_i'}$. Using this and combining the first product with the third, we obtain

$$\prod_{m_i'<m_j'} q_\lambda\left(t_{m_i} - t_{m_i'}, k_{m_1} k_{m_i}\right) \prod_{m_i<m_j'<m_i'} q_\lambda^{-1}\left(t_{m_1} - t_{m_i}, k_{m_1} k_{m_i}\right) . \tag{12.5.7}$$

Using, in the second product in (12.5.7), the change of variables

$$t_{m_1} - t_{m_i} = \left(t_{m_1} - t_{m_j'}\right) - \left(t_{m_j'} - t_{m_i}\right)$$

and the property $k_{m_1} = k_{m_j'}$, we find that (12.5.6) is equal to

$$\prod_{m_i'<m_j'} q_\lambda\left(t_{m_i} - t_{m_i'}, k_{m_1} k_{m_i}\right) \prod_{m_i<m_j'<m_i'} q_\lambda^{-1}\left(t_{m_1} - t_{m_j'}, k_{m_1} k_{m_i}\right)$$
$$\prod_{m_i<m_j'<m_i'} q_\lambda\left(t_{m_i} - t_{m_j'}, k_{m_j'} k_{m_i}\right) .$$

Substituting this into (12.5.5) we obtain

$$\sum_{j=1}^{n} \delta\left(k_{m_1} - k_{m_j'}\right) \frac{1}{\lambda^2} q_\lambda\left(t_{m_1} - t_{m_j'}, \tilde{\omega}(k_{m_1}) + k_{m_1} p\right) \tag{12.5.8}$$
$$\cdot \prod_{m_i'<m_j'} q_\lambda\left(t_{m_i} - t_{m_i'}, k_{m_1} k_{m_i}\right)$$
$$\cdot \prod_{m_i<m_j'<m_i'} q_\lambda\left(t_{m_i} - t_{m_j'}, k_{m_j'} k_{m_i}\right) \left\langle \widehat{a}_\lambda(t_{m_1}, k_{m_1}) \dots \widehat{a}_\lambda^+\left(t_{m_j'}, k_{m_j'}\right) \dots \right\rangle ,$$

where $\langle \dots \widehat{a} \dots \rangle$ means that $\widehat{a}$ is omitted in this correlation function. For the $(2n-2)$-point correlator in (12.5.8) we use (12.5.3) for $\varepsilon \setminus \{m_1, m_j'\}$, i.e.

$$\left\langle \widehat{a}_\lambda(t_{m_1}, k_{m_1}) \dots \widehat{a}_\lambda^+\left(t_{m_j'}, k_{m_j'}\right) \dots \right\rangle$$
$$= \sum_{\sigma(\varepsilon - \{m_1, m_j'\})} \prod_{h=1}^{n-1} \left[\delta\left(k_{n_h} - k_{n_h'}\right) \right.$$
$$\left. \cdot \frac{1}{\lambda^2} q_\lambda \left(\left(t_{n_h} - t_{n_h'}\right), \left(\tilde{\omega}(k_{n_h}) + k_{n_h} p + \sum_{n_\alpha < n_h < n'_\alpha} k_{n_\alpha} \cdot k_{n_h}\right)\right)\right]$$
$$\cdot \prod_{\substack{(n_j, n_j'), (n_i, n_i'); i,j=1,\dots,n-1: \\ n_j < n_i < n_j' < n_j'}} q_\lambda\left(t_{n_i} - t_{n_j'}, k_{n_i} \cdot k_{n_j'}\right) , \tag{12.5.9}$$

where $\sigma(\varepsilon\backslash\{m_1, m'_j\}) = \{(n_j < n'_j) : j = 1 \ldots n-1\}$ is the set of all partitions of $\{1 \ldots 2n\}$ associated with $\varepsilon - \{m_1, m'_j\}$ without the pair $\{m_1, m'_j\}$. Recall that the indices m_h correspond to annihilators and m'_h correspond to creators. Substituting (12.5.9) into (12.5.8) we obtain

$$\begin{aligned}
&\sum_{j=1}^{n} \delta\left(k_{m_1} - k_{m'_j}\right) \frac{1}{\lambda^2} q_\lambda \left(t_{m_1} - t_{m'_j}, \tilde{\omega}(k_{m_1}) + k_{m_1} p\right) \\
&\cdot \prod_{m'_i < m'_j} q_\lambda \left(t_{m_i} - t_{m'_i}, k_{m_1} k_{m_i}\right) \prod_{m_i < m'_j < m'_i} q_\lambda \left(t_{m_i} - t_{m'_j}, k_{m'_j} k_{m_i}\right) \\
&\cdot \sum_{\sigma(\varepsilon - \{m_1, m'_j\})} \prod_{h=1}^{n-1} \Bigg[\delta\left(k_{n_h} - k_{n'_h}\right) \\
&\quad \cdot \frac{1}{\lambda^2} q_\lambda \left(\left(t_{n_h} - t_{n'_h}\right), \left(\tilde{\omega}(k_{n_h}) + k_{n_h} p + \sum_{n_\alpha < n_h < n'_\alpha} k_{n_\alpha} \cdot k_{n_h} \right)\right)\Bigg] \\
&\cdot \prod_{\substack{(n_j, n'_j), (n_i, n'_i); i,j = 1, \ldots, n-1: \\ n_j < n_i < n'_j < n'_j}} q_\lambda \left(t_{n_i} - t_{n'_j}, k_{n_i} \cdot k_{n'_j}\right) . \qquad (12.5.10)
\end{aligned}$$

But it is easy to see that

$$\sum_{j=1}^{n} \sum_{\sigma(\varepsilon - \{m_1, m'_j\})} = \sum_{\sigma(\varepsilon)} . \qquad (12.5.11)$$

Using (12.5.11) and combining the first product in (12.5.10) with the third and the second product with the fourth we obtain (12.5.3). This finishes the proof of the theorem.

12.6 The Wick Theorem for the QED Module

Now we will prove the following module analogue of the Wick theorem, which shows that the limiting correlation functions (12.4.2–3) define a master field $b(t, k)$ satisfying the following *QED module relations*:

$$b(t, k_1) b^+(\tau, k_2) = 2\pi \delta(t - \tau) \delta(\tilde{\omega}(k_1) + k_1 p) \delta(k_1 - k_2) , \qquad (12.6.1)$$

$$b(t, k) p = (p + k) b(t, k) , \qquad (12.6.2)$$

$$b(t, k) \Phi = 0 , \qquad (12.6.3)$$

which are an operator generalization of the Boltzmannian relations.

Let ϕ be the vacuum vector and denote $\langle \cdot \rangle$ the vacuum expectation. With these notations, let us consider the following correlation function

$$\langle b^{\epsilon_N}(t_N, k_N) b^{\epsilon_{N-1}}(t_{N-1}, k_{N-1}) \ldots b^{\epsilon_1}(t_1, k_1) \rangle . \qquad (12.6.4)$$

Theorem 12.6.1.

(i) If the number of creators is not equal to the number of annihilators, then the correlator (12.6.4) is equal to zero;

(ii) if the number of creators is equal to the number of annihilators ($N = 2n$), then the correlator (12.6.4) is equal to the following:

$$\prod_{h=1}^{n} \delta\left(k_{m_h} - k_{m'_h}\right) 2\pi\delta\left(t_{m_h} - t_{m'_h}\right)$$

$$\cdot\, \delta\left(\tilde{\omega}(k_{m_h}) + k_{m_h} p + \sum_{\alpha : m_\alpha < m_h < m'_\alpha} k_{m_\alpha} \cdot k_{m_h}\right), \tag{12.6.5}$$

where $\{(m_j < m'_j) : j = 1 \dots n\}$ is the unique noncrossing pair partition of $\{1 \dots 2n\}$ associated with ε. The indices m_h correspond to annihilators, m'_h correspond to creators.

Proof 12.6.1. In Sect. 12.4 we proved that (12.6.5) is the limit of the n-point correlators of the rescaled field. Therefore they satisfy automatically all the compatibility conditions of the reconstruction theorem of [AFL82] and therefore there exist an Hilbert space, operators $b(t, k), b^+(t, k)$ and a cyclic vector Φ such that the correlators of these operators are given by (12.6.4). The structure of the master space and the commutation relations (12.6.1–3) are now a simple consequence of the induction relation (12.4.6).

12.7 The Limit White Noise Hamiltonian Equation

We investigate the limit white noise Hamiltonian equation for the evolution operator $U_t^{(\lambda)}$ in the interaction picture by means of the following generalization of the stochastic golden rule:

Using $U_{t+dt}^{(\lambda)} = U_{t,t+dt}^{(\lambda)} U_t^{(\lambda)}$ we find for $dt > 0$:

$$U_{t+dt}^{(\lambda)} = \left(1 + (-i\lambda) \int_t^{t+dt} H_{\mathrm{I}}(t_1) dt_1 \right.$$
$$\left. + (-i\lambda)^2 \int_t^{t+dt} dt_1 \int_t^{t_1} dt_2 H_{\mathrm{I}}(t_1) H_{\mathrm{I}}(t_2) + \dots \right) U_t^{(\lambda)}$$

(the dots correspond to the remaining terms of the iterated series expansion of $U_{t,t+dt}^{(\lambda)}$). Thus, writing $dU_t^{(\lambda)} = U_{t+dt}^{(\lambda)} - U_t^{(\lambda)}$, we obtain

$$dU_t^{(\lambda)} = \left((-i\lambda) \int_t^{t+dt} H_{\mathrm{I}}(t_1) dt_1 \right.$$
$$\left. + (-i\lambda)^2 \int_t^{t+dt} dt_1 \int_t^{t_1} dt_2 H_{\mathrm{I}}(t_1) H_{\mathrm{I}}(t_2) + \dots \right) U_t^{(\lambda)},$$

which, after the rescaling $t \to t/\lambda^2$, becomes

$$dU^{(\lambda)}_{t/\lambda^2} = \left[(-i) \int_t^{t+dt} dt_1 \frac{1}{\lambda} H_\mathrm{I}\left(\frac{t_1}{\lambda^2}\right) \right.$$
$$\left. + (-i)^2 \int_t^{t+dt} dt_1 \int_t^{t_1} dt_2 \frac{1}{\lambda} H_\mathrm{I}\left(\frac{t_1}{\lambda^2}\right) \frac{1}{\lambda} H_\mathrm{I}\left(\frac{t_2}{\lambda^2}\right) + \dots \right] U^{(\lambda)}_{t/\lambda^2} . \tag{12.7.1}$$

To find the limit white noise Hamiltonian equation, we need to collect all the terms of order dt in the iterated series (12.7.1). But these terms of order dt are contained only in the first two terms of the iterated series. Let us investigate the first two terms. For the first term of the perturbation theory we find

$$\int_t^{t+dt} dt_1 \frac{1}{\lambda} H_\mathrm{I}\left(\frac{t_1}{\lambda^2}\right)$$
$$= \int_t^{t+dt} dt_1 \int dk \left[\overline{g}(k)(2p+k) a_\lambda(t_1,k) + g(k) a^+_\lambda(t_1,k)(2p+k) \right] . \tag{12.7.2}$$

In the stochastic limit $\lambda \to 0$ this term gives

$$\int dk \left[\overline{g}(k)(2p+k) dB(t,k) + g(k) dB^+(t,k)(2p+k) \right] ,$$

where the stochastic differential $dB(t,k)$ is the stochastic limit of the field $a_\lambda(t,k)$ in the time interval $(t, t+dt)$:

$$dB(t,k) = \lim_{\lambda\to 0} \int_t^{t+dt} d\tau\, a_\lambda(\tau,k) = \int_t^{t+dt} d\tau\, b(\tau,k) .$$

The second term of the perturbation theory series is equal (up to terms of order $(dt)^2$) to

$$\int_t^{t+dt} dt_1 \int_t^{t_1} dt_2 \frac{1}{\lambda} H_\mathrm{I}\left(\frac{t_1}{\lambda^2}\right) \frac{1}{\lambda} H_\mathrm{I}\left(\frac{t_2}{\lambda^2}\right)$$
$$= \int_t^{t+dt} dt_1 \int_t^{t_1} dt_2 \int dk |g(k)|^2 (2p+k)^2 \frac{1}{\lambda^2} \exp\left\{ -i \frac{t_1 - t_2}{\lambda^2} \left[\omega(k) - kp + \frac{1}{2} k^2 \right] \right\}$$

due to the q-module commutation relations on the $a_\lambda(t,k)$ and p (see Sect. 12.2). Integrating over t_1, t_2 and using the formulae

$$\int_t^{t+dt} dt_1 \int_t^{t_1} dt_2 \frac{1}{\lambda^2} \exp\left(-i \frac{t_1 - t_2}{\lambda^2} x \right) = \int_t^{t+dt} dt_1 \int_{-t_1/\lambda^2}^{0} d\tau\, e^{i\tau x} ,$$

$$\int_{-\infty}^{0} dt e^{itx} = \frac{-i}{x - i0} = \pi\delta(x) - i \text{ P.P.} \frac{1}{x} ,$$

we obtain for the second term

$$-idt \int dk|g(k)|^2(2p+k)^2 \frac{1}{\omega(k) - kp + \frac{1}{2}k^2 - i0} .$$

Let us denote

$$(g|g)_-(p) = -i \int dk|g(k)|^2(2p+k)^2 \frac{1}{\omega(k) - kp + \frac{1}{2}k^2 - i0} \tag{12.7.3}$$

$$= \int dk|g(k)|^2(2p+k)^2 \left[\pi\delta\left(\omega(k) - kp + \frac{1}{2}k^2\right) - i\ \text{P.P.}\frac{1}{\omega(k) - kp + \frac{1}{2}k^2}\right] .$$

Combining together all the terms of order dt we get the following result:

Theorem 12.7.1. *The white noise Hamiltonian equation for* $U_t = \lim_{\lambda\to 0} U^{(\lambda)}_{t/\lambda^2}$ *has the form*

$$\frac{dU_t}{dt} = \left(-i \int dk \left[\overline{g}(k)(2p+k)b(t,k) + g(k)b^+(t,k)(2p+k)\right]\right)U_t \tag{12.7.4}$$

and its normally ordered form is equivalent to the quantum stochastic differential equation:

$$\begin{aligned} dU_t = -\,i\bigg(&\int dk\left[\overline{g}(k)(2p+k)dB(t,k) + g(k)dB^+(t,k)(2p+k)\right] \\ &- (g|g)_-(p)\bigg)U_t\,. \end{aligned} \tag{12.7.5}$$

Proof 12.7.1. (12.7.4) follows from the above discussion. (12.7.5) will be proved in the Remark after Theorem 12.12.1.

12.8 Free Independence of the Increments of the Master Field

In this section we prove that the stochastic differential $dB(t,k)$ and the evolution operator U_t are free independent. In fact we will prove a weaker property which, however, is sufficient for the applications discussed in Sect. 12.14 and 12.15. In the bosonic case independence follows from the commutation relation

$$[dB(t,k), U_t] = 0\,, \tag{12.8.1}$$

which implies that, for an arbitrary system observable X,

$$\langle X\, dB(t,k)U_t\rangle = 0\,, \tag{12.8.2}$$

where $\langle\ \cdot\ \rangle$ is the stochastic limit of the vacuum expectation of the boson field. In the case of Boltzmannian statistics (12.8.1) does not hold, but (12.8.2) does. This is a consequence of *free independence*: roughly speaking it means that $dB(t,k)$ kills all creators in U_t.

Lemma 12.8.1. *The stochastic differential $dB(t,k)$ and the evolution operator U_t satisfy (12.8.2).*

Proof 12.8.1. We will prove this result by analyzing the iterated series expression of U_t. We have

$$\langle X\, dB(t,k)U_t\rangle = \lim_{\lambda\to 0}\Big\langle X\int_t^{t+dt} d\tau\, a_\lambda(\tau,k)\Big[1+(-i)\int_0^t dt_1\,\frac{1}{\lambda}\,H_{\rm I}\left(\frac{t_1}{\lambda^2}\right)+\dots$$

$$+(-i)^n\int_0^t dt_1\dots\int_0^{t_{N-1}} dt_N\,\frac{1}{\lambda}\,H_{\rm I}\left(\frac{t_1}{\lambda^2}\right)\dots\frac{1}{\lambda}\,H_{\rm I}\left(\frac{t_n}{\lambda^2}\right)+\dots\Big]\Big\rangle,$$

where $\frac{1}{\lambda}\,H_{\rm I}\left(\frac{t_k}{\lambda^2}\right)$ is given by (12.8.2). The Nth term of the perturbation series is the linear combination of the following terms (we omit integration over k, k_n):

$$\left\langle X\int_t^{t+dt} d\tau\, a_\lambda(\tau,k)\int_0^t dt_1\dots\int_0^{t_{N-1}} dt_N\, a_\lambda^{\varepsilon_1}(t_1,k_1)\dots a_\lambda^{\varepsilon_N}(t_N,k_N)\right\rangle.$$

Let us shift $a_\lambda(\tau,k)$ to the right using the dynamically q-deformed relations (12.2.3). As in Sect. 12.4 we enumerate the annihilators in the product $a_\lambda^{\varepsilon_1}(t_1,k_1)\dots a_\lambda^{\varepsilon_N}(t_N,k_N)$ as $a_\lambda(t_{m_j},k_{m_j})$, $j=1\dots J$, and the creators as $a_\lambda^+(t_{m'_j},k_{m'_j})$, $j=1\dots I$, $I+J=N$. This means that if $\varepsilon_m=0$ then $a_\lambda^{\varepsilon_m}(t_m,k_m)=a_\lambda(t_{m_j},k_{m_j})$ for $m=m_j$ (and the analogous condition for $\varepsilon_m=1$).

We will use the following recurrence relation for the correlator:

$$\left\langle X\int_t^{t+dt} d\tau\int_0^t dt_1\dots\int_0^{t_{N-1}} dt_N\, a_\lambda(\tau,k)a_\lambda^{\varepsilon_1}(t_1,k_1)\dots a_\lambda^{\varepsilon_N}(t_N,k_N)\right\rangle$$

$$=\sum_{j=1}^{I}\delta\left(k-k_{m'_j}\right)\Big\langle X\int_t^{t+dt} d\tau\int_0^t dt_1\dots\int_0^{t_{N-1}} dt_N$$

$$\cdot\, a_\lambda^{\varepsilon_1}(t_1,k_1)\dots\widehat{a}_\lambda^+\left(t_{m'_j},k_{m'_j}\right)\dots a_\lambda^{\varepsilon_N}(t_N,k_N)\Big\rangle$$

$$\cdot\,\frac{1}{\lambda^2}q_\lambda\left(\tau-t_{m'_j},\omega(k)-kp+\frac{1}{2}k^2\right)\prod_{m_i>m'_j}q_\lambda^{-1}\left(\tau-t_{m'_j},kk_{m_i}\right)$$

$$\cdot\prod_{m'_i>m'_j}q_\lambda\left(\tau-t_{m'_j},kk_{m'_i}\right)\prod_{m_i<m'_j}q_\lambda^{-1}(\tau-t_{m_i},kk_{m_i})\prod_{m'_i<m'_j}q_\lambda\left(\tau-t_{m'_i},kk_{m'_i}\right),\qquad(12.8.3)$$

where as before $\widehat{a}_\lambda^+$ means that we omit the operator a_λ^+ in the product. If we could be sure that the limit of the products of the q_λ-factors is the product of the limits, it would be clear from (12.2.7) that (12.8.3) tends to zero. However, due to the exponential nature of q_λ, some cancellations might arise. The following calculation shows that this case does not happen. The right-hand side of (12.8.3) is equal to

$$\sum_{j=1}^{I} \delta\left(k - k_{m'_j}\right) \left\langle X \int_0^t dt_1 \dots \int_0^{t_{N-1}} dt_N \right.$$

$$\left. \cdot a_\lambda^{\varepsilon_1}(t_1, k_1) \dots \widehat{a}_\lambda^{+}(t_{m'_j}, k_{m'_j}) \dots a_\lambda^{\varepsilon_N}(t_N, k_N) \right\rangle$$

$$\cdot \frac{1}{\lambda^2} q_\lambda \left(-t_{m'_j}, \omega(k) - kp + \frac{1}{2}k^2\right) \prod_{m_i > m'_j} q_\lambda^{-1}\left(-t_{m'_j}, kk_{m_i}\right)$$

$$\cdot \prod_{m'_i > m'_j} q_\lambda\left(-t_{m'_j}, kk_{m'_i}\right) \prod_{m_i < m'_j} q_\lambda^{-1}(-t_{m_i}, kk_{m_i}) \prod_{m'_i < m'_j} q_\lambda\left(-t_{m'_i}, kk_{m'_i}\right)$$

$$\cdot \int_t^{t+dt} d\tau\, q_\lambda\left(\tau, \omega(k) - kp + \frac{1}{2}k^2\right) \prod_{m_i > m'_j} q_\lambda^{-1}\left(\tau, kk_{m_i}\right) \prod_{m'_i > m'_j} q_\lambda\left(\tau, kk_{m'_i}\right)$$

$$\cdot \prod_{m_i < m'_j} q_\lambda^{-1}(\tau, kk_{m_i}) \prod_{m'_i < m'_j} q_\lambda\left(\tau, kk_{m'_i}\right)$$

(where we have used the fact that q_λ is an exponential function). The first three lines of this formula do not depend on τ and the last two lines do not depend on $t_1 \dots t_N$. Therefore the stochastic limits for these values can be made independently (the limit of the product is equal to the product of the limits). It is easy to see that the stochastic limit for the multiplier that depends on τ (of the last two lines) is equal to zero because under our assumptions $\tau > t$. This finishes the proof of the lemma.

12.9 Boltzmannian White Noise Hamiltonian Equations: Normal Form

Before studying the limit evolution equation for our model we describe some general rules showing how to deal with Hamiltonian evolution equations driven by Boltzmannian white noises. Let us consider the equation

$$\partial_t U_t = -i(F b_t^+ + F^+ b_t) U_t\,, \tag{12.9.1}$$

where b_t, b_t^+ is the standard free Fock (or Boltzmannian) white noise, characterized by the relations

$$b_t b_\tau^+ = \delta(t - \tau)\,, \tag{12.9.2a}$$

$$b_t \Phi = 0 \tag{12.9.2b}$$

(Φ is the Fock vacuum). We will use the following regularization of (12.9.1):

$$b_t U_t = \lim_{\varepsilon \to 0} [c b_{t-\varepsilon} + (1-c) b_{t+\varepsilon}] U_t \,, \tag{12.9.3}$$

where c is an arbitrary complex constant, i.e. we write (12.9.1) as

$$U_t = 1 - i \lim_{\varepsilon \to 0} \int_0^{t-0} d\tau \left[F b_\tau^+ + F^+ \left(c b_{\tau-\varepsilon} + (1-c) b_{\tau+\varepsilon} \right) \right] U_\tau \,. \tag{12.9.4}$$

To make notations shorter we will not write the regularization (12.9.3–4) explicitly each time but we will use it to obtain, for example, the following relation:

$$b_t \int_0^t dt_1 b_{t_1}^+ = c \,, \tag{12.9.5}$$

which is proved as follows:

$$b_t \int_0^t dt_1 b_{t_1}^+ = \lim_{\varepsilon \to 0} (c b_{t-\varepsilon} + (1-c) b_{t+\varepsilon}) \int_0^t dt_1 b_{t_1}^+$$

$$= \lim_{\varepsilon \to 0} \int_0^t dt_1 [c \delta(t - \varepsilon - t_1) + (1-c) \delta(t + \varepsilon - t_1)] = c \,.$$

Definition 12.9.1. *The operator $P^{(-)}$ is defined on the algebra $\mathcal{F}$ of all operators A which can be represented as series in normally ordered products of the creators and annihilators. We define $P^{(-)}$ on all such formal series, but we shall use it only for convergent series. If $A \in \mathcal{F}$ then $P^{(-)} A$ is the operator obtained from the series of A by setting all the terms which contain (or, equivalently, begin with) a creator equal to zero. Moreover, by definition*

$$P^{(-)} 1 = 1 \,. \tag{12.9.6}$$

It is clear that $P^{(-)}$ maps the algebra $\mathcal{F}$ into itself; moreover it is a linear projector,

$$P^{(-)} P^{(-)} = P^{(-)} \,, \tag{12.9.7}$$

with the property that, for any $A \in \mathcal{F}$,

$$(\phi_0, P^-(A)\psi) = (\phi_0, A\psi) \,. \tag{12.9.8}$$

Theorem 12.9.1. *The white noise Hamiltonian equation (12.9.1) with the regularization (12.9.3–4) is equivalent to the normally ordered free equation:*

$$\partial_t U_t = -i [F b_t^+ U_t + F^+ b_t P^{(-)}(U_t) - i c F^+ F U_t] \,. \tag{12.9.9}$$

Proof 12.9.1. To prove the theorem, i.e. (12.9.9), one has to prove the relation

$$b_t U_t = b_t P^{(-)}(U_t) - icFU_t\,. \tag{12.9.10}$$

Let us start with the relation

$$U_t = 1 - iF^+ \int_0^t dt_1 b_{t_1} U_{t_1} - iF \int_0^t dt_1 b_{t_1}^+ U_{t_1}\,. \tag{12.9.11}$$

Multiplying (12.9.11) from the left by b_t and using (12.9.5):

$$b_t U_t = b_t - iF^+ b_t \int_0^t dt_1 b_{t_1} U_{t_1} - icFU_t\,. \tag{12.9.12}$$

By comparing (12.9.10) and (12.9.12) we see that (12.9.10) is equivalent to the identity

$$b_t P^{(-)}(U_t) = b_t - iF^+ b_t \int_0^t dt_1 b_{t_1} U_{t_1}\,. \tag{12.9.13}$$

But, by applying $P^{(-)}$ to both sides of (12.9.11) we obtain

$$P^{(-)}(U_t) = 1 - iF^+ \int_0^t dt_1 P^{(-)}(b_{t_1} U_{t_1})\,. \tag{12.9.14}$$

Multiplying (12.9.14) from the left by b_t we find:

$$b_t P^{(-)}(U_t) = b_t - iF^+ b_t \int_0^t dt_1 P^{(-)}(b_{t_1} U_{t_1})\,. \tag{12.9.15}$$

By comparing (12.9.13) and (12.9.15) we see that if we prove

$$b_t \int_0^t dt_1 b_{t_1} U_{t_1} = b_t \int_0^t dt_1 P^{(-)}(b_{t_1} U_{t_1}) \tag{12.9.16}$$

then (12.9.10) follows. To prove (12.9.16) let us consider the nth term of the expansion for U_{t_1} in the normal form:

$$\int_0^{t_1} dt_2 \int_0^{t_2} dt_3 \ldots \int_0^{t_n} dt_{n+1} F^{\varepsilon_1} F^{\varepsilon_2} \ldots F^{\varepsilon_n} b_{t_2}^{\varepsilon_2} \ldots b_{t_{n+1}}^{\varepsilon_{n+1}}\,. \tag{12.9.17}$$

If (12.9.17) contains two or more creators, i.e.

$$b_{t_2}^+ b_{t_3}^+ b_{t_y}^{\varepsilon_y} \ldots b_{t_{n+1}}^{\varepsilon_{n+1}}\,,$$

then

$$b_t \int_0^t dt_1 \int_0^{t_1} dt_2 \ldots \int_0^{t_n} dt_{n+1} b_{t_1} b_{t_2}^+ b_{t_3}^+ b_{t_4}^{\varepsilon_n} \ldots b_{t_{n+1}}^{\varepsilon_{n+1}} = 0$$

because of the non-time-consecutive pairing $\delta(t-t_3)\delta(t_1-t_2)$.

Therefore the nonvanishing contributions to the integral on the left-hand side of (12.9.16) will be given only by the scalar term in the normal expansion of U_{t_1} and by those terms of the form of (12.9.17) with exactly one creator $b^+_{t_2}$ and any number of annihilators. From this remark (12.9.16) immediately follows. Therefore the theorem is proved.

Remark 12.9.1. One may write (12.9.9) as the quantum (Boltzmannian or free) stochastic differential equation

$$dU_t = -i[FdB_t^+ + F^+dB_t - icF^+Fdt]U_t \,, \tag{12.9.18}$$

where $dB_t = b_t dt$. In fact, if we interpret dB_t as $B(\chi_{[t,t+dt]})$, then it is clear that $dB_tU_t = dB_tP^{(-)}(U_t)$ because U_t is adapted. This proves that, also in the Boltzmannian (free) case, a stochastic differential equation is a normally ordered white noise equation; however, in this case, the normal order has to be interpreted in the Boltzmannian sense.

12.10 Unitarity Conditions

Theorem 12.10.1. *For the solution U_t of (12.9.9) the following statements are equivalent:*

(i) $c = \frac{1}{2} + ix$, where x is an arbitrary real number.
(ii) U_t is isometric, i.e. one has

$$\lim_{\Delta t \to 0+} \frac{\Delta(U_t^+U_t)}{\Delta t} = 0 \,, \tag{12.10.1}$$

where the limit is meant weakly on the number vectors and, for $\Delta t > 0$,

$$\Delta(U_t^+U_t) = U_{t+\Delta t}^+U_{t+\Delta t} - U_t^+U_t \,. \tag{12.10.2}$$

Proof 12.10.1. We have

$$U_t = 1 - i\int_0^t d\sigma[Fb_\sigma^+U_\sigma + F^+b_\sigma P^{(-)}(U_\sigma) - icF^+FU_\sigma] \,, \tag{12.10.3}$$

$$U_t^+ = 1 + i\int_0^t d\tau[U_\tau^+b_\tau F^+ + P^{(-)}(U_\tau)^*b_\tau^+F + i\bar{c}U_\tau^+F^+F] \,. \tag{12.10.4}$$

Introducing for any function F_t the notation

$$\Delta F_t := F_{t+\Delta t} - F_t \quad , \qquad \Delta t > 0 \,,$$

using (12.10.3) and the algebraic identity

$$\Delta F_tG_t = (\Delta F_t)G_t + F_t(\Delta G_t) + \Delta F_t\Delta G_t \,,$$

one finds

$$\frac{\Delta(U_t^+U_t)}{\Delta t} = \frac{i}{\Delta t}\int_t^{t+\Delta t} d\tau[U_\tau^+ b_\tau + P^{(-)}(U_\tau)^* b_\tau F + i\bar{c}U_\tau^+ F^+ F]U_t$$

$$+\frac{1}{\Delta} U_t^+(-i)\int_t^{t+\Delta t} d\sigma[F b_\sigma^+ U_\sigma + F^+ b_\sigma P^{(-)}(U_\sigma) - icF^+ F U_\sigma]$$

$$+\frac{1}{\Delta t}\int_t^{t+\Delta t} d\tau U_\tau^+ F^+ b_\tau \int_t^{t+\Delta t} d\sigma b_\sigma^+ F U_\sigma + 0(\Delta t)\,. \qquad (12.10.5)$$

Now one uses the following corollary of the fact that U_t is adapted:

$$\lim_{\Delta t\to 0+} \frac{i}{\Delta t}\int_t^{t+\Delta t} d\tau U_\tau^+ F^+ b_\tau U_t = iU_t^+ F^+ b_t P^{(-)}(U_t)\,. \qquad (12.10.6)$$

Moreover, the weak continuity of the maps $t \mapsto b_t, P^{(-)}(U_t)$ implies that

$$\lim_{\Delta t\to 0+} \frac{(-i)}{\Delta t} U_t^+ \int_t^{t+\Delta t} d\sigma F^+ b_\sigma P^{(-)}(U_\sigma) = -iU_t^+ F^+ b_t P^{(-)}(U_t)\,. \quad (12.10.7)$$

Also we have, by the same argument as in (12.10.6–7),

$$\lim_{\Delta t\to 0+} \frac{i}{\Delta t}\int_t^{t+\Delta t} d\tau P^{(-)}(U_\tau)^* b_\tau^+ F U_t = iP^{(-)}(U_t)^* b_t^+ F U_t \qquad (12.10.8)$$

and

$$\lim_{\Delta t\to 0+} \frac{(-i)}{\Delta t} U_t^+ \int_t^{t+\Delta t} d\sigma F b_\sigma^+ U_\sigma = -iP^{(-)}(U_t)^* b_t^+ F U_t\,. \qquad (12.10.9)$$

In conclusion

$$\begin{aligned}
&\lim_{\Delta t\to 0+} \frac{\Delta(U_t^+U_t)}{\Delta t} \\
&= -\bar{c}U_t^+ F^+ F U_t - cU_t^+ F^+ F U_t \\
&\quad + \lim_{\Delta t\to 0+} \frac{1}{\Delta t}\int_t^{t+\Delta t} d\tau U_\tau^+ F^+ b_\tau \int_t^{t+\Delta t} d\sigma b_\sigma^+ F U_\sigma \\
&= (-\bar{c} - c + 1)U_t^+ F^+ F U_t\,.
\end{aligned}$$

The last expression vanishes if and only if $c = \frac{1}{2} + ix$. The theorem is thus proved.

12.11 Matrix Elements of the Solution

In this section we use the normally ordered form of the white noise equation (12.9.9) to compute some matrix elements of the evolution operator. We will use the following: If

$$P^{(-)}(U_t) = U_t^{(0)} + U_t^{(1)} + U_t^{(2)} + \dots , \qquad (12.11.1)$$

where $U_t^{(i)}$ contains i annihilators, then one has

$$U_t^{(0)} = \langle 0|U_t|0\rangle , \qquad (12.11.2)$$

$$\langle 0|U_t^{(1)}|f\rangle = \langle 0|U_t|f\rangle ; \qquad (12.11.3)$$

moreover,

$$\langle 0|U_t^{(n)}|f_n, f_{n-1}, \dots , f_1\rangle = \langle 0|U_t|f_n, f_{n-1}, \dots , f_1\rangle , \qquad (12.11.4)$$

where $|f_n, f_{n-1}, \dots , f_1\rangle = b^+(f_n)\dots b^+(f_1)|0\rangle$.

Lemma 12.11.1. *The vacuum expectation value of U_t is*

$$U_t^{(0)} = \langle 0|U_t|0\rangle = e^{-cF^+Ft} . \qquad (12.11.5)$$

Proof 12.11.1. Taking the vacuum expectation value of (12.9.9), we obtain

$$\frac{d\langle 0|U_t|0\rangle}{dt} = -cF^+F\langle 0|U_t|0\rangle . \qquad (12.11.6)$$

The solution of this equation gives (12.11.5).

Lemma 12.11.2. *One has*

$$\langle 0|U_t|f\rangle = -ie^{-cF^+Ft}F^+ \int_0^t dt_1 f_{t_1} . \qquad (12.11.7)$$

Proof 12.11.2. Taking the matrix element of (12.9.9) one finds

$$\frac{d}{dt}\langle 0|U_t|f\rangle = -iF^+ \left\langle 0|b_t P^{(-)}(U_t)|f\right\rangle - cF^+F\langle 0|U_t|f\rangle . \qquad (12.11.8)$$

Now

$$\langle 0|b_t P^{(-)}(U_t)|f\rangle = \left\langle 0|b_t \left[U_t^{(0)} + U_t^{(1)} + \dots\right] f\right\rangle = \langle 0|b_t|f\rangle U_t^{(0)} . \qquad (12.11.9)$$

Denoting $X_t = \langle 0|U_t|f\rangle$ and using $\langle 0|b_t|f\rangle = f_t$, one obtains from (12.11.8–9):

$$\frac{dX_t}{dt} = -iF^+U_t^{(0)}f_t - cF^+FX_t , \qquad (12.11.10)$$

where $U_t^{(0)}$ is given by (12.11.5). The solution of (12.11.10) with $X_t|_{t=0} = 1$ is

$$X_t = -ie^{-cF^+Ft}F^+ \int_0^t dt_1 f_{t_1} . \qquad (12.11.11)$$

The lemma is thus proved.

Lemma 12.11.3. *One has*

$$\langle 0|U_t|f_2, f_1\rangle = -e^{-cF^+Ft}F^{+2}\int_0^t dt_1 f_{t_1}\int_0^{t_1} dt_2 f_{t_2}\,. \tag{12.11.12}$$

Proof 12.11.3. One has

$$\begin{aligned}\langle 0|b_t P^{(-)}(U_t)|f,g\rangle &= \left\langle 0|b_t\left[U_t^{(0)} + U_t^{(1)} + U_t^{(2)} + \dots\right]|f,g\right\rangle \\ &= \langle 0|b_t U_t^{(1)}|f,g\rangle = \langle 0|b_t|g\rangle\left\langle 0\left|U_t^{(1)}\right|f\right\rangle = g_t X_t\,,\end{aligned}$$

where X_t is given by (12.11.11). Now denote

$$Y_t := \langle 0|U_t|f,g\rangle\,.$$

Using the normal form (12.9.9) of the white noise equation for U_t, we obtain

$$\frac{dY_t}{dt} = -iF^+ g_t X_t - cF^+FY_t\,. \tag{12.11.13}$$

Since X_t is given in (12.11.11), the solution of (12.11.13) is given by (12.11.12).

12.12 Normal Form of the QED Module Hamiltonian Equation

In this section we consider the evolution equation

$$\frac{dU_t}{dt} = \left(\int dk\,\left[\overline{g}(k)pb(t,k) - g(k)b^+(t,k)p\right]\right)U_t\,. \tag{12.12.1}$$

This is obtained from (12.7.4), the limit equation for the model described in Sect. 12.1, by replacing the factor $(2p+k)$ in it with p (and $-i$ with 1 in the first term and with -1 in the second). The additional term is put in normal order using the techniques developed in Sect. 12.9. Here we concentrate on the new feature arising from the nontrivial commutation relations between p and $b(t,k)$ (the module commutation relations (12.1.7–8)). We bring (12.12.1) to normal form starting from its iterated series solution

$$U(t) = 1 + \int_0^t dt_1[pb_{t_1}(g) - b_{t_1}^+(g)p] \tag{12.12.2}$$

$$+\int_0^t dt_1\int_0^{t_1} dt_2[pb_{t_1}(g) - b_{t_1}^+(g)p][pb_{t_2}(g) - b_{t_2}^+(g)p] + \dots$$

Lemma 12.12.1. *If $U(t)$ is the solution of the Boltzmannian white noise equation (12.12.2), then one has*

$$b_t(g)U(t) = b_t(g)P^{(-)}(U(t)) + (g|g)_- pU(t)\,, \tag{12.12.3}$$

where $P^{(-)}$ is the projection defined Definition 12.9.1.

Proof 12.12.1. The proof is analogous to the Proof 12.9.1.

Theorem 12.12.1. *The normally ordered form of the quantum evolution equation (12.12.1) is*

$$\partial_t U(t) = pb_t(g)P^{(-)}\big(U(t)\big) - b_t^+(g)pU(t) + p(g|g)_- pU(t)\,. \qquad (12.12.4)$$

Proof 12.12.2. It follows from Lemma 12.12.1.

Remark 12.12.1. Notice that, as remarked at the beginning of Sect. 12.8, the stochastic differential $dB(t,k)$ kills all the creators in U_t (i.e. $dB(t,k)U_t = dB(t,k)P^{(-)}U_t$). Moreover, if in (12.12.1) we replace g by ig and symmetrize in p ($pb \to pb + bp$; $b^+p \to b^+p + pb^+$), then we find precisely (12.1.5a), (12.1.4). Therefore the normally ordered form of (12.1.4) is obtained from (12.12.4) by the same replacement and symmetrization. The further replacement $b(t,k)dt = dB(t,k)$ and use of (12.1.7) now gives (12.7.5).

In other words: also in the Boltzmannian case it is true that a stochastic differential equation is equivalent to a normally ordered white noise equation.

12.13 Unitarity of the Solution: Direct Proof

In this section we prove the isometricity for the stochastic limit U_t of the evolution operator satisfying (12.7.4), i.e.

$$U_t^+ U_t = 1\,.$$

To prove this we calculate the difference

$$U^{(\lambda)+}_{t+dt/\lambda^2} U^{(\lambda)}_{t+dt/\lambda^2} - U^{(\lambda)+}_{t/\lambda^2} U^{(\lambda)}_{t/\lambda^2}$$

up to the first order in dt. Using the evolution equation, iterated up to second order as in Sect. 12.7, we express the product $U^{(\lambda)+}_{t+dt/\lambda^2} U^{(\lambda)}_{t+dt/\lambda^2}$ in the form

$$\begin{aligned}
&U^{(\lambda)+}_{t/\lambda^2}\Bigg(1 + i\int_t^{t+dt} dt_1 \int dk\, \big[\overline{g}(k)(2p+k)a_\lambda(t_1,k) + g(k)a^+_\lambda(t_s,k)(2p+k)\big] \\
&- \int_t^{t+dt} dt_1 \int_t^{t_1} dt_1' \int dk \exp\left\{+i\frac{t_1-t_1'}{\lambda^2}\left[\omega(k)-kp+\frac{1}{2}k^2\right]\right\} |g(k)|^2(2p+k)^2\Bigg) \\
&\cdot\Bigg(1 - i\int_t^{t+dt} dt_2 \int dk'\, \big[\overline{g}(k')(2p+k')a_\lambda(t_2,k') + g(k')a^+_\lambda(t_2,k')(2p+k')\big] \\
&\quad - \int_t^{t+dt} dt_2 \int_t^{t_2} dt_2' \int dk' \exp\left\{-i\frac{t_2-t_2'}{\lambda^2}\left[\omega(k')-k'p+\frac{1}{2}k'^2\right]\right\} \\
&\qquad \cdot |g(k')|^2(2p+k')^2\Bigg) U^{(\lambda)}_{t/\lambda^2}\,.
\end{aligned}$$

Therefore

$$
\begin{aligned}
&U^{(\lambda)+}_{t+dt/\lambda^2} U^{(\lambda)}_{t+dt/\lambda^2} - U^{(\lambda)+}_{t/\lambda^2} U^{(\lambda)}_{t/\lambda^2} \\
&= U^{(\lambda)+}_{t/\lambda^2} \Bigg(\int_t^{t+dt} dt_1 \int_t^{t+dt} dt_2 \int dk |g(k)|^2 (2p+k)^2 \\
&\qquad \cdot \exp\left\{ -i \frac{t_1 - t_2}{\lambda^2} \left[\omega(k) - kp + \frac{1}{2} k^2 \right] \right\} \\
&\quad - \int_t^{t+dt} dt_1 \int_t^{t_1} dt_2 \int dk |g(k)|^2 (2p+k)^2 \\
&\qquad \cdot \exp\left\{ i \frac{t_1 - t_2}{\lambda^2} \left[\omega(k) - kp + \frac{1}{2} k^2 \right] \right\} \\
&\quad - \int_t^{t+dt} dt_1 \int_t^{t_1} dt_2 \int dk |g(k)|^2 (2p+k)^2 \\
&\qquad \cdot \exp\left\{ -i \frac{t_1 - t_2}{\lambda^2} \left[\omega(k) - kp + \frac{1}{2} k^2 \right] \right\} \Bigg) U^{(\lambda)}_{t/\lambda^2} .
\end{aligned}
$$

It is easy to see that, as $\lambda \to 0$, the terms in the above difference cancel each other, because in the limit one has

$$
\begin{aligned}
&\int_t^{t+dt} dt_1 \int_t^{t_1} dt_2 \exp\left\{ i \frac{t_1 - t_2}{\lambda^2} \left[\omega(k) - kp + \frac{1}{2} k^2 \right] \right\} \\
&= \int_t^{t+dt} dt_1 \int_{t_1}^{t+dt} dt_2 \exp\left\{ -i \frac{t_1 - t_2}{\lambda^2} \left[\omega(k) - kp + \frac{1}{2} k^2 \right] \right\}
\end{aligned}
$$

and therefore the last two terms add up and cancel the first one. This proves the isometricity of U_t.

12.14 Matrix Elements of the Limit Evolution Operator

Now we start the investigation of the evolution of matrix elements of U_t. From the white noise differential equation of U_t we obtain the equation for the matrix element:

$$
\begin{aligned}
&\frac{d}{dt} \langle b(t_1, k_1) \dots b(t_n, k_n) U_t \rangle \\
&= -(g|g)_-(p + k_1 + \dots + k_n) \langle b(t_1, k_1) \dots b(t_n, k_n) U_t \rangle - i 2\pi g(k_n) \delta(t - t_n) \\
&\quad \cdot \delta\left(\omega(k_n) - k_n (p + k_1 + \dots + k_{n-1}) + \frac{1}{2} k_n^2 \right) \\
&\quad \cdot [2(p + k_1 + \dots + k_{n-1}) + k_n] \langle b(t_1, k_1) \dots b(t_{n-1}, k_{n-1}) U_t \rangle .
\end{aligned}
$$

Here $\langle \cdot \rangle$ is the vacuum expectation. We introduce the regularized operators

$$B(f) = \int dt \int dk \overline{f}(t,k) b(t,k) \,,$$

and for fixed k,

$$B_f(k) = \int dt \overline{f}(t) b(t,k) \,.$$

Theorem 12.14.1. *We obtain the following expressions for the matrix elements of the evolution operator:*

$$\begin{aligned}\langle B_f(k) U_t \rangle = & - i2\pi g(k) \delta\left(\omega(k) - kp + \frac{1}{2}k^2\right)(2p+k) \\ & \cdot \int_0^t d\tau \overline{f}(\tau) \exp\{\tau[(g|g)_-(p+k) - (g|g)_-(p)]\} \exp[-t(g|g)_-(p+k)] \,,\end{aligned} \tag{12.14.1}$$

$$\begin{aligned}\langle U_t B_f^+(k) \rangle = & - i2\pi \overline{g}(k) \delta\left(\omega(k) - kp + \frac{1}{2}k^2\right)(2p+k) \\ & \cdot \int_0^t d\tau f(\tau) \exp\{\tau[(g|g)_-(p) - (g|g)_-(p+k)]\} \exp[-t(g|g)_-(p)] \,,\end{aligned} \tag{12.14.2}$$

$$\begin{aligned}& \langle B_{f_1}(k_1) U_t B_{f_2}^+(k_2) \rangle \\ & = e^{-t(g|g)_-(p+k_1)} \Bigg[2\pi\delta(k_1 - k_2)\delta\left(\omega(k_1) - k_1 p + \frac{1}{2}k_1^2\right) \int d\tau \overline{f}_1(\tau) f_2(\tau) \\ & \quad - 2\pi g(k_1)\delta\left(\omega(k_1) - k_1 p + \frac{1}{2}k_1^2\right)(2p + k_1) \\ & \quad \cdot 2\pi \overline{g}(k_2)\delta\left(\omega(k_2) - k_2 p + \frac{1}{2}k_2^2\right)(2p + k_2) \\ & \quad \cdot \int_0^t d\tau_1 \overline{f}_1(\tau_1) \exp\{\tau_1[(g|g)_-(p + k_1) - (g|g)_-(p)]\} \\ & \quad \cdot \int_0^t d\tau_2 f_2(\tau_2) \exp\{\tau_2[(g|g)_-(p) - (g|g)_-(p + k_2)]\} \Bigg] \,.\end{aligned} \tag{12.14.3}$$

Proof 12.14.1. The expression (12.14.1) follows immediately from the evolution equation. Analogously we obtain for the evolution of the matrix element $\langle 0, U_t 1 \rangle$ the equation

$$\begin{aligned}\frac{d}{dt}\langle U_t B_f^+(k) \rangle = & \left\langle -i \int dk' \overline{g}(k')(2p + k') b(t,k') U_t B_f^+(k) \right\rangle \\ & - (g|g)_-(p) \langle U_t B_f^+(k) \rangle \,.\end{aligned}$$

Because $b(t,k')$ kills all creators in U_t, the term $b(t,k')U_t$ contains only annihilators. It is obvious that the nonzero term in $\langle b(t,k') U_t B_f^+(k) \rangle$ can arise only from the component of U_t that contains neither creators nor annihilators. Since this component is equal to $e^{-t(g|g)_-(p)}$, we get

$$\frac{d}{dt}\langle U_t B_f^+(k)\rangle = -(g|g)_-(p)\langle U_t B_f^+(k)\rangle$$

$$-i2\pi\overline{g}(k)(2p+k)\delta\left(\omega(k) - kp + \frac{1}{2}k^2\right) f(t)\exp\{-t(g|g)_-(p+k)\}\,,$$

with the initial condition $\langle U_0 B_f^+(k)\rangle = 0$. The solution of this Cauchy problem gives (12.14.2). Now let us calculate matrix elements of the form

$$\langle 1, U_t 1\rangle\,,$$

where 1 denotes an arbitrary 1-particle vector in the master space. We have

$$\begin{aligned}
\frac{d}{dt}\left\langle B_{f_1}(k_1)U_t B_{f_2}^+\right\rangle &= -\,i2\pi g(k_1)\delta\left(\omega(k_1) - k_1 p + \frac{1}{2}k_1^2\right)\\
&\quad\cdot(2p+k_1)\overline{f}_1(t)\left\langle U_t B_{f_2}^+(k_2)\right\rangle\\
&\quad-(g|g)_-(p+k_1)\left\langle B_{f_1}(k_1)U_t B_f^+(k_2)\right\rangle\\
&= -\,(g|g)_-(p+k_1)\left\langle B_{f_1}(k_1)U_t B_{f_2}^+(k_2)\right\rangle\\
&\quad-2\pi g(k_1)\delta\left(\omega(k_1) - k_1 p + \frac{1}{2}k_1^2\right)(2p+k_1)\\
&\quad\cdot 2\pi\overline{g}(k_2)\delta\left(\omega(k_2) - k_2 p + \frac{1}{2}k_2^2\right)(2p+k_2)\\
&\quad\cdot\overline{f}_1(t)\int_0^t d\tau\, f_2(\tau)\exp\{\tau[(g|g)_-(p) - (g|g)_-(p+k_2)]\}\\
&\qquad\cdot\exp[-t(g|g)_-(p)]\,.
\end{aligned}$$

Since the solution of this equation is given by (12.14.3), the theorem is proved.

The expression (12.14.3) contains the product of the δ-functions of the form

$$\delta\left(\omega(k_1)p + \frac{1}{2}k_1^2\right)\delta\left(\omega(k_2) - k_2 p + \frac{1}{2}k_2^2\right).$$

Let us analyze this product for a particular choice of $\omega(k)$, namely

$$\omega(k) = \frac{1}{2}k^2\,.$$

In this case

$$\delta\left(\omega(k) - kp + \frac{1}{2}k^2\right) = \delta\big(k(k-p)\big)\,.$$

This leads to the equation $k(k-p) = \left|k - \frac{p}{2}\right|^2 - \left(\frac{p}{2}\right)^2 = 0$, which is the equation of a sphere of radius $|p|/2$. Therefore we have

$$\delta\big(k(k-p)\big) = \frac{1}{|p|}\,\delta\left(|k| - \frac{p}{2} - \frac{|p|}{2}\right)$$
$$= \delta\left(k - \left(\frac{p}{2} + \frac{|p|}{2}\,\Omega\right)\right)\frac{1}{|p|}\left(\frac{|p|^{n-1}}{2}\right) d\Omega\,,$$

where Ω denotes a unit vector and $d\Omega$ integration over the surface of the unit sphere in $\mathbb{R}^n$ (n is the dimension of the space; in our case $n = 3$). More precisely

$$\delta\big(k_1(k_1-p)\big) \int \delta\left(k_1 - k_2 + \frac{p}{2} + \frac{|p|}{2}\,\Omega - k_1\right)\frac{1}{|p|}\left(\frac{|p|}{2}\right)^{n-1} d\Omega\,.$$

Therefore, for $\left\langle B_{f_1}(k_1)U_tB^+_{f_2}(k_2)\right\rangle$, we find

$$\begin{aligned}
&\langle B_{f_1}(k_1)U_tB^+_{f_2}(k_2)\rangle\\
&= e^{-t(g|g)_-(p+k_1)}\delta\left(\omega(k_1) - k_1p + \frac{1}{2}\,k_1^2\right)\Bigg(2\pi\delta(k_1-k_2)\int d\tau \overline{f}_1(\tau)f_2(\tau)\\
&\quad - (2\pi)^2 g(k_1)\overline{g}(k_2)(2p+k_1)(2p+k_2)\\
&\quad \cdot \int \delta\left(k_1 - k_2 - \frac{p}{2} + \frac{|p|}{2}\,\Omega - k_1\right)\frac{1}{|p|}\left(\frac{|p|}{2}\right)^{n-1} d\Omega\\
&\quad \cdot \int_0^t d\tau_1 \overline{f}_1(\tau_1)\exp\{\tau_1[(g|g)_-(p+k_1) - (g|g)_-(p)]\}\\
&\quad \cdot \int_0^t d\tau_2 f_2(\tau_2)\exp\{\tau_2[(g|g)_-(p) - (g|g)_-(p+k_2)]\}\Bigg)\,.
\end{aligned}$$

12.15 Nonexponential Decays

The aim of this section is to show that for the model of a nonrelativistic particle interacting with a quantum electromagnetic field, in particular for the polaron model, one can have not only the standard exponential decay of correlations but also some power decays may arise. The model we investigate is widely studied in elementary particle physics, solid-state physics and quantum optics (see, for example, [Bog46], [Fey72]). We consider the simplest case in which matter is represented by a single particle, say an electron, with position $q = (q_1, q_2, q_3)$ and momentum $p = (p_1, p_2, p_3)$ satisfying the commutation relations $[q_j, p_n] = i\delta_{jn}$. The electromagnetic field is described by boson operators $a(k) = \big(a_1(k), a_2(k), a_3(k)\big); a^+(k) = \big(a_1^+(k), \dots, a_3^+(k)\big)$ satisfying the canonical commutation relations

$$[a_j(k), a_n^+(k')] = \delta_{jn}\delta(k-k')\,.$$

The Hamiltonian of a free nonrelativistic atom (minimally) interacting with a quantum electromagnetic field is (neglecting polarization)

$$H = H_0 + \lambda H_{\mathrm{I}} = \int \omega(k) a^+(k) a(k) dk + \frac{1}{2} p^2 + \lambda H_{\mathrm{I}} , \tag{12.15.1}$$

where λ is a small constant, $\omega(k)$ is the dispersion law of the field and

$$H_{\mathrm{I}} = \int d^3k \left(g(k) p \cdot a^+(k) e^{-ikq} + \overline{g}(k) p \cdot a(k) e^{ikq} \right) + \text{h.c.} \tag{12.15.2}$$

Here

$$p \cdot a(k) = \sum_{j=1}^{3} p_j a_j(k) \quad , \quad p^2 = \sum_{j=1}^{3} p_j^2 ,$$

$$a^+(k) a(k) = \sum_{j=1}^{3} a_j^+(k) a_j(k) \quad , \quad kq = \sum_{j=1}^{3} k_j q_j .$$

For the polaron model the Hamiltonian has the form

$$H = \int \omega(k) a^+(k) a(k) dk + \frac{1}{2} p^2 + \lambda \int d^3k \Big(g(k) a^+(k) e^{-ikq} + \overline{g}(k) a(k) e^{ikq} \Big) .$$

This model differs from nonrelativistic QED, considered in Sect. 12.1, because of the absence of momentum p in the interaction Hamiltonian. For the analysis in this section this difference is not important. In fact the master field is described by the same commutation relations as in the nonrelativistic QED case (without the dipole approximation) (see Sect. 12.2), namely

$$b_j(t,k) p_n = (p_n - k_n) b_j(t,k) , \tag{12.15.3}$$

$$b_j(t,k) b_n^+(t',k') = 2\pi \delta(t-t') \delta \left(\omega(k) - kp + \frac{1}{2} k^2 \right) \delta(k-k') \delta_{jn} . \tag{12.15.4}$$

Let us now investigate the behaviour of the vacuum transition amplitude $\langle U_t \rangle$ using the stochastic differential equation (12.7.4). We obtain

$$\langle dU_t \rangle = \left\langle \left((-i) \int dk \overline{g}(k)(2p+k) dB(t,k) - dt \, (g|g)_-(p) \right) U_t \right\rangle .$$

Using the free independence of $dB(t,k)$ and U_t, we obtain

$$\frac{d}{dt} \langle U_t \rangle = \left\langle \frac{d}{dt} U_t \right\rangle = -(g|g)_-(p) \langle U_t \rangle .$$

Because $U_0 = 1$, we have the solution

$$\langle U_t \rangle = \exp[-t(g|g)_-(p)] .$$

In this section we calculate the matrix element $\langle X|U_t|X\rangle$, where $X = f(p)\otimes\Phi$ in the momentum representation and Φ is the vacuum vector for the master field. This matrix element is

$$\langle X|U_t|X\rangle = \int dp|f(p)|^2 \exp[-t(g|g)_-(p)]\,. \qquad (12.15.5)$$

We investigate the polaron model when $\omega(k) = 1$. For this choice of $\omega(k)$ we obtain

$$\omega(k) - kp + \frac{1}{2}k^2 = 1 - \frac{1}{2}p^2 + \frac{1}{2}(k-p)^2\,.$$

One can expect nonexponential relaxation when

$$\text{supp } f(p) \subset \{|p| < \sqrt{2}\} =: D\,, \qquad (12.15.6)$$

because if $p \in D$ then $\text{Re}(g|g)_-(p) = 0$ and there is no damping: all decay (if any) is due to interference. We assume that (12.15.6) holds, and we will use the approximation

$$\text{diam supp } g(k) \gg \text{ diam supp } f(p)\,. \qquad (12.15.7)$$

Physically this means that the particle is more localized in the momentum representation than the field. This assumption seems natural because the field's degrees of freedom are fast and the particles' ones are slow. Under this assumption we can estimate the matrix element (12.15.5). We will prove that in this case there will be polynomial decay. In fact, for $|p| < \sqrt{2}$ we obtain

$$\begin{aligned}
(g|g)_-(p) &= -i\int dk\,|g(k)|^2(2p+k)^2\frac{1}{1-\frac{1}{2}p^2+\frac{1}{2}(k-p)^2} \\
&= -2i\int dk\,|g(k)|^2 - i\,(I_1+I_2)\,, \\
I_1 &= (-2+10p^2)\int dk\,|g(k)|^2\frac{1}{1-\frac{1}{2}p^2+\frac{1}{2}(k-p)^2}\,, \\
I_2 &= 6\int dk\,|g(k)|^2 p(k-p)\frac{1}{1-\frac{1}{2}p^2+\frac{1}{2}(k-p)^2}\,.
\end{aligned}$$

Here only I_1 and I_2 depend on p and therefore can interfere. Let us find the asymptotics of $(g|g)_-(p)$ on p (we investigate the case when p is a small parameter) under the assumption that $g(k)$ is a very smooth function. More precisely

$$|g(k)|^2 = \lambda F(\lambda k) \quad , \quad F(k) > 0\,,$$

where F is a compactly supported smooth function and λ is a small parameter. By considering the Taylor expansion of F in the small parameter p:

$$\lambda F(\lambda k) = \lambda F\big(\lambda(k-p)\big) + \lambda^2\sum_i p_i\frac{\partial}{\partial k_i}F\big(\lambda(k-p)\big) + \dots\,,$$

we find that $\lambda F\big(\lambda(k-p)\big)$ is the leading term with respect to λ. Taking $\lambda \to 0$ we find that we can use $|g(k-p)|^2$ instead of $|g(k)|^2$ in the formulae for I_1 and I_2 for sufficiently smooth $g(k)$. Under the above assumptions let us calculate I_1 and I_2. We obtain

$$I_1 = \left(-2+10p^2\right)\int dk\,|g(k-p)|^2 \frac{1}{1-\frac{1}{2}p^2+\frac{1}{2}(k-p)^2}$$

$$= \left(-2+10p^2\right)\int dk\,|g(k)|^2 \frac{1}{1+\frac{1}{2}k^2} - p^2\int dk\,|g(k)|^2 \frac{1}{\left(1+\frac{1}{2}k^2\right)^2},$$

$$I_2 = pQ, \qquad Q = 6\int dk\,|g(k)|^2 k \frac{1}{1+\frac{1}{2}k^2}.$$

Therefore

$$(g|g)_-(p) = -2i\int dk\,|g(k)|^2 + 2i\int dk\,|g(k)|^2 \frac{1}{1+\frac{1}{2}k^2} - iAp^2 - ipQ,$$

$$A = 10\int dk\,|g(k)|^2 \frac{1}{1+\frac{1}{2}k^2} - \int dk\,|g(k)|^2 \frac{1}{\left(1+\frac{1}{2}k^2\right)^2}. \tag{12.15.8}$$

So, for $X(t) = \langle X|U_t|X\rangle$, we find eventually

$$X(t) = \int dp|f(p)|^2 \exp[-t(g|g)_-(p)]$$

$$= \exp\left[it2\left(\int dk\,|g(k)|^2 - \int dk\,|g(k)|^2 \frac{1}{1+\frac{1}{2}k^2}\right)\right]\int dp|f(p)|^2$$

$$\cdot \exp\left[it\left(Ap^2+pQ\right)\right].$$

Let us estimate this integral for $f(p) = e^{-Bp^2}$, $B >> 1$. Let us consider for simplicity the case $Q = 0$ (for example, $g(k)$ is spherically symmetric). In this case the integral is equal to

$$4\pi\int_0^\infty dp\,p^2 e^{-Bp^2} e^{iAtp^2} = \left(\frac{\pi}{B-iAt}\right)^{\frac{3}{2}},$$

and we find that for large t the decay of the matrix element $X(t) = \langle X|U_t|X\rangle$ is proportional to $(At)^{-\frac{3}{2}}$, where A is the functional of the cutoff function given by (12.15.8).

In summary, in the polaron model with a symmetric and very smooth cutoff function the vacuum transition amplitude with respect to any state of the particle satisfying conditions (12.15.6–7), for large t decays like $t^{-\frac{3}{2}}$ independent of the specific form of the particle state $f(p)$ (realized as a function in momentum representation). In particular, the dependence on the parameter B, corresponding to the size of the support of $f(p)$ for large t is not important.

12.16 Equilibrium States

In this and the following sections we extend the previous results, concerning the stochastic limit of QED without dipole approximation (nonrelativistic and neglecting polarization) or the polaron model, from the Fock case to the case in which the initial state of the field is a Gibbs state at a given temperature. (In fact, all the results proved in the following are valid, with the same proof for any mean zero gauge-invariant state.) We show that this leads to a new algebra,[(1)] which we call *the hot free algebra*, generalizing the QED Hilbert module algebra, discussed in Sect. 12.2, in the same sense in which the finite temperature representations of the bosonic or fermionic algebras with respect to a mean zero gauge-invariant Gaussian state, generalize the usual Fock representation.

The hot free algebra is a deformation of the free algebra in two senses:

(i) A deformation parameter appears, depending on the temperature.
(ii) The commutation relations are Hilbert module relations rather than Hilbert space relations in the sense that they cannot be realized in a usual Hilbert space, but require the introduction of a Hilbert module (see the remark at the end of Sect. 12.6).

12.17 The Master Field

The Hamiltonian we consider in the present section is exactly the same as that in Sect. 12.1. The only difference is that now the field expectation value $\langle \cdot \rangle$ is not the Fock one, but an equilibrium state $\langle \cdot \rangle$, for the free evolution, at a given inverse temperature β, i.e. the mean zero boson Gaussian state with pair correlations vanishing in the off-diagonal terms and in the diagonal ones equal to

$$\langle a_k a^+_{k'} \rangle = \frac{\delta(k-k')}{1-e^{-\beta\omega_k}}, \tag{12.17.1}$$

$$\langle a^+_{k'} a_k \rangle = \frac{\delta(k-k')}{e^{\beta\omega_k}-1} \tag{12.17.2}$$

(the other correlators can be calculated using Gaussianity). The state space of our model is the Hilbert space of the GNS representation of the field algebra with respect to $\langle \cdot \rangle$.

The first and most important step of the stochastic limit procedure is to determine the structure of the master field and the space where it lives. In our case this is given by the following:

Theorem 12.17.1. *Let the rescaled fields $a^{\pm}_{\lambda}(t,k)$ be given by (12.1.5b). Then the limit temperature correlation functions (12.4.1), i.e.*

$$\langle a^{\varepsilon_1}_{\lambda}(t_1,k_1)\dots a^{\varepsilon_N}_{\lambda}(t_N,k_N)\rangle_\beta\,, \tag{12.17.3}$$

always exist and

(i) *if the number of creators is not equal to the number of annihilators, then the above correlator is equal to zero (even before the limit);*

(ii) *if the number of creators is equal to the number of annihilators ($N = 2n$), then the limit of (12.4.1) is equal to the following sum over the noncrossing partitions:*

$$\sum_{\sigma(\varepsilon)} \prod_{h=1}^{n} \delta(k_{m'_h} - k_{m_h}) c_{m_h m'_h}(k_{m_h}) 2\pi\delta(t_{m'_h} - t_{m_h})$$

$$\cdot \delta\left(\omega(k_{m_h}) + k_{m_h} p + \sum_{\alpha} (-1)^{\varepsilon_\alpha} \chi_{(m_\alpha, m'_\alpha)}(m_h) k_{m_\alpha} \cdot k_{m_h} - \varepsilon_h k^2_{m_h}\right), \tag{12.17.4}$$

where $\{(m'_j, m_j) : j = 1 \dots n\}$ is the unique noncrossing partition of $\{1 \dots 2n\}$ associated with ε; $\chi_{(m_\alpha, m'_\alpha)}(m_h)$ is equal to 1 if m_h is between m_α and m'_α, while it is otherwise equal to 0; the indices m'_h corresponds to annihilators, m_h corresponds to creators; $(-1)^{\varepsilon_h} = 1$ for $m'_h > m_h$; $(-1)^{\varepsilon_h} = -1$ for $m'_h < m_h$;

$$c_{m_h m'_h}(k) = \frac{1}{1 - e^{-\beta\omega_k}}, \quad m'_h > m_h\,;$$

$$c_{m_h m'_h}(k) = \frac{1}{e^{\beta\omega_k} - 1}, \quad m'_h < m_h\,.$$

12.18 Proof of the Result for the Two- and Four-Point Correlators

In order to explain the main idea, we shall prove the statement of Theorem 12.17.1 in the simplest examples, i.e. the 2-point and 4-point correlators. For the 2-point correlator one has:

$$\begin{aligned}
\langle b_t(k_1) b^+_\tau(k_2)\rangle &= \lim_{\lambda\to 0} \left\langle \frac{1}{\lambda} a_{t/\lambda^2}(k_1) \frac{1}{\lambda} a^+_{\tau/\lambda^2}(k_2) \right\rangle \\
&= \lim_{\lambda\to 0} \frac{1}{\lambda^2} \left\langle \exp\left\{ \frac{it}{\lambda^2} [\omega(k_1) + k_1 p] \right\} e^{-iq(k_1 - k_2)} \right. \\
&\quad \left. \cdot \exp\left\{ -i\frac{\tau}{\lambda^2} [\omega(k_2) + k_2 p] \right\} \right\rangle \left\langle a_{k_1} a^+_{k_2} \right\rangle .
\end{aligned}$$

Using (12.17.1–2) we obtain

$$\lim_{\lambda\to 0} \frac{1}{\lambda^2} \exp\left\{ i\frac{t-\tau}{\lambda^2} [\omega(k_1) + k_1 p] \right\} \frac{\delta(k_1 - k_2)}{1 - e^{-\beta\omega(k_1)}} .$$

Using the module extension of the basic formula of the stochastic limit:

$$\lim_{\lambda\to 0}\frac{1}{\lambda^2}\exp\left\{\frac{it}{\lambda^2}\left[\omega(k)+kp\right]\right\}=2\pi\delta\big(\omega(k)+kp\big)\delta(t)\,, \tag{12.18.1}$$

we obtain the 2-point correlator

$$\langle b_t(k_1)b_\tau^+(k_2)\rangle=2\pi\delta(t-\tau)\delta\big(\omega(k_1)+k_1p\big)\cdot\frac{\delta(k_1-k_2)}{1-e^{-\beta\omega(k_1)}}\,. \tag{12.18.2}$$

Let us now investigate the following 2-point correlator

$$\langle b_\tau^+(k_2)b_t(k_1)\rangle=\lim_{\lambda\to 0}\frac{1}{\lambda^2}\left\langle a_{k_2}^+e^{ik_2q}\exp\left\{-i\frac{\tau}{\lambda^2}\left[\omega(k_2)+k_2p\right]\right\}\right.$$
$$\left.\cdot\exp\left\{\frac{it}{\lambda^2}\left[\omega(k_1)+k_1p\right]\right\}e^{-ik_1q}a_{k_1}\right\rangle.$$

Using the commutation relation for Weyl operators

$$e^{i\alpha p}e^{i\beta q}=e^{i\beta q}e^{i\alpha q}e^{i\alpha\beta}\,, \tag{12.18.3}$$

where $[p,q]=-i$, we obtain the limit relation

$$\lim_{\lambda\to 0}\frac{1}{\lambda^2}\frac{\delta(k_2-k_1)}{e^{\beta\omega(k_1)}-1}\exp\left\{i\frac{t-\tau}{\lambda^2}\left[\omega(k_1)+k_1p-k_1^2\right]\right\}.$$

Using (12.18.1) we obtain

$$\langle b_\tau^+(k_2)b_t(k_1)\rangle=2\pi\delta(t-\tau)\delta\big(\omega(k_1)+k_1p-k_1^2\big)\frac{\delta(k_2-k_1)}{e^{\beta\omega(k_1)}-1}\,. \tag{12.18.4}$$

Let us now calculate the 4-point correlator,

$$\langle b_{t_1}(k_1)b_{t_2}(k_2)b_{t_2'}^+(k_2')b_{t_1'}^+(k_1')\rangle\,. \tag{12.18.5}$$

By Gaussianity and (12.17.1–2) we obtain

$$\left\langle a_{k_1}a_{k_2}a_{k_2'}^+a_{k_1'}^+\right\rangle=\frac{1}{1-e^{-\beta\omega(k_1)}}\frac{1}{1-e^{-\beta\omega(k_2)}}$$
$$\cdot\left[\delta(k_2-k_2')\delta(k_1-k_1')+\delta(k_1-k_2')\delta(k_2-k_1')\right]. \tag{12.18.6}$$

Formula (12.18.6) for the bosonic correlator $\left\langle a_{k_1}a_{k_2}a_{k_2'}^+a_{k_1'}^+\right\rangle$ contains two terms proportional to δ-functions that correspond to two diagrams. Let us calculate the first term, which is proportional to $\delta(k_1-k_1')\delta(k_2-k_2')$. We have

$$\text{First term}=\lim_{\lambda\to 0}\frac{1}{1-e^{-\beta\omega(k_1)}}\frac{1}{1-e^{-\beta\omega(k_2)}}\delta(k_1-k_1')\delta(k_2-k_2')$$
$$\cdot\frac{1}{\lambda^4}\exp\left\{i\frac{t_1-t_1'}{\lambda^2}\left[\omega(k_1)+k_1p\right]\right\}\exp\left\{i\frac{t_2-t_2'}{\lambda^2}\left[\omega(k_2)+k_2p\right]\right\}$$
$$\cdot\exp\left[i\frac{t_2-t_2'}{\lambda^2}k_1k_2\right]. \tag{12.18.7}$$

Using (12.18.1) we obtain

$$\begin{aligned}\text{First term } &= (2\pi)^2 \frac{1}{1-e^{-\beta\omega(k_1)}}\frac{1}{1-e^{-\beta\omega(k_2)}} \\ &\cdot \delta(k_1-k_1')\delta(k_2-k_2')\delta(t_1-t_1')\delta(t_2-t_2') \\ &\cdot \delta\big(\omega(k_1)+k_1p\big)\delta\big(\omega(k_2)+k_2p+k_1k_2\big)\,. \end{aligned} \tag{12.18.8}$$

The second term of the correlator is proportional to $\delta(k_1-k_2')\delta(k_2-k_1')$. We have from (12.18.1) and the Riemann–Lebesgue lemma

$$\text{Second term } = \lim_{\lambda\to 0}\frac{1}{1-e^{-\beta\omega(k_1)}}\frac{1}{1-e^{-\beta\omega(k_2)}}\delta(k_1-k_2')\delta(k_2-k_1')$$

$$\cdot\frac{1}{\lambda^4}\exp\left\{i\frac{t_1-t_2'}{\lambda^2}\left[\omega(k_1)+k_1p\right]\right\}\exp\left\{i\frac{t_2-t_1'}{\lambda^2}\left[\omega(k_2)+k_2p\right]\right\}$$

$$\cdot\exp\left[i\frac{t_2-t_2'}{\lambda^2}k_1k_2\right]=0\,.$$

This proves that the 4-point correlator is given by (12.18.8).

12.19 The Vanishing of the Crossing Diagrams

To calculate the n-point correlators in the stochastic limit we recall that the 2-parameter family of Weyl operator $W(a,b)$ $(a,b\in\mathbb{R}^d)$ is defined by

$$W(a,b)=e^{i(a\cdot p+b\cdot q)}\,.$$

The unitary operators $W(a,b)$ satisfy

$$W(a,b)=e^{ia\cdot p}e^{ib\cdot q}e^{-i\hbar a\cdot b/2}=e^{ib\cdot q}e^{ia\cdot p}e^{i\hbar a\cdot b/2}\,,$$

$$W(a_1,b_1)W(a_2,b_2)=W(a_1+a_2,b_1+b_2)\exp\left(\frac{i\hbar}{2}(a_1\cdot b_2-a_2\cdot b_1)\right)\,, \tag{12.19.1a}$$

$$W(a_1,b_1)\dots W(a_n,b_n)=W\left(\sum_j a_j,\sum_j b_j\right)\exp\left(\frac{i\hbar}{2}\sum_{j<l}(a_j\cdot b_l-a_l\cdot b_j)\right)\,, \tag{12.19.1b}$$

$$W(a,b)^+=W(-a,-b)\,. \tag{12.19.1c}$$

Under the free system evolution we have

$$p_t=p\qquad,\qquad q_t=q+tp\,,$$

so the Weyl operators evolve as

$$e^{itp^2}W(a,b)e^{-itp^2} = e^{i(a\cdot p_t + b\cdot q_t)} = e^{i[(a+tb)p+b\cdot q]} = W(a+tb,b)\,.$$

Recalling that the rescaled field operators (12.2.2) are

$$a_\lambda(t,k) = \frac{1}{\lambda}e^{i[\omega(k)+kp]t/\lambda^2}e^{-ikq}a(k)\,, \tag{12.19.2}$$

we will consider the limit temperature correlation functions

$$\begin{aligned}&\langle b^{\epsilon_N}(t_N,k_N)b^{\epsilon_{N-1}}(t_{N-1},k_{N-1})\dots b^{\epsilon_1}(t_1,k_1)\rangle\\ &= \lim_{\lambda\to 0}\left\langle a_\lambda^{\epsilon_N}(t_N,k_N)a_\lambda^{\epsilon_{N-1}}(t_{N-1},k_{N-1})\dots a_\lambda^{\epsilon_1}(t_1,k_1)\right\rangle .\end{aligned}$$

Here $\varepsilon = \{\epsilon_N\dots\epsilon_1\}\in\{1,0\}^N$ and $\epsilon\in\{1,0\}$ ($\epsilon=0$ for a and $\epsilon=1$ for a^+). For $N=2n$ one can consider the partition $\sigma(\varepsilon)$ of ε into pairs of 0 and 1, which correspond to the Wick partition of

$$b^{\epsilon_N}(t_N,k_N)b^{\epsilon_{N-1}}(t_{N-1},k_{N-1})\dots b^{\epsilon_1}(t_1,k_1)$$

into pairs of creators and annihilators. An arbitrary partition of this kind corresponds to some Wick diagram. We will be interested in partitions that correspond to half-planar, or noncrossing, diagrams. We will call these partitions nontrivial.

Theorem 12.19.1. *The limit temperature correlation functions exist always. In addition:*

(i) If N is odd, then the above limit is equal to zero.
(ii) If $N=2n$, then the above limit, i.e. the limit

$$\lim_{\lambda\to 0}\left\langle a_\lambda^{\epsilon_{2n}}(t_{2n},k_{2n})a_\lambda^{\epsilon_{2n-1}}(t_{2n-1},k_{2n-1})\dots a_\lambda^{\epsilon_1}(t_1,k_1)\right\rangle\,, \tag{12.19.3}$$

is equal to zero if ε is trivial; it is equal to

$$\begin{aligned}&\sum_{\sigma(\varepsilon)}\prod_{h=1}^{n}\delta(k_{m'_h}-k_{m_h})c_{m_hm'_h}(k_{m_h})2\pi\delta(t_{m'_h}-t_{m_h})\\ &\cdot\delta\Big(\omega(k_{m_h})+k_{m_h}p+\hbar\sum_\alpha(-1)^{\varepsilon_\alpha}\chi_{(m_\alpha,m'_\alpha)}(m_h)k_{m_\alpha}\cdot k_{m_h}\\ &-\hbar\frac{1-(-1)^{\varepsilon_h}}{2}k_{m_h}^2\Big)\,,\end{aligned}\tag{12.19.4}$$

where $\{(m'_j,m_j)\,:\,j=1\dots n\}$ is the unique noncrossing partition of $\{1\dots 2n\}$ associated with ε. Here the index m'_h corresponds to an annihilator; m_h to a creator and

$$c_{m_hm'_h}(k) = \frac{1}{1-e^{-\beta\omega_k}},\quad m'_h > m_h\,,$$

$$c_{m_hm'_h}(k) = \frac{1}{e^{\beta\omega_k}-1},\quad m'_h < m_h\,,$$

$(-1)^{\varepsilon_h}=1$ for $m'_h>m_h$ and $(-1)^{\varepsilon_h}=-1$ for $m'_h<m_h$.

Proof 12.19.1. From (12.19.2) and the identity

$$e^{i\alpha p}e^{i\beta q} = e^{i(\alpha p+\beta q)}e^{i\frac{\hbar}{2}\alpha\beta},$$

we deduce

$$a^{\epsilon}_{t,k} \equiv \frac{1}{\lambda}\exp i(-1)^{\epsilon}\left(\frac{t}{\lambda^2}[\omega(k)+kp] - kq - \frac{\hbar}{2}\frac{t}{\lambda^2}k^2\right)a^{\epsilon}(k). \qquad (12.19.5)$$

For $\varepsilon = \{\epsilon_{2n}\dots\epsilon_1\} \in \{1,0\}^{2n}$ nontrivial, we have

$$\left\langle\prod_{j=1}^{2n} a^{\epsilon_j}_{t_j,k_j}\right\rangle$$

$$= \prod_{j=1}^{2n}\left(\frac{1}{\lambda}\exp i(-1)^{\epsilon_j}\left\{\frac{t_j}{\lambda^2}[\omega(k_j)+k_jp] - k_jq - \frac{\hbar}{2}\frac{t_j}{\lambda^2}k_j^2\right\}\right)\left\langle\prod_{h=1}^{2n} a^{\epsilon_h}(k_h)\right\rangle, \qquad (12.19.6)$$

but

$$\left\langle\prod_{h=1}^{2n} a^{\epsilon_n}(k_h)\right\rangle = \sum_{\{m'_h\neq m_h\}}\prod_{h=1}^{n}\delta\left(k_{m'_h} - k_{m_h}\right)c_{m_hm'_h}\left(k_{m_h}\right), \qquad (12.19.7)$$

that is, we sum over all possible pair contractions of annihilator–creator indices $\{(m'_h, m_h) :\ h = 1\dots n\}$. All operators in these products are ordered from the right to the left. Therefore we may write

$$\left\langle\prod_{j=1}^{2n} a^{\epsilon_j}_{t_j,k_j}\right\rangle$$

$$= \prod_{j=1}^{2n}\left[\frac{1}{\lambda}\exp i(-1)^{\epsilon_j}\left(\frac{t_j}{\lambda^2}[\omega(k_j)+k_jp] - k_jq - \frac{\hbar}{2}\frac{t_j}{\lambda^2}k_j^2\right)\right]$$

$$\cdot\sum_{\{m'_h\neq m_h\}}\prod_{h=1}^{n}\delta\left(k_{m'_h} - k_{m_h}\right)c_{m_hm'_h}\left(k_{m_h}\right). \qquad (12.19.8)$$

Now, using the rules for multiplying Weyl operators and our product convention, we have

$$\prod_{j=1}^{2n}\left[\frac{1}{\lambda}\exp i(-1)^{\epsilon_j}\left(\frac{t_j}{\lambda^2}[\omega(k_j)+k_jp] - k_jq - \frac{\hbar}{2}\frac{t_j}{\lambda^2}k_j^2\right)\right]$$

$$= \exp\left(\frac{i\hbar}{2}\sum_{1\le j<l\le 2n}(-1)^{\epsilon_j+\epsilon_l}k_j\cdot k_l\frac{t_j-t_l}{\lambda^2}\right)$$

$$\cdot\left(\frac{1}{\lambda}\right)^{2n}\exp i\sum_{j=1}^{2n}(-1)^{\epsilon_j}\left(\frac{t_j}{\lambda^2}[\omega(k_j)+k_jp] - k_jq - \frac{\hbar}{2}\frac{t_j}{\lambda^2}k_j^2\right); \qquad (12.19.9)$$

the phase factor is then

$$\frac{i\hbar}{2}\sum_{l=1}^{2n}\sum_{j<l}(-1)^{\epsilon_j+\epsilon_l}k_j\cdot k_l(t_j-t_l)\,. \tag{12.19.10}$$

Using the fact that m'_h run over half of the $2n$ indices l and m_h run over the other half, $(-1)^{\epsilon_{m'_h}}=1$ and $(-1)^{\epsilon_{m_h}}=-1$:

$$\begin{aligned}
&\frac{i\hbar}{2}\sum_{h=1}^{n}\Bigg(\sum_{1\le j<m'_h}(-1)^{\epsilon_j}k_j\cdot k_{m'_h}(t_j-t_{m'_h})\\
&\quad-\sum_{1\le j<m_h}(-1)^{\epsilon_j}k_j\cdot k_{m_h}(t_j-t_{m_h})\Bigg)\\
&=\frac{i\hbar}{2}\sum_{h=1}^{n}\Bigg(\sum_{\alpha}^{m'_\alpha<m'_h}k_{m'_\alpha}\cdot k_{m'_h}(t_{m'_\alpha}-t_{m'_h})-\sum_{\beta}^{m_\beta<m'_h}k_{m_\beta}\cdot k_{m'_h}(t_{m_\beta}-t_{m'_h})\\
&\quad-\sum_{\gamma}^{m'_\gamma<m_h}k_{m'_\gamma}\cdot k_{m_h}(t_{m'_\gamma}-t_{m_h})+\sum_{\delta}^{m_\delta<m_h}k_{m_\delta}\cdot k_{m_h}(t_{m_\delta}-t_{m_h})\Bigg)\\
&=\frac{i\hbar}{2}\sum_{h=1}^{n}(I_h+II_h)\,.
\end{aligned} \tag{12.19.11}$$

We use the fact that $k_{m_h}=k_{m'_h}$. Putting together the first term with the third and the second term with the fourth, we obtain

$$\begin{aligned}
I_h &= \sum_{\alpha}^{m'_\alpha<m'_h}k_{m_\alpha}\cdot k_{m_h}(t_{m'_\alpha}-t_{m'_h})-\sum_{\gamma}^{m'_\gamma<m_h}k_{m_\gamma}\cdot k_{m_h}(t_{m'_\gamma}-t_{m_h})\\
&=\sum_{\alpha}^{m'_\alpha<m'_h}k_{m_\alpha}\cdot k_{m_h}(t_{m'_\alpha}-t_{m_h})+\sum_{\alpha}^{m'_\alpha<m'_h}k_{m_\alpha}\cdot k_{m_h}(t_{m_h}-t_{m'_h})\\
&\quad-\sum_{\gamma}^{m'_\gamma<m_h}k_{m_\gamma}\cdot k_{m_h}(t_{m'_\gamma}-t_{m_h})\\
&=\sum_{\alpha}^{m_h<m'_\alpha<m'_h}k_{m_\alpha}\cdot k_{m_h}(t_{m'_\alpha}-t_{m_h})+\sum_{\alpha}^{m'_\alpha<m'_h}k_{m_\alpha}\cdot k_{m_h}(t_{m_h}-t_{m'_h})
\end{aligned}$$

for $m'_h>m_h$ and

$$\begin{aligned}
I_h = &-\sum_{\alpha}^{m'_h<m'_\alpha<m_h}k_{m_\alpha}\cdot k_{m_h}(t_{m'_\alpha}-t_{m_h})\\
&+\sum_{\alpha}^{m'_\alpha<m'_h}k_{m_\alpha}\cdot k_{m_h}(t_{m_h}-t_{m'_h})+k_{m_h}\cdot k_{m_h}(t_{m_h}-t_{m'_h})
\end{aligned}$$

for $m'_h < m_h$. For the sum of the second and fourth terms we obtain

$$
\begin{aligned}
-II_h &= \sum_{\beta}^{m_\beta < m'_h} k_{m_\beta} \cdot k_{m_h}(t_{m_\beta} - t_{m'_h}) - \sum_{\delta}^{m_\delta < m_h} k_{m_\delta} \cdot k_{m_h}(t_{m_\delta} - t_{m_h}) \\
&= \sum_{\beta}^{m_\beta < m'_h} k_{m_\beta} \cdot k_{m_h}(t_{m_\beta} - t_{m'_h}) - \sum_{\delta}^{m_\delta < m_h} k_{m_\delta} \cdot k_{m_h}(t_{m_\delta} - t_{m'_h}) \\
&\quad - \sum_{\delta}^{m_\delta < m_h} k_{m_\delta} \cdot k_{m_h}(t_{m'_h} - t_{m_h}) \\
&= \sum_{\beta}^{m_h < m_\beta < m'_h} k_{m_\beta} \cdot k_{m_h}(t_{m_\beta} - t_{m'_h}) + \sum_{\delta}^{m_\delta < m_h} k_{m_\delta} \cdot k_{m_h}(t_{m_h} - t_{m'_h}) \\
&\quad + k_{m_h} \cdot k_{m_h}(t_{m_h} - t_{m'_h})
\end{aligned}
$$

for $m'_h > m_h$ and

$$
-II_h = - \sum_{\beta}^{m'_h < m_\beta < m_h} k_{m_\beta} \cdot k_{m_h}(t_{m_\beta} - t_{m'_h}) + \sum_{\delta}^{m_\delta < m_h} k_{m_\delta} \cdot k_{m_h}(t_{m_h} - t_{m'_h})
$$

for $m'_h < m_h$. For (12.19.11) we find

$$
\begin{aligned}
& I_h + II_h \\
&= \sum_{\alpha}^{m_h < m'_\alpha < m'_h} k_{m_\alpha} \cdot k_{m_h}(t_{m'_\alpha} - t_{m_h}) + \sum_{\alpha}^{m'_\alpha < m'_h} k_{m_\alpha} \cdot k_{m_h}(t_{m_h} - t_{m'_h}) \\
&\quad - \sum_{\beta}^{m_h < m_\beta < m'_h} k_{m_\beta} \cdot k_{m_h}(t_{m_\beta} - t_{m'_h}) - \sum_{\beta}^{m_\beta < m_h} k_{m_\beta} \cdot k_{m_h}(t_{m_h} - t_{m'_h}) \\
&\quad - k_{m_h} \cdot k_{m_h}(t_{m_h} - t_{m'_h})
\end{aligned} \tag{12.19.12}
$$

for $m'_h > m_h$ and

$$
\begin{aligned}
& I_h + II_h \\
&= - \sum_{\alpha}^{m'_h < m'_\alpha < m_h} k_{m_\alpha} \cdot k_{m_h}(t_{m'_\alpha} - t_{m_h}) + \sum_{\alpha}^{m'_\alpha < m'_h} k_{m_\alpha} \cdot k_{m_h}(t_{m_h} - t_{m'_h}) \\
&\quad + \sum_{\beta}^{m'_h < m_\beta < m_h} k_{m_\beta} \cdot k_{m_h}(t_{m_\beta} - t_{m'_h}) - \sum_{\beta}^{m_\beta < m_h} k_{m_\beta} \cdot k_{m_h}(t_{m_h} - t_{m'_h}) \\
&\quad + k_{m_h} \cdot k_{m_h}(t_{m_h} - t_{m'_h})
\end{aligned}
$$

for $m'_h < m_h$. Let us now investigate the following term in (12.19.9):

$$\left(\frac{1}{\lambda}\right)^{2n} \exp i \sum_{j=1}^{2n} (-1)^{\epsilon_j} \left(\frac{t_j}{\lambda^2} [\omega(k_j) + k_j p] - k_j q - \frac{\hbar}{2} \frac{t_j}{\lambda^2} k_j^2 \right)$$

$$\cdot \sum_{\{m'_h \neq m_h\}} \prod_{h=1}^{n} \delta \left(k_{m'_h} - k_{m_h} \right) c_{m_h m'_h} \left(k_{m_h} \right) .$$

Note that

$$\sum_{1 \leq l \leq 2n} (-1)^{\epsilon_l} t_l k_l = - \sum_{1 \leq h \leq n} (t_{m_h} - t_{m'_h}) k_{m_h} ,$$

$$\sum_{1 \leq l \leq 2n} (-1)^{\epsilon_l} k_l q = 0 ,$$

because $k_{m_h} = k_{m'_h}$. Thus, for the term in (12.19.9) we find

$$\left(\frac{1}{\lambda}\right)^{2n} \exp -i \sum_{1 \leq h \leq n} \frac{t_{m_h} - t_{m'_h}}{\lambda^2} \left(\omega(k_{m_h}) + k_{m_h} p - \frac{\hbar}{2} k_{m_h}^2 \right)$$

$$\cdot \sum_{\{m'_h \neq m_h\}} \prod_{h=1}^{n} \delta \left(k_{m'_h} - k_{m_h} \right) c_{m_h m'_h} \left(k_{m_h} \right) .$$

With the change of variables

$$\begin{cases} u_{m_h} = t_{m_h} , \\ v_{m_h} = t_{m_h} - t_{m'_h} , \end{cases} \tag{12.19.13}$$

we obtain the following lemma.

Lemma 12.19.1. *The correlator is equal to*

$$\left\langle \prod_{j=1}^{2n} a_{t_j, k_j}^{\epsilon_j} \right\rangle = \exp \frac{i\hbar}{2} \frac{1}{\lambda^2} \sum_{h=1}^{n} \Big(I_h + II_h \Big)$$
$$\cdot \left(\frac{1}{\lambda}\right)^{2n} \exp -i \sum_{1 \leq h \leq n} \frac{v_{m_h}}{\lambda^2} \left(\omega(k_{m_h}) + k_{m_h} p - \frac{\hbar}{2} k_{m_h}^2 \right)$$
$$\cdot \sum_{\{m'_h \neq m_h\}} \prod_{h=1}^{n} \delta \left(k_{m'_h} - k_{m_h} \right) c_{m_h m'_h} \left(k_{m_h} \right) . \tag{12.19.14}$$

The phase factor in (12.19.14) is equal to

$$\sum_{\alpha}^{m_h<m'_\alpha<m'_h} k_{m_\alpha}\cdot k_{m_h}(-v_{m_\alpha}+u_{m_\alpha}-u_{m_h}) + \sum_{\alpha}^{m'_\alpha<m'_h} k_{m_\alpha}\cdot k_{m_h}v_{m_h}$$
$$-\sum_{\beta}^{m_h<m_\beta<m'_h} k_{m_\beta}\cdot k_{m_h}(v_{m_h}+u_{m_\beta}-u_{m_h})$$
$$-\sum_{\beta}^{m_\beta<m_h} k_{m_\beta}\cdot k_{m_h}v_{m_h} - k_{m_h}\cdot k_{m_h}v_{m_h} \tag{12.19.15}$$

for $m'_h > m_h$ and

$$-\sum_{\alpha}^{m'_h<m'_\alpha<m_h} k_{m_\alpha}\cdot k_{m_h}(-v_{m_\alpha}+u_{m_\alpha}-u_{m_h}) + \sum_{\alpha}^{m'_\alpha<m'_h} k_{m_\alpha}\cdot k_{m_h}v_{m_h}$$
$$+\sum_{\beta}^{m'_h<m_\beta<m_h} k_{m_\beta}\cdot k_{m_h}(v_{m_h}+u_{m_\beta}-u_{m_h})$$
$$-\sum_{\beta}^{m_\beta<m_h} k_{m_\beta}\cdot k_{m_h}v_{m_h} + k_{m_h}\cdot k_{m_h}v_{m_h}$$

for $m'_h < m_h$. The Riemann–Lebesgue lemma implies that the oscillatory factors of the type $\exp(ik^2u/\lambda^2)$ cause the associated term to vanish in the limit $\lambda \to 0$. Therefore, in this limit, a partition $\{(m_h, m'_h)\}$ survives in (12.19.14) if and only if, for each fixed $h = 1 \ldots n$ and for any α,

$$m_h < m_\alpha < m'_h \Leftrightarrow m_h < m'_\alpha < m'_h \tag{12.19.16a}$$

or

$$m_h > m_\alpha > m'_h \Leftrightarrow m_h > m'_\alpha > m'_h\,, \tag{12.19.16b}$$

i.e. if and only if it is a noncrossing partition. This means that only the non-trivial sequences $\varepsilon = \{\epsilon_{2n}\ldots\epsilon_1\} \in \{1,0\}^{2n}$ give a nontrivial contribution in the limit. Denoting $\{(m_h, m'_h)\}$ the unique pair partition associated with such a sequence, the corresponding value of the phase term (12.19.15) is

$$\sum_{\alpha}^{m_h<m'_\alpha<m'_h} k_{m_\alpha}\cdot k_{m_h}(-v_{m_\alpha}-v_{m_h}) + \sum_{\alpha}^{m'_\alpha<m'_h} k_{m_\alpha}\cdot k_{m_h}v_{m_h}$$

$$-\sum_{\beta}^{m_\beta<m_h} k_{m_\beta}\cdot k_{m_h}v_{m_h} - k_{m_h}\cdot k_{m_h}v_{m_h} \tag{12.19.17}$$

for $m'_h > m_h$ and

$$- \sum_{\alpha}^{m_h'<m'_\alpha<m_h} k_{m_\alpha} \cdot k_{m_h}(-v_{m_\alpha} - v_{m_h}) + \sum_{\alpha}^{m'_\alpha<m'_h} k_{m_\alpha} \cdot k_{m_h} v_{m_h}$$
$$- \sum_{\beta}^{m_\beta<m_h} k_{m_\beta} \cdot k_{m_h} v_{m_h} + k_{m_h} \cdot k_{m_h} v_{m_h}$$

for $m'_h < m_h$.

Let us investigate the calculated phase term. For $m_h' > m_h$ we have

$$\sum_{\alpha}^{m'_\alpha<m'_h} k_{m_\alpha} \cdot k_{m_h} v_{m_h} = \sum_{\alpha}^{m_h<m'_\alpha<m'_h} k_{m_\alpha} \cdot k_{m_h} v_{m_h} + \sum_{\alpha}^{m'_\alpha\le m_h} k_{m_\alpha} \cdot k_{m_h} v_{m_h} \, .$$

Because $m'_\alpha \neq m_h$, we have for the last term

$$\sum_{\alpha}^{m'_\alpha\le m_h} k_{m_\alpha} \cdot k_{m_h} v_{m_h} = \sum_{\alpha}^{m'_\alpha<m_h} k_{m_\alpha} \cdot k_{m_h} v_{m_h} \, .$$

Therefore the phase term is equal to

$$- \sum_{\alpha}^{m_h<m'_\alpha<m'_h} k_{m_\alpha} \cdot k_{m_h} v_{m_\alpha} + \sum_{\alpha}^{m'_\alpha<m_h} k_{m_\alpha} \cdot k_{m_h} v_{m_h}$$
$$- \sum_{\beta}^{m_\beta<m_h} k_{m_\beta} \cdot k_{m_h} v_{m_h} - k_{m_h} \cdot k_{m_h} v_{m_h} \, .$$

For the case $m'_h < m_h$, due to the noncrossing condition we have

$$- \sum_{\alpha}^{m_h'<m'_\alpha<m_h} k_{m_\alpha} \cdot k_{m_h}(-v_{m_\alpha} - v_{m_h}) = - \sum_{\alpha}^{m_h'<m_\alpha<m_h} k_{m_\alpha} \cdot k_{m_h}(-v_{m_\alpha} - v_{m_h}) \, .$$

Therefore the phase term is equal to

$$\sum_{\alpha}^{m_h'<m'_\alpha<m_h} k_{m_\alpha} \cdot k_{m_h} v_{m_\alpha} + \sum_{\alpha}^{m'_\alpha<m'_h} k_{m_\alpha} \cdot k_{m_h} v_{m_h}$$
$$- \sum_{\beta}^{m_\beta<m_h'} k_{m_\beta} \cdot k_{m_h} v_{m_h} + k_{m_h} \cdot k_{m_h} v_{m_h} \, .$$

Let us denote the phase term as

$$I_h + II_h = \Phi_h - (-1)^{\varepsilon_h} k_{m_h} \cdot k_{m_h} v_{m_h} \, .$$

Here $(-1)^{\varepsilon_h} = 1$ for $m_h' > m_h$ and $(-1)^{\varepsilon_h} = -1$ for $m_h' < m_h$. One can obtain for the phase term the following:

$$\Phi_h = - \sum_{\alpha \in (m_h, m'_h) \text{ or } (m'_h, m_h)} (-1)^{\varepsilon_h} k_{m_\alpha} \cdot k_{m_h} v_{m_\alpha}$$

$$- \sum_{\alpha : h \in (m_\alpha, m'_\alpha) \text{ or } (m'_\alpha, m_\alpha)} (-1)^{\varepsilon_\alpha} k_{m_\alpha} \cdot k_{m_h} v_{m_h} ,$$

$$\sum_{1 \le h \le n} \Phi_h = -2 \sum_{1 \le h \le n} \sum_{\substack{\alpha : h \in (m_\alpha, m'_\alpha) \\ \text{or } (m'_\alpha, m_\alpha)}} (-1)^{\varepsilon_\alpha} k_{m_\alpha} \cdot k_{m_h} v_{m_h}$$

$$= -2 \sum_{1 \le h \le n} \sum_{\alpha} (-1)^{\varepsilon_\alpha} \chi_{(m_\alpha, m'_\alpha)}(m_h) k_{m_\alpha} \cdot k_{m_h} v_{m_h} .$$

Here $\chi_{(m_\alpha, m'_\alpha)}$ is the indicator of the interval (m_α, m'_α) or (m'_α, m_α). We have proved the following lemma.

Lemma 12.19.2. *The noncrossing part of the correlator is equal to*

$$\left(\frac{1}{\lambda}\right)^{2n} \exp\Bigg(-i \sum_{1 \le h \le n} \frac{v_{m_h}}{\lambda^2} \Big[\omega(k_{m_h}) + k_{m_h} p$$

$$+ \hbar \sum_{\alpha} (-1)^{\varepsilon_\alpha} \chi_{(m_\alpha, m'_\alpha)}(m_h) k_{m_\alpha} \cdot k_{m_h} - \frac{\hbar}{2} k^2_{m_h} + \frac{\hbar}{2} (-1)^{\varepsilon_h} k^2_{m_h} \Big]$$

$$\cdot \sum_{m'_h \neq m_h} \prod_{h=1}^{n} \delta\left(k_{m'_h} - k_{m_h}\right) c_{m_h m'_h}(k_{m_h}) \Bigg) .$$

Using the Riemann–Lebesgue lemma and keeping only noncrossing partition we find the correlator from the statement of the theorem, namely that the limit

$$\lim_{\lambda \to 0} \langle a_\lambda^{\epsilon_{2n}}(t_{2n}, k_{2n}) a_\lambda^{\epsilon_{2n-1}}(t_{2n-1}, k_{2n-1}) \dots a_\lambda^{\epsilon_1}(t_1, k_1) \rangle$$

in the nontrivial case is equal to

$$\sum_{m'_h \neq m_h} \prod_{h=1}^{n} \delta\left(k_{m'_h} - k_{m_h}\right) c_{m_h m'_h}(k_{m_h}) \, 2\pi \delta\left(t_{m'_h} - t_{m_h}\right)$$

$$\cdot \, \delta\Bigg(\omega(k_{m_h}) + k_{m_h} p + \hbar \sum_{\alpha} (-1)^{\varepsilon_\alpha} \chi_{(m_\alpha, m'_\alpha)}(m_h) k_{m_\alpha} \cdot k_{m_h}$$

$$- \hbar \frac{1 - (-1)^{\varepsilon_h}}{2} k^2_{m_h} \Bigg) ,$$

where $\{(m'_j, m_j) : j = 1 \dots n\}$ is the unique noncrossing partition of $\{1 \dots 2n\}$ associated with ε. The theorem is thus proven.

12.20 The Hot Free Algebra

In analogy with the discussion of Sect. 12.1, now we want to reduce the apparently complicated expression (12.19.4) of the correlators into a simple and easy to use set of algebraic commutation rules.

Lemma 12.20.1. *The correlators of Theorem (12.17.1) are those of the field* $b_t(k)$ *defined as follows:*

$$b_t(k) = b_1(t,k) + b_2^+(t,k)\,, \tag{12.20.1}$$

where b_i *are free independent noises satisfying the following hot free algebra relations:*

$$\begin{aligned}
b_1(t,k_1)b_1^+(\tau,k_2) &= 2\pi\delta(t-\tau)\delta\big(\omega(k_1)+k_1p\big)\frac{\delta(k_1-k_2)}{1-e^{-\beta\omega(k_1)}}\,,\\
b_2(t,k_1)b_2^+(\tau,k_2) &= 2\pi\delta(t-\tau)\delta\big(\omega(k_1)+k_1(p-\hbar k_1)\big)\frac{\delta(k_1-k_2)}{e^{\beta\omega(k_1)}-1}\,,\\
b_1b_2^+ &= b_2b_1^+ = 0\,,\\
b_1(t,k)p &= (p+\hbar k)b_1(t,k)\,,\\
b_2(t,k)p &= (p-\hbar k)b_2(t,k)\,,
\end{aligned}$$

and the functional $\langle\,\cdot\,\rangle$ *is the expectation with respect to the free product of the two Fock vectors. In terms of the master field (12.20.1) this corresponds to the mean zero Gaussian field with covariance*

$$\begin{aligned}
\langle b_t^+(k)b_{t'}(k')\rangle &= \frac{1}{1-e^{-\beta\omega_k}}\,\delta(t-t')\delta(\omega(k)+kp)\delta(k-k')\,,\\
\langle b_t(k)b_{t'}^+(k')\rangle &= \frac{1}{e^{\beta\omega_k}-1}\,\delta(t-t')\delta(\omega(k)+kp-k^2)\delta(k-k')\,.
\end{aligned}$$

Proof 12.20.1. The fields b_i of the hot free algebra arise as the stochastic limit of the standard identification of the GNS representation of a boson field algebra, associated to a Gaussian equilibrium state, with the tensor product of a Fock and an anti-Fock representation. To construct such a representation we introduce (see Sect. 2.18) two independent bosonic fields $c_1(k)$, $c_2(k)$,

$$[c_i(k), c_j^+(k')] = \delta_{ij}\delta(k-k')\,,$$

such that every $c_i(k)$ acts in the Fock representation. We then consider the operators

$$\begin{aligned}
a(k) &= \sqrt{m(k)}c_1(k) + \sqrt{m(k)-1}c_2^+(k)\,,\\
a^+(k) &= \sqrt{m(k)}c_1^+(k) + \sqrt{m(k)-1}c_2(k)\,.
\end{aligned}$$

Clearly

$$[a(k), a^+(k')] = \delta(k-k')\,,$$

and for the vacuum expectation we obtain

$$\langle a(k)a^+(k')\rangle = m(k)\delta(k-k') .$$

Taking

$$m(k) = \frac{1}{1-e^{-\beta\omega_k}} ,$$

we find the thermal state (12.17.1–2) with $q = +1$.

The stochastic limit of the rescaled operator (12.19.2) will then be

$$\begin{aligned}
&\lim_{\lambda\to 0} \frac{1}{\lambda} \exp\left\{ i\frac{t}{\lambda^2}\left[\omega(k)+kp\right]\right\} e^{-ikq} a_k \\
&= \lim_{\lambda\to 0} \frac{1}{\lambda} \exp\left\{ i\frac{t}{\lambda^2}\left[\omega(k)+kp\right]\right\} e^{-ikq}\sqrt{m(k)}c_1(k) \\
&\quad + \lim_{\lambda\to 0} \frac{1}{\lambda} \exp\left\{ i\frac{t}{\lambda^2}\left[\omega(k)+kp\right]\right\} e^{-ikq}\sqrt{m(k)-1}c_2^+(k) ,
\end{aligned}$$

where now the two limits are in the Fock representation. But, from Sects. 12.5 and 12.6, we know that such limits give rise to QED Hilbert module white noises. So it is natural to expect that the master field in the temperature case shall be the sum of two such white noises $b_1(t,k)$, $b_2^+(t,k)$. In this case the above limit is equal to

$$b(t,k) = b_1(t,k) + b_2^+(t,k) ,$$

in agreement with (12.20.1). It remains to be checked that boson independence of the fields before the limit becomes free independence of the master field after the limit, i.e. $b_1b_2^+ = b_2b_1^+ = 0$. The proof is made by computing the correlation functions using the commutation relations listed above and comparing the result with (12.18.2–4). For example, using the calculations made in Sect. 12.4 for the 2-point correlators for b and (12.20.1), we have

$$\langle b_t(k)b_\tau^+(k')\rangle = \langle b_1(t,k)b_1^+(t,k')\rangle + \langle b_2^+(t,k)b_2(t,k')\rangle = \langle b_1(t,k)b_1^+(\tau,k')\rangle .$$

Therefore

$$\langle b_1(t,k_1)b_1^+(\tau,k_2)\rangle = 2\pi\delta(t-\tau)\delta\big(\omega(k_1)+k_1p\big)\frac{\delta(k_1-k_2)}{1-e^{-\beta\omega_k}} .$$

Similarly using

$$\langle b_\tau^+(k_2)b_t(k_1)\rangle = \langle b_2(\tau,k_2)b_2^+(t,k_1)\rangle ,$$

we find that

$$\langle b_2(t,k_1)b_2^+(\tau,k_2)\rangle = 2\pi\delta(t-\tau)\delta\big(\omega(k_1)+k_1(p-k_1)\big)\frac{\delta(k_2-k_1)}{e^{\beta\omega(k_1)}-1} .$$

Moreover it is easy to see that the pairings $b\left(t_{m_h'},k_{m_h'}\right)b^+\left(t_{m_h},k_{m_h}\right)$ and $b^+\left(t_{m_h},k_{m_h}\right)b\left(t_{m_h'},k_{m_h'}\right)$ give rise to the factor

$$\delta\left(k_{m'_h} - k_{m_h}\right) c_{m_h m'_h}\left(k_{m_h}\right) 2\pi\delta\left(t_{m'_h} - t_{m_h}\right)$$
$$\cdot\,\delta\left(\omega(k_{m_h}) + k_{m_h}p - \hbar\frac{1-(-1)^{\varepsilon_h}}{2}k^2_{m_h}\right),$$

and the last relation gives the term $\hbar\sum_\alpha(-1)^{\varepsilon_\alpha}\chi_{(m_\alpha,m'_\alpha)}(m_h)$ in the phase shift.

Remark 12.20.1. We conjecture that, in analogy with the result of Skeide [Ske96] for the Fock case, in this case the structure of the interacting Hilbert module defined by Lemma 12.20.1 can be reduced to the single structure of an Hilbert module by the proper choice of the left and right multiplication. This would be the finite temperature analogue of the QED Hilbert module.

12.21 Interaction of the QEM Field with a Nonfree Particle

In the previous sections we discussed the stochastic limit of quantum electrodynamics for a free particle, i.e. a particle with only kinetic energy, minimally coupled to the electromagnetic field. Now we are going to investigate the same problem in the case of a nonzero potential. The particle is not assumed to be spacially confined, so its kinetic energy spectrum is continuous. Also in this case, as $\lambda \to 0$, the rescaled propagator $U^{(\lambda)}_{t/\lambda^2}$ is approximated by a limit quantum stochastic process $U(t)$ satisfying a quantum stochastic differential equation driven by an Hilbert module boltzmannian white noise. The new feature introduced by the potential V is that this white noise is now related to the Möller wave operator of the interaction. In order to illustrate an alternative approach to the one followed in the $V = 0$ case, we shall follow the original approach of [AcLu92] and deal with the test function rather than with the distribution approach.

The free field Hamiltonian H_R and the time-zero interaction Hamiltonian will be as in Sect. 12.1, but the particle Hamiltonian is

$$H_S := p^2 + V(q)\,, \tag{12.21.1}$$

where $V \neq 0$ is a smooth potential function decreasing at infinity more rapidly than any polynomial (in fact much less stringent assumptions are sufficient). Therefore $H_{\mathrm{I}}(t)$ has the form

$$H_{\mathrm{I}}(t) = \exp[it(H_S\otimes 1 + 1\otimes H_R)]H_{\mathrm{I}}\exp[-it(H_S\otimes 1 + 1\otimes H_R)]$$
$$= \left(\int_{\mathbb{R}^d} e^{itH_S}(-ip)e^{ik\cdot q}e^{-itH_S}\otimes\overline{(S^o_t g)}(k)\;a_k\; - \;\text{h.c.}\right), \tag{12.21.2}$$

where g is a cutoff function and $S^o_t = e^{-it\omega(p)}$ is as in Sect. 12.1. In order to obtain a more explicit form for $H_{\mathrm{I}}(t)$ we use the following two lemmata:

Lemma 12.21.1. *For any bounded function or polynomial F,*

$$[e^{itH_S}, F(p)] = -i \int_0^t ds e^{isH_S}[F(p), V(q)]e^{i(t-s)H_S} . \tag{12.21.3}$$

Proof 12.21.1. Let us define

$$X(t) := [e^{itH_S}, F(p)] \ , \qquad Y(t) := -i \int_0^t ds e^{isH_S}[F(p), V(q)]e^{i(t-s)H_S} , \tag{12.21.4}$$

then

$$X(0) = Y(0) = 0$$

and

$$\frac{d}{dt} X(t) = iX(t)H_S - ie^{itH_S}[F(p), V(q)] . \tag{12.21.5}$$

On the other hand,

$$\frac{d}{dt} Y(t) = -ie^{itH_S}[F(p), V(q)] - i^2 \int_0^t ds e^{isH_S}[F(p), V(q)]e^{i(t-s)H_S} H_S$$

$$= iY(t)H_S - ie^{itH_S}[F(p), V(q)] . \tag{12.21.6}$$

The thesis follows from the uniqueness of the solution of (12.21.5) (see [Kato66]).

Lemma 12.21.2. *For any* $t \geq 0$,

$$e^{ik\cdot q} e^{itH_S} = e^{it(H_S + p\cdot k)} e^{ik\cdot q} e^{it|k|^2} . \tag{12.21.7}$$

Proof 12.21.2. One has

$$e^{ik\cdot q} e^{itH_S} e^{-ik\cdot q} = \exp\{it([e^{ik\cdot q} p e^{-ik\cdot q}]^2 + V(q)\} = \exp\{it[(p-k)^2 + V(q)]\} . \tag{12.21.8}$$

Application of Lemma 12.21.1 to the case $F(p) = ip$ shows that

$$H_{\mathrm{I}}(t) = i\Big[\int_{\mathbb{R}^d} (-i) \left(p + \int_0^t ds e^{isH_S} V'(q) e^{-isH_S}\right) e^{itH_S} e^{ik\cdot q} e^{-itH_S} \otimes \overline{(S_t g)}(k)\, a_k \; - \; \text{h.c.}\Big] . \tag{12.21.9}$$

Applying to this Lemma 12.21.2, it follows that

$$H_{\rm I}(t) = \left[\int_{\mathbb{R}^d} \left(p + \int_0^t ds e^{isH_S} V'(q) e^{-isH_S} \right) e^{itH_S} e^{-it(H_S + p\cdot k)} e^{ik\cdot q} \right.$$
$$\left. \otimes \overline{(S_t g)}(k) \; a_k \quad - \quad \text{h.c.} \right], \tag{12.21.10}$$

where now $\{S_t\}$ is given by

$$S_t := \exp\{-it[\omega(p) - |k|^2]\} S_t^o \,. \tag{12.21.11}$$

Now we recall two results from scattering theory which shall be used throughout and which justify our assumptions on the potential.

Lemma 12.21.3. *The limit*

$$\lim_{\lambda \to 0} \int_0^{r/\lambda^2} ds e^{isH_S} V'(q) e^{-isH_S} =: \Gamma_{H_S}(V'(q)) \tag{12.21.12}$$

exists for any $r > 0$ *and* $V' \in L^1(\mathbb{R}^d) \cap L^2(\mathbb{R}^d)$ *and in general is denoted by* $\Gamma_{H_S}(V'(q))$.

Proof 12.21.3. See [Kato66].

Lemma 12.21.4. *For* $V \in L^1(\mathbb{R}^d) \cap L^2(\mathbb{R}^d)$ *and any* $r > 0$*, the limits*

$$\Omega(\pm H_S, \mp p^2) := s - \lim_{\lambda \to 0} e^{\pm irH_S/\lambda^2} e^{\mp irp^2/\lambda^2} \tag{12.21.13}$$

$$\Omega(\pm p^2, \mp H_S) := s - \lim_{\lambda \to 0} e^{\pm irp^2/\lambda^2} e^{\mp irH_S/\lambda^2} \tag{12.21.14}$$

exist and are called the Möller wave operator. Moreover, all of them are complete.

Proof 12.21.4. See [Kato66].

The intertwiner property of the scattering operator implies that

$$\Omega^*(+H_S, -p^2) e^{-it_1 H_S} = e^{-it_1 p^2} \Omega^*(+H_S, -p^2) \,. \tag{12.21.15}$$

In particular

$$H_S = \Omega(+H_S, -p^2) p^2 \Omega^*(+H_S, -p^2) \,, \tag{12.21.16}$$

and since H_S is positive self-adjoint, from (12.21.16) we can deduce the identification

$$H_S = p^2_{+,-} := \Omega(+H_S, -p^2) p \Omega^*(+H_S, -p^2) \,. \tag{12.21.17}$$

12.22 The Limit Two-Point Function

As usual we begin our investigation from the rescaled evolution equation associated with the interaction (12.21.2), i.e.

$$\partial_t U^{(\lambda)}_{t/\lambda^2} = -i\frac{1}{\lambda} H_{\mathrm{I}}(t/\lambda^2) U^{(\lambda)}_{t/\lambda^2} ,$$

whose formal iterated series solution is

$$U^{(\lambda)}_{t/\lambda^2} = \sum_{n=0}^{\infty} (-i\lambda)^n \int_0^{t/\lambda^2} dt_1 \int_0^{t_1} dt_2 \cdots \int_0^{t_{n-1}} dt_n H_{\mathrm{I}}(t_1) H_{\mathrm{I}}(t_2) \cdots H_{\mathrm{I}}(t_n) . \tag{12.22.1}$$

In the following we shall find a convenient way to exchange the order of the system and field parts in the tensor product space, i.e. we write field operators on the left and system operators on the right, so that the evolved interaction Hamiltonian $H_{\mathrm{I}}(t)$ becomes

$$\int_{\mathbb{R}^d} \overline{(S_t g)}(k)\; a_k \otimes e^{itH_S}(-ip)e^{ik\cdot q}e^{-itH_S} \quad - \text{ h.c.}$$

Thus for each $n \geq 0$, one has

$$\begin{aligned} &(-i)^n H_{\mathrm{I}}(t_1) H_{\mathrm{I}}(t_2) \cdots H_{\mathrm{I}}(t_n) \\ &= \sum_{\varepsilon\in\{0,1\}^n} \int_{\mathbb{R}^{nd}} a^{\varepsilon(1)}_{k_1} \cdots a^{\varepsilon(n)}_{k_n} (S_{t_1} g)^{\varepsilon(1)}(k_1) \cdots (S_{t_n} g)^{\varepsilon(n)}(k_n) \\ &\otimes e^{it_1 H_S}\{(-ip)e^{ik_1\cdot q}\}^{\varepsilon(1)} e^{-it_1 H_S} \ldots e^{it_n H_S}\{(-ip)e^{ik_n\cdot q}\}^{\varepsilon(n)} e^{-it_n H_S} , \end{aligned} \tag{12.22.2}$$

where as usual, for any operator b,

$$b^{\varepsilon} := \begin{cases} b, \text{ if } \varepsilon = 0 , \\ b^*, \text{ if } \varepsilon = 1 , \end{cases} \tag{12.22.3a}$$

and for any function F,

$$F^{\varepsilon} := \begin{cases} F, \text{ if } \varepsilon = 0 , \\ \bar{F}, \text{ if } \varepsilon = 1 . \end{cases} \tag{12.22.3b}$$

If $n = 1$, (12.22.2) becomes

$$-iH(t) = \int_{\mathbb{R}^d} a_k \overline{(S_t g)}(k) \otimes e^{itH_S}(-ip)e^{ik\cdot q}e^{-itH_S} \; - \text{h.c.} \tag{12.22.4}$$

In order to get an idea of how to define the collective vectors, one should consider the two-point function:

$$\lambda^2 \int_0^{t/\lambda^2} dt_1 \int_0^{s/\lambda^2} dt_2 \left\langle 0 \middle| \int_{\mathbb{R}^{2d}} dk_1 dk_2 a_{k_1} \overline{(S_{t_1} f)}(k_1) a_{k_2}^+ (S_{t_2} g)(k_2) \middle| 0 \right\rangle$$
$$\cdot\, e^{it_1 H_S} (-ip) e^{ik_1 \cdot q} e^{-it_1 H_S} \cdot e^{it_2 H_S} e^{-ik_2 \cdot q} (ip) e^{-it_2 H_S}\,, \tag{12.22.5}$$

where, $f, g \in \mathcal{S}(\mathbb{R}^d)$. By the canonical commutation relations (CCR), one can know that (12.22.5) is equal to

$$\lambda^2 \int_0^{t/\lambda^2} dt_1 \int_0^{s/\lambda^2} dt_2 \int_{\mathbb{R}^d} \overline{(S_{t_1} f)}(k) (S_{t_2} g)(k)$$
$$\cdot\, e^{it_1 H_S} (-ip) e^{ik \cdot q} e^{-it_1 H_S} \cdot e^{it_2 H_S} e^{-ik \cdot q} (ip) e^{-it_2 H_S}\,. \tag{12.22.6}$$

In the product of the operators in (12.22.6), it is convenient to move the operator $e^{ik\cdot q}$ until it is erased by meeting the operator $e^{-ik\cdot q}$. This requires the following steps: by application of Lemma 12.21.2, (12.22.6) becomes

$$\lambda^2 \int_0^{t/\lambda^2} dt_1 \int_0^{s/\lambda^2} dt_2 \int_{\mathbb{R}^d} \overline{(S_{t_1} f)}(k) (S_{t_2} g)(k)$$
$$\cdot\, e^{it_1 H_S} (-ip) e^{-it_1 (H_S + p\cdot k)} \cdot e^{it_2 (H_S + p\cdot k)} (ip) e^{-it_2 H_S} e^{i(t_2 - t_1)|k|^2}\,; \tag{12.22.7}$$

by the change of variables

$$\lambda^2 t_h = s_h, \quad h = 1, 2, \qquad \tau_1 = s_1, \quad (s_2 - s_1)/\lambda^2 = \tau_2\,, \tag{12.22.8}$$

(12.22.7) becomes

$$\int_0^t d\tau_1 \int_{-\tau_1/\lambda^2}^{(s-\tau_1)/\lambda^2} d\tau_2 \int_{\mathbb{R}^d} \overline{f}(k) (S_{\tau_2} g)(k)$$
$$\cdot\, e^{i\tau_1 H_S/\lambda^2} (-ip) e^{i\tau_2 (H_S + p\cdot k)} (ip) e^{-i\tau_1 H_S/\lambda^2} e^{-i\tau_2 H_S} e^{i\tau_2 |k|^2}\,. \tag{12.22.9}$$

Now in order to exchange $e^{i\tau_1 H_S/\lambda^2}$ with $-ip$, we apply Lemma 12.21.1 to the case $F(p) = ip$ to show that (12.22.9) is equal to

$$\int_0^t d\tau_1 \int_{-\tau_1/\lambda^2}^{(s-\tau_1)/\lambda^2} d\tau_2 \int_{\mathbb{R}^d} \overline{f}(k) (S_{\tau_2} g)(k)$$
$$\cdot \left(p - i \int_0^{\tau_1/\lambda^2} ds e^{isH_S} V'(q) e^{-isH_S}\right) e^{i\tau_1 H_S/\lambda^2} e^{i\tau_2 (H_S + p\cdot k)}$$
$$\cdot\, e^{-i(\tau_1/\lambda^2 + \tau_2) H_S} \left(p + i \int_0^{-\tau_1/\lambda^2 - \tau_2} ds e^{isH_S} V'(q) e^{-isH_S}\right) e^{i\tau_2 |k|^2}\,. \tag{12.22.10}$$

By Lemma 12.21.3 and in the notations (12.21.11–12), (12.22.10) is equal to

$$o(1)+\int_0^t d\tau_1 \int_{-\tau_1/\lambda^2}^{(s-\tau_1)/\lambda^2} d\tau_2 \int_{\mathbb{R}^d} \overline{f}(k)(S_{\tau_2}g)(k) \tag{12.22.11}$$
$$\cdot \left[p - i\Gamma_{H_S}\big(V'(q)\big)\right] e^{i\tau_1 H_S/\lambda^2} e^{i\tau_2(H_S+p\cdot k)} e^{-i(\tau_1/\lambda^2+\tau_2)H_S}$$
$$\cdot \left[p - i\Gamma_{H_S}\big(V'(q)\big)\right]^* e^{i\tau_2|k|^2} .$$

With the notations (12.21.13–14) one can rewrite (12.22.11) as

$$o(1)+\int_0^t d\tau_1 \int_{-\tau_1/\lambda^2}^{(s-\tau_1)/\lambda^2} d\tau_2 \int_{\mathbb{R}^d} \overline{f}(k)(S_{\tau_2}g)(k)$$
$$\cdot \left[p - i\Gamma_{H_S}\big(V'(q)\big)\right] \Omega(+H_S,-p^2) e^{i\tau_1 p^2/\lambda^2} e^{i\tau_2(H_S+p\cdot k)} e^{-i\tau_1 p^2/\lambda^2}$$
$$\cdot \Omega^*(+H_S,-p^2) e^{-i\tau_2 H_S} \left[p - i\Gamma_{H_S}\big(V'(q)\big)\right]^* e^{i\tau_2|k|^2} . \tag{12.22.12}$$

Now we have to investigate the operator

$$e^{i\tau_1 p^2/\lambda^2} e^{i\tau_2(H_S+p\cdot k)} e^{-i\tau_1 p^2/\lambda^2}$$
$$\cdot \exp\left[i\tau_2(p^2 + e^{i\tau_1 p^3/\lambda^2 V(q)} e^{-i\tau_, p^2/\lambda^2} + p\cdot k)\right] , \tag{12.22.13}$$

and since

$$e^{inp^2} V(q) e^{-inp^2} = V(q+np) , \tag{12.22.14}$$

we know that the right-hand side of (12.22.14) is equal to

$$\exp\left\{i\tau_2[p^2 + p\cdot k + V(q+\tau_1 p/\lambda^2)]\right\} . \tag{12.22.15}$$

Now let us define a new operator

$$Y_n(t) := \exp\left\{it[p^2 + p\cdot k + V(q+np)]\right\} e^{-it(p^2+p\cdot k)} . \tag{12.22.16}$$

Lemma 12.22.1. *If the potential function V is in the Schwartz class, one has*

$$s - \lim_{n\to\infty} Y_n(t) = 1 . \tag{12.22.17}$$

Proof 12.22.1. It is easy to verify that

$$Y_n(t) = 1 + \int_0^t dt_1 V\big(q+(n+t_1)p+t_1 k\big) + \int_0^t dt_1 \int_0^{t_1} dt_2$$
$$\cdot V\big(q+(n+t_1)p+t_1 k\big) V\big(q+(n+t_2)p+t_2 k\big) + \dots$$
$$+ \int_0^t dt_1 \int_0^{t_1} dt_2 \dots \int_0^{t_{N-1}} dt_N$$
$$\cdot V\big(q+(n+t_1)p+t_1 k\big) V\big(q+(n+t_2)p+t_2 k\big) \dots$$
$$\cdot V\big(q+(n+t_N)p+t_N k\big) + \dots \tag{12.22.18}$$

Since V is bounded, in order to prove (12.22.17), it is enough to show that for any $N \geq 1$,

$$s - \lim_{n\to\infty} V\big(q + (n+t_1)p + k_1\big) V\big(q + (n+t_2)p + k_2\big) \ldots \cdot V\big(q + (n+t_N)p + k_N\big) = 0\,. \tag{12.22.19}$$

By applying a Fourier transform, one obtains, with $k_h := t_h k$,

$$\begin{aligned} &V\big(q + (n+t_1)p + k_1\big) V\big(q + (n+t_2)p + k_2\big) ... V\big(q + (n+t_N)p + k_N\big) \\ &= \int_{\mathbb{R}^{Nd}} \hat{V}(x_1)...\hat{V}(x_N) \exp\left[in\left(\sum_{1\leq h\leq N} |x_h|^2 + \sum_{1\leq\alpha<\beta\leq N} x_\alpha \cdot x_\beta\right)\right] \\ &\cdot \exp\left[\sum_{1\leq h\leq N} (k_h \cdot x_h + t_h|x_h|^2) + \sum_{1\leq\alpha<\beta\leq N} x_\alpha \cdot x_\beta t_\alpha\right] \\ &\cdot \exp\left[p \cdot \sum_{1\leq h\leq N} x_h(n+t_h)\right] \exp\left[q \cdot \sum_{1\leq h\leq N} x_h\right]. \end{aligned} \tag{12.22.20}$$

The Riemann–Lebesgue lemma gives the thesis. Using these results we are able to rewrite (12.22.12) as

$$\begin{aligned} &o(1) + \int_0^t d\tau_1 \int_{-\tau_1/\lambda^2}^{(s-\tau_1)/\lambda^2} d\tau_2 \int_{\mathbb{R}^d} \overline{f}(k)(S_{\tau_2}g)(k) \\ &\cdot \big[p - i\Gamma_{H_S}\big(V'(q)\big)\big] \Omega(+H_S, -p^2) e^{i\tau_2(p^2+p\cdot k)} \Omega^*(+H_S, -p^2) e^{-i\tau_2 H_S} \\ &\cdot \big[p - i\Gamma_{H_S}\big(V'(q)\big)\big]^* e^{i\tau_2|k|^2}\,, \end{aligned} \tag{12.22.21}$$

which as $\lambda \to 0$, tends to

$$\begin{aligned} &\langle \chi_{[0,t)}, \chi_{[0,s)} \rangle \int_{-\infty}^{\infty} dt_1 \int_{\mathbb{R}^d} \overline{f}(k)(S_{t_1}g)(k) \\ &\cdot \big[p - i\Gamma_{H_S}\big(V'(q)\big)\big] \Omega(+H_S, -p^2) e^{it_1(p^2+p\cdot k)} \Omega^*(+H_S, -p^2) e^{-it_1 H_S} \\ &\cdot \big[p - i\Gamma_{H_S}\big(V'(q)\big)\big]^* e^{it_2|k|^2}\,. \end{aligned} \tag{12.22.22}$$

The intertwiner property (12.21.14) of the scattering operator implies that

$$\Omega^*(+H_S, -p^2) e^{-it_1 H_S} = e^{-it_1 p^2} \Omega^*(+H_S, -p^2)\,;$$

thus (12.22.22) becomes

$$\begin{aligned} &\langle \chi_{[0,t)}, \chi_{[0,s)} \rangle \int_{-\infty}^{\infty} dt_1 \int_{\mathbb{R}^d} \overline{f}(k)(S_{t_1}g)(k) \\ &\cdot \big[p - i\Gamma_{H_S}\big(V'(q)\big)\big] \Omega(+H_S, -p^2) e^{it_1 p\cdot k} \Omega^*(+H_S, -p^2) \\ &\cdot \big[p - i\Gamma_{H_S}\big(V'(q)\big)\big]^* e^{it_2|k|^2}\,. \end{aligned} \tag{12.22.23}$$

Finally, Lemma 12.21.1 shows that

$$p\Omega(+H_S,-p^2) = \Omega(+H_S,-p^2)\left[p + i\Gamma_{H_S}\left(V'(q)\right)\right]\Omega(+H_S,-p^2)\,;$$

therefore, the two-point function (12.22.5) has the following limit

$$\begin{aligned}&\left\langle \chi_{[0,t)},\chi_{[0,s)}\right\rangle \int_{-\infty}^{\infty} dt_1 \int_{\mathbb{R}^d} \overline{f}(k)(S_{t_1}g)(k)\Omega(+H_S,-p^2)pe^{it_1 p\cdot k}\\ &\cdot p\Omega^*(+H_S,-p^2)e^{it_1|k|^2}\,. \end{aligned} \tag{12.22.24}$$

12.23 The Limit Four-Point Function

In order to understand which type of quantum noise can appear in our limit, let us examine the four-point functions. Essentially, we have to consider only two four-point functions which correspond to the product of creation and annihilation operators of the following types

$$aa^+aa^+\,, \qquad\qquad aaa^+a^+\,,$$

respectively. Thus we must consider the limit of

$$\begin{aligned}\Delta(\lambda,0101) &:= \lambda^4 \int_0^{T_1/\lambda^2} dt_1 \int_0^{T_2/\lambda^2} dt_2 \int_0^{T_3/\lambda^2} dt_3 \int_0^{T_4/\lambda^2} dt_4\\ &\cdot\left\langle 0\right|\int_{\mathbb{R}^{4d}} dk_1dk_2dk_3dk_4 a_{k_1}\overline{(S_{t_1}f_1)}(k_1)a^+_{k_2}(S_{t_2}f_2)(k_2)\\ &\qquad\cdot a_{k_3}\overline{(S_{t_3}f_3)}(k_3)a^+_{k_4}(S_{t_4}f_4)(k_4)\left|0\right\rangle\\ &\cdot e^{it_1H_S}(-ip)e^{ik_1\cdot q}e^{-it_1H_S}e^{it_2H_S}e^{-ik_2\cdot q}(ip)e^{-it_2H_S}e^{it_3H_S}(-ip)e^{ik_3\cdot q}e^{-it_3H_S}\\ &\cdot e^{it_4H_S}e^{-ik_4\cdot q}(ip)e^{-it_4H_S}\end{aligned} \tag{12.23.1a}$$

and

$$\begin{aligned}\Delta(\lambda,0011) &:= \lambda^4 \int_0^{T_1/\lambda^2} dt_1 \int_0^{T_2/\lambda^2} dt_2 \int_0^{T_3/\lambda^2} dt_3 \int_0^{T_4/\lambda^2} dt_4\\ &\cdot\left\langle 0\right|\int_{\mathbb{R}^{4d}} dk_1dk_2dk_3dk_4 a_{k_1}\overline{(S_{t_1}f_1)}(k_1)a_{k_2}(S_{t_2}f_2)(k_2)\\ &\qquad\cdot a^+_{k_3}\overline{(S_{t_3}f_3)}(k_3)a^+_{k_4}(S_{t_4}f_4)(k_4)\left|0\right\rangle\\ &\cdot e^{it_1H_S}(-ip)e^{ik_1\cdot q}e^{-it_1H_S}e^{it_2H_S}(-ip)e^{ik_2\cdot q}e^{-it_2H_S}e^{it_3H_S}e^{-ik_3\cdot q}(ip)e^{-it_3H_S}\\ &\cdot e^{it_4H_S}e^{-ik_4\cdot q}(ip)e^{-it_4H_S}\,.\end{aligned} \tag{12.23.1b}$$

It follows from the CCR that

$$\Delta(\lambda, 0101) := \lambda^4 \int_0^{T_1/\lambda^2} dt_1 \int_0^{T_2/\lambda^2} dt_2 \int_0^{T_3/\lambda^2} dt_3 \int_0^{T_4/\lambda^2} dt_4$$

$$\cdot \int_{\mathbb{R}^{2d}} dk_1 dk_3 \overline{(S_{t_1} f_1)}(k_1)(S_{t_2} f_2)(k_1)\overline{(S_{t_3} f_3)}(k_3)(S_{t_4} f_4)(k_3)$$

$$\cdot\, e^{it_1 H_S}(-ip)e^{ik_1\cdot q}e^{-it_1 H_S}e^{it_2 H_S}e^{-ik_1\cdot q}(ip)e^{-it_2 H_S}e^{it_3 H_S}(-ip)e^{ik_3\cdot q}e^{-it_3 H_S}$$

$$\cdot\, e^{it_4 H_S}e^{-ik_3\cdot q}(ip)e^{-it_4 H_S}\,, \tag{12.23.2}$$

which is practically a product of two objects both similar to (12.22.5). Therefore the same argument used to prove (12.22.5) gives the following result:

Lemma 12.23.1. *If the potential function V is in the Schwartz class, one has*

$$\lim_{\lambda\to 0} \Delta(\lambda, 0101) = \left\langle \chi_{[0,T_1)}, \chi_{[0,T_2)}\right\rangle \left\langle \chi_{[0,T_3)}, \chi_{[0,T_4)}\right\rangle \int_{-\infty}^{\infty} dt \int_{-\infty}^{\infty} ds$$

$$\cdot \int_{\mathbb{R}^{2d}} \overline{f_1}(k_1)(S_t f_2)(k_1)\overline{f_3}(k_2)(S_s f_4)(k_2)$$

$$\cdot\, \Omega(+H_S, -p^2)p^2 e^{itp\cdot k_1}e^{isp\cdot k_2}p^2\Omega^*(+H_S, -p^2)e^{it|k_1|^2}e^{is|k_2|^2}\,. \tag{12.23.3}$$

Proof 12.23.1. In the investigation of the two-point function, the operator p behaves like a bounded operator if one takes on the system space the partial expectation of (12.23.2) by a linear function of the form $\langle \xi, \,\cdot\, \eta\rangle$ with ξ, η in the Schwartz class. Because of this fact, one can repeat in this case the proof for the two-point function (12.22.5) and this leads to the thesis.

Now we are going to investigate the limit of the term $\Delta(\lambda, 0101)$; in the following discussion we shall always assume, without any further explicit mention of this fact, that the potential function is in the Schwartz class.

Lemma 12.23.2. *The limit, as $\lambda \to 0$, of the term $\Delta(\lambda, 0011)$, i.e. (12.23.1b), is equal to*

$$\left\langle \chi_{[0,T_1)}, \chi_{[0,T_4)}\right\rangle \left\langle \chi_{[0,T_2)}, \chi_{[0,T_3)}\right\rangle \int_{-\infty}^{\infty} dt \int_{-\infty}^{\infty} ds$$

$$\cdot \int_{\mathbb{R}^{2d}} \overline{f_1}(k_1)(S_t f_4)(k_1)\overline{f_2}(k_2)(S_s f_3)(k_2)$$

$$\cdot\, \Omega(+H_S, -p^2)p(p+k_1)e^{itp\cdot k_1}e^{isp\cdot k_2}(p+k_1)p$$

$$\cdot\, \Omega^*(+H_S, -p^2)e^{it|k_1|^2}e^{is(|k_2|^2+k_1\cdot k_2)} \tag{12.23.4}$$

Proof 12.23.2. By the CCR, one has the following expression:

$$\Delta(\lambda, 0011) := \lambda^4 \int_0^{T_1/\lambda^2} dt_1 \int_0^{T_2/\lambda^2} dt_2 \int_0^{T_3/\lambda^2} dt_3 \int_0^{T_4/\lambda^2} dt_4$$

$$\cdot \int_{\mathbb{R}^{2d}} dk_1 dk_2 \overline{(S_{t_1} f_1)}(k_1) \overline{(S_{t_2} f_2)}(k_2)(S_{t_3} f_3)(k_2)(S_{t_4} f_4)(k_1)$$

$$\cdot\ e^{it_1 H_S}(-ip)e^{ik_1\cdot q}e^{-it_1 H_S} \cdot e^{it_2 H_S}(-ip)e^{ik_2\cdot q}e^{-it_2 H_S}e^{it_3 H_S}e^{-ik_2\cdot q}(ip)e^{-it_3 H_S}$$

$$\cdot\ e^{it_4 H_S}e^{-ik_1\cdot q}(ip)e^{-it_4 H_S}\,. \tag{12.23.5}$$

Now it follows from the Weyl commutation relations that the product of the operators in (12.23.5) becomes

$$\exp(it_1 H_S)(-ip)\exp(-it_1(H_S + p\cdot k_1))\exp(it_2(H_S + p\cdot k_1))\left(-i(p+k_1)\right)$$

$$\cdot \exp(-it_2[H_S + p\cdot(k_1+k_2)])\exp(it_3[H_S + p\cdot(k_1+k_2)])\left(i(p+k_1)\right)$$

$$\cdot \exp(-it_3(H_S + p\cdot k_1))\exp(it_4(H_S + p\cdot k_1))(ip)\exp(-it_4 H_S)$$

$$\cdot \exp\left(i(t_4 - t_1)|k_1|^2\right)\exp\left(i(t_3 - t_2)(|k_2|^2 + k_1\cdot k_2)\right)\,. \tag{12.23.6}$$

The thesis follows by the same technique used to find the limit of the two-point function.

12.24 The Limit Hilbert Module

Investigation of the limit behaviour of the two- and four-point functions suggests that the limit evolution should be related to the Möller wave operator and that it should live on some Hilbert module. In this section we shall investigate the structure of this limit Hilbert module and prove that it is more general than the one used in the case with $V = 0$, which is defined by the commutation relations (12.1.8) and (12.1.10). In particular it is a Hilbert module over the $*$-algebra (not necessarily a C^*-algebra) which is the image, by the automorphism implemented by the wave operator, of the algebra of all the polynomials in the momentum operator p multiplied by a trigonometric exponential of the same operator. More precisely, define

$$\mathcal{P}_\Omega := \{\Omega(H_S, -p^2)P_n(p)e^{itp\cdot k}\Omega^*(H_S, -p^2)\}\,,\ n\in\mathbb{N},\ k\in\mathbb{R}^d,$$

$$P_n\ \{\text{is a polynomial of the } n\text{th degree}\,,\} \tag{12.24.1}$$

and it is clear that $\mathcal{P}_\Omega$ is a $*$-algebra admitting a dense subset of $L^2(\mathbb{R}^d)$ as the invariant domain. For each $f\in\mathcal{S}(\mathbb{R}^n)$, define

$$\tilde{f}_{P_n}(t) := \int_{\mathbb{R}^d} dk\,\Omega(H_S, -p^2)P_n(p)e^{itp\cdot k}\Omega^*(H_S, -p^2)(S_t f)(k) \tag{12.24.2}$$

and

$$\mathcal{F} := \left\{ \tilde{f}_a;\ f \in \mathcal{S}(\mathbb{R}^n),\ a \text{ is a polynomial in } p \right\}. \tag{12.24.3}$$

It is easy to show that $\mathcal{F}$ is a two-sided $\mathcal{P}_\Omega$ module. On the $\mathcal{P}_\Omega$ module, $\mathcal{F}$ introduce the $\mathcal{P}_\Omega$-valued inner product $(\cdot|\cdot)$:

$$(\tilde{f}_a|\tilde{g}_b) := \int_{\mathbb{R}^d} \int_{\mathbb{R}} dt\ \Omega(H_S, -p^2) a^*(p) b(p) e^{itp\cdot k} \Omega^*(H_S, -p^2) \bar{f}(k)(S_t g)(k). \tag{12.24.4}$$

It is easy to show that $\left(\mathcal{F}/(\cdot|\cdot), (\cdot|\cdot)\right)$ is a (pre)Hilbert module, and in the following we shall denote the associated Hilbert module by $\mathcal{F}$. Starting from this Hilbert module, let us consider the algebraic tensor product$L^2(\mathbb{R}_+) \odot \mathcal{F}$ on which we introduce the inner product

$$\left(\alpha \odot \tilde{f}_a | \beta \odot \tilde{g}_b\right) := \langle \alpha,\ \beta \rangle_{L^2(\mathbb{R}_+)} \cdot (\tilde{f}_a|\tilde{g}_b). \tag{12.24.5}$$

Thus we obtain a new Hilbert module. This new Hilbert module is the basic object with which we can define our limit quantum stochastic process. For each given $n \in \mathbb{N}$, on the algebraic tensor product $(L^2(\mathbb{R}_+) \odot \mathcal{F})^{\odot n}$, we introduce the inner product:

$$\begin{aligned}
&\left(\alpha_1 \odot \tilde{f}_{1,a_1} \odot \cdots \odot \alpha_n \odot \tilde{f}_{n,a_n} \middle| \beta_1 \odot \tilde{g}_{1,b_1} \odot \cdots \odot \beta_n \odot \tilde{g}_{n,b_n}\right) \\
&:= \prod_{h=1}^{n} \langle \alpha_h,\ \beta_h \rangle_{L^2(\mathbb{R}_+)} \cdot \int_{\mathbb{R}^n} du_1 \cdots du_n \int_{\mathbb{R}^{nd}} dk_1 \cdots dk_n \\
&\quad \cdot \exp\left(i \sum_{1 \le r \le h \le n-1} u_r k_r k_{h+1} \right) \Omega(H_S, -p^2) \prod_{h=1}^{n} \Big[a_h^*(p + k_1 + \cdots + k_{h-1}) \\
&\quad \cdot b_h(p + k_1 + \cdots + k_{h-1}) e^{-iu_h k_h \cdot p} \bar{f}_h(k_h)(S_{u_h} g_h)(k_h) \Big] \Omega^*(H_S, -p^2)^*,
\end{aligned} \tag{12.24.6}$$

where the equivalence classes are defined in a similar way as in the case $n = 1$ and by definition a sum extended to the set of indices $1 \le r \le h \le n-1$, with $n = 1$, is made equal to zero. Thus for each $n \in \mathbb{N}$, with the $\mathcal{P}_\Omega$-valued, $\mathcal{P}_\Omega$-right, sesquilinear form given by (12.24.6), $\left(L^2(\mathbb{R}) \odot \mathcal{F}\right)^{\odot n}$ becomes a right $\mathcal{P}_\Omega$-pre-Hilbert module and $\left(L^2(\mathbb{R}) \odot \mathcal{F}\right)^{\otimes n}$ will be used to denote it.

Since for each $n \in \mathbb{N}$, $\left(L^2(\mathbb{R}) \odot \mathcal{F}\right)^{\otimes n}$ is a $\mathcal{P}_\Omega$-pre-Hilbert module, the direct sum $\mathbb{C} \oplus \bigoplus_{n=1}^{\infty} \left(L^2(\mathbb{R}) \odot \mathcal{F}\right)^{\otimes n}$ makes sense and will be denoted by $\Gamma\left(L^2(\mathbb{R}_+) \odot \mathcal{F}\right)$ and called the $\mathcal{P}_\Omega$-Fock module over $L^2(\mathbb{R}_+) \odot \mathcal{F}$. In this pre-Hilbert module, the vector $\Psi := 1 \oplus 0 \oplus 0 \cdots$ is called the *vacuum vector*. One can easily show that

Lemma 12.24.1. *The number vector subset*

$$\Gamma := \Big\{ \left(\alpha_1 \odot \tilde{f}_1\right) \otimes \cdots \otimes \left(\alpha_n \odot \tilde{f}_n\right) \Psi ;\ n \in \mathbb{N},\ \alpha_j \in L^2(\mathbb{R}_+), \\ \tilde{f}_j \in \mathcal{F},\ j = 1 \dots n \Big\} \tag{12.24.7}$$

is a $\mathcal{P}_\Omega$*-total subset of* $\Gamma\big(L^2(\mathbb{R}_+) \odot \mathcal{F}\big)$.

Definition 12.24.1. *For each element of* $L^2(\mathbb{R}_+) \odot \mathcal{F}$*, the creator with respect to this element, denoted by* $A^+(\cdot)$*, is defined on the* $\mathcal{P}_\Omega$*-right linear span of* Γ *by* $\mathcal{P}_\Omega$*-right linearity and*

$$\begin{aligned} & A^+\left(\alpha \odot \tilde{f}\right) \left[\left(\alpha_1 \odot \tilde{f}_1\right) \otimes \cdots \otimes \left(\alpha_n \odot \tilde{f}_n\right) \Psi \right] \\ & := \left(\alpha \odot \tilde{f}\right) \otimes \left(\alpha_1 \odot \tilde{f}_1\right) \otimes \cdots \otimes \left(\alpha_n \odot \tilde{f}_n\right) \Psi , \end{aligned} \tag{12.24.8}$$

where $n \in \mathbb{N},\ \alpha,\ \alpha_j \in L^2(\mathbb{R}_+),\ \tilde{f},\ \tilde{f}_j \in \mathcal{F},\ j = 1 \dots n$.

The annihilation operator $A(\cdot)$ *is defined as the adjoint of the creation operator on the domain of number vectors.*

Remark 12.24.1. In general, $A^+\left(\alpha_1 \odot \tilde{f}_1\right) A^+\left(\alpha_2 \odot \tilde{f}_2\right)$ is not equal to $A^+\left(\alpha_2 \odot \tilde{f}_2\right) A^+\left(\alpha_1 \odot \tilde{f}_1\right)$.

The above discussion suggests we define

$$A_\lambda^+(S,T,f,a) := \lambda \int_{S/\lambda^2}^{T/\lambda^2} dt \int_{\mathbb{R}^d} dk e^{-itH_s} e^{ik\cdot p} a(p) e^{itH_s} \otimes (S_t f)(k) a_k^+ , \tag{12.24.9}$$

where $0 \le S \le T < +\infty$, $f \in \mathcal{S}(\mathbb{R}^d)$ and $a(p)$ is a polynomial in p.

The same arguments as used in the proof of Lemma 12.23.2 give the following:

Theorem 12.24.1. *For any* $N, N' \in \mathbb{N}$ *the limit of*

$$\left\langle \prod_{h=1}^{N} A_\lambda^+(S_h, T_h, f_h, a_h)\Phi, \prod_{h=1}^{N'} A_\lambda^+(S'_h, T'_h, f'_h, a'_h)\Phi \right\rangle \tag{12.24.10}$$

exists and is equal to

$$\left\langle \prod_{h=1}^{N} A^+\left(\chi_{[S_h,T_h]} \odot \tilde{f}_{h,a_h}\right)\Psi, \prod_{h=1}^{N'} A^+\left(\chi_{[S'_h,T'_h]} \odot \tilde{f}'_{h,a'_h}\right)\Psi \right\rangle . \tag{12.24.11}$$

By the same techniques used in Sects. 12.5 and 12.6, one can prove that the quantum noise is of free type in the sense that only the noncrossing diagrams contribute to its correlation functions. Therefore our Hilbert module is a deformation of the free (or Boltzmannian) Hilbert module discussed in Sect. 12.2.

12.25 The Limit Stochastic Process

Now we compute the vacuum statistics of the creation and annihilation fields on the limit Hilbert module, i.e. the expectation values

$$\left\langle \Psi, A^{\varepsilon(1)}(\alpha_1 \odot f_1) \cdots A^{\varepsilon(2n)}(\alpha_{2n} \odot f_{2n})\Psi \right\rangle , \tag{12.25.1}$$

where $n \in \mathbb{N}$, $\alpha_j \in L^2(\mathbb{R}_+), f_j \in \mathcal{F}$ $(j = 1 \dots 2n)$, $\varepsilon \in \{0,1\}^{2n}$ and

$$A^0 := A \quad , \quad A^1 := A^+$$

and where for simplicity we shall not distinguish f from $\tilde{f}$.

Moreover, it is necessary to know (12.25.1) only in the case of

$$\varepsilon(1) = 0 \quad , \quad \varepsilon(2n) = 1 . \tag{12.25.2}$$

Lemma 12.25.1. *The expectation value (12.25.2) is not equal to zero only if*

$$\sum_{h=1}^{2n} \varepsilon(h) = n . \tag{12.25.3}$$

Proof 12.25.1. This is a standard fact on free Fock modules.

Let there be given a subset $\{m_h\}_{h=1}^n \subset \{1 \dots 2n\}$ with $1 < m_1 < \cdots < m_n = 2n$. Recall from Sect. 12.4, that $\{m_h\}_{h=1}^n$ is said to admit a noncrossing pair partition if, denoting $\{m'_h\}_{h=1}^n$ an ordered version of the set $\{1 \dots 2n\} \setminus \{m_h\}_{h=1}^n$, the family $\{m'_h, m_h\}_{h=1}^n$ is a noncrossing pair partition of $\{1 \dots 2n\}$. By Sect. 12.4, if this is the case, then the choice of $\{m'_h\}$ is unique. For a given $\varepsilon \in \{0,1\}^{2n}$, denote

$$\{m_h\}_{h=1}^n := \{r \in \{1 \dots 2n\};\ \varepsilon(r) = 1\}, \qquad 1 < m_1 < \cdots < m_n = 2n . \tag{12.25.4}$$

We shall say that $\varepsilon \in \{0,1\}^{2n}$ admits a noncrossing pair partition of $\{1 \dots 2n\}$ if $\{m_h\}_{h=1}^n$ does.

Lemma 12.25.2. *The inner product (12.25.2) is equal to zero if ε does not admit a noncrossing pair partition of $\{1 \dots 2n\}$.*

Now we are led to consider (12.25.2) when ε admits a (unique) noncrossing pair partition of $\{1 \dots 2n\}$. We shall consider an expression more general than (12.25.2), namely

$$\left\langle \Psi, A^{\varepsilon(1)}(\alpha_1 \odot f_1) \cdots R(p)e^{ix \cdot p} \cdots A^{\varepsilon(2n)}(\alpha_{2n} \odot f_{2n})\Psi \right\rangle , \tag{12.25.5}$$

where $R(p)$ is a polymonial.

Since on the limit Hilbert module the left action is not defined, the most natural way to give a meaning to the quantity (12.25.5) is to define it as the limit of

$$\Big\langle \Phi, A_\lambda^{\varepsilon(1)}(S_1,T_1,f_1,a_1)\cdots A_\lambda^{\varepsilon(r)}(S_r,T_r,f_r,a_r)R(p)e^{ix\cdot p} \cdot A_\lambda^{\varepsilon(r+1)}(S_{r+1},T_{r+1},f_{r+1},a_{r+1})\cdots A_\lambda^{\varepsilon(2n)}(S_{2n},T_{2n},f_{2n},a_{2n})\Phi\Big\rangle . \tag{12.25.6}$$

Theorem 12.25.1. *The limit of (12.25.6) exists and is equal to zero if ε does not admit a noncrossing pair partition of $\{1\dots 2n\}$; however, it is equal to*

$$\prod_{h=1}^n \Big\langle \chi_{[S_{m_h},T_{m_h}]},\chi_{[S_{m'_h},T_{m'_h}]}\Big\rangle_{L^2(\mathbb{R}_+)} \cdot \int_{-\infty}^{+\infty} du_1\dots\int_{-\infty}^{+\infty} du_n \int_{\mathbb{R}^{nd}} dk_1\dots dk_n \prod_{h=1}^n (S_{u_h}f_{m_h})(k_h)\overline{f}_{m'_h}(k_h)$$
$$\cdot\Big\langle \xi, \prod_{h=1}^n\Big[e^{iu_hk_h\cdot p}a_{2h}\Big(p+\sum_{l=1}^n k_l\chi_{(m'_l,m_l)}(2h)\Big) \cdot a_{2h+1}\Big(p+\sum_{l=1}^n k_l\chi_{(m'_l,m_l)}(2h+1)\Big)\Big]R\Big(p+\sum_{l=1}^n k_l\chi_{[m'_l,m_l]}(r)\Big)e^{ix\cdot p}\eta\Big\rangle$$
$$\cdot\exp\Big[ix\cdot\sum_{l=1}^n k_l\chi_{[m'_l,m_l]}(r)\Big]\exp\Big(i\sum_{h=1}^{n-1}\sum_{r=h+1}^n u_hk_h\cdot k_r\chi_{(m'_r,m_r)}(m_h)\Big) \tag{12.25.7}$$

if ε admits a (unique) noncrossing pair partition $\{m'_h,m_h\}_{h=1}^n$.

Proof 12.25.2. The proof can be performed by putting together the argument in Sect. 12.2 and the technique used in Sect. 12.5. An interesting open problem is to determine the vacuum distribution of the classical random variable $A(\alpha\odot f_a)+A^+(\alpha\odot f_a)$ in the limit module.

12.26 The Stochastic Differential Equation

By the same technique, we are able to prove our main result, i.e. the convergence of the matrix elements, in the collective number vectors, of the wave operator at time t, $U^{(\lambda)}_{t/\lambda^2}$, to a unitary operator $U(t)$ on the Fock module described in Sect. 12.25, satisfying the stochastic differential equation

$$U(t)=1+\int_0^t\Big(dA_s^+(\tilde g_{ip})-dA_s(\tilde g_{ip})-(\tilde g_{ip}|\tilde g_{ip})_-ds\Big)U(s) \tag{12.26.1}$$

driven by the new quantum noise described in the previous section and where the quantum stochastic integrals are defined on the full $\mathcal{P}_\Omega$-Fock module described in Sect. 12.25 following the theory developed in [Lu92a], [Lu94], [Lu95]. In (12.26.1), the half-inner product $(\cdot|\cdot)_-$ is defined as in (12.7.4), and for any polynomial $g_p = g(p)$, in $\mathcal{P}_\Omega$, $\tilde{g}_{ip}$ denotes the polynomial obtained from it by replacing p by ip. In fact the estimates in Part III guarantee that in order to consider the limit

$$\left\langle \prod_{h=1}^{N} A_\lambda^+(S_h, T_h, f_h, a_h)\Phi, U_{t/\lambda^2}^{(\lambda)} \prod_{h=1}^{N'} A_\lambda^+(S'_h, T'_h, f'_h, a'_h)\Phi \right\rangle \tag{12.26.2}$$

one can replace the operator $U_{t/\lambda^2}^{(\lambda)}$ by its series expansion and take the limit (as $\lambda \to 0$) term by term. Since any term has the form (12.25.6), the same techniques as in Theorem 12.25.1 give the main result.

12.27 Notes

The results described in this chapter were obtained in [AcLu92], [AcArVo95], [AcLuVo97], [AcKoVo98], [AcKoVo98a] and [AcKoVo99].

Section (12.1)

(1) The term *entanglement* is often used to denote linear combinations (superpositions) of product vectors in a tensor product of Hilbert spaces, representing the state of a composite system. Since the term *superpositions* is already well established for these states, we prefer to use *entanglement* to denote this deeper relationship which is independent of any particular state and already emerges at a kinematical level.

Section (12.15)

The exponential decay of time correlations is a universal hallmark of unstable systems. The Fermi golden rule and the Weisskopf–Wigner approximation lead to the exponential decay law. However this law is not universal, and it is model dependent. Fock and Krylov [FoKr47], and later Khalfin [Kha57], claimed that the exponential decay law is not consistent with the general principles of quantum mechanics. Deviations from the exponential behaviour have been analysed recently in [FaPa99]. Experimental evidence for nonexponential decay in quantum tunneling is reported in [WiBh97]. The nonexponential decay for the polaron model in the stochastic limit has been derived in [AcKoVo99]. An exact formula ("ABC-formula") for matrix elements of the evolution operator, which includes in particular corrections to the stochastic limit, was obtained in [ArVo00].

Section (12.16)

(1) It is interesting to note that when the module structure is neglected the algebra we find is precisely the algebra that was found in the singleton independence central limit theorem on free groups of [Aho98a, Aho98b]. So the hot free algebra arises in a natural way not only in physics but also in mathematics.

13. The Anderson Model

As explained in Part I, the stochastic limit technique does not apply only to the open system scheme, describing a *discrete system interacting with a continuous one*, but also to single continuous systems or to two (or more) mutually interacting continuous systems. The prototype of such systems is the *Anderson model*, proposed in [And58] to explain the finite conductivity of metals. It describes a system of fermions interacting with a classical Gaussian random field, modelling the impurities of the metal. If the classical field is discrete, we are in the framework described in Part I; even degree interaction bosonization takes place as described in Sect. 11.10.

In the present chapter we consider the Anderson model, describing a non-relativistic system of fermions, in dimension $d \geq 3$, interacting with an external potential(1). In Sect. 13.2 we show that the interacting Hamiltonian becomes, in the limit, a quantum stochastic variable distributed according to a nonlinear deformation of the Wigner law.

More generally we prove that only the noncrossing diagrams in the perturbative expansion for the (connected) transition amplitude contribute to the stochastic limit. This result can be physically interpreted by saying that the noncrossing diagrams are the dominant ones when the time T is very long compared with the coupling λ (this is just the physically interesting regime). The reason why, in this limit, only the noncrossing diagrams survive is explained by Theorem 13.2.1. This theorem contains all the analytic information needed to deduce the main result of this chapter, i.e. (13.4.2). Just as in the case of quantum electrodynamics (QED) without dipole approximation, even if the limit diagrams of the correlation functions are only the noncrossing ones, the corresponding probability distribution for the time-averaged interacting Hamiltonian is not the Wigner semicircle law but a nonlinear deformation of it. Despite this nonlinearity, some properties of the free random variables still survive, and this allows us to obtain, using the combinatorial argument described in Sect. 13.6, the main result of this chapter, i.e. (13.4.2).

This is a remarkable closed nonlinear integral equation for the transition probability amplitude which, even if similar to the Schwinger–Dyson equation obtained in [Mig58] and [AGD75] for the two-point correlation function by resummation of the noncrossing Feynman graphs and to the equation for the 1-particle Green function in the $N \to \infty$ limit of the Anderson model, was not previously known in the literature. In Sect. 13.6 we find an explicit solution of this nonlinear equation.

13.1 Nonrelativistic Fermions in External Potential: The Anderson Model

The Hamiltonian of our model is

$$H = H_0 + H_{\mathrm{I}} = \int_{\Lambda} dx \tilde{\psi}^+(x) \left(\frac{\partial_x^2}{2m} - \mu \right) \tilde{\psi}^-(x) + \lambda \int_{\Lambda} dx \phi(x) \tilde{\psi}^+(x) \tilde{\psi}^-(x) , \tag{13.1.1}$$

where λ is a real number (strength of the interaction), $\Lambda \subset \mathbb{R}^d$ is a square box of side L, $\mu = p_{\mathrm{F}}^2/2m$ is the *chemical potential*, p_{F} is the *Fermi momentum*, m is the fermion mass, and $\tilde{\psi}^\varepsilon(x)$, $\epsilon = \pm 1$ is the *fermionic field* with periodic boundary condition

$$\tilde{\psi}^\varepsilon(x) = \frac{1}{L^{d/2}} \sum_k e^{i\epsilon kx} a_k^\epsilon , \tag{13.1.2}$$

where $k = \frac{2n\pi}{L}$, $n = (n_1 \dots n_d) \in Z^d$, $\varepsilon \in \{1, -1\}$,

$$\{a_k^\epsilon, a_{k'}^{-\epsilon'}\} = a_k^\epsilon a_{k'}^{-\epsilon'} + a_{k'}^{-\epsilon'} a_k^\epsilon = \delta_{\epsilon,\epsilon'} \delta_{k,k'}$$

and where for any operator X we use the notation

$$X^\varepsilon = \begin{cases} X & \text{if } \varepsilon = -1 , \\ X^+ & \text{if } \varepsilon = +1 . \end{cases}$$

The function $\phi(x)$ is an *external field*, modelled as a δ-correlated classical Gaussian random field ϕ on $\mathbb{R}^d$, with $d \geq 3$, modelling the impurities of the metal. In what follows, it will be convenient to expand the random variables $\phi(x)$ in a Fourier series,

$$\phi(x) = \frac{1}{L^d} \sum_p \phi_p e^{-ipx} , \tag{13.1.3}$$

leading to a discrete family of (complex-valued) Gaussian random variables ϕ_p with $p \in \frac{2n\pi}{L} Z^d$ and

$$E(\phi_k \phi_{k'}) = L^d \delta_{k,-k'} , \quad \phi_k^* = \phi_{-k} . \tag{13.1.4}$$

In the notations introduced above, the free evolution is characterized by the following property:

$$\tilde{\psi}^{\varepsilon}(x,t) = e^{iH_0 t}\tilde{\psi}^{\varepsilon}(x)e^{-iH_0 t} = \frac{1}{\sqrt{L^d}}\sum_k \exp\left\{i\varepsilon\left[kx + \left(\frac{|k|^2}{2m} - \mu\right)t\right]\right\} a_k^{\epsilon}$$
$$= \frac{1}{\sqrt{L^d}}\sum_k e^{i\varepsilon(kx+\omega_k t)} a_k^{\epsilon}\,, \qquad (13.1.5)$$

where $\omega_k = \left(\frac{|k|^2}{2m} - \mu\right)$. The Hamiltonian in the interaction representation is

$$\lambda H_{\mathrm{I}}(t) = \lambda e^{iH_0 t} H_{\mathrm{I}} e^{-iH_0 t} = \lambda \int_{\Lambda} dx \phi(x) \tilde{\psi}^{+}(x,t)\tilde{\psi}^{-}(x,t)\,. \qquad (13.1.6)$$

It is convenient to regularize this Hamiltonian by introducing a cutoff and normal order:

$$\lambda H_{\mathrm{I}}(t) = \lambda \int_{\Lambda} dx \phi(x) :\psi^{+}(x,t)\psi^{-}(x,t): \; + \text{h.c.}\,, \qquad (13.1.7)$$

where $\psi^{\epsilon}(x,t)$ is the regularized version of $\tilde{\psi}^{\epsilon}(x,t)$ given by

$$\psi^{\epsilon}(x,t) = \frac{1}{\sqrt{L^d}}\sum_k e^{i\varepsilon(kx+\omega_k t)} g_k a_k^{\epsilon}\,. \qquad (13.1.8)$$

g_k is a complex-valued cutoff function to be specified in the following, and $:\,\cdot\,:$ denotes normal order,

$$:a_{k+p}^{+} a_k: \, = a_{k+p}^{+} a_k - \left\langle \Phi_{\mathrm{F}}, a_{k+p,\sigma}^{+} a_{k,\sigma} \Phi_{\mathrm{F}} \right\rangle\,, \qquad (13.1.9)$$

with respect to the *ground state* of H_0,

$$\Phi_{\mathrm{F}} = \prod_{|k|\le p_{\mathrm{F}}} a_k^{+} \mid 0\rangle\,. \qquad (13.1.10)$$

The *evolution* (*wave*) *operator* at time T is defined in the usual way:

$$U_T = U_T^{(\lambda,L)} = 1 + \sum_{n=1}^{\infty}(-i)^n \lambda^n \int_0^T dt_1 \ldots \int_0^{t_{n-1}} dt_n H_{\mathrm{I}}(t_1)\ldots H_{\mathrm{I}}(t_n)\,, \qquad (13.1.11)$$

where for each finite L and T the series converges normally with respect to the fermions and as L^2 with respect to the Gaussian field.

The first excited state $\psi_k^{\epsilon}\Phi_{\mathrm{F}}$ of the free Hamiltonian can be obtained by adding or subtracting a particle to the ground state of the free Hamiltonian. The corresponding interacting state is obtained by imagining that the interaction is switched on at time $-\infty$ so that it is given by $U_{-\infty}\psi_k^{\epsilon}\Phi_{\mathrm{F}}$, where U_{-T} is given by (13.1.11) with $H_{\mathrm{I}}(-t)$ replacing $H_{\mathrm{I}}(t)$.

Remark 13.1.1. If the covariance of the ϕ field were a Schwarz function rather than a δ-function, then the discrete approximation would not be necessary and one could directly consider the mean zero Gaussian state on the canonical anti-commutation relations (CAR) algebra defined by the operator

$$Q = 2\chi - 1 ,$$

where χ is the characteristic function of the sphere centered on the origin and with a radius given by the Fermi momentum p_{F}. However in the present case the quadratic form

$$H_{\mathrm{I}} = \int dk \int dk' \overline{g}(k) g(k') \phi_{k-k'} a^+_{k'} a_k$$

cannot define an operator. In fact this would lead to

$$\langle H_{\mathrm{I}} H_{\mathrm{I}} \rangle = \int dk_1 \int dk_2 \int dk_3 \int dk_4 E(\phi_{k_1-k_2} \phi_{-(k_3-k_4)}) \cdot \langle a^+_{k_1} a_{k_2} a^+_{k_4} a_{k_3} \rangle .$$

In the expansion of this operator there is the nonconnected diagram

$$\int dk_1 \int dk_2 \int dk_3 \int dk_4 E(\phi_{k_1-k_2} \phi_{-(k_3-k_4)}) \langle a^+_{k_1} a_{k_2} \rangle \langle a^+_{k_4} a_{k_3} \rangle$$

$$= \int\int\int\int dk_1 dk_2 dk_3 dk_4 \delta(k_1 - k_2 + k_3 - k_4) \delta(k_1 - k_2) \delta(k_3 - k_4) ,$$

which is not defined because of the presence of $\delta(0)$. For this reason one restricts one's attention to the connected diagrams.

The above remark leads one to study the projection of the interacting state onto the free one from

$$\lim_{L,T\to\infty} E\left(\frac{\langle \Phi_{\mathrm{F}}, \psi_k^{-\epsilon} U_{-T} \psi_k^{\epsilon} \Phi_{\mathrm{F}} \rangle}{\langle \Phi_{\mathrm{F}}, U_{-T}, \Phi_{\mathrm{F}} \rangle} \right) \tag{13.1.12}$$

and from the known identity [AGD75]

$$\frac{\langle \Phi_{\mathrm{F}}, \psi_k^{-\epsilon} U_{-T} \psi_k^{\epsilon} \Phi_{\mathrm{F}} \rangle}{\langle \Phi_{\mathrm{F}}, U_{-T}, \Phi_{\mathrm{F}} \rangle} = \langle \Phi_{\mathrm{F}}, \psi_k^{\epsilon} U_{-T} \psi_k^{-\epsilon} \Phi_{\mathrm{F}} \rangle_{\mathrm{conn}} ,$$

where $\langle \cdot \rangle_{\mathrm{conn}}$ denotes the expectation with respect to the connected diagrams and $E(\cdot)$ denotes expectation with respect to the classical random field (13.1.3). The determination of the limit (13.1.12) is the first and most difficult step towards the determination of the 2-point correlation function (*Green function*),

$$\lim_{L,T\to\infty} E\left(\frac{\langle \Phi_{\mathrm{F}}, U_{-T} \psi_k^{-\epsilon} \psi_k^{\epsilon} U_T \Phi_{\mathrm{F}} \rangle}{\langle \Phi_{\mathrm{F}}, U_{-T} U_T \Phi_{\mathrm{F}} \rangle} \right) , \tag{13.1.13}$$

from which all the physical properties of solid-state models such as the above can be obtained[(1)].

The right insight about how the stochastic limit has to be performed in order to avoid trivialities is given by the second-order term in the expansion (13.1.11). One finds in fact that the only possibility that the limit of this term exists and is not trivial is to take the limits in the following way:

$$L \to \infty\ ,\ \lambda \to 0\ ,\ T \to \infty\ , \lambda^2 T \to \ \text{const.}$$

The above limit is equivalent to the *rescaling* $T \to T/\lambda^2$ followed by the limits

$$L \to \infty\ ,\ \lambda \to 0\ . \tag{13.1.14}$$

The limits must be taken exactly in order from left to right in (13.1.14) otherwise no limit exists. In conclusion, our goal is study the limit

$$\lim_{\lambda\to 0}\lim_{L\to\infty} E\left(\langle \Phi_{\mathrm{F}}, \psi(x) U_{T/\lambda^2} \psi^+(y) \Phi_{\mathrm{F}}\rangle_{\mathrm{conn}}\right) . \tag{13.1.15}$$

13.2 The Limit of the Connected Correlators

The first step of the stochastic limit approach is suggested by the first-order term of the iterated series (13.1.11) after the rescaling $T \to T/\lambda^2$. This is equal to the *time-averaged* interacting Hamiltonian (with $S = 0$):

$$B_\lambda(T,S) := \lambda \int_{S/\lambda^2}^{T/\lambda^2} dt\ \frac{1}{L^d} \sum_{k\neq k'} g(k)\bar{g}(k')\phi_{k'-k}\ a_k^+ a_{k'} e^{it[\omega(k)-\omega(k')]}\ . \tag{13.2.1}$$

We show that:

Theorem 13.2.1. *For each* $N \in \mathbb{N}$

$$\lim_{\lambda\to\infty}\lim_{L\to\infty} \mathrm{E}\big(\langle \phi_{\mathrm{F}}, a_{k_0} B_\lambda(T_1,S_1)\cdots B_\lambda(T_N,S_N) a_{k_0}^+ \phi_{\mathrm{F}}\rangle_{\mathrm{conn}}\big)\ , \tag{13.2.2}$$

where $\langle\cdot,\cdot\rangle_{\mathrm{conn}}$ *means expectation with respect to the connected diagrams, always exists and is equal to zero if* N *is odd; however, if* $N = 2n$*, it is equal to*

$$\sum_{\sigma\in\mathcal{S}_{1,2n}}\ \sum_{(l_h,r_h)\in\mathcal{P}nc(1,2n),h=1}^{n}\ \prod_{h=1}^{n} \left\langle \chi_{[S_{\sigma(l_h)},T_{\sigma(l_h)}]}, \chi_{[S_{\sigma(r_h)},T_{\sigma(r_h)}]}\right\rangle_{L^2(\mathbb{R})}$$

$$\cdot \int dk_{l_1}\cdots dk_{l_n}\ \chi_{B_{\mathrm{F}}^{\mathrm{c}}}(k_0) \prod_{h=1}^{n} \langle k_{l_h}, k_{r_h}\rangle\, F_\sigma(k_{l_h}) F_\sigma(k_{r_h})\ , \tag{13.2.3}$$

where

- $\mathcal{P}_{nc}(1,2n)$ *is the set of all noncrossing pair partitions* $\{l_1, r_1 \dots l_n, r_n\}$ *of the set* $\{1 \dots 2n\}$.
- $\mathcal{S}_{1,2n}$ *denotes the permutations of the set* $\{1 \dots 2n\}$.
- *Consider the set* $\{1 \dots 2n\}$ *as a 1-dimensional lattice, oriented from left to right, with lines connecting two consecutive indices. Given a noncrossing pair partition* $\{l_1, r_1 \dots l_n, r_n\}$ *of this set, to each line we associate a momentum* k_α *as follows: to the first line after a point* l_h *it is associated a momentum* k_{l_h}, *and to the first line after* r_h *it is associated a momentum* k_{r_h}, *which is a linear combination of the* $\{k_{l_h}\}_{h=1}^n$, *determined by the requirement that* $k'_{l_h} - k_{l_h} = k_{r_h} - k'_{r_h}$, *if* k'_α *is the momentum of the line preceding* α, *with* $\alpha = l_h, r_h$. *This rule simply expresses the fact that, in a vertex, the sum of the incoming momenta is equal to the sum of the outgoing ones.*
-
$$\langle k_l, k_r \rangle = \int_{-\infty}^{\infty} e^{i[\omega(k_l)-\omega(k_r)]u} g(k_l)^2 g(k_r)^2 du .$$
- *Denoting* B_{F} *the Fermi sphere,* $B_{\mathrm{F}} := \{k, |k| \le p_{\mathrm{F}}\}$, *and* $B_{\mathrm{F}}^{\mathrm{c}} = \mathbb{R}^d \setminus B_{\mathrm{F}}$ *its set-theoretical complement, the function* $F_\sigma(k_{l_h})$ *is defined by*

$$F_\sigma(k_{l_h}, l_h) := \begin{cases} \chi_{B_{\mathrm{F}}^{\mathrm{c}}}(k_{l_h}) \text{ if } \sigma(l_h) < \sigma(l_h+1) , \\ \chi_{B_{\mathrm{F}}}(k_{l_h}) \text{ if } \sigma(l_h) > \sigma(l_h+1) , \end{cases} \tag{13.2.4}$$

$$F_\sigma(k_{r_h}, r_h) := \begin{cases} \chi_{B_{\mathrm{F}}^{\mathrm{c}}}(k_{r_h}) \text{ if } \sigma(r_h) < \sigma(r_h+1) , \\ \chi_{B_{\mathrm{F}}}(k_{r_h}) \text{ if } \sigma(r_h) > \sigma(r_h+1) . \end{cases} \tag{13.2.5}$$

Note that if in (13.2.3) we neglect

- the dependence on the pair partition of the momenta k_{r_h},
- the dependence on σ of the factors $\left\langle \chi_{[S_{\sigma(l_h)}, T_{\sigma(l_h)}]}, \chi_{[S_{\sigma(r_h)}, T_{\sigma(r_h)}]} \right\rangle_{L^2(\mathbb{R})}$ and $F_\sigma(k_{l_h}) F_\sigma(k_{r_h})$ and the corresponding summation,

then (13.2.3) gives the correlators of creation and annihilation operators satisfying a Boltzmannian statistics.

13.3 The Four-Point Function

A proof of Theorem 13.2.1 will be given in Chapter 15 in a much more general situation. We consider here the case of the four-point function in order to explain in a intuitive way why in the limit only the noncrossing diagrams survive. To this goal we consider

$$\lim_{\lambda \to 0} \lim_{L \to \infty} E\left(\langle \phi_{\mathrm{F}}, a_k B_\lambda(T_1, S_1) B_\lambda(T_2, S_2) B_\lambda(T_3, S_3) B_\lambda(T_4, S_4) a_k^+ \phi_{\mathrm{F}} \rangle \right) . \tag{13.3.1}$$

Using (13.1.4) and after the limit $L \to \infty$, (13.3.1) can be written more explicitly in the form

$$\begin{aligned}
&\lambda^4 \int_{S_1/\lambda^2}^{T_1/\lambda^2} dt_1 \int_{S_2/\lambda^2}^{T_2/\lambda^2} dt_2 \int_{S_3/\lambda^2}^{T_3/\lambda^2} dt_3 \int_{S_4/\lambda^2}^{T_4/\lambda^2} dt_4 \\
&\cdot \int dx_1 dx_2 dx_3 dx_4 E\big(\phi(x_1)\phi(x_2)\phi(x_3)\phi(x_4)\big) \\
&\cdot \Big\langle \phi_{\mathrm{F}}, a_k \,:\psi^+(x_1,t_1)\psi(x_1,t_1)::\psi^+(x_2,t_2)\psi(x_2,t_2): \\
&\cdot :\psi^+(x_3,t_3)\psi(x_3,t_3)::\psi^+(x_4,t_4)\psi(x_4,t_4):\, a_k^+ \Big\rangle_{\mathrm{conn}} . \qquad (13.3.2)
\end{aligned}$$

The above fermionic expectation is given by the sum of several terms. Let us select for definiteness the following one [the sum over σ in (13.2.3) comes from taking into account all such terms]:

$$\begin{aligned}
&\langle a_k\psi^+(x_1,t_1)\rangle\langle\psi(x_1,t_1)\psi^+(x_2,t_2)\rangle\langle\psi(x_2,t_2)\psi^+(x_3,t_3)\rangle \\
&\cdot\langle\psi(x_3,t_3)\psi^+(x_4,t_4)\rangle\langle\psi(x_4,t_4)a_k^+\rangle . \qquad (13.3.3)
\end{aligned}$$

Since the expectation over the Gaussian variables in (13.3.2) is given by

$$\delta(x_1-x_4)\delta(x_2-x_3)+\delta(x_1-x_2)\delta(x_3-x_4)+\delta(x_1-x_3)\delta(x_2-x_4)\,, \qquad (13.3.4)$$

we have three terms which shall be studied separately (note that the third term corresponds to a crossing diagram, while the other ones are noncrossing). The first term is

$$\begin{aligned}
&\lambda^4 \int_{S_1/\lambda^2}^{T_1/\lambda^2} dt_1 \int_{S_2/\lambda^2}^{T_2/\lambda^2} dt_2 \int_{S_3/\lambda^2}^{T_3/\lambda^2} dt_3 \int_{S_4/\lambda^2}^{T_4/\lambda^2} dt_4 \int dx_1 dx_2 \langle a_k\psi^+(x_1,t_1)\rangle \\
&\cdot \langle\psi(x_1,t_1)\psi^+(x_2,t_2)\rangle\langle\psi(x_2,t_2)\psi^+(x_2,t_3)\rangle \\
&\cdot \langle\psi(x_2,t_3)\psi^+(x_1,t_4)\rangle\langle\psi(x_1,t_4)a_k^+\rangle \\
&=\lambda^4\frac{1}{L^{2d}}\sum_{k',k''}\int_{S_1/\lambda^2}^{T_1/\lambda^2} dt_1 \int_{S_2/\lambda^2}^{T_2/\lambda^2} dt_2 \int_{S_3/\lambda^2}^{T_3/\lambda^2} dt_3 \int_{S_4/\lambda^2}^{T_4/\lambda^2} dt_4 \\
&\cdot \exp\{-i[\omega(k)-\omega(k')](t_1-t_2)\}\exp\{-i[\omega(k)-\omega(k'')](t_3-t_4)\} \\
&\cdot \chi(|k|\ge p_{\mathrm{F}})\chi(|k'|\ge p_{\mathrm{F}})\chi(|k''|\ge p_{\mathrm{F}})[g(k)g(k')g(k)g(k')] \\
&\cdot [g(k)g(k'')g(k)g(k'')] \\
&=\frac{1}{L^{2d}}\sum_{k',k''}\int_{S_1}^{T_1} d\tau_1 \int_{(S_2-\tau_1)/\lambda^2}^{(T_2-\tau_1)/\lambda^2} d\tau_2 \int_{S_3}^{T_3} d\tau_3 \int_{(S_4-t_3)/\lambda^2}^{(T_4-t_3)/\lambda^2} d\tau_4 \\
&\cdot \exp\{-i[\omega(k)-\omega(k')]\tau_2\}\exp\{-i[\omega(k)-\omega(k'')]\tau_4\} \\
&\cdot \chi(|k|\ge p_{\mathrm{F}})\chi(|k'|\ge p_{\mathrm{F}})\chi(|k''|\ge p_{\mathrm{F}})[g(k)g(k')g(k)g(k')] \\
&\cdot [g(k)g(k'')g(k)g(k'')] . \qquad (13.3.5)
\end{aligned}$$

In the limit $\lambda \to 0, L \to \infty$, the above expression becomes

$$\langle \chi_{[S_1,T_1]}, \chi_{[S_2,T_2]} \rangle_{L^2(\mathbb{R})} \langle \chi_{[S_3,T_3]}, \chi_{[S_4,T_4]} \rangle_{L^2(\mathbb{R})}$$

$$\cdot \int dk_1 dk_2 \chi_{B_{\mathrm{F}}^{\mathrm{c}}}(k) \langle k_1, k \rangle \langle k_2, k \rangle \chi_{B_{\mathrm{F}}^{\mathrm{c}}}(k_1) \chi_{B_{\mathrm{F}}^{\mathrm{c}}}(k_2) .$$

The second term is

$$\begin{aligned}
&\lambda^4 \frac{1}{L^{2d}} \sum_{k',k''} \int_{S_1/\lambda^2}^{T_1/\lambda^2} dt_1 \int_{S_2/\lambda^2}^{T_2/\lambda^2} dt_2 \int_{S_3/\lambda^2}^{T_3/\lambda^2} dt_3 \int_{S_4/\lambda^2}^{T_4/\lambda^2} dt_4 \\
&\cdot \exp\{-i[\omega(k) - \omega(k')](t_1 - t_4)\} \exp\{-i[\omega(k') - \omega(k'')](t_2 - t_3)\} \\
&\cdot \chi(|k| \le p_{\mathrm{F}}) \chi(|k'| \le p_{\mathrm{F}}) \chi(|k''| \le p_{\mathrm{F}}) [g(k)g(k')g(k)g(k')] \\
&\cdot [g(k)g(k'')g(k)g(k'')] \\
&= \frac{1}{L^{2d}} \sum_{k',k''} \int_{S_1}^{T_1} d\tau_1 \int_{S_2}^{T_2} d\tau_2 \int_{(S_4-\tau_1)/\lambda^2}^{(T_4-t_1)/\lambda^2} d\tau_4 \exp\{-i[\omega(k) - \omega(k')]\tau_4\} \\
&\cdot \int_{(S_3-\tau_2)/\lambda^2}^{(T_3-t_2)/\lambda^2} \exp\{-i[\omega(k') - \omega(k'')]\tau_3\} \\
&\cdot \chi(|k| \ge p_{\mathrm{F}}) \chi(|k'| \ge p_{\mathrm{F}}) \chi(|k''| \ge p_{\mathrm{F}}) [g(k)g(k')g(k)g(k')] \\
&\cdot [g(k')g(k'')g(k')g(k'')] .
\end{aligned} \tag{13.3.6}$$

In the limit $L \to \infty$, the sum over k', k'' becomes an integral which is bounded by $O\left(\frac{1}{(\tau_4)^{d/2}(\tau_4-\tau_3)^{d/2}}\right)$ so that we can take the limit $\lambda \to 0$, finding

$$\langle \chi_{[S_1,T_1]}, \chi_{[S_4,T_4]} \rangle_{L^2(\mathbb{R})} \langle \chi_{[S_2,T_2]}, \chi_{[S_3,T_3]} \rangle_{L^2(\mathbb{R})}$$

$$\cdot \int dk_1 dk_2 \chi_{B_{\mathrm{F}}^{\mathrm{c}}}(k) \langle k_1, k \rangle \langle k_2, k_1 \rangle \chi_{B_{\mathrm{F}}^{\mathrm{c}}}(k_1) \chi_{B_{\mathrm{F}}^{\mathrm{c}}}(k_2) .$$

Finally the third term, corresponding to the only crossing diagram in the four-point function, is

$$\begin{aligned}
&\lambda^4 \frac{1}{L^{2d}} \sum_{k',k''} \int_{S_1/\lambda^2}^{T_1/\lambda^2} dt_1 \int_{S_2/\lambda^2}^{T_2/\lambda^2} dt_2 \int_{S_3/\lambda^2}^{T_3/\lambda^2} dt_3 \int_{S_4/\lambda^2}^{T_4/\lambda^2} dt_4 \\
&\cdot \exp\{-i[\omega(k) - \omega(k')](t_1 - t_2)\} \exp\{-i[\omega(k'') - \omega(k - k' + k'')](t_3 - t_4)\} \\
&\cdot \exp\{-i[\omega(k) - \omega(k'')]t_2\} \exp\{-i[\omega(k) - \omega(k'')]t_4\} \\
&\cdot \chi(|k| \ge p_{\mathrm{F}}) \chi(|k'| \ge p_{\mathrm{F}}) \chi(|k''| \ge p_{\mathrm{F}}) [g(k)g(k')g(k)g(k')] \\
&\cdot [g(k)g(k'')g(k)g(k'')] .
\end{aligned} \tag{13.3.7}$$

With the change of variables $\lambda^2 t_2 = \tau_2, \lambda^2 t_4 = \tau_4$, $t_1 - \tau_2/\lambda^2 = \tau_1$, $t_3 - \tau_4/\lambda^2 = \tau_3$, this becomes

$$\frac{1}{L^{2d}}\sum_{k,k'}\int_{S_2}^{T_2}d\tau_2\int_{S_4}^{T_4}d\tau_4\int_{(S_1-\tau_2)/\lambda^2}^{(T_1-\tau_2)/\lambda^2}d\tau_1\int_{(S_3-\tau_4)/\lambda^2}^{(T_3-\tau_4)/\lambda^2}d\tau_3$$
$$\cdot\exp\{-i[\omega(k)-\omega(k')]\tau_1\}\exp\{-i[\omega(k'')-\omega(k-k'+k'')]\tau_3\}$$
$$\cdot\exp\left\{-i[\omega(k)-\omega(k'')]\frac{\tau_2}{\lambda^2}\right\}\exp\left\{-i[\omega(k)-\omega(k'')]\frac{\tau_4}{\lambda^2}\right\}$$
$$\cdot\chi(|k|\le p_{\mathrm{F}})\chi(|k'|\le p_{\mathrm{F}})\chi(|k''|\le p_{\mathrm{F}})[g(k)g(k')g(k)g(k')]$$
$$\cdot[g(k)g(k'')g(k)g(k'')].$$

and, after the limit $L\to\infty$, the resulting integral vanishes in the limit $\lambda\to 0$ by dominated convergence and the Riemann–Lebesgue lemma.

13.4 The Limit of the Connected Transition Amplitude

Theorem 13.4.1. *In the notations of Theorem 13.2.1 the following identity is valid at any order of the perturbative series:*

$$\lim_{\lambda\to 0}\lim_{L\to\infty}E\left(\langle\Phi_{\mathrm{F}},a^{\varepsilon}_{k_0}U_{T/\lambda^2}a^{-\varepsilon}_{k_0}\Phi_{\mathrm{F}}\rangle_{\mathrm{conn}}\right)$$
$$=\sum_{n=0}^{\infty}\frac{1}{n!}\chi^{\varepsilon}(k_0)T^n\sum_{\substack{\{l_h,r_h\}_{h=1}^n\\ \in\mathcal{P}_{nc}(1,2n)}}\ \sum_{\substack{\sigma\in S_{1,2n}\\ \sigma(l_h)-\sigma(r_h)=\pm1\,\forall\, h\in\{1\ldots n\}}}$$
$$\int dk_{l_1}\ldots dk_{l_n}\prod_{h=1}^{n}\langle k_{l_h},k_{r_h}\rangle_{\epsilon_h}F_\sigma(k_{l_h},l_h)F_\sigma(k_{r_h},r_h)\equiv G^{\varepsilon}(k_0)\,,\qquad(13.4.1)$$

where $\epsilon_h=1$ if $\sigma(l_h)-\sigma(r_h)=+1$ and $\epsilon_h=-1$ otherwise,

$$\langle k_l,k_r\rangle_{-1}=\int_{-\infty}^{0}e^{i[\omega(k_l)-\omega(k_r)]u}g(k_l)^2g(k_r)^2\,,$$

$$\langle k_l,k_r\rangle_{1}=\int_{0}^{\infty}e^{i[\omega(k_l)-\omega(k_r)]u}g(k_l)^2g(k_r)^2\,,$$

and $\chi^{\epsilon}(k_0)=\chi_{B^c_{\mathrm{F}}}(k_0)$ if $\epsilon=1$ and $\chi^{\epsilon}(k_0)=\chi_{B_{\mathrm{F}}}(k_0)$ if $\epsilon=-1$.

The above theorem says that, in the perturbative expansion for the transition amplitude, only the noncrossing diagrams contribute in the limit.

From the explicit expression of equation (13.3.1) it follows that the functions G^{ε} satisfy the following closed set of equations:

Theorem 13.4.2. *In the same notations as in Theorem 13.4.1 one has*

$$G^{\varepsilon}(k_0)=G^{\varepsilon}_0(k_0)+TG^{\varepsilon}_0(k_0)G^{\varepsilon}(k_0)\int_{-\infty}^{+\infty}du e^{-i\omega(k_0)u}g(k_0)^2$$
$$\cdot\int_{-\infty}^{+\infty}g(k)^2e^{i\omega(k)u}[G^+(k)+G^-(k)]dk\,,\qquad(13.4.2)$$

where $G^{\varepsilon}_0(k_0)=\chi^{\epsilon}(k_0)$.

Proof 13.4.1. Let us denote $G^{\varepsilon,(n)}(k)$ the nth term of the series in (13.4.1) so that

$$G^{\varepsilon}(k) = \sum_{n=0}^{\infty} G^{\varepsilon,(n)}(k) .$$

We use a well-known property of the set $\mathcal{P}(1,2n)$ of the noncrossing pair partitions, namely

$$\mathcal{P}_{nc}(1,2n) = \bigcup_{m=1}^{n} \{(1,2m) \cup \mathcal{P}_{nc}(2,2m-1) \cup \mathcal{P}_{nc}(2m+1,2n)\} ,$$

with the convention that $\mathcal{P}_{nc}(x,y) = \emptyset$ if $x \geq y$. From this, it follows that

$$\sum_{\{l_h,r_h\}_{h=1}^{n} \in \mathcal{P}_{nc}(1,2n)} = \sum_{m=1}^{n} \sum_{\substack{\{l_h,r_h\}_{h=2}^{m-1} \\ \in \mathcal{P}_{nc}(2,2m-1)}} \sum_{\substack{\{l_h,r_h\}_{h=m+1}^{n} \\ \in \mathcal{P}_{nc}(2m+1,2n)}} .$$

Moreover, in Sect. 13.5 it is proved that, for a fixed noncrossing pair partition $(l_1, r_1 \dots l_n, r_n)$, we can write

$$\begin{aligned}
&\sum_{\substack{\sigma \in S_{1,2n} \\ \sigma(l_h)-\sigma(r_h)=\pm 1 \\ \forall h \in \{1\dots n\}}} F_{\sigma}(k_{l_h}, l_h) F_{\sigma}(k_{r_h}, r_h) = \frac{n!}{(m-1)!(n-m)!)} \\
&\cdot \chi^{\epsilon}(k_0) \Big[\chi_{B_{\mathrm{F}}}(k_{l_1}) \sum_{\substack{\sigma \in S_{2,2m-1} \\ \sigma(l_h)-\sigma(r_h)=\pm 1 \\ \forall h \in \{2\dots m-1\}, \sigma(l_2)>\sigma(l_1)}} F_{\sigma}(k_{l_h}, l_h) F_{\sigma}(k_{r_h}, r_h) \\
&+ \chi_{B_{\mathrm{F}}^{c}}(k_{l_1}) \sum_{\substack{\sigma \in S_{2,2m-1} \\ \sigma(l_h)-\sigma(r_h)=\pm 1 \\ \forall h \in \{2\dots m-1\}, \sigma(l_2)<\sigma(l_1)}} F_{\sigma}(k_{l_h}, l_h) F_{\sigma}(k_{r_h}, r_h) \Big] \\
&\cdot \sum_{\substack{\sigma \in S_{2m+1,2n} \\ \sigma(l_h)-\sigma(r_h)=\pm 1 \\ \forall h \in \{m+1\dots n\}}} F_{\sigma}(k_{l_h}, l_h) F_{\sigma}(k_{r_h}, r_h) .
\end{aligned} \tag{13.4.3}$$

Furthermore, by momentum conservation, in every connected component of the pair partition the incoming momentum is equal to that outgoing. In particular

$$k_0 = k_{r_1} = k_{r_n} .$$

Using the remarks above $G^{\varepsilon,(n)}(k_0)$ can be written, for $n \geq 1$, as

$$\sum_{m=1}^{n} \frac{T^n}{(m-1)!(n-m)!} \chi^\varepsilon(k_0) \int_{-\infty}^{+\infty} du \int \exp\{i[\omega(k_{l_1}) - \omega(k_0)]u\}$$

$$\cdot\, dug(k_0)^2 g(k_{l_1})^2 dk_{l_1} \Bigg[\chi_{B_{\mathrm{F}}}(k_{l_1}) \int dk_{l_2} ... dk_{l_{m-1}} \sum_{\substack{\{l_h,r_h\}_{h=2}^{m-1} \\ \in \mathcal{P}_{nc}(2,2m-1)}} \prod_{h=2}^{m-1} \sum_{\substack{\sigma \in \mathcal{S}_{2,2m-1} \\ \sigma(l_h)-\sigma(r_h)=\pm 1, \\ \sigma(l_2)<\sigma(l_1)}}$$

$$\cdot\, \langle k_{l_h}, k_{r_h} \rangle_{\varepsilon_h} F_\sigma(k_{l_h}) F_\sigma(k_{r_h})$$

$$+ \chi_{B_{\mathrm{F}}^{\mathrm{c}}}(k_{l_1}) \int dk_{l_2} ... dk_{l_{m-1}} \sum_{\substack{\{l_h,r_h\}_{h=m+1}^{n} \\ \in \mathcal{P}_{nc}(2,2m-1)}} \prod_{h=2}^{m-1} \sum_{\substack{\sigma \in \mathcal{S}_{2,2m-1} \\ \sigma(l_h)-\sigma(r_h)=\pm 1, \\ \sigma(l_2)>\sigma(l_1)}} \langle k_{l_h}, k_{r_h} \rangle_{\varepsilon_h} F_\sigma(k_{l_h}) F_\sigma(k_{r_h}) \Bigg]$$

$$\cdot\, \chi^\varepsilon(k_0) \int dk_{l_{m+1}} ... dk_{l_n} \sum_{\substack{\{l_h,r_h\}_{h=m+1}^{n} \\ \in \mathcal{P}_{nc}(2m+1,2n)}} \prod_{h=m+1}^{n} \sum_{\substack{\sigma \in \mathcal{S}_{2m+1,2n} \\ \sigma(l_h)-\sigma(r_h)=\pm 1}} \langle k_{l_h}, k_{r_h} \rangle_{\varepsilon_h} F_\sigma(k_{l_h}) F_\sigma(k_{r_h}) \,. \tag{13.4.4}$$

By associating the factor $(n-m)!$ to the $\mathcal{P}_{nc}(2m+1, 2n)$ sum, we see that this term reproduces $G^{\varepsilon,(n-m)}(k_0)$. Similarly by associating the factor $(m-1)!$ to the two $\mathcal{P}_{nc}(2, 2m-1)$ sums in square brackets, we see that these sums reproduce respectively $G^{+,(m-1)}(k_0)$ and $G^{-,(m-1)}(k_0)$. This shows that (13.4.4) can be written in the form

$$G^{\varepsilon,(n)}(k_0) = T \int_{-\infty}^{+\infty} du \int \exp\{i[t\omega(k_{l_1}) - \omega(k_0)]u\}\, dug(k_0)^2 g(k_{l_1})^2 dk_{l_1} \chi^\varepsilon(k_0)$$

$$\cdot \sum_{n_1+n_2=n-1} [(G^{+,(n_1)}(k_{l_1}) + G^{-,(n_1)}(k_{l_1})] G^{\varepsilon,(n_2)}(k_0)\,;$$

by summing over n, (13.4.2) is found.

Remark 13.4.1. Let $S(k, \omega)$ be the Fourier transform of the *Green function*

$$S(x-y, t-s) = \lim_{L,T\to\infty} E\left(\langle \Phi_{\mathrm{F}}, \mathcal{T} \psi^+_{x,t} \psi^-_{y,s} U_T U_{-T} \Phi_{\mathrm{F}} \rangle_{\mathrm{conn}} \right)$$

if $\mathcal{T}$ is the time order product. The closed equation found in [AGD75] is

$$S(k_0, \omega) = S_0(k_0, \omega) + S_0(k_0, \omega) S(k_0, \omega) \int_{-\infty}^{+\infty} dk u(k_0 - k) S(k, \omega)\,,$$

where $S_0(k, \omega)$ is the free Green function and $u(p)$ is a suitable cutoff function. The similarity with our (13.4.2) is striking. In fact equations of this type are a general feature of the noncrossing diagrams.

13.5 Proof of (13.4.3)

Here we prove (13.4.3). To this goal we fix the injective maps

$$l : h \in \{1 \dots n\} \to l_h \in \{1 \dots 2n\}, \tag{13.5.1}$$

$$r : l \in \{1 \dots n\} \to r_h \in \{1 \dots 2n\}$$

in such a way that $\{l_1, r_1 \dots l_n, r_n\}$ is a noncrossing pair partition of $\{1 \dots 2n\}$.

For each subset of pairs $I \subseteq \{l_1, r_1 \dots l_m, r_n\}$, define

$$I(\sigma, l) := \{l_h : (l_h, r_h) \in I\,;\ \sigma(l_h) < \sigma(l_h + 1)\}\,.$$

Similarly we can define

$$I(\sigma, r) := \{r_h : (l_h, r_h) \in I\,; \quad \sigma(r_h) < \sigma(r_h + 1)\}\,.$$

Definition 13.5.1. *Given a finite set I and a permutation σ on I, the* index of monotonicity *of σ is the map*

$$j_I(\sigma, \cdot) : x \in I \to j(\sigma, x) \in \{1, 0\}\,,$$

defined by

$$j(\sigma, x) = 1 \Leftrightarrow \sigma(x) < \sigma(x+1)\,.$$

Introducing the convention (for any set A)

$$\chi_A(k)^\varepsilon = \begin{cases} 1 & \textit{if } \varepsilon = 0\,, \\ \chi_A(k) & \textit{if } \varepsilon = 1\,, \end{cases}$$

one can write the right-hand side of (13.4.3) in the form

$$\sum_{\sigma \in \mathcal{S}'_{1,2n}} \left(\prod_{h=1}^{n} \chi_{B_{\mathrm{F}}^{\mathrm{c}}}(k_{l_h})^{j(\sigma, l_h)} \chi_{B_{\mathrm{F}}}(k_{l_h})^{1 - j(\sigma, l_h)} \right) \tag{13.5.2}$$

$$\cdot \left(\prod_{h=1}^{n} \chi_{B_{\mathrm{F}}^{\mathrm{c}}}(k_{r_h})^{j(\sigma, r_h)} \chi_{B_{\mathrm{F}}}(k_{r_h})^{1 - j(\sigma, r_h)} \right),$$

where

$$\sum_{\substack{\sigma \in S_{1,2n} \\ \sigma(l_h) - \sigma(r_h) = \pm 1\ \forall\, h \in \{1 \dots n\}}} = \sum_{\sigma \in \mathcal{S}'_{1,2n}}\,.$$

Now the set $\{l_1, r_1 \dots l_n, r_n\}$ coincides with $\{1 \dots 2n\}$. The sum (13.5.2) can be rewritten in the form

$$\sum_{\sigma \in \mathcal{S}'_{1,2n}} \prod_{\alpha=1}^{2n} \chi_{B_{\mathrm{F}}^{\mathrm{c}}}(k_\alpha)^{j(\sigma, \alpha)} \chi_{B_{\mathrm{F}}}(k_\alpha)^{1 - j(\sigma, \alpha)}\,. \tag{13.5.3}$$

Remark 13.5.1. We can see that the terms in the sum (13.5.3) depend on σ only through the function $j(\sigma,\cdot)$.

Lemma 13.5.1. *Every $\sigma \in \mathcal{S}'_{1,2n}$ can be written in a unique way in the form*

$$\sigma = \overline{\sigma}\sigma_m\sigma_{n-m}\,,$$

where

- *$\overline{\sigma}$ is characterized by the property that the sets $F \subseteq \{1\ldots 2m\}$, $F' \subseteq \{2m+1\ldots 2n\}$ exist such that*

$$\overline{\sigma}F = F'\,, \quad \overline{\sigma}F' = F\,, \quad \bar{\sigma}(x) = x \quad \text{for any} \quad x \notin F \cup F' \tag{13.5.4}$$

  and $\overline{\sigma}$ is monotone on $F \cup F'$,
- *$\sigma_m(x) = x$ for $x \notin \{1\ldots 2m\}$,*
- *$\sigma_{n-m}(x) = x$ for $x \notin \{2m+1\ldots 2n\}$.*

Proof 13.5.1. The proof of the above lemma is trivial.

Denote now

$$I = I_{\overline{\sigma}} := \overline{\sigma}^{-1}\{1, 2m\}\,,$$

then

$$I^{\mathrm{c}} = \overline{\sigma}^{-1}\{2m+1\ldots 2m\}\,.$$

Denote

$$\mathcal{S}'_I(\overline{\sigma}) = \overline{\sigma}\mathcal{S}'_{\{1,2m\}}\overline{\sigma}^{-1}\,, \quad \mathcal{S}'_{I^{\mathrm{c}}}(\overline{\sigma}) = \overline{\sigma}\mathcal{S}'_{\{2m+1,2n\}}\overline{\sigma}^{-1}\,, \tag{13.5.5}$$

and note that if σ has the form as given in Lemma 13.5.1 then

$$\sigma = (\overline{\sigma}\sigma_m\sigma_{n-m}\overline{\sigma}^{-1})\overline{\sigma} = \sigma_I\sigma_{I^{\mathrm{c}}}\overline{\sigma}\,. \tag{13.5.6}$$

Introducing the notation

$$\phi\big(k_x, j(\sigma,x)\big) = \chi_{B_{\mathrm{F}}^{\mathrm{c}}}(k_x)^{j(\sigma,x)}\chi_{B_{\mathrm{F}}}(k_x)^{1-j(\sigma,x)}$$

and using the fact that, for $x \in \{1\ldots 2m\}$,

$$\sigma_I\sigma_{I^{\mathrm{c}}}\bar{\sigma}x = \sigma_I\bar{\sigma}x$$

and the analogue identity for $\sigma_{I^{\mathrm{c}}}$, the sum (13.5.2) can be written in the form

$$\begin{aligned}
&\sum_{\overline{\sigma}}\sum_{\mathcal{S}'_I(\bar{\sigma})}\sum_{\mathcal{S}'_{I^{\mathrm{c}}}(\bar{\sigma})}\prod_{x\in\{1\ldots 2n\}}\varphi\big(k_x, j(\sigma_I\sigma_{I^{\mathrm{c}}}\overline{\sigma}, x)\big)\\
&=\sum_{\overline{\sigma}}\left(\sum_{\sigma_I\in\mathcal{S}'_I(\bar{\sigma})}\prod_{x\in\{1\ldots 2m\}}\varphi(k_x, j(\sigma_I\overline{\sigma}, x)\right)\\
&\quad\cdot\left(\sum_{\sigma_{I^{\mathrm{c}}}\in\mathcal{S}'_{I^{\mathrm{c}}}(\bar{\sigma})}\prod_{x\in\{2m+1\ldots 2n\}}\varphi\big(k_x, j(\sigma_{I^{\mathrm{c}}}\overline{\sigma}, x)\big)\right)\,.
\end{aligned} \tag{13.5.7}$$

Thus, if we prove that both the σ_I sum and the σ_{I^c} sum do not depend on $\overline{\sigma}$, the expression (13.5.7) shall be equal to

$$\binom{n}{m}\left(\sum_{\sigma\in\mathcal{S}'_{1,2m}}\prod_{x\in\{1\dots 2m\}}\varphi\big(k_x,j(\sigma,x)\big)\right)$$
$$\cdot\left(\sum_{\sigma'\in\mathcal{S}'_{2m+1,2n}}\prod_{x\in\{2m+1\dots 2n\}}\varphi\big(k_x,j(\sigma',x)\big)\right). \tag{13.5.8}$$

Lemma 13.5.2. *For every $\overline{\sigma}$ the two sums*

$$\sum_{\sigma\in\mathcal{S}'(1,2m)}\prod_{x\in\{1\dots 2m\}}\varphi\big(k_x,j(\sigma,x)\big) \tag{13.5.9}$$

and

$$\sum_{\sigma_I\in\mathcal{S}'_I(\bar\sigma)}\prod_{x\in\{1\dots 2m\}}\varphi\big(k_x,j(\sigma_I\overline{\sigma},x)\big) \tag{13.5.10}$$

coincide.

Proof 13.5.2. It is sufficient to show that, given $\overline{\sigma}$, then for each $\sigma_I\in\mathcal{S}'_I$ there exists $\sigma\in\mathcal{S}'_{2m}$ such that

$$j(\sigma_I\overline{\sigma},x)=j(\sigma,x)\ ,\quad \forall x\in\{1\dots 2m\}\,.$$

Recall that, by definition,

$$\overline{\sigma}\{1\dots 2m\}=I_{\overline{\sigma}}$$

and let $\sigma_0\in\mathcal{S}'_I$ be the unique permutation such that

$$\sigma_0 I=\sigma_0\overline{\sigma}\{1\dots 2m\}=:I_0$$

is an ordered set, i.e. the ordered version of I. Clearly the sum (13.5.10) is equal to

$$\sum_{\sigma_I\in\mathcal{S}'_I(\bar\sigma)}\prod_{x\in\{1\dots 2m\}}\varphi\big(k_x,j(\sigma_I\sigma_0\overline{\sigma},x)\big)\,. \tag{13.5.11}$$

The map

$$\beta:=\sigma_0\overline{\sigma}:\{1\dots 2m\}\to I_0$$

is the unique monotone mapping between $\{1\dots 2m\}$ and the ordered version of I.

The sum (13.5.11) then becomes

$$\sum_{\sigma_I \in \mathcal{S}'_I(\bar{\sigma})} \prod_{x \in \{1 \ldots 2m\}} \varphi\big(k_x, j(\sigma_I \beta, x)\big) . \tag{13.5.12}$$

Now denote

$$\sigma := \beta^{-1} \sigma_I \beta \tag{13.5.13}$$

and note that

$$j(\sigma_I \beta, x) = 1 \Leftrightarrow \sigma_I\big(\beta(x)\big) < \sigma_I\big(\beta(x+1)\big) .$$

However, β is monotone, so because of injectivity β^{-1} is also monotone; therefore

$$\begin{aligned} j(\sigma_I \beta, x) = 1 \quad &\Leftrightarrow \quad \beta^{-1}\sigma_I\beta(x) \leq \beta^{-1}\sigma_I\beta(x+1) \\ &\Leftrightarrow \quad \sigma(x) < \sigma(x+1) \quad \Leftrightarrow \quad j(\sigma, x) = 1 . \end{aligned}$$

This implies that

$$j(\sigma_I \beta, x) = j(\sigma, x) = 1 .$$

So we can write the sum (13.5.12) as

$$\sum_{\beta \sigma \beta^{-1} = \sigma_I \in \mathcal{S}'_{I(\bar{\sigma})}} \prod_{x \in \{1 \ldots 2m\}} \varphi\big(k_x, j(\sigma, x)\big) .$$

It is now clear that the sum over all $\sigma_I \in \mathcal{S}'_I(\bar{\sigma})$ in (13.5.9) is equivalent to the sum over all $\sigma \in \mathcal{S}'_{1,2m}$. This ends the proof of (13.5.8).

The proof of (13.4.3) follows noting that $k_0 = k_{r_m}$ and $k_{l_1} = k_{r_{m-1}}$, where (l_{m-1}, r_{m-1}) is the last pair partition "enclosed" in (l_1, r_1), so that the only dependence of the summand of (13.5.8) on $\sigma(l_1), \sigma(r_1)$ is that $\varphi(k_{l_1}) = \varphi(k_{r_{m-1}}) = \chi_{B_{\mathrm{F}}^{\mathrm{c}}}(k_{l_1})$ if $\sigma(l_2) < \sigma(l_1)$ and $\varphi(k_{l_1}) = \varphi(k_{r_{m-1}}) = \chi_{B_{\mathrm{F}}}(k_{l_1})$ otherwise.

13.6 Solution of the Nonlinear Equation (13.4.2)

If, as in the case of physical interest, $\omega(k) = |k|^2$ and $g(k)$ depends only on $|k|$, i.e. $g(k) = g(|k|)$, then the nonlinear equation (13.4.2) can be explicitly solved as follows. First note that, for $\varepsilon = 1, 2$, $G^{\varepsilon}(k_0) = 0$ if and only if $G_0^{\varepsilon}(k_0) = 0$. The "if" part is clear because $G_0^{\varepsilon}(k_0)$ multiplies the right-hand side of (13.4.2). Conversely, if $G^{\varepsilon}(k_0) = 0$ then (13.4.2) becomes $G_0^{\varepsilon}(k_0) = G^{\varepsilon}(k_0) = 0$. Second note that, by the definition of G_0^{ε}, one has $G_0^{+}(k) + G_0^{-}(k) = 1$ identically in k. Therefore, by the above remark

$$G_0^+(k) = 1 \Leftrightarrow G_0^-(k) = 0 \Leftrightarrow G^-(k) = 0\,,$$
$$G_0^-(k) = 1 \Leftrightarrow G_0^+(k) = 0 \Leftrightarrow G^+(k) = 0$$

for all k. Equivalently this means that, identically in k,

$$G_0^+ G^- = G_0^- G^+ \equiv 0\,.$$

It follows that adding the two equations (13.4.2) (for $\varepsilon = \pm$) one obtains

$$\sum_{\varepsilon=\pm} G^\varepsilon(k_0) = \sum_{\varepsilon=\pm} G_0^\varepsilon(k_0) + T \sum_{\varepsilon=\pm} G_0^\varepsilon(k_0) \sum_{\varepsilon'=\pm} G^{\varepsilon'}(k_0)$$

$$\cdot \int_{\mathbb{R}} du \int_{\mathbb{R}^d} dk \exp\{i[\omega(k) - \omega(k_0)]u\}\, |g(k_0)g(k)|^2 \sum_{\varepsilon'=\pm} G^\varepsilon(k)\,.$$

Using again that $G_0^+(k) + G_0^-(k) \equiv 1$ for all k and introducing the notation

$$G(k) := G^+(k) + G^-(k)\,,$$

we have

$$G(k_0) = 1 + TG(k_0) \int_{\mathbb{R}} du \int_{\mathbb{R}^d} dk \exp\{i[\omega(k) - \omega(k_0)]u\}\, |g(k_0)g(k)|^2 G(k)\,. \tag{13.6.1}$$

Under our assumptions on $\omega(k) = |k|^2$ and on $g(k) = g(|k|)$, the integral (13.6.1) can be explicitly evaluated, giving

$$\int_{\mathbb{R}^d} dk |g(|k_0|)g(|k|)|^2 G(k) 2\pi\delta(|k|^2 - |k_0|^2) = 2\pi |g(r_0)|^4 \cdot r_0 \int_{S_{r_0}} G(r_0, \sigma) d\sigma\,.$$

Therefore (13.6.1) is equivalent to

$$G(k_0) = \frac{1}{1 - 2\pi T |g(r_0)g(r_0)|^2 \cdot r_0 \int_{S_{r_0}} G(r_0, \sigma) d\sigma}\,, \tag{13.6.2}$$

where S_{r_0} denotes the sphere centered on the origin with radius r_0. The right-hand side of (13.6.3) tells us that $G(k_0)$ depends only on $|k_0|$, so that

$$G(r_0) = \frac{1}{1 - 2\pi T |S_{r_0}| G(r_0) |g(r_0)|^4 r_0}\,, \tag{13.6.3}$$

where $|S_{r_0}|$ is the surface of the sphere S_{r_0}. Denoting

$$a := 2\pi T |S_{r_0}| \cdot |g(r_0)|^4 r_0\,,$$

we see that $G(r_0)$ satisfies

$$G^2(r_0) a - G(r_0) + 1 = 0\,, \tag{13.6.4}$$

which, if $4a \leq 1$ (and this can be realized either by taking T to be small or by taking r_0 to be large, because g is a Schwarz function), has solutions $G(r_0) = (1 \pm \sqrt{1-4a})/2a$. But (13.6.3) implies that $G \to 1$ as $a \to 0$ and therefore

$$G(r_0) = \frac{1-\sqrt{1-4a}}{2a} = \frac{2}{1+\sqrt{1-4a}}, \tag{13.6.5}$$

which is smooth and tends to 1 as $r_0 \to \infty$.

13.7 Notes

Introduction

(1) The result described in this chapter were obtained in [AcLuMa94a,b]. The case in which the external field is replaced by a boson field is of course much more interesting and can be dealt with the techniques developed in Chap. 15.

Section 13.1

(1) The phenomenon of Anderson localization has been related to an exponential decay of the Fourier transform of equation (13.1.13), as opposed to the power law of the free case. This exponential decay is proved by [AG] for large λ or $d = 1$ as a consequence of the results of [FS83], [FSMS88] and others on the Schrödinger equation with a random potential. However it is not known what the decay is when λ is small or $d \geq 2$. In [AGD75], following an idea of Migdal [Mig58] for the boson case, it was shown that if one neglects a suitable class of contributions in the perturbative expansion for equation (13.1.13), the so-called *crossing diagrams*, one obtains a closed Schwinger–Dyson equation for the limit (13.1.13) from which the exponential decay can be deduced. However, the above-mentioned authors did not specify under which physical conditions the crossing diagrams can be neglected with respect to the noncrossing ones.
A first attempt to clarify this point was made by [Weg79], who replaced the original Hamiltonian (13.1.1) by a discrete mean field approximation in which a fictious N-valued index was added to the fermions and the classical random field was replaced by a random matrix. In the large N limit of this model only the noncrossing diagrams survive. This model was generalized by [SpNe94], [SpNe95], who replaced the large random matrix by *free independent random variables* and obtained an equation of Schwinger–Dyson type. However this result is essentially equivalent to the original assumption that only the noncrossing diagrams survive; in fact, as shown by Voiculescu [Voic91], $N \times N$ Hermitean matrices with independent Gaussian entries become, in the large N limit, free random

variables whose n-point correlation functions are described precisely by the noncrossing (or half-planar) diagrams). The proof of the Migdal conjecture, i.e. of the fact that, in the stochastic limit, only the noncrossing diagrams survive, as well as the deduction and the solution of the nonlinear integral equation were given in [AcLuMa97]. Moreover the physical meaning of this type of result is not clear because it depends in an essential way on the large N limit of a fictitious index which is absent from the original Hamiltonian. For a general consideration of the large N limit in quantum field theory see [ArVo96] and references therein.

14. Field–Field Interactions

14.1 Interacting Commutation Relations

In this chapter we begin to apply the stochastic limit to purely field theoretical Hamiltonians, i.e. fields interacting with other fields without any discrete system. The basic object to study in quantum field theory is the S-matrix introduced by Heisenberg. Bogoliubov and Shirkov developed the S-matrix formalism which includes all the quantities considered in quantum field theory [BoSch87]. The physical idea behind the S-matrix approach is that in the scattering processes there exists a characteristic time scale such that in a time regime larger then this time scale one can neglect interaction and particles evolve according to the free dynamics.

The situation in statistical physics is different because here one has not just one but several relevant time scales, and as a result we do not have here a universal method comparable with the S-matrix approach in quantum field theory. One can say that the role of the S-matrix approach in nonequilibrium statistical physics is played by various master and kinetic equations. It was the fundamental Bogoliubov idea about the existence of two time scales which led to the modern progress in the microscopic derivation of kinetic equations [Bog46]. Methods of quantum field theory, in particular Green functions, are widely used in equilibrium and nonequilibrium statistical physics [BBT61, AGD75, Pr62, Prig68, Zu71].

However there are important problems in quantum field theory where the standard S-matrix description is not very convenient or even not applicable. These include not only the investigation of bound states and spectral problems (see [BoSch87]) but also processes with unstable particles [Schw60], [GoWa64] (in fact almost all particles are unstable), atom–photon interactions [CTDG92], elementary particles in “semidressed states” with nonequilibrium proper fields [Fei72], electroweak baryogenesis and phase transitions in the early Universe and in high-energy collisions [RuSh96], quantum optics [WaMi94], etc. In the consideration of such processes we are interested in the time regime smaller than the “infinite” time when the S-matrix description becomes applicable. One can say that the consideration of such processes belongs to *nonequilibrium quantum field theory.* We believe that the S-matrix method in quantum field theory is analogous to the Gibbs distribution in equilibrium statistical physics and that there exists a general method (the stochastic limit) in nonequilibrium quantum field theory which provides a description of quantum phenomena depending on time. One of the first works on the systematic application of methods of nonequilibrium statistical physics in quantum field theory is that of Prigogine [Prig68], in which kinetic equations for the Lee model were derived.

Here we advocate the thesis that the stochastic limit method might be a general method in nonequilibrium quantum field theory.

When we say that the S-matrix method is not sufficient in nonequilibrium quantum field theory, we mean that the standard dynamical definition of the S-matrix in real time is given, for example, in terms of wave operators, including dressing [vHo55], [Fad63], [Ar73], or LSZ formalism. This definition cannot be applied immediately to the processes with unstable particles. The flexible Bogoliubov–Shirkov approach [BoSch87] to the S-matrix in principle can be applied to the description of unstable particles. There exists a phenomenological approach to the S-matrix which is not based on an Hamiltonian formalism. In this approach unstable particles are described by the Breit–Wigner complex poles of the scattering amplitudes [Che66, Bo78]. The dynamical justification of this phenomenological approach is given in the Weisskopf–Wigner resolvent method; for a discussion of the resolvent method see, for example, [GoWa64], [PriHe69]. The resolvent method is usually used for the investigation of the degrees of freedom of the system interacting with the reservoir. The stochastic limit method is conveniently used for the consideration of degrees of freedom not only of the system but also of the reservoir.

If $G(x_1 \dots x_n)$ is the Green function then we want to investigate the asymptotic behaviour of the expression $G(\mathbf{x}_1, x_1^0/\lambda^2, \dots, \mathbf{x}_n, x_n^0/\lambda^2)$ when $\lambda \to 0$. This can be performed by using the anisotropic renormalization group method [BoSch87], [ArVo94], [Mat95].

To describe and solve the dynamical equations after the stochastic limit, one has to derive the stochastic limit for the collective operators, the so-called master field. For simple models the master field is the quantum white noise whose creation and annihilation operators satisfy the relations

$$[b(t), b^+(t')] = \delta(t - t'). \tag{14.1.1}$$

In the stochastic limit one obtains for the evolution operator the equation

$$\frac{dU(t)}{dt} = [F^+(t)b(t) + b^+(t)F(t)]U(t),$$

where $b(t), b^+(t)$ are white noise operators and $F(t)$ is a regular operator function of t. This singular equation is the Hamiltonian form of a quantum stochastic differential equation and can be explicitly solved for many models.

For more complex models the master fields are more complex, and one obtains the *entangled* or *interacting commutation relations*. The program of investigation of models of quantum field theory in the stochastic limit consists of two parts. First we have to find the commutation relations for the master field; then we study the singular differential equation for the evolution operator. In recent years various modifications and deformations of the algebra of canonical commutation relations have been discussed, see, for example, [ArVo91], [Gree90], [ArVo96] and references therein. In particular, in the large N limit for $SU(N)$ invariant gauge theory (as well as for $N \times N$ matrix models), the relations

$$b(k)b^+(k') = \delta(k-k') \tag{14.1.2}$$

appear naturally [ArVo96]. Here k, k' are momentum variables. The algebra generated by the operators $b(k), b^+(k')$ satisfying (14.1.2) is called the free (or Boltzmann) algebra.

In this section we prove that the stochastic limit of interacting fields, only under the constraint of momentum conservation, leads to a new algebra of commutation relations.

We start from a model in Fock space with translation-invariant interaction H_{I}. Representing the Hamiltonian density in the momentum representation as

$$H_{\mathrm{I}}(p|k_1 \dots k_m) = H_{\mathrm{I}}^-(p|k_1 \dots k_m) + H_{\mathrm{I}}^+(p|k_1 \dots k_m),$$

we introduce entangled rescaled fields associated with the $\pm$ parts of the interaction Hamiltonian,

$$\mathcal{A}_\lambda^\pm(p|k_1 \dots k_m, t) = \frac{1}{\lambda} \exp\left(\frac{itH_0}{\lambda^2}\right) H_{\mathrm{I}}^\pm(p|k_1 \dots k_m) \exp\left(-\frac{itH_0}{\lambda^2}\right), \tag{14.1.3}$$

where e^{-itH_0} is a free evolution. The above decomposition of H_{I} has natural meaning in term of field operators.

We will prove that the stochastic limit

$$\lim_{\lambda\to 0} \mathcal{A}_\lambda^\pm(p|k_1 \dots k_m, t) = \mathcal{B}^\pm(p|k_1 \dots k_m, t) \tag{14.1.4}$$

exists in the distribution sense (see Theorem 14.2.1) and the limiting operators B^-, B^+ satisfy the relations (14.1.5–6) with energy $E(p,k)$ defined by free evolution. We obtain relations of the form

$$\begin{aligned} &B(p|k_1 \dots k_m)B^+(p'|k_1' \dots k_m') \\ &\quad = n(p)\delta\big(E(p, k_1 \dots k_m)\big)\delta(p-p')\delta(k_1-k_1') \dots \delta(k_n - k_n'), \end{aligned} \tag{14.1.5}$$

where $B^\pm(p|k_1 \dots k_m)$ is the master field, obtained as the stochastic limit of a translation-invariant interaction Hamiltonian, $n(p)$ is the operator density of particles, and $E(p, k_1 \dots k_m)$ is the energy associated with the interaction vertex. These relations are called the *entangled* (or *interacting*) *commutation relations.* Note that the equality (14.1.5) extends the Boltzmann algebra (14.1.2) because the right-hand side is an operator (in the particle space) rather than a scalar: in this sense one speaks of *Hilbert module* [AcLu96] (rather than Hilbert space) commutation relations. $B(p|k_1 \dots k_m)$ and the density $n(p)$ satisfy

$$[n(p'), B(p|k_1 \dots k_m)] = \left[\delta(p'-p) - \delta\left(p' - p + \sum k_i\right)\right] B(p|k_1 \dots k_m). \tag{14.1.6}$$

As one of the physical applications of the above ideas, we will argue that photon-splitting cascades in a magnetic field create entangled states and that photons in cascades obey not Bose statistics but a new type of statistics – infinite or quantum Boltzmann statistics. Therefore, this type of statistics has a physical meaning, since it describes photons in cascades and more generally the dominant diagrams in the long-time/weak-coupling limit in quantum field theory. The states in cascades are formed from triples of entangled photons and may be called triphons. They belong to an interacting Fock space [AcLu96], [AcLuVo97]. An interacting Cuntz algebra has also been considered in [HaSch98].

The standard definition of the stochastic limit is given as the limit of the Wightman correlation function. For some models the limit of these correlation functions is equal to zero. We show that the stochastic limit for the Green functions is nontrivial even in these cases.

In this section we prove that the stochastic limit of interacting fields, under only the constraint of momentum conservation[(1)] leads to a generalization of (14.1.2). We find that the new algebra has as its generators the master field $B(p,k)$, depending on two momenta p and k, and the operator density of particles $n(k)$, which satisfy the relations (special cases of (14.1.5))

$$B(p,k)B^+(p',k') = n(p)\delta\big(E(p,k)\big)\delta(p-p')\delta(k-k'), \tag{14.1.7}$$

$$[n(p'), B(p,k)] = \big[\delta(p'-p) - \delta(p'-p+k)\big]B(p,k), \tag{14.1.8}$$

$$[n(p), n(p')] = 0. \tag{14.1.9}$$

Here $E(p,k)$ is the energy associated with the interaction vertex.

We call the relations (14.1.7–9) *interacting commutation relations* because, on one hand, they allow one to calculate correlations of any order among the field and, on the other hand, they show that the master fields are not kinematically independent. One can obtain a generalization of the algebra (14.1.7–9) for a multiparticle master field $B(p|k_1 \dots k_n)$.

In [AcLuVo97] a generalization of the algebra (14.1.2) of the following form was obtained:

$$B(k)B^+(k') = \delta(k-k')\delta\left(\omega(k) + \hat{P}\cdot k\right), \quad [\hat{P}, B(k)] = kB(k), \tag{14.1.10}$$

(1) To avoid a discussion of the renormalization procedure, we assume that there is an ultraviolet cutoff.

with $\hat{P}$ being the operator of momenta of particles. This algebra is not realized in the usual Fock space but in the interacting Fock space. One can recover the algebra (14.1.10) from the algebra (14.1.7–9) by setting

$$\hat{P} = \int pn(p)dp, \ B(k) = \int B(p,k)dp.$$

These relations show that, contrary to what happens before the limit, the observables of the particle do not commute with the master field. In other words: before the stochastic limit the particles and the field are kinematically independent but nonlinearly related by the dynamics; after the stochastic limit the dynamics is simplified, but the particles and field are no longer kinematically independent. This is what we call *entanglement*. These new features imply that the algebra (14.1.7–9) is not realized in the usual Fock space but in the interacting Fock space [AcLu96]. This chapter is based on the papers [AcArVo97] and [AcArVo00].

14.2 The Tri-linear Hamiltonian with Momentum Conservation

Our starting point will be the Hamiltonian

$$H_\lambda = H_0 + \lambda V,$$

where the free Hamiltonian

$$H_0 = \int \varepsilon(p)c^+(p)c(p)d^3p + \int \omega(k)a^+(k)a(k)d^3k,$$
$$\{c(p), c^+(p')\} = \delta(p-p'), \ [a(k), a^+(k')] = \delta(k-k'),$$

and the interaction Hamiltonian

$$V = \int d^3kd^3pg(k,p)[c^+(p)c(p-k)a(k) + \text{h.c.}] \tag{14.2.1}$$

Here $g(k,p)$ is a test function and $\varepsilon(p)$ and $\omega(k)$ are 1-particle dispersion laws, for example, $\varepsilon(p) = p^2/2$, $\omega(k) = |k|$. The rescaled collective fields in this case have the form

$$\begin{aligned}
\mathcal{A}_\lambda(p,k,t) &= \frac{1}{\lambda} \exp\left(\frac{itH_0}{\lambda^2}\right) c^+(p)a(k)c(p-k) \exp\left(-\frac{itH_0}{\lambda^2}\right) \\
&= \frac{1}{\lambda} c^+(p)a(k)c(p-k) \exp\left(itE(p,k)/\lambda^2\right), \qquad (14.2.2) \\
\mathcal{A}_\lambda^+(p,k,t) &= \frac{1}{\lambda} \exp\left(\frac{itH_0}{\lambda^2}\right) c^+(p-k)a^+(k)c(p) \exp\left(-\frac{itH_0}{\lambda^2}\right) \\
&= \frac{1}{\lambda} c^+(p-k)a^+(k)c(p) \exp\left(-itE(p,k)/\lambda^2\right), \qquad (14.2.3)
\end{aligned}$$

where

$$E(p,k) = \epsilon(p) - \omega(k) - \epsilon(p-k) \tag{14.2.4}$$

is the corresponding energy. One has the following main theorem.

Theorem 14.2.1. *The stochastic limit*

$$\lim_{\lambda\to 0} \mathcal{A}_\lambda(p,k,t) = B^-(p,k,t), \quad \lim_{\lambda\to 0} \mathcal{A}^+_\lambda(p,k,t) = B^+(p,k,t)$$

exists in the sense of the convergence of the matrix elements ($\epsilon_i = \pm$)

$$\begin{aligned}&\lim_{\lambda\to 0} \langle 0|c(q)\mathcal{A}^{\epsilon_1}_\lambda(p_1,k_1,t_1)\dots\mathcal{A}^{\epsilon_n}_\lambda(p_n,k_n,t_n)c^+(q')|0\rangle \\ &\quad = \langle \Psi_0, c(q)B^{\epsilon_1}(p_1,k_1,t_1)\dots B^{\epsilon_n}(p_n,k_n,t_n)c^+(q')\Psi_0\rangle \end{aligned}\tag{14.2.5}$$

as distributions, and the limiting operators $B^- = B$ *and* B^+ *satisfy the entangled commutation relations*

$$B(p,k,t)B^+(p',k',t') = 2\pi\delta(t-t')\delta(p-p')\delta(k-k')\cdot\delta\big(E(p,k)\big)n(p), \tag{14.2.6}$$

$$[n(p'), B^\mp(p,k,t)] = (\pm)(\delta(p'-p) - \delta(p'-p+k))B^\mp(p,k,t), \tag{14.2.7}$$

$$[n(p), n(p')] = 0. \tag{14.2.8}$$

Here Ψ_0 *is the vacuum in the new Hilbert space characterized by the relations*

$$B(p,k,t)\prod_i c^+(q_i)\Psi_0 = 0.$$

We use the same notations for the creation and annihilation operators of c-particles in the original and in the new Hilbert spaces. $n(p)$ *is the operator density of the c-particles,* $n(p) = c^+(p)c(p)$.

Remark 14.2.1. If we set

$$B(p,k,t) = b(t)\otimes B(p,k),$$

where

$$b(t)b^+(t') = 2\pi\delta(t-t'),$$

then we obtain the relations (14.2.5),

$$B(p,k)B^+(p',k') = n(p)\delta\big(E(p,k)\big)\delta(p-p')\delta(k-k').$$

The theorem will be proved in the next section. Note that for the right-hand side of (14.2.6) to be nonzero we have to chose suitable dispersion relations, so that there are nontrivial solutions of $E(p,k) = 0$.

14.3 Proof of Theorem 14.2.1

Let us consider the matrix elements

$$\langle 0|c(q)\prod_{i=1}^{v}\mathcal{A}_{\lambda}^{\epsilon_i}(p_i,k_i,t_i)c^+(q')|0\rangle. \tag{14.3.1}$$

To evaluate (14.3.1) we apply Wick's theorem [BoSch87] and consider only the corresponding connected diagrams. Each vertex contains three lines which are characterized by two momenta (k_i, p_i). We find the momentum corresponding to the third line using momentum conservation. If a diagram contains L loops, then only $L+A+B-1$ momenta are independent (L momenta for the loop variables and $A+B-1$ momenta for the external lines). For every vertex there is the corresponding energy exponent. These energies depend on the momenta of the lines that enter into the given vertex,

$$E_i^{\pm} = E_i^{\pm}(k_i, p_i),$$

via the dispersion laws

$$E_i^{\pm} = [\epsilon(p_i) - \epsilon(p_i \pm k_i) \pm \omega(k_i)]. \tag{14.3.2}$$

In the proof of the theorem we will use Proposition 1.2.1.

Let us consider a connected diagram corresponding to the matrix element (14.3.1). The proof of the theorem consists of three parts. First we prove that only diagrams consisting of pairs of conjugated vertices do not vanish in the limit $\lambda \to 0$. Next we prove that such diagrams are in fact the noncrossing or half-planar diagrams. Finally we show that the noncrossing diagrams are described by the entangled commutation relations.

Generally, the sets of momenta corresponding to different vertices are different. However, it may happen that *the same set of momenta corresponds to two different vertices.* More precisely, momenta which come in the first vertex come out from the second one and vice versa.

Definition 14.3.1. *We say that two incident vertices of a given connected diagram are* conjugated *if the momenta coming in the first vertex come out from the second vertex, i.e. the vertices have the same momenta but with opposite orientation.*

If the i-vertex has a conjugated vertex, then we denote the latter by $\hat{i}$. A typical example of diagrams containing at least one pair of conjugated vertices is a diagram with a mass insertion such that this insertion contains a line that does not cross others lines of the diagram. With these premises, let us prove the main

Lemma 14.3.1. *If a connected diagram does not consist only of pairs of conjugated vertices, then it vanishes in the limit $\lambda \to 0$ (in the sense of distributions).*

Proof 14.3.1. To a given diagram, representing a matrix element of the form (14.3.1) integrated over $t_1 \dots t_v$ with some test functions, corresponds an expression that schematically can be written as

$$\frac{1}{\lambda^v} \int \exp\left(i \sum_{i=1}^{v} E_i t_i/\lambda^2\right) \phi(t,p,q) \prod_{a}^{A+B-1} dp_a \prod_{l=1}^{L} dq_l \prod_{i=1}^{v} dt_i; \tag{14.3.3}$$

here by t we mean $t_1 \dots t_v$, by p we mean $p_1 \dots p_A$, $p'_1 \dots p'_B$ and q denotes the set of independent momenta associated with the diagram under consideration. E_i are given by (14.3.2) and $\phi(t,p,q)$ is a test function.

To evaluate the asymptotic behaviour of this expression when $\lambda \to 0$, we will make a change of variables corresponding to the conjugated vertices. Note that for the nonvanishing diagrams, the number of vertices v should be even, $v = 2n$. Suppose there are n_0 pairs of conjugated vertices, which we denote $\{i_1, \hat{i}_1, \dots, i_{n_0}, \hat{i}_{n_0}\}$. Let us divide the set of all vertices $\{1 \dots 2n\}$ into two disjoint subsets $\{i_1, i_2, \dots, i_{n_0}, i_{n_0+1}, \dots, i_n\}$ and $\{\hat{i}_1, \hat{i}_2, \dots, \hat{i}_{n_0}, i_{n+n_0+1}, \dots i_{2n}\}$ so that in every subset there are no conjugated vertices. We denote the corresponding set of time variables $\{t_{i_1} \dots t_{i_n}\}$ by $t^{(1)}$ and the set $\{t_{\hat{i}_1}, \dots, t_{\hat{i}_{n_0}}, t_{i_{n+n_0+1}}, \dots, t_{i_{2n}}\}$ by $t^{(2)}$. Now we perform the following change of variables

$$\left(t^{(1)}, t^{(2)}\right) \to \left(\tau, t^{(2)}\right), \quad t^{(1)} = t^{(2)} + \lambda^2 \tau, \tag{14.3.4}$$

or more precisely

$$\begin{aligned} \left(t^{(1)}, t^{(2)}\right) &= (t_1 \dots t_{2n}) \to \left(\tau, t^{(2)}\right) \\ &= (\tau_1 \dots \tau_n;\ t_{\hat{i}_j}, j = 1 \dots n_0;\ t_{i_{n+r}},\ r = n_0 + 1 \dots n), \end{aligned} \tag{14.3.5}$$

$$t_{i_j} = t_{\hat{i}_j} + \lambda^2 \tau_j\ , \quad 1 \le j \le n_0, \tag{14.3.6}$$

$$t_{i_j} = t_{i_{n+j}} + \lambda^2 \tau_j\ , \quad n_0 < j \le n. \tag{14.3.7}$$

After this the integral (14.3.3) takes the following form

$$\begin{aligned} &\int \exp\left(i \sum_{j=1}^{n} \tau_j E_{i_j}\right) \exp\left(i \sum_{j=1}^{n_0} (E_{i_j} + E_{\hat{i}_j}) t_{\hat{i}_j}/\lambda^2\right) \\ &\cdot \exp\left(i \sum_{j=n_0+1}^{n} (E_{i_j} + E_{i_{n+j}}) t_{i_{n+j}}/\lambda^2\right) \\ &\cdot \phi\left(t^{(2)} + \lambda^2 \tau, t^{(2)}, p, q\right) \cdot \prod_{j=1}^{n} d\tau_j \prod_{j=1}^{n_0} dt_{\hat{i}_j} \prod_{j=n_0+1}^{n} dt_{i_{n+j}} \prod dp \cdot \prod dq. \end{aligned} \tag{14.3.8}$$

By the definition of conjugated vertices, $E_{i_j} + E_{\hat{i}_j} = 0$, and we are left with

$$\int \exp\left(i\sum_{j=1}^{n} \tau_j E_{i_j}\right) \exp\left[i \sum_{j=n_0+1}^{n} (E_{i_j} + E_{i_{n+j}}) t_{i_{n+j}}/\lambda^2\right]$$
$$\cdot\, \phi(t^{(2)} + \lambda^2\tau, t^{(2)}, p, q) \cdot \prod_{j=1}^{n} d\tau_j \prod_{j=1}^{n_0} dt_{\hat{i}_j} \prod_{j=n_0+1}^{n} dt_{i_{n+j}} \prod dp \cdot \prod dq. \tag{14.3.9}$$

Here $t^{(2)} + \lambda^2\tau$ schematically represents the dependence of half of the t variables on λ. When $\lambda \to 0$ we can neglect the dependence of ϕ on λ and the integration over τ gives a product of $\delta(E_{i_j})$:

$$\int \prod_{j=1}^{n} \delta(E_{i_j}) \exp\left[i \sum_{h=n_0+1}^{n} (E_{i_h} + E_{i_{n+h}}) t_{i_{n+h}}/\lambda^2\right] \phi(t, t, p, q)$$
$$\cdot \prod_{k=n_0+1}^{n} dt_{i_{n+k}} \prod dp \prod dq. \tag{14.3.10}$$

Note that the second exponent in (14.3.9) vanishes since the energies in the conjugated vertices are equal.

Suppose that the diagram contains nonconjugated vertices, i.e. that $n \neq n_0$. When $\lambda \to 0$ (14.3.10) goes to zero, since according to our assumption the set of momenta in the vertices i_j and i_{n+j}, $n_0 < j \leq n$, do not coincide; hence the functions $E_{i_j} + E_{i_{n+j}}$ as functions of momenta do not vanish, and therefore according the Riemann–Lebesgue lemma we obtain zero in the limit $\lambda \to 0$.

In the case when $n = n_0$ the exponent in (14.3.10) vanishes and generally we obtain the nonzero answer:

$$\int \left(\prod_{i=1}^{n} \delta(E_{i_j})\right) \phi\left(\{t_{i_j}\}, \{t_{i_j}\}, p, q\right) \prod dp \prod dq. \tag{14.3.11}$$

The lemma is thus proved.

Lemma 14.3.2. *If a connected diagram with two external lines consists only of pairs of conjugated vertices, then it is half-planar, i.e. it can be drawn in the half-plane without self-intersections.*

We will not present here the simple proof of this lemma.

Lemma 14.3.3. *The limiting expression (14.3.11) is equal to the right-hand side of the relation (14.2.5).*

The proof is obtained by using the algebra (14.2.6–8).

The theorem follows from Lemmata 14.3.1–3.

Remark 14.3.1. We have considered only matrix elements with two external particles, so they do not vanish only if v is even: $v = 2n$. In the general case of matrix elements with an arbitrary number of external lines, the number of vertices v can be odd. For the case of odd v ($v = 2n - 1$), we once again select n vertices and make the following change of variables:

$$(t_1 \dots t_{2n-1}) \to (\tau_1 \dots \tau_n; t_{\hat{i}_j}, j = 1 \dots n_0; t_{i_{n+r}}, r = 1 \dots n - 1 - n_0), \tag{14.3.12}$$

with τ_i as before [see (14.3.6)].

$$\begin{aligned} t_{i_j} &= t_{\hat{i}_j} + \lambda^2 \tau_j, \quad 1 \le j \le n_0, \\ t_{i_j} &= t_{i_{n+j}} + \lambda^2 \tau_j, \quad n_0 + 1 \le j \le n - 1, \\ t_{i_n} &= \lambda^2 \tau_n. \end{aligned}$$

The difference with the case of even v is that we obtain an extra factor λ in front of the integral:

$$\begin{aligned} \lambda \int \exp\left(i \sum_{j=1}^{n} \tau_j E_{i_j} \right) &\exp\left[i \sum_{j=1}^{n_0} (E_{i_j} + E_{\hat{i}_j}) t_{\hat{i}_j} / \lambda^2 \right] \\ &\cdot \exp\left[i \sum_{j=n_0+1}^{n-1} (E_{i_j} + E_{i_{n+j}}) t_{i_{n+j}} / \lambda^2 \right] \\ &\cdot \phi(t + \lambda\tau, t, p, q) \prod dp dq \prod_{j=1}^{n} dt_j \prod_{j=n_0}^{n_0} dt_{i_{n+j}}. \end{aligned} \tag{14.3.13}$$

Here we use the same schematical notations as in (14.3.8). When $\lambda \to 0$ we neglect the λ dependence of ϕ and we get a product of δ-functions. The second exponent disappears. The third exponent disappears in the case when $n_0 = n - 1$, but since we have an extra factor λ, (14.3.8) always goes to zero as $\lambda \to 0$.

Remark 14.3.2. The theorem admits a generalization to the case of an arbitrary number of external particles,

$$\left\langle 0 | \prod_i c(q_i) \mathcal{A}_\lambda^{\epsilon_1}(p_1, k_1, t_1) \dots \mathcal{A}_\lambda^{\epsilon_n}(p_n, k_n, t_n) \prod_j c^+(q'_j) | 0 \right\rangle.$$

Remark 14.3.3. For the composite operators

$$\begin{aligned} \mathcal{A}_\lambda(t, p, k_1 \dots k_m) = &\frac{1}{\lambda} c^+(p) c(p - k_1) a^{\epsilon_1}(k_1) \dots a^{\epsilon_m}(k_m) \\ &\cdot \exp\left[itE(p, k_1 \dots k_m)/\lambda^2 \right] \end{aligned}$$

in the stochastic limit, one obtains the entangled commutation relations (14.3.3).

14.4 Example: Four Internal Lines

Let us illustrate using the example of four internal lines the fact that in the stochastic limit with momentum conservation only diagrams with noncrossing lines survive. To an arbitrary diagram with four internal lines corresponds an expression which schematically can be written as

$$\frac{1}{\lambda^4}\int \exp\left(\frac{i}{\lambda^2}\sum(\pm)E^{(j)}t_j\right)\phi(p,k(q))\,dqdp\prod dt_j .$$

Here ϕ accumulates all the form-factors and test functions. The $E^{(j)}$ are as in (14.3.11); $k^{(j)}$ is the momentum of the line coming into the i-vertex; $p^{(j)}$, $p^{(j)} \pm k^{(j)}$ the momenta of the lines coming out from the i-vertex; $\pm$ corresponds to a vertex with two or one incoming particle, respectively. For diagrams with two crossing lines the following relations between momenta hold:

$$\begin{gathered} k^{(1)}=k^{(3)}, \quad k^{(2)}=k^{(4)}, \\ p^{(1)}-k^{(1)}=p^{(2)}, \quad p^{(2)}-k^{(2)}=p^{(3)}-k^{(3)}, \quad p^{(4)}=p^{(3)}+k^{(4)}. \end{gathered} \tag{14.4.1}$$

This diagram corresponds to two independent momenta; we denote them q. The sets of momenta corresponding to different vertices are different. Now let us consider the diagram with noncrossing lines. In this case,

$$\begin{gathered} k^{(1)}=k^{(4)}, \quad k^{(2)}=k^{(3)}, \\ p^{(1)}-k^{(1)}=p^{(2)}, \quad p^{(2)}=p^{(3)}, \quad p^{(1)}=p^{(4)}. \end{gathered} \tag{14.4.2}$$

We see that there are two vertices such that the momenta coming into the first vertex come out from the second one and vice versa. We call these vertices *conjugated* (conjugated vertices are denoted by a hat, i.e. $\hat{1}=4$ and $\hat{2}=3$). Only diagrams consisting of pairs of conjugated vertices survive in the limit $\lambda \to 0$. Indeed, making the change of variables,

$$\begin{gathered} (t_1,t_2,t_{\hat{1}},t_{\hat{2}}) \to (\tau_1,\tau_2,t_{\hat{1}},t_{\hat{2}}), \\ \tau_1=\frac{t_1-t_{\hat{1}}}{\lambda^2}, \quad \tau_2=\frac{t_2-t_{\hat{2}}}{\lambda^2}, \end{gathered}$$

we see that the relations (14.4.2) give contributions containing the following factors:

$$\delta\left(E^{(1)}_{c\to c+a}\right)\delta\left(E^{(2)}_{c\to c+a}\right)\delta(t_1-t_{\hat{1}})\delta(t_2-t_{\hat{2}}),$$

where the energy factors $E^{(1)}$, $E^{(2)}$ are not independent due to momentum conservation. For the relations (14.4.1) a similar change of variables,

$$\begin{gathered} (t_1,t_2,t_3,t_4) \to (\tau_1,\tau_2,\tau_3,\tau_4), \\ \tau_1=\frac{t_1-t_3}{\lambda^2}, \quad \tau_2=\frac{t_2-t_3}{\lambda^2}, \end{gathered}$$

does not remove the dependence of the exponent on λ, and we are left with

$$\int e^{it(E_3)/\lambda^2} e^{it_4+(E_4)/\lambda^2} \delta(E_1^i)\delta(E_2)\phi(p,k(q))\,dpdq;$$

which vanishes due to the generalized Riemann–Lebesgue lemma.

14.5 The Stochastic Limit for Green Functions

The master field in the standard formulation of the stochastic limit is defined by the limit, as $\lambda \to 0$, of the Wightman correlation functions

$$\left\langle 0|A_\lambda^{(1)}(t_1)\ldots A_\lambda^{(n)}(t_n)|0\right\rangle . \tag{14.5.1}$$

This limit defines the Hilbert space in which the master field lives. For some models the limit of these correlation functions is trivial (equal to zero). However this does not mean that the stochastic limit of such a model is trivial, because we can consider another natural formulation of the stochastic limit which is defined by the convergence of the chronologically ordered correlation functions (Green functions):

$$\left\langle 0|T\left(A_\lambda^{(1)}(t_1)\ldots A_\lambda^{(n)}(t_n)\right)|0\right\rangle . \tag{14.5.2}$$

For the evolution operator one has to consider

$$\left\langle 0|T\left(A_\lambda^{(1)}(t_1)\ldots A_\lambda^{(n)}(t_n)U(t)\right)|0\right\rangle . \tag{14.5.3}$$

Here the T-product is defined as

$$T\left(A_\lambda^{(1)}(t_1)\ldots A_\lambda^{(n)}(t_n)\right) = A_\lambda^{(i_1)}(t_{i_1})\ldots A_\lambda^{(i_n)}(t_{i_n})$$

if $t_{i_1} \geq \ldots \geq t_{i_n}$. The limit of the Green functions can be nontrivial even if the limit of the Wightman functions vanishes. In the simplest case in the first formulation we use

$$\lim_{\lambda\to 0} \frac{1}{\lambda^2} e^{itx/\lambda^2} = 2\pi\delta(t)\delta(x), \tag{14.5.4}$$

and in the second

$$\lim_{\lambda\to 0} \frac{1}{\lambda^2}\theta(t) e^{itx/\lambda^2} = i\delta(t)\frac{1}{x+i0}. \tag{14.5.5}$$

Here

$$\theta(t) = 1, t \geq 0, \quad \theta(t) = 0, t < 0.$$

In particular if

$$A_\lambda(t,p) = \frac{1}{\lambda} e^{it\omega(p)/\lambda^2} a(p),$$

where $[a(p), a^+(p')] = \delta(p-p'),\ p, p' \in R^3$, then the stochastic limit of the Wightman functions is

$$\begin{aligned}\langle 0|A_\lambda(t,p)A_\lambda^+(t',p')|0\rangle &= \frac{1}{\lambda^2} e^{i(t-t')\omega(p)/\lambda^2}\delta(p-p') \\ &\to 2\pi\delta(t-t')\delta\big(\omega(p)\big)\delta(p-p')\end{aligned}$$

and the stochastic limit of the Green functions is

$$\begin{aligned}\langle 0|T(A_\lambda(t,p)A_\lambda^+(t',p'))|0\rangle &= \frac{1}{\lambda^2}\theta(t-t')e^{i(t-t')\omega(p)/\lambda^2}\delta(p-p') \\ &\to i\delta(t-t')\frac{1}{\omega(p)+i0}\delta(p-p').\end{aligned}$$

It is important to notice that in the latter formula one obtains a nontrivial limit even if $\delta\big(\omega(p)\big) = 0$, while in the former the limit is trivial.

14.6 Second Quantized Representation of the Nonrelativistic QED Hamiltonian

In this section we introduce an interaction between two boson fields which, when restricted to the 1-particle subspace of one of the two, reproduces the standard nonrelativistic QED Hamiltonian. To this goal let us consider the Hamiltonian

$$\begin{aligned}H = &\int \varepsilon(p)c^+(p)c(p)dp + \int \omega(k)a^+(k)a(k)dk \\ &+ \int [f(k,p)a^+(k)c^+(p-k)c(p) + \overline{f}(k,p)a(k)c^+(p)c(p-k)]dkdp,\end{aligned} \tag{14.6.1}$$

where $c(p)$, $c^+(p)$ (a_k, a_k^+) act on the Fock space $\mathcal{F}_c$ ($\mathcal{F}_a$) and

$$[a(k), a^+(k')] = \delta(k-k'), \quad [c(p), c^+(p')] = \delta(p-p').$$

The Hamiltonian (14.6.1) acts on the Fock space $\mathcal{F} = \mathcal{F}_a \otimes \mathcal{F}_c$. Here $f(k,p)$ is a test function (form-factor). Defining the *total momentum operator* by

$$P := \int pc^+(p)c(p)dp + \int ka^+(k)a(k)dk,$$

we have the conservation of momenta: $[H, P] = 0$. Let us take the restriction of (14.6.1) to the 1-particle subspace generated by the vectors $|p\rangle = c^+(p)|0\rangle$. We have

$$
\begin{aligned}
H|p\rangle =& \varepsilon(p)|p\rangle + \int \omega(k)a^+(k)a(k)dk|p\rangle \\
&+ \int dkf(k,p)a^+(k)|p-k\rangle + \int dk\overline{f}(k,k+p)a(k)|k+p\rangle. \quad (14.6.2)
\end{aligned}
$$

The number vectors $|p\rangle$ have to be understood in the sense of distributions

$$
|\varphi\rangle = \int \varphi(p)|p\rangle dp, \quad \langle\varphi|p\rangle = \varphi(p). \quad (14.6.3)
$$

This gives

$$
\int \varphi(p)|p+k\rangle dp = \int \varphi(p'-k)|p'\rangle dp' = \int (e^{ikq}\varphi)(p')dp',
$$

because in the momentum representation

$$
q = i\frac{\partial}{\partial p}.
$$

Using the duality (14.6.3) we have

$$
(e^{ikq}\varphi)(p') = \langle e^{ikq}\varphi|p'\rangle = \langle\varphi|e^{-ikq}p'\rangle,
$$

and we can write

$$
\int dkf(k,p)a^+(k)|p-k\rangle = \int dkf(k,p)e^{-ikq}a^+(k)|p\rangle.
$$

Therefore we can rewrite (14.6.2) as

$$
\begin{aligned}
H|p\rangle = \Big(&\varepsilon(p) + \int \omega(k)a^+(k)a(k)dk \\
&+ \int dk[f(k,p)e^{-ikq}a^+(k) + e^{ikq}f(k,p)a(k)]\Big)|p\rangle.
\end{aligned}
$$

If one takes

$$
f(k,p) = pg(k),
$$

then H, restricted to the 1-particle subspace of the c field, will take the form

$$
H|_{1,c} = \varepsilon(p) + \int dk\omega(k)a^+(k)a(k) + \int dk[pg(k)e^{-ikq}a^+(k) + \overline{g}(k)e^{ikq}pa(k)],
$$

which is the standard form of the Hamiltonian describing a particle interacting with a quantum field.

14.7 Interacting Commutation Relations and Nonrelativistic QED Module Algebra

In the previous section we showed that from the point of view of the stochastic limit, the QED Hamiltonian is a particular case of the trilinear interaction considered in Sect. 14.3. In this section we show that the limit algebra is also a generalization of the QED module algebra discussed in Sect. 12.2. Let us consider the entangled commutation relations

$$B^-(p|k_1 \dots k_m)B^+(p'|k_1' \dots k_m') = \delta(p-p')\delta(k-k')\delta\big(E(p,k_1 \dots k_m)\big)n(p), \tag{14.7.1}$$

$$[n(p), B^{\mp}(p'|k_1 \dots k_m)] = [\delta(p-p') - \delta(p'-p \mp \Sigma k_i)]B^{\mp}(p'|k_1 \dots k_n), \tag{14.7.2}$$

$$[n(p), n(p')] = 0. \tag{14.7.3}$$

These relations describe entanglement of a quantum system already at the kinematical ("free") level. Here

$$n(p) = c^+(p)c(p).$$

Let us denote

$$\mathcal{P} = \int pn(p)dp \equiv \int pdE(p) \tag{14.7.4}$$

the second quantized momenta operator and

$$\mathcal{B}^{\pm}(k_1 \dots k_m) = \int B^{\pm}(p|k_1 \dots k_m)dp. \tag{14.7.5}$$

Now if we multiply (14.7.2) by p and integrate over p and p', we obtain

$$[\mathcal{P}, \mathcal{B}^{\mp}(k_1 \dots k_m)] = (\Sigma k_i)\mathcal{B}^{\mp}(k_1 \dots k_m). \tag{14.7.6}$$

Integrating (14.7.1) over p and p' we obtain

$$\mathcal{B}^-(k_1 \dots k_m)\mathcal{B}^+(k_1' \dots k_m') = \delta(k-k')\delta(E(\mathcal{P}, k_1 \dots k_m)), \tag{14.7.7}$$

where we use notations of operator calculus,

$$f(\mathcal{P}) = \int f(p)n(p)dp \equiv \int f(p)dE(p). \tag{14.7.8}$$

Now if we restrict ourselves to the case $m = 1$ and the 1-particle subspace in (14.7.4) then (14.7.6) and (14.7.7) become

$$[P, b^{\mp}(k)] = kb^{\pm}(k) \tag{14.7.9}$$

or

$$Pb^{\mp}(k) = b^{\mp}(k)(P \pm k)$$

and

$$b^-(k)b^+(k') = \delta(k-k')\delta\big(E(\mathcal{P},k)\big). \tag{14.7.10}$$

The relations (14.7.9) and (14.7.10) reduce to those found in the nonrelativistic QED module algebra if we take

$$E(p,k) = \frac{p^2}{2} - \frac{(p-k)^2}{2} - |k| = pk - \frac{k^2}{2} - |k|.$$

14.8 Decay and the Universality Class of the QED Hamiltonian

In the formulation of Theorem 14.2.1 we assumed the special form (14.2.1) of the interaction Hamiltonian. In local quantum field theory the typical Hamiltonian is more complicated than (14.2.1). In particular, for the Yukawa interaction of fields ψ and ϕ the Hamiltonian has the form

$$H_\lambda = H_0 + \lambda V, \tag{14.8.1}$$

where H_0 is the sum of the free Hamiltonians for the fermionic field ψ, and for the scalar field ϕ with relativistic dispersion laws

$$\omega_a(k) = \sqrt{m_a^2 + k^2}, \qquad \omega_b(k) = \sqrt{m_b^2 + k^2}, \tag{14.8.2}$$

$$\begin{aligned} V &= \int d^3x g\bar{\psi}\psi\phi \\ &= \int d^3k d^3p g(k,p)[c^+(p)c(p-k)a(k) + c^+(p)c^+(k-p)a(k) \\ &\qquad\qquad + c^+(p)a^+(k)c^+(-p-k) + \text{h.c.}]. \end{aligned} \tag{14.8.3}$$

Here $g(k,p)$ is a test function. This expression is not a well-defined operator in the Fock space, but it defines a bilinear form. We have the following collective operators:

$$\frac{1}{\lambda} c^+(p)c(p-k)a(k) \exp\left(\frac{it}{\lambda^2}\left[\omega_c(p) - \omega_c(p-k) - \omega_a(k)\right]\right), \tag{14.8.4}$$

$$\frac{1}{\lambda} c^+(p)c^+(k)a(p+k) \exp\left(\frac{it}{\lambda^2}\left[\omega_c(p) + \omega_c(k) - \omega_a(p+k)\right]\right), \tag{14.8.5}$$

$$\frac{1}{\lambda} c^+(p)c^+(k)a^+(-p-k) \exp\left(\frac{it}{\lambda^2}\left[\omega_c(p) + \omega_c(k) + \omega_a(p+k)\right]\right). \tag{14.8.6}$$

In the limit $\lambda \to 0$ the operator (14.8.6) vanishes because in the correlation functions one obtains

$$\delta\big(\omega_c(p) + \omega_a(k) + \omega_c(p+k)\big) = 0$$

due to the positivity of energy. The limit of the operator (14.8.4) is also zero, since for the relativistic dispersion laws (14.8.2) one has

$$\delta\big(\omega_c(p) + \omega_a(k) - \omega_c(p+k)\big) = 0.$$

Only the operator (14.8.5) has the chance to be nonzero. In this case we have to find nontrivial solutions of

$$\omega_c(p) + \omega_c(k-p) = \omega_a(k). \tag{14.8.7}$$

There are solutions if

$$m_a^2 > 2m_c^2,$$

i.e. if there is decay. Therefore we find that the relativistic interaction Hamiltonian (14.8.3) is within the same stochastic *universality class* as the interaction Hamiltonian (14.2.1)

$$V = \int d^3k d^3p g(k,p)\big(c^+(p)c^+(k-p)a(k) + \text{h.c.}\big). \tag{14.8.8}$$

Remark 14.8.1. The above statement, and more generally the results of Sects. 14.2–14.5 are not in contradicton with Theorem 11.5.1 (the universality class principle). In fact this theorem holds when one considers:

(i) all diagrams;
(ii) arbitrary correlations;
(iii) interactions without momentum conservation.

In the present chapter we are considering:

(i) only connected diagrams;
(ii) interactions with momentum conservation
(iii) (sometimes) only time-ordered correlations.

In these situations also the N-type blocks can give nontrivial contributions to the stochastic limit.

In QED the interaction has the form

$$V = \int d^3x \overline{\psi}\gamma^\mu \psi A_\mu.$$

If one neglects the spinor and polarization indices then the free Hamiltonian has the form (14.8.2), and the interaction Hamiltonian has the form (14.8.3). We see that the stochastic limit of the QED Hamiltonian is reduced to the form of the above-discussed Hamiltonian, and as a result we obtain a trivial limit, in agreement with the fact that there is no decay in the standard QED. However, if we consider QED in an external field we can obtain a nonzero result due to the change in the dispersion law. One such example will be considered in the next section.

14.9 Photon Splitting Cascades and New Statistics

In the previous section we have seen that the stochastic limit of the Wightman functions in relativistic QED is trivial and that it can be nontrivial in the presence of external fields. It is known that in an external magnetic field a photon splits into two. This splitting is one of the most interesting manifestations of the nonlinearity of Maxwell's equations with radiative corrections. In a constant uniform field, this process occurs with conservation of energy and momentum. The process was considered by Adler et al. in the early 1970s by using the Heisenberg–Euler effective Lagrangian [ABCR70], [Adl71], [BeLi82]. Photon splitting was considered as a possible mechanism for the production of linearly polarized gamma-rays in a pulsar field. Recently the splitting of photons has found astrophysical applications in the study of annihilation line suppression in gamma-ray pulsars and spectral formation of gamma-ray bursts from neutron stars [BHG97], [HBG96]. Photon splitting cascades have also been used in models of soft gamma-ray repeaters, where they soften the photon spectrum [Bar95]. The process of photon splitting is potentially important in applications for a possible explanation of the origin of high-energy cosmic rays from Active Galactic Nuclei [Pro96]. A recalculation of the amplitude for photon splitting in a strong magnetic field has been performed recently in [AdSc96], [HeyHe97].

We will start from a discussion of the theory of photon splitting cascades, show the emergence of infinite statistics in this theory and then discuss its connection with the stochastic limit of quantum field theory. In the decay of a photon with momenta k into photons with momenta k_1 and k_2, we have conservation of momentum and energy: $k = k_1 + k_2$, $\omega(k) = \omega_1(k_1) + \omega_2(k_2)$. For photons in vacuum, in the absence of external fields, $\omega = \omega_1 = \omega_2 = k$, and although these two equations have a solution, the decay is forbidden by the invariance under charge conjugation (Furry's theorem).

In a constant uniform magnetic field B_0 there are only two decay processes kinematically allowed: $\gamma_{\|} \to \gamma_{\perp} + \gamma_{\perp}$ and $\gamma_{\|} \to \gamma_{\|} + \gamma_{\perp}$. Here the subscripts $\perp$ and $\|$ denote polarizations of the photon with respect to the vector B_0. More precisely, in the presence of a magnetic field one has a distinctive plane, namely the kB_0 plane. One takes the linear polarizations of the magnetic field of the photon parallel and orthogonal to this plane as the two independent polarizations of the photon, $\|$ and $\perp$, respectively.

The vacuum in the presence of the field B_0 acquires an index of refraction n, and the photon dispersion relation is modified from $k/\omega = 1$ to $k/\omega = n$. The indices of refraction $n_{\|,\perp}$ can be calculated from the Heisenberg–Euler effective Lagrangian. Adler has shown that for subcritical fields in the limit of weak vacuum dispersion, only the splitting mode $\| \to \perp + \perp$ operates below the pair production threshold. For weak dispersion $n_{\perp} = 1 + \frac{7}{90}\beta$ and $n_{\|} = 1 + \frac{2}{45}\beta$, where $\beta = \frac{e^4 h}{m^4 c^7} B_0^2 \sin^2\theta$ and θ is the angle between k and B_0. It is mentioned by Harding et al. that in magnetar models of soft gamma-ray repeaters [HBG96], where supercritical fields are employed, moderate vacuum dispersion arises. In such a regime, it is not clear whether Adler's selection rules still endure, since in his analysis higher-order contributions to the vacuum polarization are omitted. In [HBG96] photon cascades are considered for the case where all three photon splitting modes allowed by CP invariance are operating. Baier et al. [BaMiSh96] have found that there is only one allowed transition ($\| \to \perp + \perp$) for any magnetic field. They suggested that a photon cascade could develop only if the magnetic field changes its direction. It seems that the question of the validity of Adler's rule for a nonweak vacuum dispersion deserves further study. In this section we consider photon cascades when both kinematically allowed modes ($\| \to \perp + \perp$ and $\| \to \| + \perp$) operate. The interaction operator for the decay $\| \to \perp + \perp$ is known to be [BeLi82]

$$V_1(t) = \lambda_1 \int (B_0 E_1)(B_0 E_2)(B_0 B) d^3 x, \tag{14.9.1}$$

where the coupling constant $\lambda_1 = 13e^6/315\pi^2 m^8$ and magnetic and electric parts of photon field are

$$\begin{aligned} B &= i(4\pi)^{1/2} k \times E_{\|} e^{-i(kr-\omega t)} a_{\|}(k), \\ E_1 &= -i(4\pi)^{1/2} \omega_1 E_{\perp} e^{i(k_1 r - \omega_1 t)} a_{\perp}^{+}(k_1); \end{aligned} \tag{14.9.2}$$

similarly for E_2, $\omega = \omega_{\|}(k)$ and $\omega_i = \omega_{\perp}(k_i)$, $i = 1, 2$. Here B, B_0, E_i are three-dimensional fields, and k_i are three dimensional vectors.

For the decay $\| \to \| + \perp$ one has a similar interaction operator with the operator structure

$$\mathcal{A}^{+}(t) = \lambda a_{\|}^{+}(p-k) a_{\perp}^{+}(k) a_{\|}(p) e^{-itE}, \tag{14.9.3}$$

where $E = \omega_{\|}(p) - \omega_{\|}(p-k) - \omega_{\perp}(k)$. The coupling constant λ in this case can be estimated to be $\lambda/\lambda_1 = \alpha(B_0/B_{\text{cr}})^2$, where α is the fine structure constant, $\alpha = e^2/\hbar c$ and $B_{\text{cr}} = m^2 c^3/e\hbar \simeq 4.4 \times 10^{13}$ Gauss.

Let us consider a photon cascade created by a photon with momentum p and polarization $\|$. The photon splits as $\gamma_{\|}(p) \to \gamma_{\perp}(k_1) + \gamma_{\|}(p-k_1)$. Then one has the next generation of splitting: $\gamma_{\|}(p-k_1) \to \gamma_{\perp}(k_2) + \gamma_{\|}(p-k_1-k_2)$ etc. After N generations of splitting one has a cascade with N photons with $\perp$ polarization and momenta $k_1, k_2, \ldots, k_N$ and also one photon with $\|$ polarization and momentum $k - k_1 - \cdots - k_N$.

Our goal is to consider cascades with real photons (i.e. on the mass shell) including the intermediate states (compare with what one sees in the Wilson camera). The corresponding diagram is not a Feynman one because all the lines (including an intermediate one) correspond to real particles on the mass shell and not to virtual states. More precisely all the lines in the diagram are "dressed" lines on the mass shell, and the initial photon $\gamma_{\|}(k)$ is prepared in a special way such that it undergoes the decay in a finite time. So we cannot use the standard S-matrix approach and the standard Feynman diagram technique to describe this process. The diagram is also not a diagram in the noncovariant diagram technique because we have the conservation of energy at every vertex. A cascade may be intuitively described by the following state:

$$|\psi(p,k_1,k_2)\rangle = f(p,k_1)f(p-k_1,k_2)a_{\parallel}^{+}(p-k_1-k_2)a_{\perp}^{+}(k_2)a_{\perp}^{+}(k_1)|0\rangle, \tag{14.9.4}$$

where momentum conservation is built in creation and annihilation operators and energy conservation is accounted for by the factor $f(p,k) = f\big(\omega_{\parallel}(p) - \omega_{\perp}(k) - \omega_{\parallel}(p-k)\big)$, where $f(\omega)$ is a function with support at $\omega = 0$. As we shall see below this is not the δ-function but roughly speaking its "square root". Indeed, the transition amplitude between two cascade states is given by the scalar product

$$\begin{aligned}&\langle\psi(p',k_1',k_2')|\psi(p,k_1,k_2)\rangle\\ &\qquad=|f(p,k_1)|^2|f(p-k_1,k_2)|^2\delta(p-p')\delta(k_1-k_1')\delta(k_2-k_2').\end{aligned} \tag{14.9.5}$$

Note that in the scalar product (14.9.5) only the noncrossing diagram contributes. In fact the contribution from the crossing diagram is negligible because of conservation of energy and momentum. This is the crucial point where the difference between our diagrams describing real particles in intermediate states and the Feynman diagrams having virtual particles in intermediate states is manifested. In the Feynman diagram technique the amplitude of emission of the two photons is represented by a sum of two diagrams differing by the order in which the two photons are emitted. Here we have only one diagram.

Now let us observe that if in (14.9.4) we replace the operators $a_{\perp}^{+}(k_1)$ and $a_{\perp}^{+}(k_2)$ by the quantum Boltzmann operators $b_{\perp}(k_1)$ and $b_{\perp}(k_2)$ satisfying the relations (14.1.2), i.e. $b_{\perp}(k)b_{\perp}^{+}(p) = \delta(k-p)$, then it will be automatically guaranteed that only the noncrossing diagrams survive. Therefore it is natural to describe cascade wave functions in terms of these operators. It is well known that standard free photons are bosons. Therefore to see the quantum Boltzmann statistics, we have to prepare a special state depending on the interaction. In fact it is natural to expect that the cascades with physical intermediate states occur at a time scale slower than the one occurring in the standard S-matrix approach to multiparticle production. A natural method leading to this result is suggested by the stochastic limit technique.

Now let us consider the question of how one can prepare a state which exhibits the new statistics for photons in cascade. If we would deal with the scattering of 2-particle states at infinite time (S-matrix), we simply have to consider two Feynman diagrams to take into account the Bose statistics of photons. However, in the cascade, we deal with evolution in finite time, and the states of photons $\gamma_{\perp}(k_1)$ and $\gamma_{\perp}(k_2)$ are prepared in a special way because they are emitted at times t_1 and t_2, respectively. Therefore, there is a reason not to add the second diagram. A special procedure which is adequate to this situation is the stochastic limit technique. In our case the master field is given by the asymptotic behaviour of the following collective operator:

$$\lim_{\lambda\to 0}\frac{1}{\lambda}a^+_{\|}(p-k)a^+_{\perp}(k)a_{\|}(p)e^{-itE/\lambda^2}=\mathcal{B}^+(p,k), \tag{14.9.6}$$

where E is the same as in (14.9.3) and the limit is meant in the sense of the Wightman correlation functions.

It follows from Theorem 14.2.1 that the master field $\mathcal{B}^{\pm}(p,k)$ satisfies the commutation relations

$$\mathcal{B}(p,k,t)\mathcal{B}^+(p',k',t')=2\pi\delta(t-t')\delta(E)\delta(k-k')\delta(p-p')b^+_{\|}(k)b_{\|}(k), \tag{14.9.7}$$

and $b_{\|}(p), b^+_{\|}(p')$ satisfy the relations

$$b_{\|}(p)b^+_{\|}(p')=\delta(p-p'). \tag{14.9.8}$$

The presence of the $\delta(E)$-factor $[E=\omega_{\|}(p)-\omega_{\|}(p-k)-\omega_{\perp}(k)]$ has two important physical consequences. First, the commutation relations for the $B^{\#}$ are not a consequence of the corresponding relations for $b^{\#}_{\|}$ and $b^{\#}_{\perp}$: the three photons are entangled into a single new object that we call a *triphon*. Second, the triphon creation and annihilation operators $B^{\#}$ operate not on the usual Fock space but in the interacting Fock space. By introducing the auxiliary creation and annihilation operators $b(t), b^+(t),\ b_{\perp}(p), b^+_{\perp}(p)$, satisfying the quantum Boltzmann relations

$$b(t)b^+(t')=\delta(t-t'), \tag{14.9.9}$$

$$b_{\perp}(p)b^+_{\perp}(p')=\delta(p-p'), \tag{14.9.10}$$

and introducing the symbolic relation

$$\mathcal{B}^+(p,k,t)=b^+(t)b^+_{\|}(k_1)b^+_{\perp}(k_2)b_{\|}(k)(2\pi)^{1/2}\delta_{1/2}(E), \tag{14.9.11}$$

we can disentangle the master field by expressing it as a product of individual Boltzmannian fields. Here the notation $\delta_{1/2}(E)$ is purely symbolic; it simply means that, since the new commutation relation (14.9.7) is quadratic in the master creation and annihilation operators, if in the above symbolic relation we consider the symbol $\delta_{1/2}(E)$ as a scalar function satisfying the formal relation $[\delta_{1/2}(E)]^2=\delta(E)$, and if we use the right-hand side of (14.9.11) to express the left-hand side of (14.9.7), then the standard Boltzmannian relations (14.9.10) and (14.9.8) will reduce (14.9.7) to an identity. The intuitive understanding of the "disentangling" relation (14.9.11) is that the triphon master field $\mathcal{B}^+(p,k,t)$ can be expressed as the product of three "Boltzmannian photons": one can consider that each photon of the triple has its own (Boltzmannian) creation and annihilation operators depending on its own momentum; however in the master field these operators can only appear in the combination given by the right-hand side of (14.9.11), and there is a constraint among the three momenta expressed by energy conservation.

Note the Boltzmannian white noise relation (14.9.9), which makes our model particularly suitable for Monte Carlo simulations. The origin of these new commutation relations lies in the fact that the crossing diagrams in the computation of the matrix element (14.9.4) are suppressed in the weak-coupling/long-time limit.

A photon splits into two not only in a magnetic field but also in a nonlinear medium. In fact such processes are well known in nonlinear quantum optics (see, for example, [WaMi94]). In the nonlinear process of parametric down conversion a high-frequency photon splits into two photons with frequencies such that their sum equals that of the high-energy photon. The two photons produced in this process possess quantum correlations and have identical intensity fluctuations.

In conclusion, in this section we have argued that photon cascades in a strong magnetic field might create a new type of entangled states (triphons) which obey not Bose but quantum Boltzmann statistics. This prediction is based on the assumption that both kinematically allowed photon splitting modes operate. Given the validity of this assumption, we prove that in the stochastic regime the intermediate photons in a cascade are real and virtual particles. The dominating contributions to the dynamics come from these entangled triples of photons which behave like single new particles, *triphons*, whose statistics can be experimentally observed by counting the emitted photons in the corresponding cascades.

As explained in the introduction, the time scale in which our predictions are true is long compared to the strength of the coupling but short if compared to the time scale of the S-matrix approach. This remark should be kept into account in a possible experimental verification of these predictions. A better theoretical understanding of the photon splitting with nonweak dispersion is required. From the experimental side new more precise devices such as the planned Integral mission [BHG97] might significantly advance our understanding of the fundamental problem of photon statistics.

Part III

Estimates and Proofs

15. Analytical Theory of Feynman Diagrams

Part III is devoted to some analytical results useful in the consideration of the stochastic limit. As explained in the Preface, we will not include the detailed proofs of all the equations deduced in the text. Instead of this we have tried to condense into a few mathematical theorems, the basic estimates which can be applied to a multiplicity of models. We have organized the material as follows:

In Chap. 15, we give some basic analytical results. These essentially amount to an analytical representation and estimates of the Feynman diagrams. These estimates allow one to take the term-by-term limit of the iterated series expansion of the evolution operator.

Chapter 16 is devoted to prove the convergence, in the sense of term-by-term convergence of the iterated series expansion, of the time evolution operator U_{t/λ^2} and the Heisenberg evolutions to the solution of a Langevin equation, i.e. $U_{t/\lambda^2}(1 \otimes X)U^*_{t/\lambda^2}$ and $U^*_{t/\lambda^2}(1 \otimes X)U_{t/\lambda^2}$, where X is a *good* (e.g. bounded) linear operator on the system space. We prove the result in the Bose and in the Fermi case for a general gauge-invariant Gaussian initial state of the reservoir (in particular for the Fock and the finite temperature equilibrium states). This is done by:

(i) simplifying the interaction Hamiltonian;
(ii) stating and proving several principles by which the weak-coupling limit (WCL) works;
(iii) giving and proving the precise form of the limit space on which the above limits live;
(iv) calculating the limit, as $\lambda \to 0$, of the matrix elements of U_{t/λ^2}, $U_{t/\lambda^2}(1 \otimes X)U^*_{t/\lambda^2}$ and $U^*_{t/\lambda^2}(1 \otimes X)U_{t/\lambda^2}$ with respect to collective vectors; these limits are calculated in the sense of term-by-term convergence of their iterated series expansion.

We work in the open system or system–reservoir scheme described in Chap. 5, and we keep the notations of this chapter. In particular we denote

(i) $\mathcal{H}_S$ the state space of the system S, and $\mathcal{H}_1$ the 1-particle state space of the reservoir R. Both are Hilbert spaces.

(ii)

$$\Gamma(\mathcal{H}_1) \otimes \mathcal{H}_S$$

the state space of the (S, R) system. Often in the previous chapters we adopted the notation $\mathcal{H}_S \otimes \Gamma(\mathcal{H}_1)$, but for some purposes the present notation is more convenient. Any operator L on $\mathcal{H}_S$ $[\Gamma(\mathcal{H}_1)]$ will be identified (if no confusion can arise) to $1 \otimes L$ $[L \otimes 1]$;

(iii) $1 \otimes H_S$ the Hamiltonian of the system S, and $H_R \otimes 1$ the Hamiltonian of the reservoir; H_{I} the interaction Hamiltonian;

(iv) λ the coupling constant between the system and the reservoir, and

$$1 \otimes H_S + H_R \otimes 1 + \lambda H_{\mathrm{I}} =: H_0 + \lambda H_{\mathrm{I}} =: H_{\text{full}}$$

the total Hamiltonian (in this book we deal mainly with the weak-coupling regime);

(v) $\varphi(\cdot) := \langle \Phi, \cdot \Phi \rangle$ the initial state of the reservoir.

Let there be given a pre-Hilbert space $\mathcal{H}$. Unless otherwise specified, the boson-Fock space $\Gamma(\mathcal{H})$, over $\mathcal{H}$, is defined as

$$\mathbb{C} \oplus \bigoplus_{n=1}^{\infty} \mathcal{H}^{\oplus n} .$$

For $n = 0$, $\mathcal{H}^{\otimes n} = \mathbb{C} \cdot \Phi$, where Φ is the Fock vacuum and, for $1 \leq n \in \mathbb{N}$, as a linear space, $\mathcal{H}^{\otimes n}$ is the subspace of the tensor product of n copies of $\mathcal{H}$ consisting of all the elements invariant under arbitrary permutations of the factors in the tensor product. An element of $\mathcal{H}^{\otimes n}$ is called an n-particle vector; the symmetrized tensor product of $f_1 \cdots f_n \in \mathcal{H}$ will be denoted by $f_1 \otimes \cdots \otimes f_n$, and the scalar product between two n-particle vectors is defined by

$$\langle f_1 \otimes \cdots \otimes f_n, g_1 \otimes \cdots \otimes g_n \rangle := \sum_{\sigma \in \mathcal{S}_n} \prod_{k=1}^{n} \langle f_{\sigma(k)}, g_k \rangle ,$$

where, here and in the following, $\mathcal{S}_n$ denotes the set of all permutations on $\{1 \ldots n\}$. For any $f \in \mathcal{H}$, the creator with the "test function" f is defined by

$$A^+(f) f_1 \otimes \cdots \otimes f_n = f \otimes f_1 \otimes \cdots \otimes f_n .$$

If $f \in \mathcal{H}$, the identity (with $g_1 := f$)

$$\left\langle f_1 \otimes \cdots \otimes f_n, A^+(f) g_2 \otimes \cdots \otimes g_n \right\rangle = \sum_{\sigma \in \mathcal{S}_n} \prod_{k=1}^{n} \left\langle f_{\sigma(k)}, g_k \right\rangle$$

$$= \langle f_1 \otimes \cdots \otimes f_n, g_1 \otimes \cdots \otimes g_n \rangle = \sum_{k=1}^{n} \langle f_k, g_1 \rangle \sum_{\substack{\sigma \in \mathcal{S}_n \\ \sigma(1)=k}} \prod_{k=2}^{n} \langle f_{\sigma(k)}, g_k \rangle$$

$$= \left\langle \sum_{k=1}^{n} \langle f, f_k \rangle f_1 \otimes \cdots \otimes \widehat{f_k} \otimes \cdots \otimes f_n, g_2 \otimes \cdots \otimes g_n \right\rangle$$

shows that the adjoint of the creator $A^+(f)$ is well defined on the n-particle vectors and that it has the form

$$A(f)\Phi := 0\,,$$

$$A(f)\, f_1 \otimes \cdots \otimes f_n := \sum_{k=1}^{n} \langle f, f_k \rangle\, f_1 \circ \cdots \otimes \widehat{f_k} \otimes \cdots \otimes f_n\,,$$

where, as usual, $\widehat{g}$ means that g is absent. $A(f)$ is called the annihilator with the "test function" f. From the identities

$$A(f)\, A^+(g_{n+1})\, g_1 \otimes \cdots \otimes g_n = \sum_{k=1}^{n+1} \langle f_{n+1}, g_k \rangle\, g_1 \otimes \cdots \otimes \widehat{g_k} \otimes \cdots \otimes g_{n+1}\,,$$

$$A^+(g_{n+1})\, A(f)\, g_1 \otimes \cdots \otimes g_n = \sum_{k=1}^{n} \langle f_{n+1}, g_k \rangle\, g_1 \otimes \cdots \otimes \widehat{g_k} \otimes \cdots \otimes g_{n+1}\,,$$

subtracting the second from the first, one finds the *canonical commutation relations* (CCR):

$$\left[A(f), A^+(g)\right] = \langle f, g \rangle\,.$$

Similarly, one can introduce the fermion Fock space and creation–annihilation operators. Instead of the CCR one now has the *canonical anti-commutation relations* (CAR).

15.1 The Connected Component Theorem

Unless otherwise specified throughout this chapter we shall use the following notations:

- $n, N \in \mathbb{N}$ are natural integers such that $n \geq 2$, $N \geq n/2$;
- $S_1, T_1, \cdots, S_n, T_n \in \mathbb{R}$ are real numbers such that $S_k \leq T_k$, $\forall k = 1 \cdots n$;
- $F_1 \cdots F_N \in L^1(\mathbb{R})$ are complex valued functions;
- $P_1 \cdots P_N$ are real valued locally bounded functions on $\mathbb{R}$.

The main purpose of this section is to estimate integrals of the form

$$\int_{S_1}^{T_1} dt_1 \cdots \int_{S_n}^{T_n} dt_n \prod_{h=1}^{N} \left| F_h\left(\frac{t_{p_h} - t_{q_h} + P_h(\lambda)}{\lambda^2}\right)\right|\,.$$

Definition 15.1.1. *Let there be a product of the form*

$$\prod_{h=1}^{N} F_h\left(t_{p_h} - t_{q_h}\right)\,, \tag{15.1.1}$$

where $1 \leq p_h \neq q_h \leq n$. *For* $1 \leq \alpha, \beta \leq n$ *we say that* t_α *and* t_β *are* connected *with respect to (15.1.1) if there exist an integer* m *and indices* $1 \leq h_1 < h_2 < \cdots < h_m \leq N$ *such that in (15.1.1) there is a subproduct*

$$\prod_{j=1}^{m} F_{h_j}\left(t_{p_{h_j}} - t_{q_{h_j}}\right),$$

with

$$\alpha \in \{p_{h_1}, q_{h_1}\}, \ \beta \in \{p_{h_m}, q_{h_m}\}, \tag{15.1.2a}$$

$$\left\{p_{h_j}, q_{h_j}\right\} \cap \left\{p_{h_{j+1}}, q_{h_{j+1}}\right\} \neq \emptyset \quad , \quad \forall\, j = 1, 2, \cdots, m-1. \tag{15.1.2b}$$

We shall use the notation $t_\alpha \rightleftharpoons t_\beta$ *to denote that* t_α *,* t_β *are connected. By definition any* t_h *is connected to itself, i.e.* $t_h \rightleftharpoons t_h$ $(h \in \{1, 2, \cdots, N\})$*. Clearly* to be connected *is an equivalence relation. Its equivalence classes are called* connected components. *They uniquely determine a partition of the set* $\{1, 2, \cdots, n\}$.

Definition 15.1.2.

(i) *We say that the variables* $\{t_1, t_2, \cdots, t_n\}$ *are* totally connected *if for any* $h, r \in \{1, 2, \cdots, n\}$, $t_h \rightleftharpoons t_r$; *this is equivalent to saying that all variables* $t_1 \cdots t_n$ *belong to the same connected component.*
(ii) *We say* t_α, t_β *are* directly connected *if there is an* $h = 1, 2, \cdots, N$ *such that*

$$\{\alpha, \beta\} = \{p_h, q_h\}.$$

Sometimes, we will omit the time index t*, and we will say that* $\{1, 2, \cdots, n\}$ *are* totally connected *if* $\{t_1, t_2, \cdots, t_n\}$ *are* totally connected.

Remark 15.1.1. The intuitive notion of a connected component can be described by the following example: Consider the following vacuum expectation value of four annihilators and four creators:

$$\langle A(S_{t_1} f_1)^2 A(S_{t_2} f_2)^2 A^+(S_{t_3} f_3)^2 A^+(S_{t_4} f_4)^2 \rangle. \tag{15.1.3}$$

By Gaussianity this is a sum of products of factors,

$$\prod_{j \in \{1,2\}, k \in \{3,4\}} \langle A(S_{t_j} f_j) A^+(S_{t_k} f_k) \rangle. \tag{15.1.4}$$

In these pairings we distinguish two situations:

(i) each block $A^{\varepsilon_i}(S_{t_j} f_j)^2$ is entirely paired with another block;
(ii) at least one block is paired with more than one other blocks.

In case (i), if the block $A(S_{t_1}f_1)^2$ is paired with $A^+(S_{t_k}f_k)^2$, then (15.1.3) must be the sum for $\{k,l\}=\{3,4\}$ of

$$\langle A(S_{t_1}f_{t_1}),A^+(S_{t_k}f_k)\rangle^2\langle A(S_{t_2},f_2),A^+(S_{t_l}f_l)\rangle^2\,. \tag{15.1.5}$$

In this example the pairs $\{t_1,t_k\}$ and $\{t_2,t_l\}$ are connected with respect to (15.1.5), but the pairs $\{t_1,t_l\}$, $\{t_2,t_k\}$ are not connected.

In case (ii), the only possibility is that (15.1.4) contains a subproduct of the form

$$\begin{gathered}\langle A(S_{t_1}f_1)A^+(S_{t_3}f_3)\rangle\\ \langle A(S_{t_1}f_1)A^+(S_{t_4}f_4)\rangle\\ \langle A(S_{t_2}f_2)A^+(S_{t_3}f_3)\rangle\\ \langle A(S_{t_2}f_2)A^+(S_{t_4}f_4)\rangle\,.\end{gathered} \tag{15.1.6}$$

Note that (15.1.4) contains both factors of the block $A(S_{t_1}f_1)^2$ but only one factor of the blocks $A^+(S_{t_3}f_3)^2$, $A^+(S_{t_4}f_4)$. So in this case we have only one connected component and every t_α is connected with every other t_β $(\alpha\neq\beta)$.

In the following, for any function F, we denote by $F^{(\lambda)}$ the function

$$F^{(\lambda)}(x):=F\left(\frac{x}{\lambda^2}\right)\,.$$

Lemma 15.1.1. *Let*

- $S\leq T$;
- $G\in L^1(\mathbb{R})$ *positive;*
- *locally bounded functions* $P_1\cdots P_N$.

Then the integral

$$\int_S^T dt_1\cdots\int_S^T dt_n\prod_{h=1}^N G\left(\frac{t_{p_h}-t_{q_h}+P_h(\lambda)}{\lambda^2}\right) \tag{15.1.7}$$

is of the order $O\left(\lambda^{2n-2}\right)$ *for any* $n\geq 2$, $N\geq n/2$ *if the following three conditions are satisfied:*

(i) $\prod_{h=1}^N G\left(\frac{t_{p_h}-t_{q_h}}{\lambda^2}\right)$ *has totally connected variables;*
(ii) for any $1\leq h\neq k\leq n$, $\{p_h,q_h\}\neq\{p_k,q_k\}$;
(iii) for any $r,h,k\in\{1,2,\cdots,N\}$ *such that* $h\neq k$, *for any* $j=1,2,\cdots,N$ *and for any distinct pairs* $\{t_{p_h},t_{q_h}\}\neq\{t_{p_k},t_{q_k}\}$, *by deleting the connecting variable from one pair one cannot obtain another pair, i.e.*

$$\{p_h,q_h,p_k,q_k\}\setminus\{p_r\}\neq\{p_j,q_j\}\ \text{ and }\ \{p_h,q_h,p_k,q_k\}\setminus\{q_r\}\neq\{p_j,q_j\}\,.$$

Remark 15.1.2. In order to understand the meaning of condition (iii), let us consider an example in which this condition is not verified (see also the comments before Lemma 15.1.2). Take $N = n = 3,\ p_1 = 1, q_1 = 2,\ p_2 = 1, q_2 = 3,\ p_3 = 2,\ q_3 = 3$. Then

$$\{p_1, q_1, p_3, q_3\} \setminus \{q_1\} = \{1,2,3\} \setminus \{2\} = \{1,3\} = \{p_2, q_2\} .$$

If condition (iii) is verified, for any subproduct of

$$\prod_{h=1}^{N} G\left(\frac{t_{p_h} - t_{q_h} + P_h(\lambda)}{\lambda^2}\right)$$

but not itself, there is at least one variable t_j disappearing in the subproduct.

Proof 15.1.1. If $n = 2$, by the change of variables

$$s_1 := t_1 , \qquad s_2 := \frac{t_2 - s_1 + P(\lambda)}{\lambda^2} ,$$

(15.1.7) becomes

$$\int_S^T dt_1 \int_S^T dt_2 G\left(\frac{t_2 - t_1 + P(\lambda)}{\lambda^2}\right)$$
$$= \lambda^2 \int_S^T ds_1 \int_{[S-s_1+P(\lambda)]/\lambda^2}^{[T-s_1+P(\lambda)]/\lambda^2} G(s_2) ds_2 \le \lambda^2 \|\chi_{(S,T)}\|_1 \|G\|_1 ,$$

which is a quantity of order $O(\lambda^2) = O\left(\lambda^{2n-2}\right)$. Suppose that the conclusion of Lemma 15.1.1 is true for $k \le n$ and denote by $\{\alpha_2 \cdots \alpha_m\}$ the subset of $\{2 \cdots n+1\}$ such that the variable t_1 is directly connected to the variable t_{α_j}. By renaming the variables, we can assume that $t_{\alpha_j} = t_j$ for any $j = 2, 3, \cdots, m$. Thus, (15.1.7) becomes

$$\int_S^T dt_1 \cdots \int_S^T dt_m \prod_{k=2}^{m} G\left(\frac{t_k - t_1 + \tilde{P}_k(\lambda)}{\lambda^2}\right) \tag{15.1.8}$$
$$\cdot \int_S^T dt_{m+1} \cdots \int_S^T dt_{n+1} \prod_{\substack{1 \le h \le N \\ q_h \ge 2, m+1 \le p_h \le n+1}} G\left(\frac{t_{p_h} - t_{q_h} + \tilde{Q}_h(\lambda)}{\lambda^2}\right) ,$$

where $\tilde{P}_k, \tilde{Q}_h \in \{P_1 \cdots P_N\}$. By the change of variables

$$s_k := \frac{t_k - t_1 - \tilde{P}_k(\lambda)}{\lambda^2} , \qquad k = 2, 3, \cdots, m ,$$

the first factor of (15.1.8) becomes

$$\lambda^{2m-2} \int_S^T dt_1 \int_{[S-t_1-\tilde{P}_k(\lambda)]/\lambda^2}^{[T-t_1-\tilde{P}_k(\lambda)]/\lambda^2} ds_2 \cdots \int_{[S-t_1-\tilde{P}_k(\lambda)]/\lambda^2}^{[T-t_1-\tilde{P}_k(\lambda)]/\lambda^2} ds_m \tag{15.1.9a}$$

$$\cdot \prod_{k=2}^{m} G(s_k) ,$$

and the second factor of (15.1.8) becomes

$$\int_S^T dt_{m+1} \cdots \int_S^T dt_{n+1} \prod_{\substack{1 \le h \le N \\ m+1 \le q_h, p_h \le n+1}} G\left(\frac{t_{p_h} - t_{q_h} + \tilde{Q}_h(\lambda)}{\lambda^2}\right)$$

$$\prod_{\substack{1 \le h \le N \\ q_h = j \in \{2 \cdots m\}, p_h \le n+1}} G\left(\frac{t_{p_h} - t_1 + \tilde{Q}_h(\lambda) - \tilde{P}_j(\lambda) - s_j\lambda^2}{\lambda^2}\right) . \tag{15.1.9b}$$

Since

$$S - T - \tilde{P}_k(\lambda) \le S - t_1 - \tilde{P}_k(\lambda) \le s_j\lambda^2 \le T - t_1 - \tilde{P}_k(\lambda) \le T - S - \tilde{P}_k(\lambda)$$

for all $S \le t_1 \le T$ and λ, it follows that, in the second product of (15.1.9b), the function $\tilde{Q}_h(\lambda) - \tilde{P}_j(\lambda) - s_j\lambda^2$ is locally bounded. Moreover

$$\int_{[S-t_1-\tilde{P}_k(\lambda)]/\lambda^2}^{[T-t_1-\tilde{P}_k(\lambda)]/\lambda^2} ds_2 \cdots \int_{[S-t_1-\tilde{P}_k(\lambda)]/\lambda^2}^{[T-t_1-\tilde{P}_k(\lambda)]/\lambda^2} ds_m \prod_{k=2}^{m} G(s_k)$$

is bounded, uniformly in t_1 and λ, by $\|G\|_1^{m-1}$. So (15.1.8) is less than or equal to

$$\|G\|_1^{m-1} \lambda^{2m-2} \cdot \int_S^T dt_1 \,[\text{the quantity (15.1.9b)}] .$$

Since the quantity $\int_S^T dt_1$ [the quantity (15.1.9b)] involves in fact $n+1-(m-1)$ integrals and $m \ge 2$ implies that $n+1-(m-1) \le n$, it follows from the induction assumption that

$$\int_S^T dt_1 \,[\text{the quantity (15.1.9b)}] = O\left(\lambda^{2(n+2-m)-2}\right) .$$

So (15.1.8) is of the order $O(\lambda^{2m-2}) \cdot O(\lambda^{2(n+2-m)+2}) = O(\lambda^{2n}) = O\left(\lambda^{2(n+1)-2}\right)$. This completes the proof.

Now let us investigate the role played by conditions (ii) and (iii) in Lemma 15.1.1. If condition (ii) is not satisfied, we could meet a term such as

$$\int_S^T dt_1 \int_S^T dt_2 G\left(\frac{t_2 - t_1}{\lambda^2}\right) G\left(\frac{t_2 - t_1}{\lambda^2}\right),$$

corresponding to the case $N = 2$ and $\{p_1, q_1\} = \{p_2, q_2\} = \{1, 2\}$. If condition (iii) is not satisfied, we could meet a term such as

$$\int_S^T dt_1 \int_S^T dt_2 \int_S^T dt_2 G\left(\frac{t_2 - t_1}{\lambda^2}\right) G\left(\frac{t_3 - t_1}{\lambda^2}\right) G\left(\frac{t_3 - t_2}{\lambda^2}\right),$$

(note that $\{3, 1, 3, 2\} \setminus \{3\} = \{1, 2\}$) and we need a stronger integrability condition on G to deal with these quantities, for example, that G belongs to $L^p(\mathbb{R})$ for every p or that G belongs to $L^1(\mathbb{R}) \cap L^\infty(\mathbb{R})$. This suggests that if the function G is also bounded we can generalize Lemma 15.1.1 by dropping conditions (ii) and (iii). This is done in the following.

Lemma 15.1.2. *Let there be given for any $n \geq 2$, $N \geq n/2$,*

- *$S \leq T$;*
- *$G \in L^1 \cap L^\infty(\mathbb{R})$ positive;*
- *locally bounded functions $P_1 \cdots P_N$.*

Then (15.1.7) is of order $O(\lambda^{2n-2})$ if the product $\prod_{h=1}^N G\left(\frac{t_{p_h} - t_{q_h}}{\lambda^2}\right)$ has totally connected variables.

Proof 15.1.2. If the condition (ii) in Lemma 15.1.1 is not satisfied, i.e. if there are integers m and $k_1 \cdots k_m$ such that, for some h with $1 \leq h \neq k_j \leq n$, $\{p_h, q_h\} = \{p_{k_j}, q_{k_j}\}$, then a product of G-factors will appear with the same variable $t_{p_h} - t_{q_h}$; we reduce such a product to a single factor by majorizing the other ones in $\|\cdot\|_\infty$-norm. Thus (15.1.7) is dominated by

$$\|G\|_\infty^m \int_S^T dt_1 \cdots \int_S^T dt_n \prod_{\substack{1 \leq h \leq N \\ h \notin \{k_1 \cdots k_m\}}} G\left(\frac{t_{p_h} - t_{q_h} + P_h(\lambda)}{\lambda^2}\right),$$

and now the product $\prod_{\substack{1 \leq h \leq N \\ h \notin \{k_1 \cdots k_m\}}} G\left(\frac{t_{p_h} - t_{q_h} + P_h(\lambda)}{\lambda^2}\right)$ has totally connected variables. By repeating this procedure, we reduce the situation to the case when condition (ii) in Lemma 15.1.1 is verified.

If condition (iii) in Lemma 15.1.1 is not satisfied, i.e. if for some $1 \leq r, j, h, k \leq N$ with h, k distinct, $\{p_h, q_h, p_k, q_k\} \setminus \{p_r\} = \{p_j, q_j\}$, then $\{p_h, q_h\} \cap \{p_k, q_k\}$ must be equal to $\{p_r\}$. Without a loss of generality, let us assume that $q_h = p_k = p_r$. Thus $\{p_h, q_k\} = \{p_j, q_j\}$. By majorizing in norm the function $G\left(\frac{t_{p_j}-t_{q_j}+P_j(\lambda)}{\lambda^2}\right)$, (15.1.7) is dominated by

$$\|G\|_\infty \int_S^T dt_1 \cdots \int_S^T dt_n \prod_{\substack{1\le h\le N \\ h\neq j}} G\left(\frac{t_{p_h}-t_{q_h}+P_h(\lambda)}{\lambda^2}\right).$$

By iterating this procedure we reduce ourselves to the case in which also condition (iii) holds. The thesis now follows from Lemma 15.1.1.

Lemma 15.1.3. *Let there be given, for any $n \ge 2$ and $N \ge n/2$,*

- $S_1, T_1, \cdots, S_n, T_n \in \mathbb{R}$ *such that* $S_k \le T_k$, $\forall k = 1 \cdots n$;
- *functions* $F_1 \cdots F_N \in L^1(\mathbb{R}) \cap L^\infty(\mathbb{R})$;
- *locally bounded functions* $P_1 \cdots P_N$.

Then, if the product $\prod_{h=1}^N F_h\left(t_{p_h} - t_{q_h}\right)$ *has totally connected variables, the integral*

$$\int_{S_1}^{T_1} dt_1 \cdots \int_{S_n}^{T_n} dt_n \prod_{h=1}^N \left| F_h\left(\frac{t_{p_h}-t_{q_h}+P_h(\lambda)}{\lambda^2}\right)\right| \tag{15.1.10}$$

is of order $O\left(\lambda^{2n-2}\right)$.

Proof 15.1.3. The difference between the assumptions of Lemma 15.1.3 and Lemmata 15.1.1–2 is that in this case we are dealing with different S_k, T_k and F_k and we are not assuming that F_k are symmetric functions. By defining

$$S := \min_{1\le k\le n}(S_k) \quad , \quad T := \max_{1\le k\le n}(T_k) \quad , \quad G(x) := \max_{1\le k\le N}|F_k(x)|$$

and applying Lemmata 15.1.1–2 the thesis follows.

Theorem 15.1.1. *(The first connected component theorem). Let*

- $S_1, T_1 \cdots S_n, T_n \in \mathbb{R}$ *such that* $S_h \le T_h$, $\forall h = 1 \cdots n$;
- *functions* $F_1 \cdots F_N \in L^1(\mathbb{R}) \cap L^\infty(\mathbb{R})$;
- *locally bounded functions* $P_1 \cdots P_N$.

For any $n \ge 2$ and $N \ge n/2$, if, with respect to the product $\prod_{h=1}^N F_h(t_{p_h} - t_{q_h})$, *the variables* $t_1 \cdots t_n$ *can be divided into k connected components, then the integral*

$$\int_{S_1}^{T_1} dt_1 \cdots \int_{S_n}^{T_n} dt_n \prod_{h=1}^N \left| F_h\left(\frac{t_{p_h}-t_{q_h}+P_h(\lambda)}{\lambda^2}\right)\right| \tag{15.1.11}$$

is of order $O(\lambda^{2n-2k})$.

Proof 15.1.4. By assumption the variables $t_1 \cdots t_n$ can be divided into k connected components. Let m_j $(j = 1, 2, \cdots, k)$ be the number of variables in the jth component. Since one variable t_α can be contained only in one connected components, (15.1.11) factorizes in a product of multiple integrals: one for each connected component. Thus the proof can be completed just by applying Lemma 15.1.3 to each of the integrals in the product.

Corollary 15.1.1. (The second connected component theorem). *Let there be given for any $n \geq 2$ and $N \geq n/2$,*

- $S_1, T_1 \cdots S_n, T_n \in \mathbb{R}$ *such that* $S_h \leq T_h$, $\forall h = 1 \cdots n$;
- *functions* $F_1 \cdots F_N \in L^1(\mathbb{R}) \cap L^\infty(\mathbb{R})$;
- *locally bounded functions* $P_1 \cdots P_N$.

If, with respect to the product $\prod_{h=1}^N F_h\left(t_{p_h} - t_{q_h}\right)$, the variables $t_1 \cdots t_n$ can be divided into k connected components, then the integral

$$\int_{S_1/\lambda^2}^{T_1/\lambda^2} dt_1 \cdots \int_{S_n/\lambda^2}^{T_n/\lambda^2} dt_n \prod_{h=1}^N \left| F_h\left(t_{p_h} - t_{q_h} + P_h(\lambda)/\lambda^2\right)\right| \tag{15.1.12}$$

is of order $O(\lambda^{-2k})$.

Proof 15.1.5. By the change of variables

$$t_h \hookrightarrow \lambda^2\, t_h, \quad h = 1, 2, \cdots, n\,, \tag{15.1.13}$$

(15.1.12) becomes

$$\lambda^{-2n} \int_{S_1}^{T_1} dt_1 \cdots \int_{S_n}^{T_n} dt_n \prod_{h=1}^N \left| F_h\left(\frac{t_{p_h} - t_{q_h} + P_h(\lambda)}{\lambda^2}\right)\right|, \tag{15.1.14}$$

which is of the order $\lambda^{-2n} \cdot O(\lambda^{2n-2k}) = O(\lambda^{-2k})$ by the first connected component theorem.

15.2 The Factorization Theorem

In this section $\mathcal{H}$ will be a given Hilbert space and S_t a 1-parameter unitary group acting on it. $\Gamma(\mathcal{H})$ will denote the Fock space over $\mathcal{H}$, Φ the *vacuum state*, and $A(f)$ and $A^+(f)$ respectively the annihilator and creator with test function f. We shall not distinguish between Bose and Fermi Fock space because in the present section the commutation relations will not play a role and both cases will be dealt with by the same method. Recall our convention that, for an element O of a $*$-algebra, we use the notation

$$O^\varepsilon := \begin{cases} O, & \text{if } \varepsilon = 0\,, \\ O^*, & \text{if } \varepsilon = 1\,. \end{cases}$$

The main objects to be investigated in this section are integrals of the form

$$\lambda^{-n-m}\int_{S_1}^{T_1}dt_1\cdots\int_{S_n}^{T_n}dt_n\left\langle\Phi,B_1^{\varepsilon(1)}\left(\frac{t_1}{\lambda^2}\right)\cdots B_n^{\varepsilon(n)}\left(\frac{t_n}{\lambda^2}\right)\Phi\right\rangle \tag{15.2.1}$$

or equivalently

$$\lambda^{n-m}\int_{S_1/\lambda^2}^{T_1/\lambda^2}dt_1\cdots\int_{S_n/\lambda^2}^{T_n/\lambda^2}dt_n\left\langle\Phi,B_1^{\varepsilon(1)}\left(t_1\right)\cdots B_n^{\varepsilon(n)}\left(t_n\right)\Phi\right\rangle,$$

where

- $S_1,T_1\cdots S_n,T_n\in\mathbf{R}$, $S_j\le T_j$, for all $j=1,2,\cdots,n$;
- each $B_j(t)$ is either of $\mathcal{A}$-type, i.e.

$$A\left(S_tf_1^{(j)}\right)\cdots A\left(S_tf_{m_j}^{(j)}\right),$$

  or of $\mathcal{N}$-type, i.e.

$$A^+\left(S_tf_1^{(j)}\right)\cdots A^+\left(S_tf_{k_j}^{(j)}\right)A\left(S_tg_1^{(j)}\right)\cdots A\left(S_tg_{h_j}^{(j)}\right);$$

- m is the number of B_j of $\mathcal{N}$-type.

The main result of this section is the so-called *factorization theorem.* Essentially, this theorem says that, up to an $o(1)$, the quantity (15.2.1) has the form

$$\sum\int_{S_1}^{T_1}dt_1\cdots\int_{S_n}^{T_n}dt_n\prod\langle\Phi,\text{(a product of annihilators)}$$
$$\cdot\,\text{(a product of some }\mathcal{N}\text{-blocks)}\cdot\text{(a product of creators)}\Phi\rangle\,.$$

Let us fix an $\varepsilon\in\{0,1\}^n$; obviously, the scalar product in (15.2.1) is equal to zero or to a sum of products of 2-point functions. Define, for any $j=1,2,\cdots,n$,

$$\begin{aligned}r_j:=&\left(\text{the number of creators in }B_j^{\varepsilon(j)}\right)\\&-\left(\text{the number of annihilators in }B_j^{\varepsilon(j)}\right).\end{aligned}$$

Then the scalar product in (15.2.1) can be nonzero only if

$$\sum_{j=1}^{n}r_j=0\,. \tag{H15.2.1}$$

Moreover the scalar product in (15.2.1) can be nonzero only if

$$B_1^{\varepsilon(1)}\text{ is }\mathcal{A}\text{-type and }B_n^{\varepsilon(n)}\text{ is }\mathcal{A}^+\text{-type}\,. \tag{H15.2.2}$$

Denote by N_+ (N_-) the number of creators (annihilators) in the product $B_1^{\varepsilon(1)}\left(\frac{t_1}{\lambda^2}\right)\cdots B_n^{\varepsilon(n)}\left(\frac{t_n}{\lambda^2}\right)$, then condition (H15.2.1) is equivalent to

$$N_+ = N_- .$$

In the following we shall study (15.2.1) under the hypotheses (H15.2.1) and (H15.2.2), and we shall denote

$$N := N_+ = N_- .$$

Under (H15.2.1) the scalar product in (15.2.1) is equal to a sum of the form

$$\sum_{\substack{1\le p_h<q_h\le n\\ \{p_h,q_h\}_{h=1}^N=\{1,2,\cdots,n\}}} \prod_{h=1}^N c_h \left\langle S_{t_{p_h}/\lambda^2} f_h, S_{t_{q_h}/\lambda^2} g_h \right\rangle , \tag{15.2.2}$$

where

- any f_h is one of the test functions of some annihilator in the block with the time index t_{p_h};
- any g_h is one of the test functions of some creator in the block with the time index t_{q_h};
- c_h is equal to 1 in the boson case and to either $+1$ or -1 in the fermion case.

Remark 15.2.1. It is clear that each scalar product involves an annihilator and a creator, i.e.

$$\left\langle S_{t_{p_h}/\lambda^2} f_h, S_{t_{q_h}/\lambda^2} g_h \right\rangle = \left\langle \Phi, A\left(S_{t_{p_h}/\lambda^2} f_h\right) A^+\left(S_{t_{q_h}/\lambda^2} g_h\right) \Phi \right\rangle .$$

In the following, we shall say that *the annihilator* $A\left(S_{t_{p_h}/\lambda^2} f_h\right)$ *is paired with the creator* $A^+\left(S_{t_{q_h}/\lambda^2} g_h\right)$ *to produce a scalar product.*

As suggested by the above discussion, we are going to consider the integral

$$\int_{S_1}^{T_1} dt_1 \cdots \int_{S_n}^{T_n} dt_n \prod_{h=1}^N c_h \left\langle S_{t_{p_h}/\lambda^2} f_h, S_{t_{q_h}/\lambda^2} g_h \right\rangle \tag{15.2.3}$$

for an arbitrary choice of $1 \le p_h < q_h \le n$ such that $\{p_h, q_h\}_{h=1}^N = \{1, 2, \cdots, n\}$. Define

$$F_h(s) := c_h \cdot \langle f_h, S_s g_h \rangle .$$

By the first connected component theorem, we know the following:

Theorem 15.2.1. *The quantity (15.2.3) is of order $O(\lambda^{2n-2k})$, where k is the number of connected components of the variables $t_1 \cdots t_n$ with respect to $\prod_{h=1}^{N} F_h(t_{p_h} - t_{q_h})$.*

Now let us estimate the minimum number of connected components and analyze the case in which this minimum is achieved.

Lemma 15.2.1. *The number k of connected components is less than or equal to $n/2$.*

Proof 15.2.1. As said before, for a fixed product $\prod_{h=1}^{N} F_h\left(t_{p_h} - t_{q_h}\right)$, "$\rightleftharpoons$" is an equivalence relation. Therefore the variables $t_1 \cdots t_n$ can be divided into disjoint classes with respect to this relation. Since each component involves at least 2 variables, if the variables $t_1 \cdots t_n$ can be divided into k classes, then k should be less than or equal to $n/2$.

Lemma 15.2.2. *If n is odd then one has, with k as in Lemma 15.2.1,*

$$2n - 2k \geq n + 1\,.$$

Proof 15.2.2. If n is odd then $k \leq \frac{n}{2}$ is equivalent to $k \leq \frac{n-1}{2}$. So

$$2n - 2k \geq 2n - (n-1) = n + 1\,.$$

Lemma 15.2.3. *If, for some $j \in \{2, 3, \cdots, n-1\}$, the block $B_j^{\varepsilon(j)}$ is of $\mathcal{N}$-type, then the connected component which includes the time variable t_j involves at least three time variables.*

Proof 15.2.3. By the CCR (or CAR), any creator in $B_j^{\varepsilon(j)}$ must be paired with an annihilator with time variable t_x to produce a scalar product and any annihilator A in $B_j^{\varepsilon(j)}$ must be paired with a creator with time variable t_y to produce a scalar product. Since clearly $x < j < y$, it follows that the time variables t_x, t_j, t_y are included in the same connected component.

Theorem 15.2.2. *If among the blocks $B_1^{\varepsilon(1)}(t_1) \cdots B_n^{\varepsilon(n)}(t_n)$, there are m blocks of $\mathcal{N}$-type, then with k as in Lemma 15.2.1, one has*

$$k \leq \frac{n-m}{2} \tag{15.2.4}$$

or equivalently

$$2n - 2k \geq n + m\,. \tag{15.2.5}$$

In particular, if n is odd,

$$2n - 2k \geq n + m + 1\,. \tag{15.2.6}$$

Remark 15.2.2. Note that (H15.2.2) implies that $m \leq n - 2$.

Proof 15.2.4. By the same argument as in Lemma 15.2.2 we need only to prove (15.2.4) [or equivalently (15.2.5)]. If $m = 0$ (15.2.4) has been proved in Lemma 15.2.1. So we can assume that $m \geq 1$. Now we complete the proof by induction on n. First of all, note that the condition $m \geq 1$ together with (H15.2.2) implies that n must be greater than or equal 3. If $n = 3$ then m must be equal to 1, since $1 \leq m \leq n-2 = 1$. In this case Lemma 15.2.2 gives (15.2.5). Suppose that (15.2.4) is true for all $n = 3, 4, \cdots, r$ and consider the case $n = r + 1$.

Denote $t_{\alpha_1} \cdots t_{\alpha_m}$ the time variables of the $\mathcal{N}$-blocks, and let us consider the connected component which includes the time variable t_{α_1}. If this component includes all the time variables, then $t_1 \cdots t_n$ are totally connected with respect to the product $\prod_{h=1}^{N} F_h(t_{p_h} - t_{q_h})$; therefore $k = 1$ and, since $m \leq n - 2$,

$$\frac{n-m}{2} \geq \frac{n-(n-2)}{2} = 1 = k\,.$$

If the component does not include all the time variables, we relabel $s_1 \cdots s_j$ (certainly $j \geq 3$ by Lemma 15.2.3) the time variables which are connected with t_{α_1}, and $\tau_1 \cdots \tau_{r+1-j}$ the other time variables. Suppose that there are x $\mathcal{N}$-blocks with the time variable belonging to $\{s_1 \cdots s_j\}$ (thus there shall be $m - x$ $\mathcal{N}$-blocks with time variable belonging to $\{\tau_1 \cdots \tau_{r+1-j}\}$). By the induction assumption the number of connected components in $\{\tau_1 \cdots \tau_{r+1-j}\}$ must be less than or equal to

$$\frac{r+1-j-(m-x)}{2} = \frac{r+1+x-j-m}{2}\,.$$

So the full number of connected components k must be less than or equal to

$$1 + \frac{r+1+x-j-m}{2} = \frac{r+1+2+x-j-m}{2}\,.$$

As shown in Lemma 15.2.1, j must be greater than or equal to $x + 2$, and therefore

$$k \leq \frac{r+1+2+x-j-m}{2} = \frac{r+1-(j-x-2)-m}{2} \leq \frac{r+1-m}{2}\,.$$

Thus we have completed the proof.

Now let us return to (15.2.1). Using the argument to reach (15.2.2), we reduce the study of the limit, as $\lambda \to 0$, of (15.2.1) to the consideration of the limit of integrals of the form

$$\lambda^{-n-m} \int_{S_1}^{T_1} dt_1 \cdots \int_{S_n}^{T_n} dt_n \prod_{h=1}^{N} F_h\left(\frac{t_{p_h} - t_{q_h}}{\lambda^2}\right) \tag{15.2.7}$$

for an arbitrary choice of $1 \leq p_h < q_h \leq n$ such that $\{p_h, q_h\}_{h=1}^{N} = \{1, 2, \cdots, n\}$. On the other hand, as a corollary of the first connected component theorem, we have the following:

Lemma 15.2.4. *The quantity (15.2.7) is of order $O(\lambda^{n-m-2k})$ where k is as in Lemma 15.2.1.*

Proof 15.2.5. By applying the first connected component theorem, we have

$$\int_{S_1}^{T_1} dt_1 \cdots \int_{S_n}^{T_n} dt_n \prod_{h=1}^{N} F_h\Big(\frac{t_{p_h} - t_{q_h}}{\lambda^2}\Big) = O(\lambda^{2n-2k}),$$

and so (15.2.7) has order $\lambda^{-n-m} \cdot O(\lambda^{2n-2k}) = O(\lambda^{n-m-2k})$.

Fix any choice of $\{p_h, q_h\}_{h=1}^{N}$ such that $1 \le p_h < q_h \le n$ and $\{p_1, q_1, \cdots, p_N, q_N\} = \{1, 2, \cdots, n\}$; suppose that the time variables $t_{j_1} \cdots t_{j_l}$ form a connected component and without loss of generality; suppose also that $t_{j_1} < \cdots < t_{j_l}$. Then $B_{j_1}^{\varepsilon(j_1)}$ ($B_{j_l}^{\varepsilon(j_l)}$) is a product of annihilators (creators). With these choices, we have the following:

Lemma 15.2.5. *If either of the following two conditions is satisfied then $k < \frac{n-m}{2}$:*

- *there is an $r = 2, 3, \cdots, l-1$ such that B_{j_r} is* not *$\mathcal{N}$-type;*
- *$n - m$ is odd*

Moreover, the limit, as $\lambda \to 0$, of (15.2.7) is equal to zero.

Proof 15.2.6. Since among the B_j, there are m blocks of $\mathcal{N}$-type so among the B_j, there are $n - m$ blocks of $\mathcal{A}^+$-type or $\mathcal{A}$-type. As a consequence, there are at most $(n-m)/2$ connected components. Moreover, one can have exactly $(n-m)/2$ connected components only if $(n-m)/2$ is even and in each connected component there are exactly two time variables corresponding to $\mathcal{A}^+$-block or $\mathcal{A}$-block. Thus we have proved the first statement.

Moreover, if $k < (n-m)/2$, we have clearly $k \le (n-m)/2 - 1$ and from Lemma 15.2.4 it follows that

$$\text{the quantity (15.2.7)} = O(\lambda^{n-m-2k}),$$

which is a fortiori of order $O\left(\lambda^2\right)$ since

$$n - m - 2k \ge n - m - (n - m - 2) = 2 .$$

Now we are ready to state and prove the main result of this section:

Theorem 15.2.3. *In (15.2.1), for each fixed ε, denote*

$$\begin{aligned} \mathcal{T}_- &:= \{j = 1, 2, \cdots, n : \ B_j \text{ is a product of annihilators}\}, \\ \mathcal{T}_+ &:= \{j = 1, 2, \cdots, n : \ B_j \text{ is a product of creators}\}, \\ \mathcal{T}_0 &:= \{1, 2, \cdots, n\} \setminus (\mathcal{T}_- \cup \mathcal{T}_+) , \end{aligned}$$

and let

$$m := \textit{the cardinality of } \mathcal{T}_0, \quad K := \frac{n-m}{2}.$$

Then, as $\lambda \to 0$, (15.2.1), up to an $o(1)$ term, is equal to

$$\chi_{2\mathbb{N}}(n-m) \sum_{\substack{0 \le r_1 \cdots r_K \le m \\ r_1 + \cdots + r_K = m}} \sum_{\substack{\alpha_{1,1} < \cdots < \alpha_{r_1,1} \cdots \alpha_{1,K} < \cdots < \alpha_{r_K,K} \\ \alpha_{x,h} \neq \alpha_{y,j},\ \forall h \neq j,\ \textit{and}\ 1 \le x \le r_h,\ 1 \le y \le r_j}}$$

$$\cdot \sum_{\substack{1 \le y_1 \cdots y_K \le n \\ y_h \in \mathcal{T}_+, y_h > x_h + r_h, h = 1 \cdots K}} C\lambda^{n-m} \int_{S_1/\lambda^2}^{T_1/\lambda^2} dt_1 \cdots \int_{S_n/\lambda^2}^{T_n/\lambda^2} dt_n$$

$$\cdot \prod_{h=1}^{(n-m)/2} \left\langle \Phi, \mathcal{A}(t_{x_h})(\text{a product of } r_h \text{ blocks of } \mathcal{N}\textit{-type}) \mathcal{A}^+(t_{y_h})\Phi \right\rangle , \tag{15.2.8}$$

where

- *by $\mathcal{N}(s)$ [$\mathcal{A}(s)$ and $\mathcal{A}^+(s)$], we mean the $\mathcal{N}$-block [annihilation block and creation block] with the time variable s, and we do not specify the precise form and the corresponding test functions if it is not necessary;*
- *C is equal to 1 in the boson case and to a certain power of -1, depending on the order of the blocks, in the fermion case;*
- *$x_1 < x_2 < \cdots < x_K$ are determined by the following rule: $x_1 = 1$ and, having defined x_h, x_{h+1} is the index of the first annihilator block, counting from left to right, not yet used to produce the scalar product.*

Proof 15.2.7. Since the factor C is obtained, as in the statement, from the CCR (or CAR), in the following without loss of generality we shall forget this factor.

By definition, m is the number of $\mathcal{N}$-blocks and we have shown that the first block must be of $\mathcal{A}$-type and the last of $\mathcal{A}^+$-type. Moreover, it follows from Lemma 15.2.5 that, up to an $o(1)$ term, $n-m$ must be even.

Consider now the connected component which includes the time variable $t_1 = t_{x_1}$. If we denote by $t_1, t_{\alpha_{1,1}}, \cdots, t_{\alpha_{r_1,1}}, t_{y_1}$ the time variables which are connected with t_1 and $1 < \alpha_{1,1} < \cdots < \alpha_{r_1,1} < y_1$, then obviously $B_{y_1}^{\varepsilon(y_1)}$ must be an $\mathcal{A}^+$-type block. Moreover, Lemma 15.2.5 guarantees that, up to an $o(1)$, any $B_{\alpha_{x,1}}^{\varepsilon(\alpha_{x,1})}$ $(x = 1 \cdots r_1)$ must be an $\mathcal{N}$-block.

Many choices of $y_1 \in \mathcal{T}_+$ are possible, but in any case y_1 must be larger than $x_1 + r_1$. Therefore, up to an $o(1)$ term, (15.2.8) is equal to

$$\chi_{2\mathbb{N}}(n-m) \sum_{0\le r_1\le m} \sum_{\substack{y_1\\ x_1+r_1<y_1, y_1\in\mathcal{T}_+}} \lambda^{n-m} \int_{S_1/\lambda^2}^{T_1/\lambda^2} dt_1$$

$$\cdot \int_{S_n/\lambda^2}^{T_n/\lambda^2} dt_n \left\langle \Phi, \prod_{\substack{1\le h\le n\\ h\notin\{x_1,\alpha_{1,1},\cdots,\alpha_{r_1,1},y_1\}}} B_h^{\varepsilon(h)}(t_h)\Phi \right\rangle$$

$$\cdot \langle \Phi, \mathcal{A}(t_{x_1})\mathcal{N}(t_{\alpha_{1,1}})\cdots\mathcal{N}(t_{\alpha_{r_1,1}})\mathcal{A}^+(t_{y_1})\Phi\rangle . \tag{15.2.9}$$

Define x_2 as in the statement of Theorem 15.2.3, then the first scalar product of (15.2.9) is not equal to zero only if $x_2 \in \mathcal{T}_-$, that is the operator $B_{x_2}^{\varepsilon(x_2)}$ is an $\mathcal{A}$-block.

Denote by $t_{x_2}, t_{\alpha_{1,2}}, \cdots, t_{\alpha_{r_2,2}}, t_{y_2}$ the time variables connected to t_{x_2} with the order $x_2 < \alpha_{1,2} < \cdots < \alpha_{r_2,2} < y_2$, then Lemma 15.2.5 shows that, up to an $o(1)$ term, $B_{y_2}^{\varepsilon(y_2)}$ is an $\mathcal{A}^+$-block and any $B_{\alpha_{x,2}}^{\varepsilon(\alpha_{x,2})}$ is an $\mathcal{N}$-block. Moreover, $0 \le r_2 \le m - r_1$, because r_1 blocks of $\mathcal{N}$-type have been used in the preceding connected component. So, up to an $o(1)$ term, (15.2.9) is

$$\chi_{2\mathbb{N}}(n-m) \sum_{0\le r_1\le m} \sum_{\substack{y_1\\ x_1+r_1<y_1,\ y_1\in\mathcal{T}_+}} \sum_{\substack{r_2\ge 0\\ r_1+r_2\le m}} \sum_{\substack{y_2\\ x_2+r_2<y_2,\ y_2\in\mathcal{T}_+}}$$

$$\cdot \lambda^{n-m} \int_{S_1/\lambda^2}^{T_1/\lambda^2} dt_1 \cdots \int_{S_n/\lambda^2}^{T_n/\lambda^2} dt_n \left\langle \Phi, \prod_{\substack{1\le h\le n\\ h\notin\bigcup_{j=1}^2\{x_j,\alpha_{1,j},\cdots,\alpha_{r_j,j},y_j\}}} B_h^{\varepsilon(h)}(t_h)\Phi \right\rangle$$

$$\cdot \left\langle \Phi, \mathcal{A}(t_{x_1})\mathcal{N}(t_{\alpha_{1,1}})\cdots\mathcal{N}(t_{\alpha_{r_1,1}})\mathcal{A}^+(t_{y_1})\Phi \right\rangle$$

$$\cdot \left\langle \Phi, \mathcal{A}(t_{x_2})\mathcal{N}(t_{\alpha_{1,2}})\cdots\mathcal{N}(t_{\alpha_{r_2,2}})\mathcal{A}^+(t_{y_2})\Phi \right\rangle . \tag{15.2.10}$$

By iterating this procedure and noticing that if the jth connected component involves r_j operators of $\mathcal{N}$-type then $r_1 + \cdots + r_K$ must be equal to m, we finish the proof.

Remark 15.2.3. Analytically, the construction of x_h can been given as follows:

$$x_1 = 1$$

$$x_2 := \min\left\{h :\ h = 1,2,\cdots,n,\ h \notin \{x_1,\alpha_{1,1},\cdots,\alpha_{r_1,1},y_1\}\right\},$$

$$x_3 := \min\left\{h :\ h = 1,2,\cdots,n,\right.$$

$$\left. h \notin (\{x_1,\alpha_{1,1},\cdots,\alpha_{r_1,1},y_1\} \cup \{x_2,\alpha_{1,2},\cdots,\alpha_{r_2,2},y_2\})\right\},$$

$$\cdots \qquad \cdots$$

$$x_K := \min\left\{h:\ h=1,2,\cdots,n,\ h\notin\bigcup_{h=1}^{K-1}\{x_h,\alpha_{1,h},\cdots,\alpha_{r_h,h},y_h\}\right\}.$$

In order to have a better intuition of the above result, we give the following:

Corollary 15.2.1. *If $m=0$, as $\lambda\to 0$, (15.2.1) is equal, up to an $o(1)$ term, to*

$$\chi_{2\mathbb{N}}(n) \sum_{\substack{1\le y_1\cdots y_{n/2}\le n\\ y_h\in T_+,y_h>x_h,h=1\cdots n/2}} C\lambda^n \int_{S_1/\lambda^2}^{T_1/\lambda^2} dt_1 \cdots \int_{S_n/\lambda^2}^{T_n/\lambda^2} dt_n \prod_{h=1}^{n/2}\left\langle \Phi, \mathcal{A}(t_{x_h})\mathcal{A}^+(t_{y_h})\Phi\right\rangle . \qquad (15.2.11)$$

Remark 15.2.4. Note that, in the case of $m=0$, if an annihilator with the time variable t_j is paired with a creator with the time variable t_h to produce a scalar product, it must be true that, up to an $o(1)$ term:

(i) The number of annihilators with time variable t_j must be equal to the number of creators with the time variable t_h.
(ii) Any annihilator with the time variable t_j must be paired with a creator with the time variable t_h, and any creator with the time variable t_h must be paired with an annihilator with the time variable t_j.

Also the principles of linearization and bosonization are consequences of the above results (see Sect. 16.2).

Remark 15.2.5. At the beginning of this section we supposed that all the S_h, T_h $(h=1\cdots n)$ are constants. However this assumption is not essential, since all the results stated concern orders of magnitudes and since in the proofs we have majorized these quantities respectively by $T := \max\{T_1\cdots T_n\}$ and $S := \min\{S_1\cdots S_n\}$. In fact the only assumptions needed for the validity of the above results are

- S_1, T_1 are constants;
- For $h=2,3,\cdots,n$, S_h, T_h are bounded functions of the variables S_j, T_j, t_j, λ with $j\in\{1,2,\cdots,h-1\}$.

This fact is very useful for the stochastic limit.

Finally it is worth mentioning that all the results and conclusions in this section are still true if we replace (15.2.1) by

$$\lambda^{-n-m} \int_{S_1}^{T_1} dt_1 \cdots \int_{S_n}^{T_n} dt_n \left\langle \Phi, B_1^{\varepsilon(1)}\left(\frac{t_1}{\lambda^2}\right) \cdots B_{k-1}^{\varepsilon(1)}\left(\frac{t_{k-1}}{\lambda^2}\right) \right.$$
$$\cdot \left[B_k^{\varepsilon(k)}\left(\frac{t_k}{\lambda^2}\right), B_{k+1}^{\varepsilon(k+1)}\left(\frac{t_{k+1}}{\lambda^2}\right) \cdots B_{k+m}^{\varepsilon(k+m)}\left(\frac{t_{k+m}}{\lambda^2}\right) \right]$$
$$\left. \cdot B_{k+m+1}^{\varepsilon(k+m+1)}\left(\frac{t_{k+m+1}}{\lambda^2}\right) \cdots B_n^{\varepsilon(n)}\left(\frac{t_n}{\lambda^2}\right) \Phi \right\rangle \tag{15.2.12}$$

since (15.2.12) is the difference between two terms both having a form like (15.2.1).

15.3 The Case of Many Independent Fields

In the present section we show that all the results of the preceding section, where we have dealt with a single field, remain valid without any change in their proofs for interactions which are polynomials in several independent fields. In fact, in all the proofs in the previous section, the only property of the expectation functional we have used is its Gaussianity, i.e. the fact that the scalar product in (15.2.1) is a sum of products of 2-point functions. Therefore all the results in this section are still valid if we replace each creation or annihilation operator (and the corresponding cyclic vector) by a product of independent copies. More precisely we will show that, if we replace

- the vacuum (cyclic vector) Φ by a tensor product of a certain number, say M, of vacua (cyclic vectors), $\Phi_1 \otimes \Phi_2 \otimes \cdots \otimes \Phi_M$,
- any $\mathcal{A}$-type ($\mathcal{A}^+$-type) block by a tensor product of M $\mathcal{A}$-type ($\mathcal{A}^+$-type) blocks (different blocks may have a different number of factors),
- any $\mathcal{N}$-type block by a M-tuple tensor product in which each factor is either a $\mathcal{N}$-type block or the identity,

then all the results of the previous section continue to hold. The extension of the estimates is straightforward. The only point that requires some care is the precise calculation of the weight that arises in the calculation of vacuum expectation values of blocks of Fermi operators. The remaining of this section will be devoted to this topic.

15.4 The Fermi Block Theorem

Let there be given $N \in \mathbb{N}$, $\varepsilon \in \{0,1\}^{2N}$ and $2N$ blocks of Fermi creation–annihilation operators on a Fermi Fock space, say $\Gamma_{\text{Fermi}}(\mathcal{H})$. We shall consider the vacuum expectation

$$\left\langle \Phi, \mathcal{A}^{\varepsilon(1)}(1,n_1)\cdots \mathcal{A}^{\varepsilon(2N)}(2N,n_{2N})\Phi \right\rangle , \tag{15.4.1}$$

where, for any $h = 1, 2, \cdots, 2N$, if $\varepsilon(h) = 0$, $\mathcal{A}^{\varepsilon(h)}(h, n_h) := \mathcal{A}(n_h)$ is a product of n_h Fermi annihilators and if $\varepsilon(h) = 1$, $\mathcal{A}^{\varepsilon(h)}(h, n_h) := \mathcal{A}^+(n_h)$ is a product of n_h Fermi creators.

The vacuum expectation (15.4.1) can be different from zero only if $\varepsilon(1) = 0$ and $\varepsilon(2N) = 1$. In the following, we assume this without any special comment. Moreover, we shall consider only the case in which

(i) among the product of blocks in (15.4.1), there are exactly N $\mathcal{A}$-blocks and N $\mathcal{A}^+$-blocks i.e.

$$\sum_{h=1}^{2N} \varepsilon(h) = N$$

(ii) for any $j = 1, 2, \cdots, 2N$, the cardinality of the set $\{k : 1 \le k \le j, \varepsilon(k) = 0\}$ is greater than or equal to the cardinality of the set $\{k : 1 \le k \le j, \varepsilon(k) = 1\}$.

Remark 15.4.1. Condition (ii) above means that for any $j = 1, 2, \cdots, 2N$, among the operators $\mathcal{A}^{\varepsilon(1)}(1, n_1) \cdots \mathcal{A}^{\varepsilon(j)}(j, n_j)$, there are more $\mathcal{A}$-blocks than $\mathcal{A}^+$-blocks.

As usual we use p_h to label the annihilator indices,

$$\{p_h\}_{h=1}^N := \{j : \ 1 \le j \le 2N, \ \varepsilon(j) = 0\} ,$$

and we fix their order to be increasing: $p_1 < p_2 < \cdots < p_N$ (note that p_1 must be equal to 1). Thus, by applying the CAR, we are able to rewrite (15.4.1) in the form

$$\begin{aligned}&\sum_{\substack{1<q_1,q_2,\cdots,q_N\le 2N\\ p_h<q_h,\ h=1,2,\cdots,N,\{q_h\}_{h=1}^N=\{1,\cdots,2N\}\setminus\{p_h\}_{h=1}^N}} c(\{p_h, q_h\}_{h=1}^N)\\ &\cdot \prod_{h=1}^N \langle \Phi, \mathcal{A}(p_h, n_{p_h})\mathcal{A}^+(q_h, n_{q_h})\Phi\rangle\\ &+ \text{ OTHER TERMS} .\end{aligned} \tag{15.4.2}$$

Our goal in this section is not to estimate the correction term in (15.4.1) but only to evaluate the factor $c(\{p_h, q_h\}_{h=1}^N)$.

Since $\langle \Phi, \cdot \Phi\rangle$ is the vacuum state, (15.4.2) is equal to

$$\begin{aligned}&\sum_{\substack{1<q_1,q_2,\cdots,q_N\le 2N\\ p_h<q_h,\ h=1,2,\cdots,N,\{q_h\}_{h=1}^N=\{1\cdots 2N\}\setminus\{p_h\}_{h=1}^N}} c(\{p_h, q_h\}_{h=1}^N)\\ &\cdot \prod_{h=1}^N \langle \Phi, \mathcal{A}(p_h, n_{p_h})\mathcal{A}^+(q_h, n_{p_h})\Phi\rangle\\ &+ \text{ OTHER TERMS} .\end{aligned} \tag{15.4.3}$$

We know from the CAR the following:

(i) The factor $c\left(\{p_h, q_h\}_{h=1}^N\right)$ comes from the action of bringing the last annihilation block $\mathcal{A}(p_N, n_{p_N})$ to the nearest left of the creation block $\mathcal{A}^+(q_N, n_{q_N})$; the annihilation block $\mathcal{A}(p_{N-1}, n_{p_{N-1}})$ to the nearest left of the creation block $\mathcal{A}^+(q_{N-1}, n_{q_{N-1}})$; $\cdots$; and the annihilation block $\mathcal{A}(p_1, n_{p_1})$ to the nearest left of the creation block $\mathcal{A}^+(q_1, n_{p_1})$.

(ii) The factor $c\left(\{p_h, q_h\}_{h=1}^N\right)$ has the form $(-1)^{\nu\left(\{p_h, q_h\}_{h=1}^N\right)}$ where $\nu(\{p_h,\, q_h\}_{h=1}^N) \in \mathbb{N}$ is a natural number which counts how many commutations must be done if we want to carry out the action described in (i) with the agreement that we shall not commute operators inside a given block.

The number $\nu\left(\{p_h, q_h\}_{h=1}^N\right)$ can be calculated using the following steps:

In order to bring each annihilator in the block $\mathcal{A}(p_N, n_{p_N})$ to the nearest left of the creation block $\mathcal{A}^+(q_N, n_{q_N})$, we have to commute it with all creators in those blocks $\mathcal{A}^+(q_j, n_{q_j})$ satisfying $p_N < q_j < q_N$. Therefore $\sum_{j:\ p_N<q_j<q_N} n_{q_j}$ commutations are needed. Thus to bring all annihilators in the block $\mathcal{A}(p_N, n_{p_N})$ to the nearest left of the creation block $\mathcal{A}^+(q_N, n_{q_N})$ one needs $n_{p_N} \sum_{j:\ p_N<q_j<q_N} n_{q_j}$ commutations.

In the second step we bring the annihilators in the block $\mathcal{A}(p_{N-1}, n_{p_{N-1}})$ to the nearest left of the creation block $\mathcal{A}^+(q_{N-1}, n_{q_{N-1}})$. But we do not know if $q_N > q_{N-1}$. If $q_N < q_{N-1}$, we need to commute each annihilator in the block $\mathcal{A}(p_{N-1}, n_{p_{N-1}})$ not only with all creators in the blocks $\mathcal{A}^+(q_j, n_{q_j})$ satisfying $p_{N-1} < q_j < q_{N-1}$ but also with all annihilators in the annihilation block $\mathcal{A}(p_N, n_{p_N})$ and so we have to do

$$n_{p_{N-1}} \left(\sum_{j:\ p_{N-1}<q_j<q_{N-1}, j\neq N} n_{q_j} + 2n_{p_N} \right)$$

commutations. If $q_N > q_{N-1}$, we have to commute each annihilator in the block $\mathcal{A}(p_{N-1}, n_{p_{N-1}})$ only with all creators in the blocks $\mathcal{A}^+(q_j, n_{q_j})$ satisfying $p_{N-1} < q_j < q_{N-1}$, and therefore we have to do

$$n_{p_{N-1}} \sum_{j:\ p_{N-1}<q_j<q_{N-1}, j\neq N} n_{q_j}$$

commutations. Thus, to bring the annihilators in the block $\mathcal{A}(p_{N-1}, n_{p_{N-1}})$ to the nearest left of the creation block $\mathcal{A}^+(q_{N-1}, n_{q_{N-1}})$, one needs

$$n_{p_{N-1}} \sum_{j:\ p_{N-1}<q_j<q_{N-1}, j\neq N} n_{q_j} + E_{N-1}$$

commutations, where E_{N-1} is zero or an even number.

By the same argument as in the second step, to bring the annihilators in the block $\mathcal{A}(p_{N-2}, n_{p_{N-2}})$ to the nearest left of the creation block $\mathcal{A}^+(q_{N-2}, n_{q_{N-2}})$, one needs

$$n_{p_{N-2}} \sum_{j:\ p_{N-2}<q_j<q_{N-2}, j\notin\{N-1,N\}} n_{q_j} + E_{N-2}$$

commutations, where E_{N-2} is zero or an even number.

Iterating this procedure we find that to bring the annihilators in the block $\mathcal{A}(p_2, n_{p_2})$ to the nearest left of the creation block $\mathcal{A}^+(q_2, n_{q_2})$ one needs

$$n_{p_2} \sum_{j:\ p_2<q_j<q_2, j\notin\{3,\cdots,N\}} n_{q_j} + E_2$$

commutations, where E_2 is zero or an even number.

Finally, notice that $p_1 = 1$ and $p_1 < p_2 < \cdots < p_N$. So if $p_2 > q_1$, it must be true that $q_1 = 2$, i.e. the annihilation block $\mathcal{A}(p_1, n_{p_1})$ is already at the nearest left of the creation block $\mathcal{A}^+(q_1, n_{q_1})$; if $p_2 < q_1$, by the same argument as before, to bring the annihilators in the block $\mathcal{A}(q_1, n_{p_1})$ to the nearest left of the creation block $\mathcal{A}^+(q_1, n_{q_1})$ one needs E_1 commutations and E_1 is an even number.

In summary, we have obtained the following theorem.

Theorem 15.4.1. *(The first* Fermi block *theorem). The factor* $c(\{p_h, q_h\}_{h=1}^N)$ *is equal to*

$$(-1)^{\sum_{h=2}^N n_{p_h} \sum_{j:\ p_h<q_j<q_h,\ j\notin\{h+1\cdots N\}} n_{q_j}}, \tag{15.4.4}$$

where by definition $\{N+1, N\} := \emptyset$.

Now we try to further simplify the exponential in (15.4.4). To this goal note that if n_{p_h} is even the product

$$n_{p_h} \sum_{j:\ p_h<q_j<q_h,\ j\notin\{h+1,\cdots,N\}} n_{q_j}$$

will always be even, and its contribution to (15.4.4) is trivial. So (15.4.4) is equal to (-1) elevated to the exponent

$$\sum_{\substack{2\le h\le N,\\ n_{p_h} \text{ is odd}}} n_{p_h} \sum_{j:\ p_h<q_j<q_h,\ j\notin\{h+1\cdots N\}} n_{q_j}\,. \tag{15.4.5}$$

Similarly we can take away from (15.4.5) all the terms that multiply an even n_{q_j}. After that the exponent (15.4.5) of (-1) becomes

$$\sum_{\substack{2\le h\le N\\ n_{p_h} \text{ is odd}}} n_{p_h} \sum_{\substack{j:\ p_h<q_j<q_h,\ j\notin\{h+1\cdots N\},\\ n_{q_j} \text{ is odd}}} n_{q_j}$$

and therefore (15.4.4) is equal to

$$\prod_{\substack{2\le h\le N,\\ n_{p_h} \text{ is odd}}} (-1)^{n_{p_h} \sum_{j:\ p_h<q_j<q_h,\ j\notin\{h+1\cdots N\},\ n_{q_j} \text{ is odd}} n_{q_j}}$$

$$= \prod_{\substack{2\le h\le N,\\ n_{p_h} \text{ is odd}}} \left((-1)^{n_{p_h}}\right)^{\sum_{j:\ p_h<q_j<q_h,\ j\notin\{h+1\cdots N\},\ n_{q_j} \text{ is odd}} n_{q_j}} \tag{15.4.6}$$

But since n_{p_h} is odd, $(-1)^{n_{p_h}}$ is equal to -1. Therefore the right-hand side of (15.4.6) is equal to

$$\prod_{\substack{2\le h\le N,\\ n_{p_h} \text{ is odd}}} \prod_{\substack{j:\ p_h<q_j<q_h,\ j\notin\{h+1\cdots N\},\\ n_{p_j} \text{ is odd}}} (-1)^{n_{p_j}}$$

$$= \prod_{\substack{2\le h\le N,\\ n_{p_h} \text{ is odd}}} \prod_{\substack{j:\ p_h<q_j<q_h,\ j\notin\{h+1\cdots N\},\\ n_{p_j} \text{ is odd}}} (-1)$$

$$= \prod_{\substack{2\le h\le N,\\ n_{p_h} \text{ is odd}}} (-1)^{\sum_{j:\ p_h<q_j<q_h,\ j\notin\{h+1\cdots N\},\ n_{q_j} \text{ is odd}} 1}$$

$$= \prod_{\substack{2\le h\le N,\\ n_{p_h} \text{ is odd}}} (-1)^{\left|\{j:\ p_h<q_j<q_h,\ j\notin\{h+1\cdots N\},\ n_{q_j} \text{ is odd}\}\right|}, \tag{15.4.7}$$

where $|\ \cdot\ |$ denotes cardinality. This gives:

Theorem 15.4.2. *(The second* Fermi block *theorem). The factor* $c\left(\{p_h, q_h\}_{h=1}^N\right)$ *in (15.4.2) is equal to*

$$\prod_{\substack{2\le h\le N\\ n_{p_h} \text{ is odd}}} (-1)^{\left|\{j:\ p_h<q_j<q_h,\ j\notin\{h+1\cdots N\},\ n_{q_j} \text{ is odd}\}\right|}.$$

In summary, the Fermi block tells us that

(i) even power blocks do not contribute anything to the factor $c\left(\{p_h, q_h\}_{h=1}^N\right)$;

(ii) any odd power block gives to the factor $c\left(\{p_h, q_h\}_{h=1}^N\right)$ the same contribution that would be given by a first-order block, i.e. a linear term.

15.5 Non-Time-Consecutive Terms: The First Vanishing Theorem

In this section we study integrals of the form

$$\lambda^{2n} \int_{S_1/\lambda^2}^{T_1/\lambda^2} dt_1 \cdots \int_{S_{2n}/\lambda^2}^{T_{2n}/\lambda^2} dt_{2n} \prod_{h=1}^{n} F_h(t_{p_h} - t_{q_h}), \tag{15.5.1}$$

where

- $n \in \mathbb{N}$, $S_1, T_1 \in \mathbb{R}$;
- S_2, T_2 are bounded functions of S_1, T_1, t_1; S_3, T_3 are bounded functions of $S_1, T_1, t_1, S_2, T_2, t_2, \cdots, S_{2n}, T_{2n}$ are bounded functions of $\{S_h, T_h, t_h\}_{h=1}^{2n-1}$; moreover $S_h \le T_h$, $h = 1, 2, \cdots, 2n$;
- for any $h = 1, 2, \cdots, n$, the function F_h belongs to $\mathbf{L}^1(\mathbb{R})$;
- $\{p_h, q_h\}_{h=1}^{n}$ is a pair partition of $\{1, 2, \cdots, 2n\}$.

In the following, for any $m \in \mathbb{N}$, Δ_m denotes the m-simplex:

$$\Delta_m := \{(x_1, x_2, \cdots, x_m) : x_1 \ge x_1 \ge \cdots \ge x_m \ge 0\}.$$

Lemma 15.5.1. *For any $F \in L^1(\mathbb{R})$ positive and $S \le T \in \mathbb{R}$ with $T \ge 0$, the limit, as $\lambda \to 0$, of*

$$\lambda^4 \int_{[S/\lambda^2, T/\lambda^2]^4} dt_1 dt_2 ds_1 ds_2 F(t_1 - t_2) F(s_1 - s_2) \chi_{\Delta_3}(t_1, s_1, t_2) \tag{15.5.2}$$

is equal to zero.

Proof 15.5.1. By exchanging the second and the third integrals, one knows that (15.5.2) is less than or equal to

$$\lambda^4 \int_{S/\lambda^2}^{T/\lambda^2} ds_2 \int_0^{T/\lambda^2} dt_1 \int_0^{t_1} dt_2 \int_{t_2}^{t_1} ds_1 F(t_1 - t_2) F(s_1 - s_2). \tag{15.5.3}$$

By the change of variables $\lambda^2 t_1 \hookrightarrow t_1$, $\quad \lambda^2 t_2 \hookrightarrow t_2$, $\quad \lambda^2 s_1 \hookrightarrow s_1$, (15.5.3) becomes

$$\lambda^{-2} \int_{S/\lambda^2}^{T/\lambda^2} ds_2 \int_0^{T} dt_1 \int_0^{t_1} dt_2 \int_{t_2}^{t_1} ds_1 F\left(\frac{t_1 - t_2}{\lambda^2}\right) F\left(\frac{s_1}{\lambda^2} - s_2\right). \tag{15.5.4}$$

With the further change of variables

$$s_1 = s_1, \qquad s := s_2 - \frac{s_1}{\lambda^2}, \qquad t_1 = t_1, \qquad t := \frac{t_2 - t_1}{\lambda^2},$$

(15.5.4) can be rewritten as

$$\int_0^T dt_1 \int_{-t_1/\lambda^2}^{0} dt F(-t) \int_{t_1 + \lambda^2 t}^{t_1} ds_1 \int_{(S - s_1)/\lambda^2}^{(T - s_1)/\lambda^2} ds F(-s). \tag{15.5.5}$$

By introducing

$$G_\lambda(t_1,t) := \chi_{[-t_1/\lambda^2,0]}(t)\int_{t_1+\lambda^2 t}^{t_1} ds_1 \int_{(S-s_1)/\lambda^2}^{(T-s_1)/\lambda^2} dsF(-s),\ t_1\in[0,T],$$

(15.5.5) becomes

$$\int_0^T dt_1 \int_{-\infty}^0 dtF(-t)G_\lambda(t_1,t)\ .$$

Since

- for any $t\in[-t_1/\lambda^2,0]$, $t_1\in[0,T]$, the quantity $t_1+\lambda^2 t$ always belongs to $[0,T]$,
- F is positive and integrable,

we know that G_λ is uniformly bounded. So (15.5.5) is less than or equal to

$$\int_0^T dt_1 \int_{-\infty}^0 dt\,|F(-t)|\cdot|G_\lambda(t_1,t)|\ . \tag{15.5.6}$$

Since for each (t_1,t),

$$|G_\lambda(t_1,t)|\le\lambda^2 t\longrightarrow 0,\ \text{ as }\lambda\to 0\,, \tag{15.5.7}$$

by the Lebesgue-dominated convergence theorem, we complete the proof.

Now we shall prove the *first vanishing theorem* (or *non-time-consecutive theorem*)

Theorem 15.5.1. *(The first vanishing theorem).In the assumptions listed at the beginning of this section, for any $n\ge 2$, $j\ne r\in\{1,2,\cdots,n\}$, the limit, as $\lambda\to 0$, of*

$$\lambda^{2n}\int_{S_1/\lambda^2}^{T_1/\lambda^2} dt_1\cdots\int_{S_{2n}/\lambda^2}^{T_{2n}/\lambda^2} dt_{2n}\prod_{h=1}^n |F_h(t_{p_h}-t_{q_h})|$$

$$\cdot\Big[\chi_{\Delta_3}(t_{p_j},t_{p_r},t_{q_j})+\chi_{\Delta_3}(t_{p_j},t_{q_r},t_{q_j})+\chi_{\Delta_3}(t_{q_j},t_{p_r},t_{p_j})+\chi_{\Delta_3}(t_{q_j},t_{q_r},t_{p_j})\Big] \tag{15.5.8}$$

is equal to zero.

Proof 15.5.2. For $n=2$, Lemma (15.5.1) gives our thesis and therefore we shall assume that $n\ge 3$. Denote

$$S:=\min\{S_1,S_2,\cdots,S_{2n}\}\quad,\quad T:=\max\{T_1,T_2,\cdots,T_{2n}\}\vee 0\quad,$$
$$F(x):=\max_{1\le h\le n}\big(|F_h(x)|+|F_h(-x)|\big)\,,$$

then (15.5.8) is less than or equal to

$$\lambda^{2n}\int_{S/\lambda^2}^{T/\lambda^2} dt_1\cdots\int_{S/\lambda^2}^{T/\lambda^2} dt_{2n}\prod_{h=1}^{n} F(t_{p_h}-t_{q_h})$$
$$\cdot\Big[\chi_{\Delta_3}(t_{p_j},t_{p_r},t_{q_j})+\chi_{\Delta_3}(t_{p_j},t_{q_r},t_{q_j})$$
$$+\chi_{\Delta_3}(t_{q_j},t_{p_r},t_{p_j})+\chi_{\Delta_3}(t_{q_j},t_{q_r},t_{p_j})\Big]. \tag{15.5.9}$$

In order to apply Lemma 15.5.1 we rewrite the first term of (15.5.9) as

$$\lambda^4\int_{[S/\lambda^2,T/\lambda^2]^4} dt_{p_j}dt_{p_r}dt_{q_j}dt_{q_r}F(t_{p_j}-t_{q_j})F(t_{p_r}-t_{q_r})\chi_{\Delta_3}(t_{p_j},t_{p_r},t_{q_j})$$
$$\cdot\lambda^{2n-4}\int_{[S/\lambda^2,T/\lambda^2]^{2n-4}} dt_1\cdots\widehat{dt_{p_j}}\cdots\widehat{dt_{p_r}}\cdots\widehat{dt_{q_j}}\cdots\widehat{dt_{q_r}}\cdots dt_n\prod_{\substack{1\le h\le n\\ h\notin\{j,r\}}} F(t_{p_h}-t_{q_h}). \tag{15.5.10}$$

By the change of variables $\lambda^2 t_\alpha \hookrightarrow t_\alpha,\ \ \alpha\notin\{p_j,p_r,q_j,q_r\}$, the second factor becomes

$$\lambda^{-2n+4}\int_{[S,T]^{2n-4}} dt_1\cdots\widehat{dt_{p_j}}\cdots\widehat{dt_{p_r}}\cdots\widehat{dt_{q_j}}\cdots\widehat{dt_{q_r}}\cdots dt_n\prod_{\substack{1\le h\le n\\ h\notin\{j,r\}}} F\left(\frac{t_{p_h}-t_{q_h}}{\lambda^2}\right),$$

where, with respect to the product $\prod_{\substack{1\le h\le n\\ h\notin\{j,r\}}} F(t_{p_h}-t_{q_h})$, all variables involved are totally connected. So, by the first connected component theorem, this is a quantity of order $\lambda^{-2n+4}O(\lambda^{2(2n-4)-2}) = O\left(\lambda^{2n-6}\right)$, which is bounded since $n\ge 3$. Moreover, by Lemma 15.5.1, the first factor of (15.5.10) goes to zero as $\lambda\to 0$. The other terms in (15.5.9) can be shown to go to zero as $\lambda\to 0$ in a similar way. This proves the thesis.

Corollary 15.5.1. *For any $n\ge 2,\ j\ne r\in\{1,2,\cdots,n\}$, as $\lambda\to 0$, the limit of*

$$\lambda^{2n}\int_{S_1/\lambda^2}^{T_1/\lambda^2} dt_1\cdots\int_{S_{2n}/\lambda^2}^{T_{2n}/\lambda^2} dt_{2n}\prod_{h=1}^{n}|F_h(t_{p_h}-t_{q_h})|$$
$$\cdot\left[\chi_{\left(t_{p_j},t_{q_j}\right)}(t_{p_r})+\chi_{\left(t_{p_j},t_{q_j}\right)}(t_{q_r})+\chi_{\left(t_{q_j},t_{p_j}\right)}(t_{p_r})+\chi_{\left(t_{q_j},t_{p_j}\right)}(t_{q_r})\right] \tag{15.5.11}$$

is equal to zero.

Proof 15.5.3. The difference between (15.5.11) and (15.5.8) is the fact that we do not assume that $t_{p_r}, t_{q_j}, t_{q_j}, t_{p_j}$ are positive. Clearly we need to study only the first term, which is less than or equal to

$$\lambda^{2n} \int_{[S/\lambda^2, T/\lambda^2]^{2n}} dt_1 \cdots dt_{2n} \prod_{h=1}^{n} \left|F_h(t_{p_h} - t_{q_h})\right| \chi_{\left(t_{p_j}, t_{q_j}\right)}(t_{p_r}) .$$

By the change of variables

$$s_k := \lambda^2 t_k, \; k = 1, 2, \cdots, 2n ,$$

the above quantity becomes

$$\lambda^{-2n} \int_{[S,T]^{2n}} ds_1 \cdots ds_{2n} \prod_{h=1}^{n} \left|F_h\left(\frac{s_{p_h} - s_{q_h}}{\lambda^2}\right)\right| \chi_{\left(s_{p_j}, s_{q_j}\right)}(s_{p_r}) . \qquad (15.5.12)$$

By the change of variables

$$u_k := s_k - S, \quad k = 1, 2, \cdots, 2n ,$$

(15.5.12) becomes

$$\lambda^{-2n} \int_{[0,T-S]^{2n}} du_1 \cdots du_{2n} \prod_{h=1}^{n} \left|F_h\left(\frac{u_{p_h} - u_{q_h}}{\lambda^2}\right)\right| \chi_{\left(u_{p_j}, u_{q_j}\right)}(u_{p_r}) ,$$

which is obviously equal to

$$\lambda^{2n} \int_{[0,(T-S)/\lambda^2]^{2n}} du_1 \cdots du_{2n} \prod_{h=1}^{n} \left|F_h(u_{p_h} - u_{q_h})\right| \chi_{\left(u_{p_j}, u_{q_j}\right)}(u_{p_r}) .$$

By Theorem 15.5.1 this completes the proof.

15.6 Non-Time-Consecutive Terms: The Second Vanishing Theorem

Now we prove the *second vanishing theorem* (whose main use will be in the low-density limit [LDL]).

Theorem 15.6.1. *(The second vanishing theorem). Suppose that*

- $n, m = 2, 3, 4, ..., \; S, T \in \mathbb{R}, \; S < T$;
- *the function* F *belongs to* $L^1(\mathbb{R}) \cap L^\infty(\mathbb{R})$;
- $\alpha \in \{1, 2, ..., m\}$, $1 \le p \ne q \le n$.

Then, as $\lambda \to 0$*, the quantity*

$$\lambda^4 \int_{[S/\lambda^2, T/\lambda^2]^{n+m}} dt_1 \cdots dt_n ds_1 \cdots ds_m \, \chi_{\Delta_3}(t_p, s_\alpha, t_q) F(t_1 - t_n) F(s_1 - s_m) \cdot \prod_{h=1}^{n-1} F(t_h - t_{h+1}) \prod_{j=1}^{m-1} F(s_j - s_{j+1}) \tag{15.6.1}$$

goes to zero.

Proof 15.6.1. First of all, the following facts are clear:

- we can assume that $T > 0$ since the emergence of $\chi_{\Delta_3}(t_p, s_\alpha, t_q)$ shows that (15.6.1) is equal to zero if $T < 0$;
- it is sufficient to show that (15.6.1) goes to zero for F positive;
- p can be assumed to be less than q.

Thus, with the change of variables

$$\lambda^2 t_h \hookrightarrow t_h \ , \ h = 1, 2, ..., n, \qquad \lambda^2 s_j \hookrightarrow s_j \ , \ j = 1, 2, ..., m \, ,$$

(15.6.1) can be rewritten as

$$\lambda^{4-2(n+m)} \int_{[S,T]^{n+m-3}} \int_0^T dt_p \int_0^{t_p} ds_\alpha \int_0^{s_\alpha} dt_q dt_1 \cdots \widehat{dt_p} \cdots \widehat{dt_q} \cdots dt_n ds_1 \cdots \widehat{ds_\alpha} \cdots ds_m \cdot F\left(\frac{t_n - t_1}{\lambda^2}\right) F\left(\frac{s_m - s_1}{\lambda^2}\right) \prod_{h=1}^{n-1} F\left(\frac{t_h - t_{h+1}}{\lambda^2}\right) \prod_{j=1}^{m-1} F\left(\frac{s_j - s_{j+1}}{\lambda^2}\right) . \tag{15.6.2}$$

With the new change of variables

$$x_1 := t_1, \ x_{h+1} := \frac{t_{h+1} - t_h}{\lambda^2} \ , \ h = 1, 2, ..., n - 1 \, ,$$

the integral $\int_{[S,T]^{p-1}} dt_1 \cdots dt_{p-1} \int_0^T dt_p \int_0^{t_p} ds_\alpha$ becomes

$$\int_S^T dx_1 \int_{\frac{(S-x_1)}{\lambda^2}}^{\frac{(T-x_1)}{\lambda^2}} dx_2 \int_{\frac{(S-x_1)}{\lambda^2} - x_2}^{\frac{(T-x_1)}{\lambda^2} - x_2} dx_3 \cdots \int_{\frac{(S-x_1)}{\lambda^2} - x_2 - \cdots - x_{p-2}}^{\frac{(T-x_1)}{\lambda^2} - x_2 - \cdots - x_{p-2}} dx_{p-1} \cdot \int_{\frac{(S-x_1)}{\lambda^2} - x_2 - \cdots - x_{p-1}}^{\frac{(T-x_1)}{\lambda^2} - x_2 - \cdots - x_{p-1}} dx_p \int_0^{\lambda^2(x_p + x_{p-1} + \cdots + x_2) + x_1} ds_\alpha \, ,$$

and where $\lambda^2(x_p + x_{p-1} + \cdots + x_2) + x_1$ is in fact t_p belonging to $[0, T]$, the integral $\int_0^{s_\alpha} dt_q$ becomes $\int_{-\frac{x_1}{\lambda^2} - x_2 - \cdots - x_{q-1}}^{\frac{(s_\alpha - x_1)}{\lambda^2} - x_2 - \cdots - x_{q-1}} dx_q$. Obviously, any x_j is run over a certain subset of $\mathbb{R}$, and so (15.6.2) will be controlled by

$$\int_S^T dx_1 \int_{\frac{(S-x_1)}{\lambda^2}}^{\frac{(T-x_1)}{\lambda^2}} dx_2 \int_{\frac{(S-x_1)}{\lambda^2}-x_2}^{\frac{(T-x_1)}{\lambda^2}-x_2} dx_3 \cdots \int_{\frac{(S-x_1)}{\lambda^2}-x_2-\cdots-x_{p-2}}^{\frac{(T-x_1)}{\lambda^2}-x_2-\cdots-x_{p-2}} dx_{p-1}$$
$$\cdot \int_{\frac{(S-x_1)}{\lambda^2}-x_2-\cdots-x_{p-1}}^{\frac{(T-x_1)}{\lambda^2}-x_2-\cdots-x_{p-1}} dx_p$$
$$\cdot \lambda^{2-2m} \int_0^{\lambda^2(x_p+x_{p-1}+\cdots+x_2)+x_1} \chi_{[0,T]}(s_\alpha)\, ds_\alpha \int_{[S,T]^{m-1}} ds_1 \cdots \widehat{ds_\alpha} \cdots ds_m$$
$$\cdot F\left(\frac{s_m - s_1}{\lambda^2}\right) \prod_{j=1}^{m-1} F\left(\frac{s_j - s_{j+1}}{\lambda^2}\right)$$
$$\cdot \int_{\mathbb{R}^{n-p-1}} dx_{p+1} \cdots \widehat{dx_q} \cdots dx_n$$
$$\cdot \int_{-\frac{x_1}{\lambda^2}-x_2-\cdots-x_{q-1}}^{\frac{(s_a-x_1)}{\lambda^2}-x_2-\cdots-x_{q-1}} dx_q F(x_n + \cdots + x_2) \prod_{h=1}^{n-1} F(-x_{h+1}), \tag{15.6.3}$$

where since the function F is integrable, the integrals with respect to $dx_1 \cdots dx_n$ are bounded. In order to complete the proof, we introduce the following change of variables for s_j:

$$\tau_1 := s_\alpha, \quad \tau_\alpha := s_1, \quad \tau_j := s_j, \quad j = \{1, 2, ..., m\} \setminus \{1, \alpha\},$$
$$y_1 := \tau_1, \quad y_j := \frac{\tau_{j+1} - \tau_j}{\lambda^2}, \quad j = 2, ..., m-1.$$

With these change of variables, we have

$$\lambda^{2-2m} \int_0^{\lambda^2(x_p+x_{p-1}+\cdots+x_2)+x_1} ds_\alpha \int_{[S,T]^{m-1}} ds_1 \cdots \widehat{ds_\alpha} \cdots ds_m$$
$$\cdot F\left(\frac{s_m - s_1}{\lambda^2}\right) \prod_{j=1}^{m-1} F\left(\frac{s_j - s_{j+1}}{\lambda^2}\right)$$
$$= \lambda^{2-2m} \int_0^{\lambda^2(x_p+x_{p-1}+\cdots+x_2)+x_1} d\tau_1 \int_{[S,T]^{m-1}} d\tau_2 \cdots d\tau_m$$
$$\cdot F\left(\frac{\tau_m - \tau_\alpha}{\lambda^2}\right) F\left(\frac{\tau_\alpha - \tau_2}{\lambda^2}\right) \prod_{j=2}^{m-1} F\left(\frac{\tau_j - \tau_{j+1}}{\lambda^2}\right)$$
$$= \int_0^{\lambda^2(x_p+x_{p-1}+\cdots+x_2)+x_1} dy_1 \int_{\frac{(S-y_1)}{\lambda^2}}^{\frac{(T-y_1)}{\lambda^2}} dy_2 \int_{\frac{(S-y_1)}{\lambda^2}-y_2}^{\frac{(T-y_1)}{\lambda^2}-y_2} dy_3$$
$$\cdots \int_{\frac{(S-y_1)}{\lambda^2}-y_2-\cdots-y_{m-1}}^{\frac{(T-y_1)}{\lambda^2}-y_2-\cdots-y_{m-1}} dy_m$$

$$\cdot F(y_m + \cdots + y_{\alpha+1}) F(y_\alpha + \cdots + y_3) \prod_{j=2}^{m-1} F(-y_{j+1})$$

$$\leq \int_0^{\lambda^2(x_p+x_{p-1}+\cdots+x_2)+x_1} dy_1 \int_{\mathbb{R}} dy_2 \int_{\mathbb{R}} dy_3 \cdots \int_{\mathbb{R}} dy_m F(y_m + \cdots + y_{\alpha+1})$$

$$\cdot F(y_\alpha + \cdots + y_3) \prod_{j=2}^{m-1} F(-y_{j+1}) .$$

By applying this to (15.6.3), one knows that (15.6.3) is less than or equal to

$$\int_S^T dx_1 \int_{\mathbb{R}^{n-2}} dx_2 \cdots \widehat{dx_q} \cdots dx_n \int_{-\frac{x_1}{\lambda^2}-x_2-\cdots-x_{q-1}}^{\frac{(y_1-x_1)}{\lambda^2}-x_2-\cdots-x_{q-1}} dx_q$$

$$\cdot \int_0^{\lambda^2(x_p+x_{p-1}+\cdots+x_2)+x_1} \chi_{[0,T]}(y_1)\, dy_1 \int_{\mathbb{R}} dy_2 \int_{\mathbb{R}} dy_3 \cdots \int_{\mathbb{R}} dy_m$$

$$\cdot F(x_n + \cdots + x_2) \cdot \prod_{h=1}^{n-1} F(-x_{h+1})$$

$$\cdot F(y_m + \cdots + y_{\alpha+1}) \cdot F(y_\alpha + \cdots + y_3) \cdot \prod_{j=1}^{m-1} F(-y_{j+1})$$

$$= \int_{\mathbb{R}^{n+m}} dx_1 \cdots dx_n dy_1 \cdots dy_m \chi_{\left[-\frac{x_1}{\lambda^2}-x_2-\cdots-x_{q-1}, \frac{(y_1-x_1)}{\lambda^2}-x_2-\cdots-x_{q-1}\right]}(x_q)$$

$$\cdot \chi_{\left[0,\lambda^2(x_p+x_{p-1}+\cdots+x_2)+x_1\right]}(y_1)$$

$$\cdot \chi_{[0,T]}(y_1) \cdot \chi_{[S,T]}(x_1)\, F(x_n + \cdots + x_2) \cdot \prod_{h=1}^{n-1} F(-x_{h+1})$$

$$\cdot F(y_m + \cdots + y_{\alpha+1}) \cdot F(y_\alpha + \cdots + y_3) \cdot \prod_{j=1}^{m-1} F(-y_{j+1}) . \tag{15.6.4}$$

Since the product of all functions in the integral is less than or equal to

$$3 \|F\|_\infty \cdot \chi_{[0,T]}(y_1) \cdot \chi_{[S,T]}(x_1) \cdot \prod_{h=1}^{n-1} F(-x_{h+1}) \cdot \prod_{j=1}^{m-1} F(-y_{j+1}) ,$$

which belongs to $\mathbf{L}^1(\mathbb{R}^{n+m})$, and since the function

$$\chi_{\left[-\frac{x_1}{\lambda^2}-x_2-\cdots-x_{q-1}, \frac{(y_1-x_1)}{\lambda^2}-x_2-\cdots-x_{q-1}\right]}(x_q) \cdot \chi_{\left[0,\lambda^2(x_p+x_{p-1}+\cdots+x_2)+x_1\right]}(y_1)$$

goes to zero as $\lambda \to 0$ almost always, the dominated convergence theorem gives the thesis.

Corollary 15.6.1. *Let the function F and the indices α, $n, m = 2, 3, 4, \ldots$, be as in Theorem 15.6.1 and suppose that $S_h < T_h$, for all $h = 1, 2, \cdots, m+n$.*

As $\lambda \to 0$, the quantity

$$\lambda^4 \int_{S_1/\lambda^2}^{T_1/\lambda^2} dt_1 \cdots \int_{S_n/\lambda^2}^{T_n/\lambda^2} dt_n \int_{S_{n+1}/\lambda^2}^{T_{n+1}/\lambda^2} ds_1 \cdots \int_{S_{n+m}/\lambda^2}^{T_{n+m}/\lambda^2} ds_{n+m} \; \chi_{(t_p,t_q)}(s_\alpha)$$

$$\cdot F(t_1 - t_n) F(s_1 - s_m) \prod_{h=1}^{n-1} F(t_h - t_{h+1}) \prod_{j=1}^{m-1} F(s_j - s_{j+1}) \tag{15.6.5}$$

goes to zero.

Proof 15.6.2. The proof can be made by just

(i) noticing that the absolute value of (15.6.5) is less than or equal to

$$\lambda^4 \int_{[S/\lambda^2, T/\lambda^2]} dt_1 \cdots dt_n ds_1 \cdots ds_{n+m} \; \chi_{(t_p,t_q)}(s_\alpha)$$

$$\cdot \left| F(t_1 - t_n) F(s_1 - s_m) \prod_{h=1}^{n-1} F(t_h - t_{h+1}) \prod_{j=1}^{m-1} F(s_j - s_{j+1}) \right| ;$$

where $S := \min\{S_1, \cdots, S_{n+m}\}$, $\quad T := \max\{T_1, \cdots, T_{n+m}\}$.

(ii) making the same change of variables as we did in the proof of Corollary (15.5.3);

(iii) applying Theorem 15.6.1.

15.7 The Type-I Term Theorem

In this section we shall study the following type of integrals:

$$\int_{S_1/\lambda^2}^{T_1/\lambda^2} ds_1 \cdots \int_{S_N/\lambda^2}^{T_N/\lambda^2} ds_N \int_{S_1'/\lambda^2}^{T_1'/\lambda^2} du_1 \cdots \int_{S_M'/\lambda^2}^{T_M'/\lambda^2} du_M \int_0^{t/\lambda^2} dt_1 \int_0^{t_1} dt_2$$

$$\cdots \int_0^{t_{n-1}} dt_n \cdot F_{\lambda,\epsilon}(s_1 \cdots s_N, t_1 \cdots t_n, u_1 \cdots u_M) \tag{15.7.1}$$

where, $F_{\lambda,\epsilon}(s_1 \cdots s_N, t_1 \cdots t_n, u_1 \cdots u_M)$ has the form

$$\lambda^{N+k_-+M+k_++2k+2m+\epsilon} \delta_{N-k_-, M-k_+}$$

$$\cdot \prod_{h=1}^{m} \left(f_h^{(-1)}(t_{r_h} - t_{l_h}) \prod_{j=1}^{r_h - l_h} f_{j,h}^{(1)}(t_{l_h+j} - t_{l_h+j-1}) \right) \prod_{h=1}^{k} f_h^{(0)}(t_{p_h+1} - t_{p_h})$$

$$\cdot \prod_{h=1}^{k_-} f_h^{(-)}(t_{a_h} - s_{\sigma(h)}) \prod_{h=1}^{k_+} f_h^{(+)}(u_{\tau(h)} - t_{b_h}) \prod_{h=1}^{N-k_-} f_h^{(\pm)}(u_{\tau(k_++h)} - s_{\sigma(k_-+h)}),$$

$$\text{and } \prod_{h=1}^{n}(\cdots) := 1 \quad \text{if } n = 0. \tag{15.7.2}$$

Remark 15.7.1. This very general form of $F_{\lambda,\epsilon}$ covers simultaneously several types of integral arising in the polynomial WCL and in the LDL. Each of the functions f accounts for some type of pairing (line) occurring in the normally ordered form of matrix elements (between two collective vectors) of some terms of the iterated series of the evolution operator. More precisely:

- $f^{(-1)}$ occurs in the Fock part of the LDL;
- $f^{(1)}$ occurs in the anti-Fock part of the LDL;
- $f^{(0)}$ accounts for the internal lines in the WCL;
- $f^{(-)}$ accounts for the external lines to the left vector in the WCL and LDL;
- $f^{(+)}$ accounts for the external lines to the right vector in the WCL and the LDL;
- $f^{(\pm)}$ accounts for the external lines which pair two vectors in the WCL and LDL.

Remark 15.7.2. The power ϵ, in the exponent of λ in (15.7.2), also has its origins in the LDL and is due to the asymmetry between the Fock and anti-Fock case. In fact the test functions of the anti-Fock creators and annihilators are multiplied by the factor λ, while those of the Fock creators and annihilators are of order 1. For this reason the terms of the form $1 \otimes A^+A$ will produce pairings which are always multiplied by the factor λ^2 (these factors are taken into account by the exponent ϵ) and therefore vanish in the limit. While the factors of the form $A^+A \otimes 1$ are of finite order and therefore in the limit give a nontrivial contribution.

Moreover, we will suppose the following:

- $\epsilon \geq 0$, $0 \leq m \leq n/3$;
- $1 \leq l_1 < r_1 < l_2 < r_2 < \cdots < l_m < r_m \leq n$ satisfy $r_h - l_h > 1$ for all $h = 1, \cdots, m$;
- $k \leq \frac{1}{2}[n - 1 - \sum_{h=1}^m (r_h - l_h)]$ and $1 \leq p_1 < \cdots < p_k < n$ satisfy $\{p_h, p_h + 1\}_{h=1}^k \subset \cup_{h=1}^m \{l_h, l_h + 1, \cdots, r_h\}$;
- $\sigma \in \mathcal{S}_N$, $\tau \in \mathcal{S}_M$;
- $k_-, k_+ \in \mathbb{N}$ such that $k_- + k_+ = n - \sum_{h=1}^m (r_h - l_h + 1) - 2k$;
- $1 \leq a_1 < \cdots < a_{k_-} \leq n$ are such that

$$\{a_h\}_{h=1}^{k_-} \subset \{1, 2, \cdots, n\} \setminus \left(\cup_{h=1}^m \{l_h, l_h + 1, \cdots, r_h\} \cup \{p_h, p_h + 1\}_{h=1}^k\right) ;$$

- $1 \leq b_1 < \cdots < b_{k_+} \leq n$ are such that

$$\{b_h\}_{h=1}^{k_+} \subset \{1, 2, \cdots, n\} \setminus \Big(\cup_{h=1}^m \{l_h, l_h + 1, \cdots, r_h\} \cup \{p_h, p_h + 1\}_{h=1}^k \cup \{a_h\}_{h=1}^{k_-}\Big) ;$$

- all functions $f_{j,h}^{(1)}$, $f_h^{(0)}$, $f_h^{(-1)}$, $f_h^{(-)}$, $f_h^{(+)}$ and $f_h^{(\pm)}$ for all possible choices of j, h are in $L^1(\mathbb{R}) \cap L^\infty(\mathbb{R})$.

With the above assumptions the variables

$$\{s_h\}_{h=1}^N \cup \{t_h\}_{h=1}^n \cup \{u_h\}_{h=1}^N$$

are divided into $m + k + k_- + k_+ + N - k_-$ connected components. So the first connected component theorem shows that (15.7.1), as $\lambda \to 0$, has order

$$\begin{aligned}&\lambda^{N+k_-+M+k_++2k+2m+\epsilon} \cdot O(\lambda^{-2(m+k+k_++N)}) \\ =&\lambda^{2N+2k_++2k+2m+\epsilon} \cdot O(\lambda^{-2(m+k+k_++N)}) = O(\lambda^\epsilon),\end{aligned}$$

which is nontrivial only if $\epsilon = 0$. That is, we need to study (15.7.1) only for $\epsilon = 0$. In the following for simplicity we call each connected component involving more than 2 (or exactly 2) time variables a *long chain* (*short chain*). The first step is to study (15.7.1) in the case $m = 0$, i.e. there are no long chains. In the WCL only such quantities arise. In the following we use the notation

$$\Delta_r(t) := \{(s_1 \cdots s_r) : 0 \le s_r \le \cdots \le s_1 \le t\}.$$

Theorem 15.7.1. *(The type-I term theorem). As $\lambda \to 0$, the limit of (15.7.1) exists and is equal to zero if $\epsilon > 0$; if $\varepsilon = 0$, it is equal to*

$$\prod_{h=1}^m \int_{-\infty}^0 dv_1 \cdots \int_{-\infty}^0 dv_{r_h-l_h} f_h^{(-1)}(v_1 + v_2 + \cdots + v_{r_h-l_h}) \prod_{j=1}^{r_h-l_h} f_{j,h}^{(1)}(v_j)$$

$$\cdot \prod_{h=1}^k \int_{-\infty}^0 dv f_h^{(0)}(v) \int_{\Delta_{k_-+k_++k+m}(t)} dt_1 \cdots \widehat{dt_{p_1+1}} \cdots \widehat{dt_{l_1+1}} \cdots \widehat{dt_{r_1+1}}$$

$$\cdots \widehat{dt_{p_k+1}} \cdots \widehat{dt_{l_m+1}} \cdots \widehat{dt_{r_m+1}} \cdots dt_n$$

$$\cdot \prod_{h=1}^{k_-} \chi_{[S_{\sigma(h)},T_{\sigma(h)}]}(t_{a_h}) \int_{\mathbb{R}} dv f_h^{(-)}(v) \prod_{h=1}^{k_+} \chi_{[S_{\tau(h)},T_{\tau(h)}]}(t_{b_h}) \int_{\mathbb{R}} dv f_h^{(+)}(v)$$

$$\cdot \prod_{h=1}^{N-k_-} \left\langle \chi_{[S_{\sigma(k_-+h)},T_{\sigma(k_-+h)}]}, \chi_{[S_{\tau(k_++h)},T_{\tau(k_++h)}]} \right\rangle \int_{\mathbb{R}} dv f_h^{(\pm)}(v). \quad (15.7.3)$$

Proof 15.7.1. As already stated, (15.7.1) has order $O(\lambda^\epsilon)$; therefore, as $\lambda \to 0$, it goes to zero if $\epsilon > 0$. So we need only consider the case when $\epsilon = 0$. The first step is trivial: with the change of variables

$$\lambda^2 t_h \hookrightarrow t_h, \quad h = 1, 2, \cdots, n, \qquad \lambda^2 s_h \hookrightarrow s_h, \quad h = 1, 2, \cdots, N,$$
$$\lambda^2 u_h \hookrightarrow u_h, \quad h = 1, 2, \cdots, M,$$

(15.7.1) becomes

$$\int_{S_1}^{T_1} ds_1 \cdots \int_{S_N}^{T_N} ds_N \int_{S'_1}^{T'_1} du_1 \int_{S'_M}^{T'_M} du_M \int_0^t dt_1 \int_0^{t_1} dt_2 \cdots \int_0^{t_{n-1}} dt_n$$
$$\cdot \lambda^{-2(N+M+n)} F_{\lambda,0}\big(s_1/\lambda^2 \cdots s_N/\lambda^2, t_1/\lambda^2$$
$$\cdots t_n/\lambda^2, u_1/\lambda^2 \cdots u_M/\lambda^2\big)\,. \tag{15.7.4}$$

In the second step we change the variables

$$y_{l_j+1} := \frac{t_{l_j+1} - t_{l_j}}{\lambda^2} \tag{15.7.5a}$$

and keep t_{l_j} invariant. Thus (15.7.4) becomes

$$\lambda^{-2(N+M+n)} \lambda^{N+k_-+M+k_++2k+2m} \delta_{N-k_-,M-k_+}$$
$$\int_{S_1}^{T_1} ds_1 \cdots \int_{S_N}^{T_N} ds_N \int_{S'_1}^{T'_1} du_1 \int_{S'_M}^{T'_M} du_M$$
$$\cdot \int_0^t dt_1 \cdots \int_0^{t_{l_1-1}} dt_{l_1} \int_{-t_{l_1}/\lambda^2}^0 dy_{l_1+1} \int_0^{t_{l_1}+y_{l_1+1}\lambda^2} dt_{l_1+2}$$
$$\cdots \int_0^{t_{r_1-1}} dt_{r_1} \int_0^{t_{r_1}} dt_{r_1+1} \cdots \int_0^{t_{l_m-1}} dt_{l_m} \int_{(t_{l_m+1}-t_{l_m})/\lambda^2}^0 dy_{l_m+1}$$
$$\cdot \int_0^{t_{l_m}+y_{l_m+1}\lambda^2} dt_{l_m+2} \cdots \int_0^{t_{r_m-1}} dt_{r_m}$$
$$\cdot \int_0^{t_{r_m}} dt_{r_m+1} \cdots \int_0^{t_{n-1}} dt_n \prod_{h=1}^m \lambda^2 f_{1,h}^{(1)}(y_{l_h+1})$$
$$\cdot \prod_{h=1}^m \left[f_h^{(-1)} \left(\frac{t_{r_h} - t_{l_h}}{\lambda^2} \right) \prod_{j=2}^{r_h - l_h} f_{j,h}^{(1)} \left(\frac{t_{l_h+j} - t_{l_h+j-1}}{\lambda^2} \right) \right]$$
$$\cdot \prod_{h=1}^k f_h^{(0)} \left(\frac{t_{p_h+1} - t_{p_h}}{\lambda^2} \right) \prod_{h=1}^{k_-} f_h^{(-)} \left(\frac{t_{a_h} - s_{\sigma(h)}}{\lambda^2} \right)$$
$$\cdot \prod_{h=1}^{k_+} f_h^{(+)} \left(\frac{u_{\tau(h)} - t_{b_h}}{\lambda^2} \right) \prod_{h=1}^{N-k_-} f_h^{(\pm)} \left(\frac{u_{\tau(k_++h)} - s_{\sigma(k_-+h)}}{\lambda^2} \right). \tag{15.7.6}$$

Now we make the successive change of variables

$$y_{l_j+2} := \frac{t_{l_j+2} - t_{l_j+1}}{\lambda^2} = \frac{t_{l_j+2} - t_{l_j}}{\lambda^2} - y_{l_j+1}\,, \tag{15.7.5b}$$

$$y_{l_j+3} := \frac{t_{l_j+3} - t_{l_j+2}}{\lambda^2} = \frac{t_{l_j+3} - t_{l_j}}{\lambda^2} - y_{l_j+1} - y_{l_j+2}\,, \tag{15.7.5c}$$

and eventually

$$y_{r_j} = y_{l_j + r_j - l_j} := \frac{t_{r_j} - t_{r_j - 1}}{\lambda^2} = \frac{t_{r_j} - t_{l_j}}{\lambda^2} - y_{l_j+1} - y_{l_j+2} - \cdots - y_{r_j - 1} \, . \tag{15.7.5d}$$

Noticing that

$$\frac{t_{r_j} - t_{l_j}}{\lambda^2} = \sum_{x=1}^{r_j - l_j} \frac{t_{l_j + x} - t_{l_j + x.1}}{\lambda^2} = \sum_{x=1}^{r_j - l_j} y_{l_j + x} \, ,$$

we are able to rewrite (15.7.6) as

$$\begin{aligned}
&\lambda^{-2(N+M+n)} \lambda^{N+k_- + M + k_+ + 2k + 2m} \lambda^{2 \sum_{h=1}^{m} (r_h - l_h)} \delta_{N-k_-, M-k_+} \\
&\cdot \int_{S_1}^{T_1} ds_1 \cdots \int_{S_N}^{T_N} ds_N \int_{S_1'}^{T_1'} du_1 \cdots \int_{S_M'}^{T_M'} du_M \\
&\cdot \int_0^t dt_1 \cdots \int_0^{t_{l_1 - 1}} dt_{l_1} \int_{-t_{l_1}/\lambda^2}^{0} dy_{l_1+1} \int_{-t_{l_1}/\lambda^2 - y_{l_1+1}}^{0} dy_{l_1+2} \\
&\cdots \int_{-t_{l_1}/\lambda^2 - y_{l_1+1} - \cdots - y_{r_1 - 1}}^{0} dy_{r_1} \int_0^{t_{l_1} + \lambda^2 (y_{l_1+1} + \cdots + y_{r_1})} dt_{r_1+1} \\
&\cdots \int_0^{t_{l_m - 1}} dt_{l_m} \int_{-t_{l_m}/\lambda^2}^{0} dy_{l_m+1} \int_{-t_{l_m}/\lambda^2 - y_{l_m+1}}^{0} dy_{l_m+2} \\
&\cdots \int_{-t_{l_m}/\lambda^2 - y_{l_m+1} - \cdots - y_{r_m - 1}}^{0} dy_{r_m} \int_0^{t_{r_m} + \lambda^2 (y_{l_m+1} + \cdots + y_{r_m})} dt_{r_1+1} \\
&\cdots \int_0^{t_{n-1}} dt_n \prod_{h=1}^{m} \left[f_h^{(-1)} \left(\sum_{j=1}^{r_h - l_h} y_{l_h + j} \right) \prod_{j=1}^{r_h - l_h} f_{1,h}^{(1)} (y_{l_h + j}) \right] \\
&\cdot \prod_{h=1}^{k} f_h^{(0)} \left(\frac{t_{p_h + 1} - t_{p_h}}{\lambda^2} \right) \prod_{h=1}^{k_-} f_h^{(-)} \left(\frac{t_{a_h} - s_{\sigma(h)}}{\lambda^2} \right) \\
&\cdot \prod_{h=1}^{k_+} f_h^{(+)} \left(\frac{u_{\tau(h)} - t_{b_h}}{\lambda^2} \right) \prod_{h=1}^{N-k_-} f_h^{(\pm)} \left(\frac{u_{\tau(k_+ + h)} - s_{\sigma(k_- + h)}}{\lambda^2} \right) .
\end{aligned} \tag{15.7.7}$$

The third step is to change the variables

$$x_h := (t_{p_h+1} - t_{p_h}) / \lambda^2, \quad h = 1 \cdots k \, , \tag{15.7.8}$$

and keep t_j invariant for any $j \in \{1, 2, \cdots, n\} \setminus \{p_h + 1\}_{h=1}^{k}$. Thus, (15.7.7) can be written as

$$
\begin{aligned}
&\lambda^{-2(N+M+n)}\lambda^{N+k_-+M+k_++2k+2m}\lambda^{2\sum_{h=1}^{m}(r_h-l_h)+2k}\delta_{N-k_-,M-k_+}\\
&\cdot\int_{S_1}^{T_1}ds_1\cdots\int_{S_N}^{T_N}ds_N\int_{S_1'}^{T_1'}du_1\cdots\int_{S_M'}^{T_M'}du_M\\
&\cdot\int_0^{t}dt_1\cdots\int_0^{t_{l_1-1}}dt_{l_1}\int_{-t_{l_1}/\lambda^2}^{0}dy_{l_1+1}\int_{-t_{l_1}/\lambda^2-y_{l_1+1}}^{0}dy_{l_1+2}\\
&\cdots\int_{-t_{l_1}/\lambda^2-y_{l_1+1}-\cdots-y_{r_1-1}}^{0}dy_{r_1}\int_0^{t_{l_1}+\lambda^2(y_{l_1+1}+\cdots+y_{r_1})}dt_{r_1+1}\\
&\cdots\int_0^{t_{p_1-1}}dt_{p_1}\int_{-t_{p_1}/\lambda^2}^{0}dx_1\int_0^{t_{p_1}+\lambda^2x_1}dt_{p_1+2}\cdots\int_0^{t_{l_m-1}}dt_{l_m}\int_{-t_{l_m}/\lambda^2}^{0}dy_{l_m+1}\\
&\cdot\int_{-t_{l_m}/\lambda^2-y_{l_m+1}}^{0}dy_{l_m+2}\cdots\int_{-t_{l_m}/\lambda^2-y_{l_m+1}-\cdots-y_{r_m-1}}^{0}dy_{r_m}\\
&\cdot\int_0^{t_{r_m}+\lambda^2(y_{l_m+1}+\cdots+y_{r_m})}dt_{r_1+1}\cdots\int_0^{t_{p_k-1}}dt_{p_k}\\
&\cdot\int_{-t_{p_k}/\lambda^2}^{0}dx_k\int_0^{t_{p_k}+\lambda^2x_k}dt_{p_k+2}\cdots\int_0^{t_{n-1}}dt_n\\
&\cdot\prod_{h=1}^{m}\left[f_h^{(-1)}\left(\sum_{j=1}^{r_h-l_h}y_{l_h+j}\right)\prod_{j=1}^{r_h-l_h}f_{1,h}^{(1)}(y_{l_h+j})\right]\\
&\cdot\prod_{h=1}^{k}f_h^{(0)}(x_h)\prod_{h=1}^{k_-}f_h^{(-)}\left(\frac{t_{a_h}-s_{\sigma(h)}}{\lambda^2}\right)\\
&\cdot\prod_{h=1}^{k_+}f_h^{(+)}\left(\frac{u_{\tau(h)}-t_{b_h}}{\lambda^2}\right)\prod_{h=1}^{N-k_-}f_h^{(\pm)}\left(\frac{u_{\tau(k_++h)}-s_{\sigma(k_-+h)}}{\lambda^2}\right). \qquad (15.7.9)
\end{aligned}
$$

The fourth step is to change the variables

$$
s_h' := \frac{s_{\sigma(h)}-t_{a_h}}{\lambda^2},\quad h=1\cdots k_-;\qquad u_h' := \frac{u_{\tau(h)}-t_{b_h}}{\lambda^2},\quad h=1,\cdots,k_+\,, \qquad (15.7.10)
$$

and keep invariant t_j, $u_{\tau(k_++h)}$, $s_{\sigma(k_-+h')}$ for all $j\in\{1,2,\cdots,n\}\backslash\{p_h+1\}_{h=1}^{k}$, $h=1,2,\cdots,N-k_-$ and $h'=1,2,\cdots,M-k_+$. With this change of variables, (15.7.9) becomes

$$
\begin{aligned}
&\lambda^{-2(N+M+n)}\lambda^{N+k_-+M+k_++2k+2m} \\
&\cdot\lambda^{2\sum_{h=1}^{m}(r_h-l_h)+2k+2k_-+2k_+}\delta_{N-k_-,M-k_+} \\
&\cdot\int_0^t dt_1\cdots\int_0^{t_{l_1-1}}dt_{l_1}\int_{-t_{l_1}/\lambda^2}^0 dy_{l_1+1} \\
&\cdot\int_{-t_{l_1}/\lambda^2-y_{l_1+1}}^0 dy_{l_1+2}\cdots\int_{-t_{l_1}/\lambda^2-y_{l_1+1}-\cdots-y_{r_1-1}}^0 dy_{r_1} \\
&\cdot\int_0^{t_{l_1}+\lambda^2(y_{l_1+1}+\cdots+y_{r_1})}dt_{r_1+1}\cdots\int_0^{t_{p_1-1}}dt_{p_1} \\
&\cdot\int_{-t_{p_1}/\lambda^2}^0 dx_1\int_0^{t_{p_1}+\lambda^2x_1}dt_{p_1+2}\cdots\int_0^{t_{l_m-1}}dt_{l_m}\int_{-t_{l_m}/\lambda^2}^0 dy_{l_m+1} \\
&\cdot\int_{-t_{l_m}/\lambda^2-y_{l_m+1}}^0 dy_{l_m+2}\cdots\int_{-t_{l_m}/\lambda^2-y_{l_m+1}-\cdots-y_{r_m-1}}^0 dy_{r_m} \\
&\cdot\int_0^{t_{r_m}+\lambda^2(y_{l_m+1}+\cdots+y_{r_m})}dt_{r_1+1}\cdots\int_0^{t_{p_k-1}}dt_{p_k}\int_{-t_{p_k}/\lambda^2}^0 dx_k \\
&\cdot\int_0^{t_{p_k}+\lambda^2x_k}dt_{p_k+2}\cdots\int_0^{t_{n-1}}dt_n \\
&\cdot\prod_{h=1}^k f_h^{(0)}(x_h)\prod_{h=1}^m\left[f_h^{(-1)}\left(\sum_{j=1}^{r_h-l_h}y_{l_h+j}\right)\prod_{j=1}^{r_h-l_h}f_{1,h}^{(1)}(y_{l_h+j})\right] \\
&\cdot\prod_{h=1}^{k_-}\int_{(S_{\sigma(h)}-t_{a_h})/\lambda^2}^{(T_{\sigma(h)}-t_{a_h})/\lambda^2}ds f_h^{(-)}(-s)\prod_{h=1}^{k_+}\int_{(S'_{\tau(h)}-t_{b_h})/\lambda^2}^{(T_{\tau(h)}-t_{b_h})/\lambda^2}du f_h^{(+)}(u) \\
&\cdot\int_{S_{\sigma(k_-+1)}}^{T_{\sigma(k_-+1)}}ds_{\sigma(k_-+1)}\cdots\int_{S_{\sigma(N)}}^{T_{\sigma(N)}}ds_{\sigma(N)}\int_{S'_{\tau(k_++1)}}^{T'_{\tau(k_++1)}}du_{\tau(k_++1)} \\
&\cdots\int_{S'_{\tau(M)}}^{T'_{\tau(M)}}du_{\tau(M)}\cdot\prod_{h=1}^{N-k_-}f_h^{(\pm)}\left(\frac{u_{\tau(k_++h)}-s_{\sigma(k_-+h)}}{\lambda^2}\right).
\end{aligned}
\tag{15.7.11}
$$

The fifth step is to change the variables

$$
z_h := \frac{u_{\tau(k_++h)}-s_{\sigma(k_-+h)}}{\lambda^2},\quad h=1,2,\cdots,N-k_-=M-k_+\,, \tag{15.7.12}
$$

and keep invariant $s_{\sigma(k_-+h')}$ for all $h'=1,2,\cdots,M-k_+$. So, we are able to rewrite (15.7.11) as

$$
\begin{aligned}
&\lambda^{-2(N+M+n)}\lambda^{N+k_-+M+k_++2k+2m} \\
&\cdot\lambda^{2\sum_{h=1}^{m}(r_h-l_h)+2k+2k_-+2k_++2(N-k_-)}\delta_{N-k_-,M-k_+} \\
&\cdot\int_0^t dt_1\cdots\int_0^{t_{l_1-1}}dt_{l_1}\int_{-t_{l_1}/\lambda^2}^0 dy_{l_1+1}\int_{-t_{l_1}/\lambda^2-y_{l_1+1}}^0 dy_{l_1+2} \\
&\cdots\int_{-t_{l_1}/\lambda^2-y_{l_1+1}-\cdots-y_{r_1-1}}^0 dy_{r_1}\int_0^{t_{l_1}+\lambda^2(y_{l_1+1}+\cdots+y_{r_1})} dt_{r_1+1} \\
&\cdots\int_0^{t_{p_1-1}}dt_{p_1}\int_{-t_{p_1}/\lambda^2}^0 dx_1\int_0^{t_{p_1}+\lambda^2x_1}dt_{p_1+2} \\
&\cdots\int_0^{t_{l_m-1}}dt_{l_m}\int_{-t_{l_m}/\lambda^2}^0 dy_{l_m+1} \\
&\cdot\int_{-t_{l_m}/\lambda^2-y_{l_m+1}}^0 dy_{l_m+2}\cdots\int_{-t_{l_m}/\lambda^2-y_{l_m+1}-\cdots-y_{r_m-1}}^0 dy_{r_m} \\
&\cdot\int_0^{t_{r_m}+\lambda^2(y_{l_m+1}+\cdots+y_{r_m})} dt_{r_1+1}\cdots\int_0^{t_{p_k-1}}dt_{p_k} \\
&\cdot\int_{-t_{p_k}/\lambda^2}^0 dx_k\int_0^{t_{p_k}+\lambda^2x_k}dt_{p_k+2}\cdots\int_0^{t_{n-1}}dt_n \\
&\cdot\prod_{h=1}^{k}f_h^{(0)}(x_h)\prod_{h=1}^{m}\left[f_h^{(-1)}\left(\sum_{j=1}^{r_h-l_h}y_{l_h+j}\right)\prod_{j=1}^{r_h-l_h}f_{1,h}^{(1)}(y_{l_h+j})\right] \\
&\cdot\prod_{h=1}^{k_-}\int_{(S_{\sigma(h)}-t_{a_h})/\lambda^2}^{(T_{\sigma(h)}-t_{a_h})/\lambda^2}ds f_h^{(-)}(-s)\prod_{h=1}^{k_+}\int_{(S'_{\tau(h)}-t_{b_h})/\lambda^2}^{(T_{\tau(h)}-t_{b_h})/\lambda^2}du f_h^{(+)}(u) \\
&\cdot\int_{S_{\sigma(k_-+1)}}^{T_{\sigma(k_-+1)}}ds_{\sigma(k_-+1)}\cdots\int_{S_{\sigma(N)}}^{T_{\sigma(N)}}ds_{\sigma(N)} \\
&\cdot\prod_{h=1}^{N-k_-}\int_{\left(S'_{\tau(k_++1)}-s_{\sigma(k_-+h)}\right)/\lambda^2}^{\left(T'_{\tau(k_++1)}-s_{\sigma(k_-+h)}\right)/\lambda^2}du f_h^{(\pm)}(u)\,.
\end{aligned}
\tag{15.7.13}
$$

Since $n = 2k + k_- + k_+ + \sum_{h=1}^{m}(r_h - l_h + 1)$ and $N - k_- = M - k_+$, we have

$$
\begin{aligned}
&N+k_-+M+k_++2k+2m\\
&+2\sum_{h=1}^{m}(r_h-l_h)-2n-2N-2M+2k+2k_-+2k_++2(N-k_-)\\
=&N+k_-+M+k_++2k+2m+2\sum_{h=1}^{m}(r_h-l_h)-2(2k+k_-+k_+\\
&+\sum_{h=1}^{m}(r_h-l_h+1)-2N-2M+2k+2k_-+2k_++2(N-k_-)=0\,.
\end{aligned}
\tag{15.7.14}
$$

So the factor

$$\lambda^{-2(N+M+n)}\lambda^{N+k_-+M+k_++2k+2m}\lambda^{2\sum_{h=1}^{m}(r_h-l_h+1)+2k+2k_-+2k_+2(N-k_-)}$$

in (15.7.13) is equal to 1. Thanks to this fact and to the Lebesgue-dominated convergence theorem, we complete the proof.

15.8 The Double Integral Lemma

The following result is very important for constructing the master space and for proving that the limit of the time-rescaled evolution operator verifies a quantum stochastic differential equation.

Lemma 15.8.1. (The double integral lemma). *Let $F \in \mathbf{L}^1(\mathbb{R})$ and let, for each $\lambda \in \mathbb{R}_+$, $G_\lambda : \mathbb{R} \to \mathbb{C}$ be a continuous function such that, for some constant $C < +\infty$*

$$\sup\nolimits_{(\lambda,t)\in\mathbb{R}_+\times\mathbb{R}} |G_\lambda(t)| \le C\,, \tag{15.8.1a}$$

$$\lim\nolimits_{\lambda\to 0} G_\lambda(t+\lambda^2 r) = G_0(t)\,, \tag{15.8.1b}$$

for any $r \in \mathbb{R}$. Then, for any $S, T, S', T' \in \mathbb{R}$ such that $S < T,\ S' < T'$,

$$
\begin{aligned}
&\lim_{\lambda\to 0}\frac{1}{\lambda^2}\int_S^T dt_1\int_{S'}^{T'} dt_2 F\left(\frac{t_1-t_2}{\lambda^2}\right)G_\lambda(t_1)\\
&=\int_{\mathbb{R}} dt G_0(t)\chi_{[S,T]}(t)\,\chi_{[S',T']}(t)\cdot\int_{-\infty}^{\infty}F(s)ds\,.
\end{aligned}
\tag{15.8.2}
$$

Moreover, if $F \in \mathbf{L}^1(\mathbb{R}) \cap \mathbf{L}^\infty(\mathbb{R})$, the limit is locally uniform for S, T, S', T', i.e. uniform for S, T, S', T' running over a bounded set of $\mathbb{R}$.

Proof 15.8.1. By the change of variables

$$t_2 = t_2,\quad s_1 := \frac{t_1-t_2}{\lambda^2}\,, \tag{15.8.3}$$

one has

$$\begin{aligned}
&\frac{1}{\lambda^2}\int_S^T dt_1 \int_{S'}^{T'} dt_2 F\left(\frac{t_1-t_2}{\lambda^2}\right) G_\lambda(t_1)\\
&=\int_{S'}^{T'} dt_2 \int_{(S-t_2)/\lambda^2}^{(T-t_2)/\lambda^2} ds_1 F(s_1)\, G_\lambda\left(t_2+\lambda^2 s_1\right)\\
&=\int_{S'}^{T'} dt_2 \int_{\mathbb{R}} ds_1 \chi_{[(S-t_2)/\lambda^2,(T-t_2)/\lambda^2]}(s_1)\, F(s_1)\, G_\lambda\left(t_2+\lambda^2 s_1\right)\\
&=\int_{\mathbb{R}} dt_2 \int_{\mathbb{R}} ds_1\, \chi_{(S',T')}(t_2)\, \chi_{[(S-t_2)/\lambda^2,(T-t_2)/\lambda^2]}(s_1)\, F(s_1)\, G_\lambda\left(t_2+\lambda^2 s_1\right) .
\end{aligned} \tag{15.8.4}$$

So, by the Lebesgue-dominated convergence theorem, one obtains

$$\begin{aligned}
&\lim_{\lambda\to 0}\frac{1}{\lambda^2}\int_S^T dt_1 \int_{S'}^{T'} dt_2 F\left(\frac{t_1-t_2}{\lambda^2}\right) G_\lambda(t_1)\\
&=\int_{\mathbb{R}} dt_2 \int_{\mathbb{R}} ds_1\, \chi_{(S',T')}(t_2)\, F(s_1)\, G_0(t_2) \lim_{\lambda\to 0}\chi_{[(S-t_2)/\lambda^2,(T-t_2)/\lambda^2]}(s_1) .
\end{aligned} \tag{15.8.5}$$

For each $t_2 \in (S',T')$ and $s_1 \in \mathbb{R}$, the limit $\lim_{\lambda\to 0}\chi_{[(S-t_2)/\lambda^2,(T-t_2)/\lambda^2]}(s_1)$ is possibly nonzero only if $S-t_2 \le 0$ and $T-t_2 \ge 0$, that is if $t_2 \in [S,T]$. Moreover, since $\chi_{\{S,T\}}$ is equal to zero a.e., we conclude that

$$\begin{aligned}
&\lim_{\lambda\to 0}\frac{1}{\lambda^2}\int_S^T dt_1 \int_{S'}^{T'} dt_2 F\left(\frac{t_1-t_2}{\lambda^2}\right) G_\lambda(t_1)\\
&=\int_{\mathbb{R}} dt_2 \int_{\mathbb{R}} ds_1 \chi_{(S',T')}(t_2)\, \chi_{[S,T]}(t_2) \left(1-\chi_{\{S,T\}}\right)(t_2)\, F(s_1)\, G_0(t_2)\\
&\cdot \lim_{\lambda\to 0}\chi_{[(S-t_2)/\lambda^2,(T-t_2)/\lambda^2]}(s_1)\\
&=\int_{\mathbb{R}} dt_2 \int_{\mathbb{R}} ds_1\, \chi_{(S',T')}(t_2)\, \chi_{(S,T)}(t_2)\, F(s_1)\, G_0(t_2)\\
&\cdot \lim_{\lambda\to 0}\chi_{[(S-t_2)/\lambda^2,(T-t_2)/\lambda^2]}(s_1) .
\end{aligned} \tag{15.8.6}$$

Obviously, for each $t_2 \in (S,T)$ and $s_1 \in \mathbb{R}$,

$$\lim_{\lambda\to 0}\chi_{[(S-t_2)/\lambda^2,(T-t_2)/\lambda^2]}(s_1) = 1 . \tag{15.8.7}$$

So, we have

$$\begin{aligned}
&\lim_{\lambda\to 0}\frac{1}{\lambda^2}\int_S^T dt_1 \int_{S'}^{T'} dt_2 F\left(\frac{t_1-t_2}{\lambda^2}\right) G_\lambda(t_1)\\
&=\int_{\mathbb{R}} dt_2 \chi_{(S',T')}(t_2)\, \chi_{(S,T)}(t_2)\, G_0(t_2) \int_{\mathbb{R}} ds_1 F(s_1) ,
\end{aligned} \tag{15.8.8}$$

and this is exactly (15.8.2). To prove the uniformity of the convergence, it is sufficient to consider separately the following two cases:

(i) $[S,T] = [S',T']$;
(ii) $[S,T] \cap [S',T'] = \emptyset$.

In fact, in the general case, the quantity

$$\frac{1}{\lambda^2}\int_S^T dt_1 \int_{S'}^{T'} dt_2 F\left(\frac{t_1 - t_2}{\lambda^2}\right) G_\lambda(t_1)$$

can be reduced to a sum of at most four terms, and in each of these terms the situation is like either case (i) or case (ii). In case (i) we have

$$\begin{aligned}
&\left|\frac{1}{\lambda^2}\int_S^T dt_1 \int_S^T dt_2 F\left(\frac{t_1 - t_2}{\lambda^2}\right) G_\lambda(t_1)\right. \\
&\left. - \int_{\mathbb{R}} dt_2 \chi_{(S,T)}(t_2)\, \chi_{(S,T)}(t_2)\, G_0(t_2) \int_{\mathbb{R}} F(t)\, dt\right| \\
&= \left|\int_S^T dt_2 \int_{(S-t_2)/\lambda^2}^{(T-t_2)/\lambda^2} ds_1 F(s_1)\, G_\lambda\left(t_2 + \lambda^2 s_1\right) - \int_S^T dt_2 G_0(t_2) \int_{\mathbb{R}} F(t)\, dt\right| \\
&\le \left|\int_S^T dt_2 \int_{(S-t_2)/\lambda^2}^{(T-t_2)/\lambda^2} ds_1 F(s_1)\, G_\lambda\left(t_2 + \lambda^2 s_1\right)\right. \\
&\left. - \int_S^T dt_2 \int_{(S-t_2)/\lambda^2}^{(T-t_2)/\lambda^2} ds_1 F(s_1)\, G_0(t_2)\right| \\
&+ \left|\int_S^T dt_2 G_0(t_2) \int_{(S-t_2)/\lambda^2}^{(T-t_2)/\lambda^2} ds_1 F(s_1) - \int_S^T dt_2 G_0(t_2) \int_{\mathbb{R}} F(t)\, dt\right| \\
&\le \int_S^T dt_2 \int_{\mathbb{R}} ds_1 \left|F(s_1)\right| \cdot \left|G_\lambda\left(t_2 + \lambda^2 s_1\right) - G_0(t_2)\right| \\
&+ \int_S^T dt_2 \left|G_0(t_2)\right| \left|\int_{(S-t_2)/\lambda^2}^{(T-t_2)/\lambda^2} ds_1 F(s_1) - \int_{\mathbb{R}} F(t)\, dt\right| . \qquad (15.8.9)
\end{aligned}$$

By dominated convergence,

$$\begin{aligned}
&\lim_{\lambda\to 0}\int_S^T dt_2 \int_{\mathbb{R}} ds_1 \left|F(s_1)\right| \cdot \left|G_\lambda\left(t_2 + \lambda^2 s_1\right) - G_0(t_2)\right| \\
&= \int_S^T dt_2 \int_{\mathbb{R}} ds_1 \left|F(s_1)\right| \cdot \lim_{\lambda\to 0}\left|G_\lambda\left(t_2 + \lambda^2 s_1\right) - G_0(t_2)\right| = 0 , \qquad (15.8.10)
\end{aligned}$$

and it is obvious that the convergence is locally uniform for S, T. Moreover,

$$\int_S^T dt_2 \left|G_0(t_2)\right| \left| \int_{(S-t_2)/\lambda^2}^{(T-t_2)/\lambda^2} ds_1 F(s_1) - \int_{\mathbb{R}} F(t)\, dt \right|$$

$$\leq \int_S^T dt_2 \left|G_0(t_2)\right| \left(\int_{(T-t_2)/\lambda^2}^{\infty} dt \left|F(t)\right| + \int_{-\infty}^{(S-t_2)/\lambda^2} dt \left|F(t)\right| \right),$$

which, as $\lambda \to 0$, tends to zero locally uniformly in S, T. In case (ii), assuming, without loss of generality, that $0 \leq S' \leq T' \leq S \leq T$ and choosing $\epsilon > 0$, arbitrarily small, one has

$$\left| \frac{1}{\lambda^2} \int_S^T dt_1 \int_S^T dt_2 F\left(\frac{t_1 - t_2}{\lambda^2}\right) G_\lambda(t_1) \right| \leq C \int_{S'}^{T'} dv \int_{(S-v)/\lambda^2}^{(T-v)/\lambda^2} du \left|F(u)\right|$$

$$= \left(\int_{S'}^{T'-\epsilon} + \int_{T'-\epsilon}^{T'} \right) dv \int_{(S-v)/\lambda^2}^{(T-v)/\lambda^2} du \left|F(u)\right| \leq 2\epsilon \cdot \sup_t \left|F(t)\right|$$

$$+ \int_{S'}^{T'} dv \int_{(S-v)/\lambda^2}^{(T-v)/\lambda^2} du \left|F(u)\right|. \tag{15.8.11}$$

Since, for any $v \in [S', T' - \epsilon]$, $S - v > 0$ and $T - v > 0$, the right-hand side of (15.8.11) can be made uniformly small for S, T, S', T' running over a bounded set of $\mathbb{R}$.

Corollary 15.8.1. *For any $F \in \mathbf{L}^1(\mathbb{R})$, for any $S, T, S', T' \in \mathbb{R}$ such that $S < T$, $S' < T'$, the limit, as $\lambda \to 0$, of*

$$\lambda^2 \int_{S/\lambda^2}^{T/\lambda^2} dt_1 \int_{S'/\lambda^2}^{T'/\lambda^2} dt_2 F(t_1 - t_2)$$

exists and is equal to

$$\langle \chi_{[S,T]}, \chi_{[S',T']} \rangle \cdot \int_{-\infty}^{\infty} F(t) dt,$$

where the scalar product of the characteristic functions is meant in $\mathbf{L}^2(\mathbb{R})$. Moreover, if $F \in \mathbf{L}^1(\mathbb{R}) \cap \mathbf{L}^\infty(\mathbb{R})$, the limit is locally uniform in S, T, S', T'.

In addition, by just copying the proof of Lemma 15.8.1, we have the following:

Lemma 15.8.2. *Let $F \in \mathbf{L}^1(\mathbb{R})$ and let, for each $\lambda \in \mathbb{R}_+$, $G_\lambda : \mathbb{R} \to \mathbb{C}$ be a continuous function verifying the conditions of Lemma 15.8.1. Then, for any $S, T \in \mathbb{R}$ such that $S < T$,*

$$\lim_{\lambda \to 0} \frac{1}{\lambda^2} \int_S^T dt_1 \int_S^{t_1} dt_2 F\left(\frac{t_1 - t_2}{\lambda^2}\right) G_\lambda(t_1)$$

$$= \int_{\mathbb{R}} dt G_0(t) \chi_{[S,T]}(t) \cdot \int_0^{\infty} F(s) ds. \tag{15.8.12}$$

Moreover, if $F \in \mathbf{L}^1(\mathbb{R}) \cap \mathbf{L}^\infty(\mathbb{R})$, the limit is locally uniform for S, T.

15.9 The Multiple-Simplex Theorem

In this section we will study integrals of the form

$$\lambda^{-2M-2}\int_0^t dt_1\int_0^{t_1} dt_2\cdots\int_0^{t_{n_1-1}} dt_{n_1}\int_0^t dt_{n_1+1}\int_0^{t_{n_1+1}} dt_{n_1+2}$$
$$\cdots\int_0^{t_{n_2-1}} dt_{n_2}\cdots\int_0^t dt_{n_{m-1}+1}\int_0^{t_{n_{m-1}+1}} dt_{n_{m-1}+2}$$
$$\cdots\int_0^{t_{n_m-1}} dt_{n_m}\int_{S_1}^{T_1} dt_{n_m+1}\cdots\int_{S_N}^{T_N} dt_{n_m+N}$$
$$\cdot F_0\left(\frac{t-t_{q_0}}{\lambda^2}\right)\cdot\prod_{h=1}^{M} F_h\left(\frac{t_{p_h}-t_{q_h}}{\lambda^2}\right), \tag{15.9.1}$$

where $m, N, n_k \in \mathbb{N}, k = 1, 2, \cdots, m, 1 \le n_1 < \cdots < n_m$ and $N+n_m = 2M+1$ is odd; $\{F_h\}_{h=0}^{M} \subset \mathbf{L}^1(\mathbb{R})$; $\{p_h, q_h\}_{h=1}^{M} \cup \{q_0\} = \{1, 2, \cdots, 2M+1\}$.

First of all, without loss of generality, we can assume that $p_h < q_h$ for all $h = 1, 2, \cdots, M$. In fact, if for some h, $q_h < p_h$, one can exchange q_h with p_h and replace $F_h(\cdot)$ by $F_h(-\cdot)$. Moreover, we introduce the notation $n_0 := 0$.

Theorem 15.9.1. *(The multiple-simplex theorem). If there are* $0 \le k \le m$ *and* $j \in \{2 \cdots n_{k+1} - n_k\}$ *such that* $q_0 = n_k + j$*, then as* $\lambda \to 0$*, the quantity (15.9.1) goes to zero. Moreover, the quantity obtained by replacing all the functions in (15.9.1) by their moduli goes to zero.*

Remark 15.9.1. This result means that the limit (15.9.1) is not equal to zero only if $q_0 \in \{n_k + 1\}_{k=0}^{m-1} \cup \{n_m + h\}_{h=1}^{N}$.

Proof 15.9.1. Without loss of generality, one can assume that all the functions F_h are positive and $k = 0$. Clearly $n_1 - n_0 = n_1$ must be greater than or equal to 2 and $q_0 = j$. Thus it is sufficient to consider integrals of the form

$$\int_0^{t_{n_1-1}} dt_{n_1}\int_0^t dt_{n_1+1}\int_0^{t_{n_1+1}} dt_{n_1+2}$$
$$\cdots\int_0^{t_{n_2-1}} dt_{n_2}\cdots\int_0^t dt_{n_{m-1}+1}\cdot\int_0^{t_{n_{m-1}+1}} dt_{n_{m-1}+2}\cdots\int_0^{t_{n_m-1}} dt_{n_m}$$
$$\cdot\int_{S_1}^{T_1} dt_{n_m+1}\cdots\int_{S_N}^{T_N} dt_{n_m+N} F_0\left(\frac{t-t_j}{\lambda^2}\right)$$
$$\cdot\prod_{h=1}^{M} F_h\left(\frac{t_{p_h}-t_{q_h}}{\lambda^2}\right). \tag{15.9.2}$$

By changing the order of the integrals, one obtains

$$\int_0^t dt_1 \int_0^{t_1} dt_2 \cdots \int_0^{t_{n_1-1}} dt_{n_1} = \int_0^t dt_j \int_{t_j}^t dt_1 \int_{t_j}^{t_1} dt_2 \int_{t_j}^{t_2} dt_3 \cdots \int_{t_j}^{t_{j-2}} dt_{j-1}$$
$$\cdot \int_0^{t_j} dt_{j+1} \cdots \int_0^{t_{n_1-1}} dt_{n_1} \, . \tag{15.9.3}$$

By the change of variables

$$s := \frac{t_j - t}{\lambda^2}, \; t_k \text{ remain invariant for all } k \in \{1, 2, \cdots, 2M+1\} \setminus \{j\} \, ,$$

and applying (15.9.3), (15.9.2) becomes

$$\int_{-\infty}^0 ds F_0(-s) \chi_{[-t/\lambda^2, 0]}(s) \cdot \lambda^{-2M}$$
$$\cdot \int_{t+\lambda^2 s}^t dt_1 \int_{t+\lambda^2 s}^{t_1} dt_2 \int_{t+\lambda^2 s}^{t_2} dt_3 \cdots \int_{t+\lambda^2 s}^{t_{j-2}} dt_{j-1}$$
$$\cdot \int_0^{t+\lambda^2 s} dt_{j+1} \cdots \int_0^{t_{n_1-1}} dt_{n_1} \cdot \int_0^t dt_{n_1+1} \int_0^{t_{n_1+1}} dt_{n_1+2}$$
$$\cdots \int_0^{t_{n_2-1}} dt_{n_2} \cdots \int_0^t dt_{n_{m-1}+1} \int_0^{t_{n_{m-1}+1}} dt_{n_{m-1}+2}$$
$$\cdots \int_0^{t_{n_m-1}} dt_{n_m} \cdot \int_{S_1}^{T_1} dt_{n_m+1} \cdots \int_{S_N}^{T_N} dt_{n_m+N} \prod_{h=2}^M F_h \left(\frac{t_{p_h} - t_{q_h}}{\lambda^2} \right) . \tag{15.9.4}$$

Note that for any $r = 1, \cdots, j-1$,

$$0 \le t + \lambda^2 s \le t_r \le t_{r-1} \le t \, ,$$

hence (15.9.4) is dominated by

$$\int_{-\infty}^0 ds F_0(-s) \chi_{[-t/\lambda^2, 0]}(s) \cdot \lambda^{-2M} \int_{[S,T]^{2M}} dt_1 \cdots \widehat{dt_j} \cdots dt_{n_m+N}$$
$$\cdot \prod_{h=2}^M F_h \left(\frac{t_{p_h} - t_{q_h}}{\lambda^2} \right) , \tag{15.9.5}$$

where

$$S := \min \{0, S_h : h = 1, 2, \cdots, N\}, T := \max \{t, T_h : h = 1, 2, \cdots, N\} \, .$$

In the integral

$$\int_{[S,T]^{2M}} dt_1 \cdots \widehat{dt_j} \cdots dt_{n_m+N} \prod_{h=2}^{M} F_h\left(\frac{t_{p_h}-t_{q_h}}{\lambda^2}\right), \tag{15.9.6}$$

each $\{t_{p_h}, t_{q_h}\}$ forms a connected component. Therefore, by the first connected component theorem, (15.9.6) is a quantity of order $O\left(\lambda^{2(2M)-2M}\right) = O\left(\lambda^{2M}\right)$; therefore, (15.9.5) has the form

$$\int_{-\infty}^{0} ds F_0(-s)\, \chi_{[-t/\lambda^2,0]}(s) \cdot O(1)\,,$$

and by dominated convergence one can perform the $\lim_{\lambda\to0}$ inside the integral $\int_{-\infty}^{0} ds$. So, to complete the proof, it is sufficient to show that the ds-integral in (15.9.4) tends to zero as $\lambda \to 0$ or, equivalently, that the limit of

$$\begin{aligned}
&\chi_{[-t/\lambda^2,0]}(s) \cdot \lambda^{-2M} \int_{t+\lambda^2 s}^{t} dt_1 \int_{t+\lambda^2 s}^{t_1} dt_2 \int_{t+\lambda^2 s}^{t_2} dt_3 \\
&\cdots \int_{t+\lambda^2 s}^{t_{j-2}} dt_{j-1} \int_0^{t+\lambda^2 s} dt_{j+1} \cdots \int_0^{t_{n_1-1}} dt_{n_1} \\
&\cdot \int_0^{t} dt_{n_1+1} \int_0^{t_{n_1+1}} dt_{n_1+2} \cdots \int_0^{t_{n_2-1}} dt_{n_2} \\
&\cdots \int_0^{t} dt_{n_{m-1}+1} \int_0^{t_{n_{m-1}+1}} dt_{n_{m-1}+2} \cdots \int_0^{t_{n_m-1}} dt_{n_m} \\
&\cdot \int_{S_1}^{T_1} dt_{n_m+1} \cdots \int_{S_N}^{T_N} dt_{n_m+N} \prod_{h=2}^{M} F_h\left(\frac{t_{p_h}-t_{q_h}}{\lambda^2}\right)
\end{aligned} \tag{15.9.7}$$

is equal to zero for any s. In order to do this one can assume, without loss of generality, that $p_1 = 1$. With S, T introduced after (15.9.5), (15.9.7) is dominated by

$$\begin{aligned}
&\chi_{[-t/\lambda^2,0]}(s) \cdot \lambda^{-2M} \int_{t+\lambda^2 s}^{t} dt_1 \int_{[S,T]^{2M-1}} dt_2 \cdots \widehat{dt_j} \cdots dt_{n_m+N} \\
&\cdot \prod_{h=1}^{M} F_h\left(\frac{t_{p_h}-t_{q_h}}{\lambda^2}\right).
\end{aligned} \tag{15.9.8}$$

By the change of variables

$$\tau := \frac{t_{q_1}-t_1}{\lambda^2}, \quad t_k \text{ are invariant for all } k \neq q_1\,,$$

(15.9.8) becomes

$$\chi_{[-t/\lambda^2,0]}(s)\cdot\int_{t+\lambda^2 s}^{t} dt_1 \int_{(S-t_1)/\lambda^2}^{(T-t_1)/\lambda^2} d\tau\,|F_1(-\tau)|$$
$$\cdot\lambda^{-2M+2}\int_{[S,T]^{2M-2}} dt_2\cdots\widehat{dt_j}\cdots\widehat{dt_{q_1}}\cdots dt_{n_m+N}\prod_{h=2}^{M}F_h\left(\frac{t_{p_h}-t_{q_h}}{\lambda^2}\right), \tag{15.9.9}$$

which, applying again the first connected component theorem, is seen to be of the form

$$\chi_{[-t/\lambda^2,0]}(s)\cdot\int_{t+\lambda^2 s}^{t} dt_1 \int_{(S-t_1)/\lambda^2}^{(T-t_1)/\lambda^2} d\tau F_1(-\tau)\cdot O(1)$$
$$=\chi_{[-t/\lambda^2,0]}(s)\cdot O(1)\cdot\int_{t+\lambda^2 s}^{t} dt_1\,,$$

which, as $\lambda\to 0$, goes to zero for any s.

15.10 The Multiple Integral Lemma

Now we are interested in knowing the limit of (15.9.1) in the case of $N=0$ (no external lines), i.e.

$$\lambda^{-2M-2}\int_0^t dt_1\int_0^{t_1} dt_2\cdots\int_0^{t_{n_1-1}} dt_{n_1}\int_0^t dt_{n_1+1}\int_0^{t_{n_1+1}} dt_{n_1+2}$$
$$\cdots\int_0^{t_{n_2-1}} dt_{n_2}\cdots\int_0^t dt_{n_{m-1}+1}\int_0^{t_{n_{m-1}+1}} dt_{n_{m-1}+2}$$
$$\cdots\int_0^{t_{n_m-1}} dt_{n_m}F_0\left(\frac{t-t_{q_0}}{\lambda^2}\right)\cdot\prod_{h=1}^{M}F_h\left(\frac{t_{p_h}-t_{q_h}}{\lambda^2}\right) \tag{15.10.1}$$

when $q_0\in\{n_h+1\}_{h=0}^m$. One can also assume that $q_0=1$.

Lemma 15.10.1. (The multiple integral lemma). *As $\lambda\to 0$, the limit of*

$$\lambda^{-2M-2}\int_0^t dt_1\int_0^{t_1} dt_2\cdots\int_0^{t_{n_1-1}} dt_{n_1}\int_0^t dt_{n_1+1}\int_0^{t_{n_1+1}} dt_{n_1+2}$$
$$\cdots\int_0^{t_{n_2-1}} dt_{n_2}\cdots\int_0^t dt_{n_{m-1}+1}\int_0^{t_{n_{m-1}+1}} dt_{n_{m-1}+2}$$
$$\cdots\int_0^{t_{n_m-1}} dt_{n_m}F_0\left(\tfrac{t-t_1}{\lambda^2}\right)\cdot\prod_{h=1}^{M}F_h\left(\tfrac{t_{p_h}-t_{q_h}}{\lambda^2}\right) \tag{15.10.2a}$$

is equal to the limit of

$$\int_{-\infty}^0 dsF_0(-s)\cdot\lambda^{-2M}\int_0^t dt_2\cdots\int_0^{t_{n_1-1}} dt_{n_1}\int_0^t dt_{n_1+1}\cdots\int_0^{t_{n_2-1}} dt_{n_2}$$
$$\cdots\int_0^t dt_{n_{m-1}+1}\int_0^{t_{n_{m-1}+1}} dt_{n_{m-1}+2}\cdots\int_0^{t_{n_m-1}} dt_{n_m}\prod_{h=1}^{M}F_h\left(\tfrac{t_{p_h}-t_{q_h}}{\lambda^2}\right) \tag{15.10.2b}$$

in the sense that if one of the limits exists then so does the other and the two limits are equal.

Proof 15.10.1. By the change of variable

$$s := \frac{t_1 - t}{\lambda^2},$$

(15.10.2a) becomes

$$\int_{-t/\lambda^2}^{0} ds F_0(-s) \cdot \lambda^{-2M} \int_0^{t_1} dt_2 \cdots \int_0^{t_{n_1-1}} dt_{n_1} \int_0^{t} dt_{n_1+1} \int_0^{t_{n_1+1}} dt_{n_1+2}$$
$$\cdots \int_0^{t_{n_2-1}} dt_{n_2} \cdots \int_0^{t} dt_{n_{m-1}+1} \int_0^{t_{n_{m-1}+1}} dt_{n_{m-1}+2}$$
$$\cdots \int_0^{t_{n_m-1}} dt_{n_m} \prod_{h=1}^{M} F_h\left(\frac{t_{p_h} - t_{q_h}}{\lambda^2}\right).$$

By the first connected component theorem, the quantity

$$\lambda^{-2M} \int_0^{t_1} dt_2 \cdots \int_0^{t_{n_1-1}} dt_{n_1} \int_0^{t} dt_{n_1+1} \int_0^{t_{n_1+1}} dt_{n_1+2} \cdots \int_0^{t_{n_2-1}} dt_{n_2}$$
$$\cdots \int_0^{t} dt_{n_{m-1}+1} \int_0^{t_{n_{m-1}+1}} dt_{n_{m-1}+2} \cdots \int_0^{t_{n_m-1}} dt_{n_m} \prod_{h=1}^{M} F_h\left(\frac{t_{p_h} - t_{q_h}}{\lambda^2}\right)$$

is bounded. So there is a constant C, such that

$$|\text{the quantity (15.10.2a)} - \text{the quantity (15.10.2b)}| \le C$$
$$\cdot \int_{-\infty}^{0} ds \left|F_0(-s)\right| \chi_{(-\infty, -t/\lambda^2)}(s),$$

which tends to zero.

15.11 The Second Multiple-Simplex Theorem

Let us consider a quantity of the form

$$\lambda^{-2M} \int_0^{t} dt_1 \int_0^{t_1} dt_2 \cdots \int_0^{t_{n_1-1}} dt_{n_1} \int_0^{t} dt_{n_1+1} \int_0^{t_{n_1+1}} dt_{n_1+2}$$
$$\cdots \int_0^{t_{n_2-1}} dt_{n_2} \cdots \int_0^{t} dt_{n_{m-1}+1} \int_0^{t_{n_{m-1}+1}} dt_{n_{m-1}+2} \cdots \int_0^{t_{n_m-1}} dt_{n_m}$$
$$\cdot \int_{S_1}^{T_1} dt_{n_m+1} \cdots \int_{S_N}^{T_N} dt_{n_m+N} \prod_{h=1}^{M} F_h\left(\frac{t_{p_h} - t_{q_h}}{\lambda^2}\right). \tag{15.11.1}$$

It is worth to notice that (15.11.1) is in a certain sense an integral form of the quantity (15.9.1). We shall assume as usual that $p_h < q_h$ for all h. One can suppose that, for all j, $n_j > 1$. In fact, if for some j $n_j = 1$, then we define $S_{N+1} := 0$ and $T_{N+1} := t$. Denote (15.11.1) by

$$L_\lambda\left(t; \{p_h, q_h\}_{h=1}^M, \{n_j\}_{j=1}^m\right).$$

As argued in Proof 15.10.1, one is sure that

$$L_\lambda\left(t; \{p_h, q_h\}_{h=1}^M, \{n_j\}_{j=1}^m\right)$$

is bounded. Moreover, one has the following:

Lemma 15.11.1. *$L_\lambda\left(t; \{p_h, q_h\}_{h=1}^M, \{n_j\}_{j=1}^m\right)$ tends to zero as $\lambda \to 0$ if there is an $h = 1, 2, \cdots, M$ and a $j = 0, 1, \cdots, m$ such that $n_j + 1 \le p_h, q_h < n_{j+1}$, $|p_h - q_h| > 1$.*

Proof 15.11.1. Without loss of generality, we assume that $p_h < q_h$; then there is at least one $k \in \{1, 2, \cdots, M\}$, such that

$$\text{either } n_j + 1 \le p_h < p_k < q_h < n_{j+1} \quad \text{or} \quad n_j + 1 \le p_h < q_k < q_h < n_{j+1}.$$

So one can replace the function $\prod_{h=1}^M F_h\left(\frac{t_{p_h} - t_{q_h}}{\lambda^2}\right)$ by

$$\text{either} \quad \chi_{(t_{p_h}, t_{q_h})}(t_{p_k}) \prod_{h=1}^M F_h\left(\frac{t_{p_h} - t_{q_h}}{\lambda^2}\right)$$
$$\text{or} \quad \chi_{(t_{p_h}, t_{q_h})}(t_{q_k}) \prod_{h=1}^M F_h\left(\frac{t_{p_h} - t_{q_h}}{\lambda^2}\right).$$

The thesis then follows from the first vanishing theorem.

Theorem 15.11.1. *As $\lambda \to 0$, the limit of $L_\lambda\left(t; \{p_h, q_h\}_{h=1}^M, \{n_j\}_{j=1}^m\right)$ always exists.*

Proof 15.11.2. The proof shall be done by induction. In the case $M = 2$, there are two possible cases to be considered:

$$\lambda^{-2}\int_0^t ds_1 \int_0^{s_1} ds_2 F\left(\frac{s_1 - s_2}{\lambda^2}\right), \quad \lambda^{-2}\int_S^T ds_1 \int_{S'}^{T'} ds_2 F\left(\frac{s_1 - s_2}{\lambda^2}\right).$$

By Corollary 15.8.1 and Lemma 15.8.2, we know that as $\lambda \to 0$, the limits of the above terms exist. Suppose that as $\lambda \to 0$, the limit of (15.11.1) exists for $n_m + N = 2M$, and let us show that the quantity

$$\begin{aligned}
&L_\lambda\left(t;\{p_h,q_h\}_{h=1}^{M+1},\{n_j\}_{j=1}^{m}\right)\\
&:=\lambda^{-2M-2}\int_0^t dt_1\int_0^{t_1}dt_2\cdots\int_0^{t_{n_1-1}}dt_{n_1}\int_0^t dt_{n_1+1}\int_0^{t_{n_1+1}}dt_{n_1+2}\\
&\cdots\int_0^{t_{n_2-1}}dt_{n_2}\cdots\int_0^t dt_{n_{m-1}+1}\cdot\int_0^{t_{n_{m-1}+1}}dt_{n_{m-1}+2}\cdots\int_0^{t_{n_m-1}}dt_{n_m}\\
&\cdot\int_{S_1}^{T_1}dt_{n_m+1}\cdots\int_{S_N}^{T_N}dt_{n_m+N}\prod_{h=1}^{M+1}F_h\left(\frac{t_{p_h}-t_{q_h}}{\lambda^2}\right)
\end{aligned}\tag{15.11.2}$$

has a limit as $\lambda\to 0$, where $n_m+N = 2(M+1)$, $\{p_h,q_h\}_{h=1}^{M+1} = \{1,2,\cdots,2M+2\}$ and $\{F_h\}_{h=1}^{M+1}\subset\mathbf{L}^1(\mathbb{R})$.

Step 1: Clearly, for each fixed λ, as a function of the variable t, L_λ is absolutely continuous. So

$$\begin{aligned}
&L_\lambda\left(t;\{p_h,q_h\}_{h=1}^{M+1},\{n_j\}_{j=1}^{m}\right)=\int_0^t\frac{d}{ds}L_\lambda\left(s;\{p_h,q_h\}_{h=1}^{M+1},\{n_j\}_{j=1}^{m}\right)ds\\
&=\lambda^{-2M-2}\int_0^t ds\frac{d}{ds}\int_0^s dt_1\int_0^{t_1}dt_2\cdots\int_0^{t_{n_1-1}}dt_{n_1}\int_0^s dt_{n_1+1}\int_0^{t_{n_1+1}}dt_{n_1+2}\\
&\cdots\int_0^{t_{n_2-1}}dt_{n_2}\cdots\int_0^s dt_{n_{m-1}+1}\int_0^{t_{n_{m-1}+1}}dt_{n_{m-1}+2}\cdots\int_0^{t_{n_m-1}}dt_{n_m}\\
&\cdot\int_{S_1}^{T_1}dt_{n_m+1}\cdots\int_{S_N}^{T_N}dt_{n_m+N}\prod_{h=1}^{M+1}F_h\left(\frac{t_{p_h}-t_{q_h}}{\lambda^2}\right).
\end{aligned}\tag{15.11.3}$$

On the right-hand side of (15.11.3), the derivative $\frac{d}{ds}(\cdots)$ gives a sum of m terms with a similar form; let us analyze the first one (the other terms are similar), i.e.

$$\begin{aligned}
&\lambda^{-2M-2}\int_0^t ds\int_0^s dt_2\cdots\int_0^{t_{n_1-1}}dt_{n_1}\int_0^s dt_{n_1+1}\int_0^{t_{n_1+1}}dt_{n_1+2}\cdots\int_0^{t_{n_2-1}}dt_{n_2}\\
&\cdots\int_0^s dt_{n_{m-1}+1}\int_0^{t_{n_{m-1}+1}}dt_{n_{m-1}+2}\cdots\int_0^{t_{n_m-1}}dt_{n_m}\int_{S_1}^{T_1}dt_{n_m+1}\\
&\cdots\int_{S_N}^{T_N}dt_{n_m+N}\prod_{h=1}^{M+1}F_h\left(\frac{t_{p_h}-t_{q_h}}{\lambda^2}\right),
\end{aligned}\tag{15.11.4}$$

where t_1 means s. Since $p_h<q_h$ for all $h=1,2,\cdots,M+1$, the index 1 should be some p_h, we assume, without loss of generality, that $p_1=1$. So (15.11.4) has the form

$$\int_0^t ds\ \lambda^{-2M-2} \int_0^s dt_2 \cdots \int_0^{t_{n_1-1}} dt_{n_1} \int_0^s dt_{n_1+1} \int_0^{t_{n_1+1}} dt_{n_1+2}$$
$$\cdots \int_0^{t_{n_2-1}} dt_{n_2} \cdots \int_0^s dt_{n_{m-1}+1} \int_0^{t_{n_{m-1}+1}} dt_{n_{m-1}+2} \cdots \int_0^{t_{n_m-1}} dt_{n_m}$$
$$\cdot \int_{S_1}^{T_1} dt_{n_m+1} \cdots \int_{S_N}^{T_N} dt_{n_m+N} F_1\left(\frac{s-t_{q_1}}{\lambda^2}\right) \prod_{h=2}^{M+1} F_h\left(\frac{t_{p_h}-t_{q_h}}{\lambda^2}\right)$$
$$=: \int_0^t ds L_\lambda\left(s;\{p_h,q_h\}_{h=2}^{M+1}\right) . \tag{15.11.5}$$

Denoting

$$S := \min\{0, S_h : h=1,2,\cdots,N\},\quad T := \max\{t, T_h : h=1,2,\cdots,N\}\ ,$$

one finds that

$$\left|L_\lambda\left(s;\{p_h,q_h\}_{h=2}^{M+1}\right)\right|$$
$$\leq \lambda^{-2M-2}\int_{[S,T]^{n_m+N-1}} dt_2\cdots dt_{n_m+N}\left|F_1\left(\frac{s-t_{q_1}}{\lambda^2}\right)\right|\prod_{h=2}^{M+1}\left|F_h\left(\frac{t_{p_h}-t_{q_h}}{\lambda^2}\right)\right| . \tag{15.11.6}$$

By the change of variable

$$\tau := \frac{t_{q_1}-s}{\lambda^2}\ , \tag{15.11.7}$$

the right-hand side of (15.11.6) becomes

$$\int_{(S-s)/\lambda^2}^{(T-s)/\lambda^2} d\tau\,|F_1(-\tau)|\cdot\lambda^{-2M}\int_{[S,T]^{n_m+N-2}} dt_2 dt_3\cdots\widehat{dt_{q_1}}\cdots dt_{n_m+N}$$
$$\cdot\prod_{h=2}^{M+1}\left|F_h\left(\frac{t_{p_h}-t_{q_h}}{\lambda^2}\right)\right| . \tag{15.11.8}$$

Since we are dealing with the case of $n_m+N=2(M+1)$, one knows that $n_m+N-2=2M$ and $\{p_h,q_h\}_{h=2}^{M+1}$ is nothing other than $\{2,3,\cdots,q_1-1,q_1+1,$ $\cdots,\ n_m+N\}$. It follows from the first connected component theorem that (15.11.8) is a quantity of order

$$\lambda^{-2M}\cdot O\left(\lambda^{2(2M)-2M}\right)=O(1)\ .$$

In other words, $L_\lambda\left(s;\{p_h,q_h\}_{h=2}^{M+1}\right)$ is bounded. By dominated convergence, if the limit, as $\lambda\to 0$, of $L_\lambda\left(s;\{p_h,q_h\}_{h=2}^{M+1}\right)$ exists, then so does the limit of (15.11.5); moreover,

$$\lim_{\lambda\to 0}\int_0^t L_\lambda\left(s;\{p_h,q_h\}_{h=2}^{M+1}\right)ds=\int_0^t ds\lim_{\lambda\to 0}L_\lambda\left(s;\{p_h,q_h\}_{h=2}^{M+1}\right).$$

Therefore we need to show the existence of $\lim_{\lambda\to 0}L_\lambda\left(s;\{p_h,q_h\}_{h=2}^{M+1}\right)$ for any $s\in[0,t]$.

Step 2: Since the quantity $L_\lambda\left(s;\{p_h,q_h\}_{h=2}^{M+1}\right)$ involves $2(M+1)-1=2M+1$ integrals, Theorem 15.9.1 guarantees that its limit is zero if $q_1\in\{3,4,\cdots,n_1\}\cup\bigcup_{k=2}^m\{n_{k-1}+2,\cdots,n_k\}$. So

$$\lim_{\lambda\to 0}\int_0^t L_\lambda\left(s;\{p_h,q_h\}_{h=2}^{M+1}\right)ds=\int_0^t ds\lim_{\lambda\to 0}L_\lambda\left(s;\{p_h,q_h\}_{h=2}^{M+1}\right)=0.$$

Step 3: Now let us prove the existence of $\lim_{\lambda\to 0}L_\lambda\left(s;\{p_h,q_h\}_{h=2}^{M+1}\right)$ in the case when $q_1=2$. By the change of variable (15.11.7), $L_\lambda\left(s;\{p_h,q_h\}_{h=2}^{M+1}\right)$ becomes,

$$\begin{aligned}
&\int_{-s/\lambda^2}^0 F_1(-\tau)\,d\tau\cdot\lambda^{-2M}\int_0^s dt_{n_1+1}\int_0^{t_{n_1+1}}dt_{n_1+2}\cdots\int_0^{t_{n_2-1}}dt_{n_2}\\
\cdots\ &\int_0^s dt_{n_{m-1}+1}\int_0^{t_{n_{m-1}+1}}dt_{n_{m-1}+2}\cdots\int_0^{t_{n_m-1}}dt_{n_m}\\
\cdot\ &\int_{S_1}^{T_1}dt_{n_m+1}\cdots\int_{S_N}^{T_N}dt_{n_m+N}\prod_{h=2}^{M+1}F_h\left(\frac{t_{p_h}-t_{q_h}}{\lambda^2}\right)\\
=\ &\int_{-\infty}^0\chi_{[-s/\lambda^2,0]}(\tau)F_1(-\tau)\,d\tau\cdot\lambda^{-2M}\\
\cdot\ &\int_0^s dt_{n_1+1}\int_0^{t_{n_1+1}}dt_{n_1+2}\cdots\int_0^{t_{n_2-1}}dt_{n_2}\\
\cdots\ &\int_0^s dt_{n_{m-1}+1}\int_0^{t_{n_{m-1}+1}}dt_{n_{m-1}+2}\cdots\int_0^{t_{n_m-1}}dt_{n_m}\int_{S_1}^{T_1}dt_{n_m+1}\\
\cdots\ &\int_{S_N}^{T_N}dt_{n_m+N}\cdot\prod_{h=2}^{M+1}F_h\left(\frac{t_{p_h}-t_{q_h}}{\lambda^2}\right);
\end{aligned}\tag{15.11.9a}$$

if $n_1=2$ (then q_1 must be equal to n_2); if $n_1>2$

$$\begin{aligned}
&\int_{-s/\lambda^2}^0 F_1(-\tau)\,d\tau\lambda^{-2M}\int_0^{s+\lambda^2\tau}dt_3\cdots\int_0^{t_{n_1-1}}dt_{n_1}\\
&\cdot\int_0^s dt_{n_1+1}\cdots\int_0^{t_{n_2-1}}dt_{n_2}\cdots\int_0^s dt_{n_{m-1}+1}\\
&\cdot\int_0^{t_{n_{m-1}+1}}dt_{n_{m-1}+2}\cdots\int_0^{t_{n_m-1}}dt_{n_m}\int_{S_1}^{T_1}dt_{n_m+1}\cdots\int_{S_N}^{T_N}dt_{n_m+N}\\
&\cdot\prod_{h=2}^{M+1}F_h\left(\frac{t_{p_h}-t_{q_h}}{\lambda^2}\right)=\int_{-\infty}^0\chi_{[-s/\lambda^2,0]}(\tau)F_1(-\tau)\,d\tau\cdot\lambda^{-2M}\int_0^s dt_3\\
&\cdots\int_0^{t_{n_1-1}}dt_{n_1}\int_0^s dt_{n_1+1}\int_0^{t_{n_1+1}}dt_{n_1+2}\cdots\int_0^{t_{n_2-1}}dt_{n_2}
\end{aligned}$$

$$\cdots \int_0^s dt_{n_{m-1}+1} \int_0^{t_{n_{m-1}+1}} dt_{n_{m-1}+2} \cdots \int_0^{t_{n_m-1}} dt_{n_m}$$

$$\cdot \int_{S_1}^{T_1} dt_{n_m+1} \cdots \int_{S_N}^{T_N} dt_{n_m+N} \cdot \prod_{h=2}^{M+1} F_h \left(\frac{t_{p_h} - t_{q_h}}{\lambda^2} \right)$$

$$+ \int_{-\infty}^0 \chi_{[-s/\lambda^2,0]}(\tau) F_1(-\tau) \, d\tau \cdot \lambda^{-2M} \int_s^{s+\lambda^2\tau} dt_3$$

$$\cdots \int_0^{t_{n_1-1}} dt_{n_1} \int_0^s dt_{n_1+1} \int_0^{t_{n_1+1}} dt_{n_1+2} \cdots \int_0^{t_{n_2-1}} dt_{n_2}$$

$$\cdots \int_0^s dt_{n_{m-1}+1} \int_0^{t_{n_{m-1}+1}} dt_{n_{m-1}+2} \cdots \int_0^{t_{n_m-1}} dt_{n_m}$$

$$\cdot \int_{S_1}^{T_1} dt_{n_m+1} \cdots \int_{S_N}^{T_N} dt_{n_m+N} \cdot \prod_{h=2}^{M+1} F_h \left(\frac{t_{p_h} - t_{q_h}}{\lambda^2} \right) . \quad (15.11.9b)$$

By the first connected component theorem, the Lebesgue-dominated convergence theorem and the induction assumption, the limits, as $\lambda \to 0$, of the quantity in (15.11.9a) and of the first factor in (15.11.9b) exist and are equal to

$$\int_{-\infty}^0 F_1(-\tau) \, d\tau \cdot \lim_{\lambda \to 0} \lambda^{-2M} \int_0^s dt_{n_1+1} \int_0^{t_{n_1+1}} dt_{n_1+2} \cdots \int_0^{t_{n_2-1}} dt_{n_2}$$

$$\cdots \int_0^s dt_{n_{m-1}+1} \int_0^{t_{n_{m-1}+1}} dt_{n_{m-1}+2} \cdots \int_0^{t_{n_m-1}} dt_{n_m}$$

$$\cdot \int_{S_1}^{T_1} dt_{n_m+1} \cdots \int_{S_N}^{T_N} dt_{n_m+N} \cdot \prod_{h=2}^{M+1} F_h \left(\frac{t_{p_h} - t_{q_h}}{\lambda^2} \right) \quad (15.11.10a)$$

and

$$\int_{-\infty}^0 F_1(-\tau) \, d\tau \cdot \lim_{\lambda \to 0} \lambda^{-2M} \int_0^s dt_3 \cdots \int_0^{t_{n_1-1}} dt_{n_1} \int_0^s dt_{n_1+1} \cdots \int_0^{t_{n_2-1}} dt_{n_2}$$

$$\cdots \int_0^s dt_{n_{m-1}+1} \int_0^{t_{n_{m-1}+1}} dt_{n_{m-1}+2} \cdots \int_0^{t_{n_m-1}} dt_{n_m} \int_{S_1}^{T_1} dt_{n_m+1}$$

$$\cdots \int_{S_N}^{T_N} dt_{n_m+N} \cdot \prod_{h=2}^{M+1} F_h \left(\frac{t_{p_h} - t_{q_h}}{\lambda^2} \right) \quad (15.11.10b)$$

respectively. Now we show that the second term in (15.11.9b) goes to zero. In fact, since for any $\tau \in \left[-s/\lambda^2, 0\right]$, $s + \lambda^2\tau \le s$, the absolute value of

$$\lambda^{-2M}\int_s^{s+\lambda^2\tau}dt_3\cdots\int_0^{t_{n_1-1}}dt_{n_1}\int_0^s dt_{n_1+1}\int_0^{t_{n_1+1}}dt_{n_1+2}\cdots\int_0^{t_{n_2-1}}dt_{n_2}$$
$$\cdots\int_0^s dt_{n_{m-1}+1}\int_0^{t_{n_{m-1}+1}}dt_{n_{m-1}+2}\cdots\int_0^{t_{n_m-1}}dt_{n_m}\int_{S_1}^{T_1}dt_{n_m+1}$$
$$\cdots\int_{S_N}^{T_N}dt_{n_m+N}\prod_{h=2}^{M+1}F_h\left(\frac{t_{p_h}-t_{q_h}}{\lambda^2}\right)$$

is less than or equal to

$$\lambda^{-2M}\int_{s+\lambda^2\tau}^s dt_3\int_{[S,T]^{n_m+N-3}}dt_4\cdots dt_{n_m+N}\prod_{h=2}^{M+1}\left|F_h\left(\frac{t_{p_h}-t_{q_h}}{\lambda^2}\right)\right|. \tag{15.11.11}$$

Without loss of generality, let us assume that $p_2=3$. By the change of the variable

$$u:=\frac{t_{q_2}-t_3}{\lambda^2},$$

(15.11.11) becomes

$$\int_{s+\lambda^2\tau}^s dt_3\int_{(S-t_3)/\lambda^2}^{(T-t_3)/\lambda^2}du\,|F_h(-u)|\cdot\lambda^{-2(M-1)}$$
$$\cdot\int_{[S,T]^{n_m+N-4}}dt_4\cdots\widehat{dt_{q_2}}\cdots dt_{n_m+N}\cdot\prod_{h=3}^{M+1}\left|F_h\left(\frac{t_{p_h}-t_{q_h}}{\lambda^2}\right)\right|. \tag{15.11.12}$$

By the first connected component theorem, there is a constant C such that the quantity (15.11.12) is less than or equal to

$$C\lambda^2\,|\tau|\,.$$

So, the module of the second term of the right-hand side of (15.11.9b) is dominated by

$$C\int_{-\infty}^0 d\tau\,|F_1(-\tau)|\cdot\chi_{[-s/\lambda^2,0]}(\tau)\,\lambda^2\,|\tau|\,.$$

Since

$$\chi_{[-s/\lambda^2,0]}(\tau)\,\lambda^2\,|\tau|\le s,\quad\lim_{\lambda\to0}\chi_{[-s/\lambda^2,0]}(\tau)\,\lambda^2\,|\tau|=0\,,$$

one gets the result.

In summary, in the case of $q_1=2$, $\lim_{\lambda\to0}L_\lambda\left(s;\{p_h,q_h\}_{h=2}^{M+1}\right)$ exists and is equal to the sum of (15.11.10a) and (15.11.10b).

Step 4: By the same arguments as in Step 2, one can be sure that $\lim_{\lambda\to 0} L_\lambda\left(s;\{p_h,q_h\}_{h=2}^{M+1}\right)$ exists if $q_1 \in \{n_k+1\}_{k=1}^{m-1}$.

Step 5: In order to complete the proof, we need to show that

$$\lim_{\lambda\to 0} L_\lambda\left(s;\{p_h,q_h\}_{h=2}^{M+1}\right)$$

exists in the case of $q_1 \geq n_m+1$. Without loss of generality, we shall assume that $q_1 = n_m + N$. By the change of variable (15.11.7), $L_\lambda\left(s;\{p_h,q_h\}_{h=2}^{M+1}\right)$ becomes

$$\begin{aligned}&\lambda^{-2M}\int_0^s dt_2\cdots\int_0^{t_{n_1-1}} dt_{n_1}\int_0^s dt_{n_1+1}\int_0^{t_{n_1+1}} dt_{n_1+2}\cdots\int_0^{t_{n_2-1}} dt_{n_2}\\ &\cdots\int_0^s dt_{n_{m-1}+1}\int_0^{t_{n_{m-1}+1}} dt_{n_{m-1}+2}\cdots\int_0^{t_{n_m-1}} dt_{n_m}\\ &\cdot\int_{S_1}^{T_1} dt_{n_m+1}\cdots\int_{S_{N-1}}^{T_{N-1}} dt_{n_m+N-1}\prod_{h=2}^{M+1} F_h\left(\frac{t_{p_h}-t_{q_h}}{\lambda^2}\right)\cdot\int_{(S_N-s)/\lambda^2}^{(T_N-s)/\lambda^2} d\tau F_1(-\tau)\,.\end{aligned}$$

Then the existence of the limit of $L_\lambda\left(s;\{p_h,q_h\}_{h=2}^{M+1}\right)$ as $\lambda\to 0$ is guaranteed by the induction assumption and the following easily checked fact:

$$\lim_{\lambda\to 0}\int_{(S_N-s)/\lambda^2}^{(T_N-s)/\lambda^2} d\tau F_1(-\tau) = \begin{cases}\int_{-\infty}^{\infty} d\tau F_1(\tau)\,, & \text{if } s\in(S_N,T_N)\,,\\ \int_{-\infty}^{0} d\tau F_1(\tau)\,, & \text{if } s=S_N\,,\\ \int_0^{\infty} d\tau F_1(\tau)\,, & \text{if } s=T_N\,,\\ 0, & \text{if } s\notin[S_N,T_N]\,.\end{cases}$$

15.12 Some Combinatorial Facts and the Block Normal Ordering Theorem

Throughout this book, for each $n\in\mathbb{N}$ and a set E, we use the notation E^n to denote the set of all functions defined on the set $\{1,2,\cdots,n\}$ and taking values in the set E. If $n\in\mathbb{N}$ is a natural number, a family $\{p_h,q_h\}_{h=1}^n$ is called a *pair partition* of $\{1,2,\cdots,2n\}$ if

(i) $\{p_1,q_1,\cdots,p_n,q_n\}=\{1,2,\cdots,2n\}$;
(ii) $1\le p_1<\cdots<p_n\le 2n$ (or $1\le q_1<\cdots<q_n\le 2n$);
(iii) $p_h<q_h$ for any $h=1,\cdots,n$;
A family $\{p_h,q_h\}_{h=1}^n$ is called a noncrossing pair partition (or admissible pair partition) if it satisfies (i–iii) and
(iv) for any $1\le h<j\le n$, $p_h<p_j<q_h$ if and only if $p_h<q_j<q_h$.

Now we give a geometrical interpretation of pair partitions and noncrossing pair partitions.

Let us associate the number 1 with the n elements of the set $\{1, 2, \cdots, 2n\}$ (1-points) and the number 0 with the remaining n elements (0-points). Each such arrangement corresponds to a choice of a function in $\varepsilon \in \{0,1\}^{2n}$ such that $\sum_{k=1}^{2n} \varepsilon(k) = n$. In the following, in order to indicate the fact that such arrangement corresponds to an $\varepsilon \in \{0,1\}^{2n}$ such that $\sum_{k=1}^{2n} \varepsilon(k) = n$, we shall say that the arrangement is *balanced.*

For a fixed balanced arrangement we draw a line, starting from a 0-point and ending at a 1-point, so that the 0-point is placed on the left of the 1-point. If we require that any point be connected to another point by at most one line, then clearly on the $2n$ points with a certain balanced arrangement, one can draw at most n lines. In fact, each balanced arrangement permitting n lines corresponds to an $\varepsilon \in \{0,1\}^{2n}$ that satisfies

(i) $\sum_{k=1}^{2n} \varepsilon(k) = n$;
(ii) $\forall j = 1, 2, \cdots, n$.

$$|\{h : h = j, j+1, \cdots, 2n, \varepsilon(k) = 1\}| \geq |\{h : h = j, j+1, \cdots, 2n, \varepsilon(k) = 0\}| .$$

In the following we shall use $\{0,1\}_{+}^{2n}$ to denote the subset of all $\varepsilon \in \{0,1\}^{2n}$ satisfying (i) and (ii).

Obviously, for each $\varepsilon \in \{0,1\}_{+}^{2n}$, the n lines, associated with ε as mentioned before, give a pair partition on the set $\{1, 2, \cdots, 2n\}$, say $\{p_h, q_h\}_{h=1}^{n}$, where $\{p_h\}_{h=1}^{n}$ ($\{q_h\}_{h=1}^{n}$) is the set of the left (right) ends points of these n lines. On the other hand, any pair partition $\{p_h, q_h\}_{h=1}^{n}$ on the set $\{1, 2, \cdots, 2n\}$ gives a naturally balanced arrangement and n lines as mentioned before.

A pair partition $\{p_h, q_h\}_{h=1}^{n}$ is noncrossing if and only if the n above mentioned lines do not cross.

We can introduce also pair partitions and non-crossing pair partitions on any set $\{a_1, a_2, \cdots, a_{2n}\}$, where the $2n$ points are assumed to be distinct. In fact, define

$$J(a_k) = k, \ k = 1, 2, \cdots, 2n .$$

$\{p_h, q_h\}_{h=1}^{n}$ is called a pair partition (noncrossing pair partition) on the set $\{a_1, a_2, \cdots, a_{2n}\}$ if $\{J(p_h), J(q_h)\}_{h=1}^{n}$ is a pair partition (noncrossing pair partition) on the set $\{1, 2, \cdots, 2n\}$. The following results are well known [$|\cdot|$ denotes cardinality].

Lemma 15.12.1.

$$|\{\textit{pair partitions on the set } \{1, 2 \cdots, 2n\}\}| = \int_{-\infty}^{\infty} \frac{x^{2n}}{\sqrt{2\pi}} e^{-\frac{x^2}{2}} dx = (2n-1)!!$$

Lemma 15.12.2.

$$|\{\textit{noncrossing pair partitions on the set } \{1,2\cdots,2n\}\}| = \int_{-2}^{2} x^{2n}\frac{2}{\pi}\sqrt{1-\frac{x^2}{4}}dx = \frac{1}{n+1}\binom{2n}{n}$$

is exactly the nth Catalan number.

Lemma 15.12.3. *Each* $\varepsilon \in \{0,1\}_+^{2n}$ *determines possibly many pair partitions but exactly one noncrossing pair partition.*

Lemma 15.12.4. *Given a noncrossing pair partition* $\{l_h, r_h\}_{h=1}^n$ *on the set* $\{1,2,\cdots,2n\}$*, by taking out a pair* $\{l_j, r_j\}$*, the remaining part* $\{l_h, r_h\}_{1\le h\le n,\, h\ne j}$ *is still a noncrossing pair partition of the set*

$$\{1,2,\cdots,l_j-1,\widehat{l_j},l_j+1,\cdots,r_j-1,\widehat{r_j},r_j+1,\cdots,2n\}.$$

Lemma 15.12.5. *(General noncrossing principle). Let us assume we are given* $n\in\mathbb{N}$ *and* $\varepsilon\in\{0,1\}^n$ *such that for any* $j=1,2,\cdots,n$

$$|\{h:\ h\ge j,\ \varepsilon(h)=1\}| \ge |\{h:\ h\ge j,\ \varepsilon(h)=0\}|\ .$$

Denote $m := |\{h:\varepsilon(h)=0\}|$ *and* $\{l_j\}_{j=1}^m = \{h:\varepsilon(h)=0\}$ *with the order* $l_1<\cdots<l_m$*. Then for any* $\{a_j\}_{j=1}^m \subset \{h:\varepsilon(h)=1\}$ *satisfying the conditions:*

(i) $|\{a_j\}_{j=1}^m| = m$ *(i.e. the* $a_1,\cdots,a_m$ *are distinct),*
(ii) for any $j=1,2,\cdots,m, |\{h:a_h\le j\}| \le |\{h:l_h\le j\}|$,

there exists a unique non-crossing pair partition $\{l_j,a_j\}_{j=1}^m$.

Theorem 15.12.1. *(Normal ordering theorem). For any Fock space* $\Gamma(\mathcal{H})$*, for any* $n\in\mathbb{N}$ *and* $g_1\cdots g_n\in\mathcal{H}$*, the normally (or Wick) ordered form of the product*

$$A^{\varepsilon(1)}(g_1)\cdots A^{\varepsilon(n)}(g_n) \tag{15.12.1}$$

is equal, denoting $k := n-\sum_{h=1}^n \varepsilon(h)$*, to*

$$\sum_{m=0}^{k\wedge(n-k)} \sum_{\substack{1\le p_1<\cdots<p_m\le n\\ \varepsilon(p_h)=0,h=1,2,\cdots,m}} \sum_{\substack{1\le q_1\cdots q_m\le n\\ p_h<q_h,\varepsilon(q_h)=1,h=1,2,\cdots,m}} C\prod_{h=1}^m \langle g_{p_h}, g_{q_h}\rangle$$
$$\cdot \prod_{\substack{1\le\alpha\le n\\ \varepsilon(\alpha)=1,\alpha\notin\{q_h\}_{h=1}^m}} A^+(g_\alpha) \prod_{\substack{1\le\beta\le n\\ \varepsilon(\beta)=0,\beta\notin\{p_h\}_{h=1}^m}} A(g_\beta)\,, \tag{15.12.2}$$

where in the Boson case $C=1$ *and in the Fermion case* $C\in\{1,-1\}$.

Proof 15.12.1. Clearly it is sufficient to deal only with the boson case. Using the CCR

$$A(g_k)\,A^+(g_h) = A^+(g_h)\,A(g_k) + \langle g_k, g_h\rangle \tag{15.12.3}$$

to bring (15.12.1) to its normally ordered form, a certain number m of scalar products are produced. Since each scalar product uses a creator and an annihilator, the number m must be less than or equal to the minimum between $n-k$ [the number of creators in (15.12.1)] and k [the number of annihilators in (15.12.1)]. In other words, we have to sum for all possible $m = 0, 1, \cdots, k \wedge (n-k)$.

The number m indicates only how many creators–annihilators pairs are used to produce the scalar product and does not give any information on which creators–annihilators are used to produce the scalar products. Therefore, we must, for any fixed m, also take into account all possible creators–annihilators pairs. By $\{p_1 \cdots p_m\}$ ($\{q_1 \cdots q_m\}$) we denote the indices of all annihilators (creators) used to produce scalar products; then one of the two sets can be assumed to be ordered, and we shall assume that $p_1 < \cdots < p_m$. Moreover, it must be true that $\varepsilon(p_h) = 0, \varepsilon(q_h) = 1,\ p_h < q_h$ for all $h = 1, 2, \cdots, m$. So (15.12.1) is equal to

$$\sum_{m=0}^{k\wedge(n-k)} \sum_{\substack{1\le p_1<\cdots<p_m\le n \\ \varepsilon(p_h)=0, h=1,2,\cdots,m}} \sum_{\substack{1\le q_1,\cdots,q_m\le n \\ p_h<q_h, \varepsilon(q_h)=1, h=1,2,\cdots,m}} \prod_{h=1}^{m} \langle g_{p_h}, g_{q_h}\rangle$$

$$\cdot(\text{product of all creators not used to produce scalar product})$$
$$\cdot(\text{product of all annihilators not used to produce scalar product})\,.$$

But this is exactly the quantity (15.12.2).

By combining together the arguments of the normal ordering theorem and the factorization principle, we easily obtain the following:

Theorem 15.12.2. *(The block normal ordering theorem). For $N', N'' \in \mathbb{N}$ let there be given two vectors $\Phi'_\lambda, \Phi''_\lambda$ of the form*

$$\Phi'_\lambda = \prod_{h=1}^{N'} \lambda \int_{S'_h/\lambda^2}^{T'_h/\lambda^2} \mathcal{A}^+(s, x_h)\, ds\Phi, \quad \Phi''_\lambda = \prod_{h=1}^{N''} \lambda \int_{S''_h/\lambda^2}^{T''_h/\lambda^2} \mathcal{A}^+(s, y_h)\, ds\Phi\,,$$

where $\mathcal{A}^+(s,x)$ is a product of x creators. Then for any $n \in \mathbb{N}$ and $\varepsilon \in \{0,1\}^n$, with

$$k = n - \sum_{h=1}^{n} \varepsilon(h)\,,$$

the matrix element

$$\left\langle \Phi'_\lambda, \prod_{h=1}^{n} \lambda \int_{S_h/\lambda^2}^{T_h/\lambda^2} dt_h \mathcal{A}^{\varepsilon(j)}\left(t_h, z_h\right) \Phi''_\lambda \right\rangle \tag{15.12.4}$$

is equal to

$$\begin{aligned} & o(1) + C \sum_{m=0}^{k\wedge(n-k)} \sum_{\substack{1\le p_1<\cdots<p_m\le n \\ \varepsilon(p_h)=0, h=1,2,\cdots,m}} \sum_{\substack{1\le q_1,\cdots,q_m\le n \\ p_h<q_h, \varepsilon(q_h)=1, h=1,2,\cdots,m}} \\ & \cdot \prod_{h=1}^{m} \left\langle \lambda \int_{S_{p_h}/\lambda^2}^{T_{p_h}/\lambda^2} dt_{p_h} \mathcal{A}^{\varepsilon(p_h)}\left(t_{p_h}, z_{p_h}\right), \lambda \int_{S_{q_h}/\lambda^2}^{T_{q_h}/\lambda^2} dt_{q_h} \mathcal{A}^{\varepsilon(q_h)}\left(t_{q_h}, z_{q_h}\right) \right\rangle \\ & \cdot \left\langle \Phi'_\lambda, \prod_{\substack{1\le\alpha\le n \\ \varepsilon(\alpha)=1, \alpha\notin\{q_h\}_{h=1}^m}} \lambda \int_{S_a/\lambda^2}^{T_\alpha/\lambda^2} dt_\alpha \mathcal{A}^{+}\left(t_\alpha, z_\alpha\right) \right. \\ & \cdot \left. \prod_{\substack{1\le\beta\le n \\ \varepsilon(\beta)=0, \beta\notin\{p_h\}_{h=1}^m}} \lambda \int_{S_\beta/\lambda^2}^{T_\beta/\lambda^2} dt_\beta \mathcal{A}\left(t_\beta, z_\beta\right) \Phi''_\lambda, \right\rangle \end{aligned} \tag{15.12.5}$$

where C is a constant belonging to $\{1, -1\}$ in the Fermi case, $C = 1$ in the Boson case, and the term $o(1)$ is meant asymptotically as $\lambda \to 0$.

The next thing to be considered in this section is n-tuple commutator.

Lemma 15.12.6. *Let there be given $n+1$ elements $X, B_1 \cdots B_n$ $(n \in \mathbb{N})$ of an algebra. The n-tuple commutator*

$$\left[B_1, \cdots, \left[B_{n-1}, \left[B_n, X\right]\right] \cdots\right] \tag{15.12.6}$$

has the form

$$\sum_{d=0}^{n} \sum_{\sigma\in\mathcal{S}_n} \theta(d,\sigma) B_{\sigma(1)} \cdots B_{\sigma(d)} X B_{\sigma(d+1)} \cdots B_{\sigma(n)}, \tag{15.12.7}$$

where, for any $d \in \{0, 1, 2, \cdots, n\}$, $\sigma \in \mathcal{S}_n$, and the factor $\theta(d, \sigma)$ takes values in the set $\{0, 1, -1\}$. Moreover,

(i) the set $\{\sigma \in \mathcal{S}_n : \theta(n, \sigma) \neq 0\}$ includes only the identity;
(ii) the set $\{\sigma \in \mathcal{S}_n : \theta(0, \sigma) \neq 0\}$ includes only the permutation which brings $\{1, 2, \cdots, n\}$ to $\{n, \cdots, 2, 1\}$.

Proof 15.12.2. For $n = 1$, (15.12.6) is equal to $B_1X - XB_1$ which has the form (15.12.7) with $\theta(0, \sigma) = -1$ and $\theta(1, \sigma) = 1$. Supposing that, for $n = m$, (15.12.6) has the form (15.12.7), we find the following for the case $n = m+1$:

$$
\begin{aligned}
&\Big[B_1, \cdots, \Big[B_{m-1}, \big[B_m, [B_{m+1}, X]\big]\Big] \cdots\Big] \\
&= \Big[B_1, \cdots, \big[B_{m-1}, [B_m, B_{m+1}X]\big] \cdots\Big] - \Big[B_1, \cdots, \big[B_{m-1}, [B_m, XB_{m+1}]\big] \cdots\Big] \\
&= \sum_{d=0}^{m} \sum_{\sigma \in \mathcal{S}_m} \theta(d, \sigma) B_{\sigma(1)} \cdots B_{\sigma(d)} B_{m+1} X B_{\sigma(d+1)} \cdots B_{\sigma(m)} \\
&- \sum_{d=0}^{m} \sum_{\sigma \in \mathcal{S}_m} \theta(d, \sigma) B_{\sigma(1)} \cdots B_{\sigma(d)} X B_{m+1} B_{\sigma(d+1)} \cdots B_{\sigma(m)} ,
\end{aligned}
$$

which obviously has the form

$$
\sum_{j=0}^{m+1} \sum_{\tau \in \mathcal{S}_{m+1}} \theta(j, \tau) B_{\tau(1)} \cdots B_{\tau(j)} X B_{\tau(j+1)} \cdots B_{\tau(m+1)} .
$$

For $n = 1$, (i) is clear. Suppose that (i) is true for $n = m$. Then for $n = m+1$ we find

$$
\begin{aligned}
&\Big[B_1, \cdots, \Big[B_{m-1}, \big[B_m, [B_{m+1}, X]\big]\Big] \cdots\Big] \\
&= B_1 \Big[B_2, \cdots, \Big[B_{m-1}, \big[B_m, [B_{m+1}, X]\big]\Big] \cdots\Big] \\
&\quad - \Big[B_2, \cdots, \Big[B_{m-1}, \big[B_m, [B_{m+1}, X]\big]\Big] \cdots\Big] B_1 ,
\end{aligned}
$$

and only the first term on the right-hand side can give a term like

$$
B_1 \cdot (\text{a product of } B_2, \cdots, B_{m+1}) \cdot X .
$$

By applying the induction assumption to

$$
\Big[B_2, \cdots, \Big[B_{m-1}, \big[B_m, [B_{m+1}, X]\big]\Big] \cdots\Big] ,
$$

we complete the proof for (i). Similarly, one can prove (ii).

16. Term-by-Term Convergence

In this chapter the initial state of the reservoir will be the vacuum state and we study the limits, as $\lambda \to 0$, of matrix elements of the evolution operator in the collective vectors,

$$\left\langle \prod_{h=1}^{N} \lambda \int_{S_h/\lambda^2}^{T_h/\lambda^2} \mathcal{C}_h(s_h) ds_h \Phi, U_{t/\lambda^2} \prod_{h=1}^{N'} \lambda \int_{S'_h/\lambda^2}^{T'_h/\lambda^2} \mathcal{C}'_h(s'_h) ds'_h \Phi \right\rangle \tag{16.0.1}$$

and of the corresponding matrix elements for the Heisenberg evolution

$$\left\langle \prod_{h=1}^{N} \lambda \int_{S_h/\lambda^2}^{T_h/\lambda^2} \mathcal{C}_h(s_h) ds_h \Phi, U_{t/\lambda^2} (1 \otimes X) U^*_{t/\lambda^2} \prod_{h=1}^{N'} \lambda \int_{S'_h/\lambda^2}^{T'_h/\lambda^2} \mathcal{C}'_h(s'_h) ds'_h \Phi \right\rangle \tag{16.0.2a}$$

and

$$\left\langle \prod_{h=1}^{N} \lambda \int_{S_h/\lambda^2}^{T_h/\lambda^2} \mathcal{C}_h(s_h) ds_h \Phi, U^*_{t/\lambda^2} (1 \otimes X) U_{t/\lambda^2} \prod_{h=1}^{N'} \lambda \int_{S'_h/\lambda^2}^{T'_h/\lambda^2} \mathcal{C}'_h(s'_h) ds'_h \Phi \right\rangle \tag{16.0.2b}$$

In (16.0.1) and (16.0.2a,b):

(i) $N, N' \in \mathbf{N}$;
(ii) $\mathcal{C}_h, \mathcal{C}'_j$ are:
- a sum of products of creators, i.e. a sum of $\mathcal{A}^+$-type terms;

or

- a sum of [products of creators] $\cdot$ [products of annihilators], i.e. a sum of $\mathcal{N}$-type terms;

(iii) X is a linear operator on the system space.

Replacing formally, in (16.0.1) and (16.0.2a,b), the evolution operator by its iterated series, (16.0.1) becomes equal to a sum of terms of the form

$$\lambda^N \int_{S_1/\lambda^2}^{T_1/\lambda^2} ds_1 \cdots \int_{S_N/\lambda^2}^{T_N/\lambda^2} ds_N \, (-i\lambda)^n \int_0^{t/\lambda^2} dt_1 \int_0^{t_1} dt_2 \cdots \int_0^{t_{n-1}} dt_n \lambda^{N'} \int_{S'_1/\lambda^2}^{T'_1/\lambda^2} ds'_1 \cdots \int_{S'_{N'}/\lambda^2}^{T'_{N'}/\lambda^2} ds'_{N'} \cdot \left\langle \prod_{h=1}^{N} \mathcal{C}_h(s_h)\Phi, H_{\mathrm{I}}(t_1) \cdots H_{\mathrm{I}}(t_n) \prod_{h=1}^{N'} \mathcal{C}'_h(s'_h)\Phi \right\rangle, \quad (16.0.3)$$

(16.0.2a) is equal to the sum of terms of the form

$$\lambda^N \int_{S_1/\lambda^2}^{T_1/\lambda^2} ds_1 \cdots \int_{S_N/\lambda^2}^{T_N/\lambda^2} ds_N (-i\lambda)^n \int_0^{t/\lambda^2} dt_1 \int_0^{t_1} dt_2 \cdots \int_0^{t_{n-1}} dt_n \cdot (i\lambda)^m \int_0^{t/\lambda^2} d\tau_1 \int_0^{\tau_1} d\tau_2 \cdots \int_0^{\tau_{m-1}} d\tau_m \lambda^{N'} \int_{S'_1/\lambda^2}^{T'_1/\lambda^2} ds'_1 \cdots \int_{S'_{N'}/\lambda^2}^{T'_{N'}/\lambda^2} ds'_{N'} \cdot \left\langle \prod_{h=1}^{N} \mathcal{C}_h(s_h)\Phi, H_{\mathrm{I}}(t_1) \cdots H_{\mathrm{I}}(t_n)(1 \otimes X) H_{\mathrm{I}}(\tau_m) \cdots H_{\mathrm{I}}(\tau_1) \prod_{h=1}^{N'} \mathcal{C}'_h(s'_h)\Phi \right\rangle \quad (16.0.4a)$$

and (16.0.2b) is equal to the sum of terms of the form

$$\lambda^N \int_{S_1/\lambda^2}^{T_1/\lambda^2} ds_1 \cdots \int_{S_N/\lambda^2}^{T_N/\lambda^2} ds_N (i\lambda)^m \int_0^{t/\lambda^2} d\tau_1 \int_0^{\tau_1} d\tau_2 \cdots \int_0^{\tau_{m-1}} d\tau_m \cdot (-i\lambda)^n \int_0^{t/\lambda^2} dt_1 \int_0^{t_1} dt_2 \cdots \int_0^{t_{n-1}} dt_n \lambda^{N'} \int_{S'_1/\lambda^2}^{T'_1/\lambda^2} ds'_1 \cdots \int_{S'_{N'}/\lambda^2}^{T'_{N'}/\lambda^2} ds'_{N'} \cdot \left\langle \prod_{h=1}^{N} \mathcal{C}_h(s_h)\Phi, H_{\mathrm{I}}(\tau_m) \cdots H_{\mathrm{I}}(\tau_1)(1 \otimes X) H_{\mathrm{I}}(t_1) \cdots H_{\mathrm{I}}(t_n) \prod_{h=1}^{N'} \mathcal{C}'_h(s'_h)\Phi \right\rangle . \quad (16.0.4b)$$

In all cases the indices m, n of the summation run over the natural integers $\{0, 1, 2, \cdots\}$.

If the initial state is a temperature state, then the basic objects to be studied in the weak-coupling limit (WCL) are essentially the same as (16.0.3) and (16.0.4a,b), but with the following modifications:

(i) the vacuum vector Φ should be replaced by the tensor product of the vacuum and the anti-vacuum vector: $\Phi \otimes \Phi^\iota$;

(ii) $\mathcal{C}_h, \mathcal{C}'_j$ are

- a sum of products of creators on the first space;

or

- a sum of products of creators on the second space;

or

- a sum of products of
  [creators on the first space] $\otimes$ [creators on the second space] $\cdot$[annihilators on the first space] $\otimes$ [annihilators on the second space];

such a term will also be called a $\mathcal{N}$-type block.

Remark 16.0.1. If we take $N = N' = 0$, then (16.0.1) and (16.0.2a,b) become just the expectation with respect to the state determined by the vector Φ.

The present chapter is devoted to the consideration of the limits (16.0.3) and (16.0.4a,b) for fixed n, m. Such limits will be called simply the term-by-term limits. We shall investigate the term-by-term limits not only for the vacuum case but also for the finite temperature case.

16.1 The Universality Class Principle and Effective Interaction Hamiltonians

We will use the following notations: $\mathcal{A}(t;x)$:= a product of x annihilators with time variable t,

$$\mathcal{A}^+(t;x) := \left(\mathcal{A}(t;x)\right)^*, \tag{16.1.1}$$

$$\mathcal{N}(t;x,y) := \mathcal{A}^+(t;x)\mathcal{A}(t;y), \tag{16.1.2}$$

$$\mathcal{C}_j(t;x,y) := \begin{cases} \mathcal{A}(t;y), & \text{if } j = 0, \\ \mathcal{A}^+(t;x), & \text{if } j = 1, \\ \mathcal{N}^+(t;x,y), & \text{if } j = 2. \end{cases} \tag{16.1.3}$$

From Chap. 11 we know that the standard form of the time-evolved interaction Hamiltonian, in the Fock case, is a sum of terms of the following type:

$$-i\lambda[\mathcal{A}^+(t,x) \otimes D_x e^{i\omega t} + \mathcal{N}(t;y,z) \otimes L_{y,z}(t) - \text{ h.c.}], \tag{16.1.4a}$$

where D and $L(t)$ are linear operators on the system space. Moreover, just by moving the frequencies term $e^{it\omega}$ into the block $\mathcal{A}^+(t,x)$ and changing the unitary group S_t^0 into $S_t^0 e^{it\omega/x}$, one can rewrite the standard form of the time-evolved interaction Hamiltonian as a sum of terms of the following type

$$-i\lambda[\mathcal{A}^+(t,x) \otimes D_x + \mathcal{N}(t;y,z) \otimes L_{y,z}(t) - \text{ h.c.}], \tag{16.1.4b}$$

and we will use this notation when it is not necessary to specify the frequency term. Similarly, in the standard form of the interaction Hamiltonian, we have not made explicit the dependence on the test functions. In fact we will prove that this information is irrelevant for the determination of the universality class.

In the temperature case the state space of the reservoir has the form

$$\Gamma(\mathcal{H}_1) \otimes \Gamma(\iota\mathcal{H}_1),$$

where $\mathcal{H}_1$ is the 1-particle space (ket vectors) and $\iota\mathcal{H}_1$ is its conjugate space (bra vectors) and, if Φ is the Fock vacuum and Φ^ι the anti-Fock vacuum, the initial state is

$$\langle \Phi \otimes \Phi^\iota, \Phi \otimes \Phi^\iota \rangle .$$

We will use the following notations:
$\mathcal{A}_1(t;x)$:= a product of x annihilators on the first space with the time variable t,

$$\mathcal{A}_1^+(t;x) := \left(\mathcal{A}_1(t;x)\right)^*,$$

$\mathcal{A}_2(t;x)$:= a product of x annihilators on the second space with the time variable t,

$$\mathcal{A}_2^+(t;x) := \left(\mathcal{A}_2(t;x)\right)^*,$$

$$\mathcal{A}(t;x,y) := \mathcal{A}_1(t;x) + \mathcal{A}_2(t;y) \qquad , \quad \mathcal{A}^+(t;x,y) := \left(\mathcal{A}(t;x,y)\right)^*$$

and

$$\begin{aligned}\mathcal{N}(t;u,u',v,v',z,z',w,w') :=\, & a\ \mathcal{A}_1^+(t;u)\mathcal{A}_1(t;u') + b\ \mathcal{A}_2^+(t;v)\mathcal{A}_2(t;v') \\ & + c\ [\mathcal{A}_1^+(t;z)\mathcal{A}_1(t;z')] \otimes [\mathcal{A}_2^+(t;w)\mathcal{A}_2(t;w')],\end{aligned}$$

where a, b, c are constants. As in the Fock case we did not make explicit the dependence on the test functions and on the frequency terms.

With these notations the standard form of the time-evolved interaction Hamiltonian is a sum of terms of the following type:

$$\begin{aligned}-i\lambda[\mathcal{A}^+(t,x,y) \otimes D_{x,y} + \mathcal{N}(t;u,u',v,v',z,z',w,w') \\ \otimes L_{u,u',v,v',z,z',w,w'}(t) - \text{ h.c.}] .\end{aligned} \tag{16.1.5}$$

In the following (both in the Fock case and in the temperature case) if we are interested only in the time variable t, we will simply write $\mathcal{A}^+(t)$ $[\mathcal{A}(t), \mathcal{N}(t)]$ to denote the $\mathcal{A}^+$ ($\mathcal{A}$, $\mathcal{N}$)-blocks for this time variable. In these notations (16.1.3) becomes

$$\mathcal{C}_j(t) := \begin{cases} \mathcal{A}(t), & \text{if } j = 0, \\ \mathcal{A}^+(t), & \text{if } j = 1, \\ \mathcal{N}^+(t), & \text{if } j = 2. \end{cases} \tag{16.1.6}$$

Then by using the standard form (16.1.4b) and (16.1.5) of $H_{\mathrm{I}}(t)$, the basic quantities to be estimated, i.e. (16.0.3), (16.0.4a) and (16.0.4b), can always be expressed as a finite sum of quantities of the form

$$\lambda^n \int_{S_1/\lambda^2}^{T_1/\lambda^2} dt_1 \cdots \int_{S_n/\lambda^2}^{T_n/\lambda^2} dt_n G(t_1 \cdots t_n) \left\langle \tilde{\Phi}, \prod_{h=1}^n \mathcal{C}_{j_h}(t_h)\tilde{\Phi} \right\rangle, \tag{16.1.7}$$

where $G(t_1 \cdots t_n)$ is a bounded function and

$$\tilde{\Phi} := \begin{cases} \Phi, & \text{if the initial state is the vacuum,} \\ \Phi \otimes \Phi^{\iota}, & \text{if the initial state is finite–temperature.} \end{cases}$$

For example, the quantity (16.0.3) can be rewritten as

$$\sum \lambda^{N+N'+n} \int_{S_1/\lambda^2}^{T_1/\lambda^2} ds_1 \cdots \int_{S_N/\lambda^2}^{T_N/\lambda^2} ds_N \int_0^{t/\lambda^2} dt_1 \int_0^{t/\lambda^2} dt_2 \cdots \int_0^{t/\lambda^2} dt_n$$
$$\cdot \int_{S'_1/\lambda^2}^{T'_1/\lambda^2} ds'_1 \cdots \int_{S'_{N'}/\lambda^2}^{T'_{N'}/\lambda^2} ds'_{N'} \, (-i)^n \chi_{\Delta_n}(t_1 \cdots t_n)$$
$$\cdot \langle \Phi, C^*_N(s_N) \cdots C^*_1(s_1) C_{j_1}(t_1) \cdots C_{j_n}(t_n) C'_1(s'_1) \cdots C'_{N'}(s'_{N'}) \Phi \rangle \, .$$

Theorem 16.1.1. *For any $n \in \mathbb{N}$, for any $S_1, T_1 \cdots S_n, T_n \in \mathbf{R}$ such that $S_h \leq T_h \; \forall h = 1, 2 \cdots n$, for any bounded function $G(t_1 \cdots t_n)$, the limit, as $\lambda \to 0$, of*

$$\lambda^n \int_{S_1/\lambda^2}^{T_1/\lambda^2} dt_1 \cdots \int_{S_n/\lambda^2}^{T_n/\lambda^2} dt_n \; G(t_1 \cdots t_n) \left\langle \tilde{\Phi}, \prod_{h=1}^{n} C_{j_h}(t_h; x_h, y_h) \tilde{\Phi} \right\rangle \tag{16.1.8}$$

is equal to zero if either of the following conditions holds:

(i) *the first block is not a product of annihilators, i.e. $j_1 \neq 0$;*
(ii) *the last block is not a product of creators, i.e. $j_n \neq 1$;*
(iii) *in the product $\prod_{h=1}^{n} C_{j_h}(t_h; x_h, y_h)$, the number of the creators is different from that of annihilators;*
(iv) *n is odd;*
(v) *among the blocks $C_{j_1}(t_1; x_1, y_1) \cdots C_{j_h}(t_h; x_n, y_n)$, there is some $\mathcal{N}$-type term, i.e. there is some $h \in \{1, 2 \cdots N\}$ such that $j_h = 2$.*

Proof 16.1.1. As argued in Chap. 15, just by applying the canonical commutation relations (CCR) (or canonical anti-commutation relations CAR) one knows that, if at least one of the conditions (i), (ii) or (iii) is verified, then (16.1.8) is equal to zero; otherwise (16.1.8) is equal to a quantity of the form

$$\lambda^n \int_{S_1/\lambda^2}^{T_1/\lambda^2} dt_1 \cdots \int_{S_n/\lambda^2}^{T_n/\lambda^2} dt_n G(t_1 \cdots t_n) \sum_{p_h, q_h} C\left(\{p_h, q_h\}_{h=1}^{N}\right) \prod_{h=1}^{N} F_h(t_{p_h} - t_{q_h}) \, , \tag{16.1.9}$$

where

(i) N is the number of creators in the product $\prod_{h=1}^{n} C_{j_h}(t_h; x_h, y_h)$ (which must be the same as the number of annihilators in the same product);
(ii) $\{p_1, q_1 \cdots p_N, q_N\} = \{1, 2 \cdots n\}$ and $p_h < q_h$ for any $h = 1, 2 \cdots N$;

(iii) any function F_h has the form

$$F_h(t) = \left\langle S_t^0 f_h, g_h \right\rangle ,$$

where f_h (g_h) are test functions coming from the annihilators (creators) in the product $\prod_{h=1}^n C_{j_h}(t_h; x_h, y_h)$;

(iv) the factor $C\left(\{p_h, q_h\}_{h=1}^N\right)$ is equal to 1 if the reservoir is bosonic, but otherwise is equal to 1 or -1.

Because of (iii), we know that the conditions of the first connected component theorem are satisfied. Therefore, by applying that theorem, we conclude that (16.1.9) is a quantity of the order

$$\lambda^n \cdot O(\lambda^{-2k}) = O(\lambda^{n-2k}) ,$$

where k is the number of connected components of the variables $t_1 \cdots t_n$ with respect to the product $\prod_{h=1}^N F_h\left(t_{p_h} - t_{q_h}\right)$. Moreover, as argued in Sect. 16.2, $k \le \frac{n}{2}$. So if n is odd, one has $k \le \frac{n-1}{2}$, and therefore the quantity (16.1.9) has order $O(\lambda)$, and therefore it tends to zero as $\lambda \to 0$.

Finally, if there exists $h \in \{1, 2 \cdots n\}$ such that $j_h = 2$, then the hth block is an $\mathcal{N}$-block. So the connected component which includes the time variable t_h includes at least 3 time variables. As a consequence of this fact, the number k of connected components is less than or equal to

$$1 + \frac{n-3}{2} = \frac{n-1}{2} .$$

Thus (16.1.9), which a priori is a quantity of the order $O(\lambda^{n-2k})$, is in fact of the order $O(\lambda)$; hence it tends to zero as $\lambda \to 0$.

Just by repeating the above proof, one can obtain the following:

Theorem 16.1.2. *The conclusion of Theorem 16.1.1 is still true if we assume that*

(i) S_1, T_1 *are constants;*

(ii) For $h = 2, 3 \cdots n$, S_h, T_h *are bounded functions of the variables* S_j, T_j, t_j, *with* $j \in \{1, 2 \cdots h-1\}$.

Using these results, we are able to prove *the universality class principle*: *$\mathcal{N}$-type blocks can be neglected in the time-evolved interaction Hamiltonian and collective vectors.* In particular the effective time-evolved interaction Hamiltonian $H_{\mathrm{I}}^{\mathrm{e}}(t)$ has the form (11.5.4) and (11.5.5).

16.2 Block and Orthogonalization Principles

In this section we try to understand the role played by the different terms in the effective time-evolved Hamiltonian $H_{\mathrm{I}}^{\mathrm{e}}(t)$. In order to do this, we consider quantities such as

$$\lambda^{2n} \int_{S_1/\lambda^2}^{T_1/\lambda^2} dt_1 \cdots \int_{S_{2n}/\lambda^2}^{T_{2n}/\lambda^2} dt_{2n}\; G(t_1 \cdots t_{2n}) \left\langle \tilde{\Phi}, \prod_{h=1}^{2n} \mathcal{A}^{\varepsilon(h)}(t_h; j_h)\tilde{\Phi} \right\rangle , \tag{16.2.1}$$

where here and in the following, $\varepsilon \in \{0,1\}^n$,

$$\mathcal{A}^{\varepsilon}(t;j) := \begin{cases} \text{a product of } j \text{ annihilators with time variable } t, & \text{if } \varepsilon = 0\,, \\ \text{a product of } j \text{ creators with time-variable } t, & \text{if } \varepsilon = 1\,, \end{cases}$$

and $G(t_1 \cdots t_n)$ is a bounded function.

As shown in Theorem 16.1.1, the quantity (16.2.1) is different from zero only if

(i) $\varepsilon(1) = 0$;
(ii) $\varepsilon(n) = 1$;
(iii) $\sum_{k=1}^{2n} \varepsilon(k) = n$;
(iv) in the product $\prod_{h=1}^{n} \mathcal{A}^{\varepsilon(h)}(t_h; j_h)$ the number of creators is the same as that of annihilators.

In the following, the four conditions above shall be called the *nontriviality* conditions, and in our discussion we will assume that these conditions are satisfied. Denote by $\{p_h\}_{h=1}^{n} := \{k : \varepsilon(k) = 0\}$ (annihilator indices) with the order $1 \le p_1 < \cdots p_n \le 2n$. From Theorem 15.2.3 and its corollary, one easily obtains the following:

Theorem 16.2.1. *As* $\lambda \to 0$, *(16.2.1) is equal to*

$$o(1) + \sum_{\substack{1 \le q_1 \cdots q_n \le 2n \\ p_h < q_h, \forall h = 1,2 \cdots n}} c(\{p_h, q_h\}_{h=1}^{n}) \prod_{h=1}^{n} \lambda^2 \int_{S_{p_h}/\lambda^2}^{T_{p_h}/\lambda^2} dt_{p_h}$$

$$\cdot \int_{S_{q_h}/\lambda^2}^{T_{q_h}/\lambda^2} dt_{q_h} \left\langle \tilde{\Phi}, \mathcal{A}(t_{p_h}, j_{p_h}) \mathcal{A}^{+}(t_{q_h}, j_{q_h}) \tilde{\Phi} \right\rangle , \tag{16.2.2}$$

where as usual $c(\{p_h, q_h\}_{h=1}^{n})$ *is equal to* ± 1 *and, if the reservoir is bosonic, it is always equal to* 1.

16.3 The Stochastic Resonance Principle

In this section we deal with the contribution of the frequency terms to the WCL. The basic object to be considered is

$$\lambda^2 \int_{S/\lambda^2}^{T/\lambda^2} dt \int_{S'/\lambda^2}^{T/\lambda^2} d\tau \left\langle \tilde{\Phi}, \mathcal{A}(t,j) e^{-it\omega} \mathcal{A}^{+}(\tau, j) e^{i\tau\omega'} \tilde{\Phi} \right\rangle . \tag{16.3.1}$$

The stochastic resonance principle says that

Theorem 16.3.1. *As $\lambda \to 0$, the limit of (16.3.1) is equal to zero if $\omega \neq \omega'$.*

Proof 16.3.1. Obviously, for any type of reservoir (bosonic or fermionic), for any initial state (vacuum or finite temperature), for any $j \in \mathbf{N}$, the scalar product

$$\langle \tilde{\Phi}, \mathcal{A}(t,j)\mathcal{A}^+(\tau,j)\tilde{\Phi}\rangle$$

has the form $F(t-\tau)$ for some function $F \in L^1(\mathbf{R}) \cap L^\infty(\mathbf{R})$. Therefore, (16.3.1) has the form

$$\lambda^2 \int_{S/\lambda^2}^{T/\lambda^2} dt \int_{S'/\lambda^2}^{T'/\lambda^2} d\tau e^{-it\omega} e^{i\tau\omega'} F(t-\tau)\,. \tag{16.3.2}$$

By the change of variables

$$v := \lambda^2\,\tau, \quad u := t - \tau = t - v/\lambda^2\,,$$

(16.3.2) becomes

$$\int_{S'}^{T'} e^{i(\omega'-\omega)v/\lambda^2} dv \int_{(S-v)/\lambda^2}^{(T-v)/\lambda^2} du F(u) e^{-iu\omega}\,. \tag{16.3.3}$$

Since $F \in L^1(\mathbf{R}) \cap L^\infty(\mathbf{R})$ [so is $e^{-iu\omega}F(u)$], the function

$$G_\lambda(v) := \int_{(S-v)/\lambda^2}^{(T-v)/\lambda^2} du F(u) e^{-iu\omega}$$

is in $L^1(\mathbf{R}) \cap L^\infty(\mathbf{R})$ and $|G_\lambda(v)| \leq \|F\|_1$. By the following lemma, we know that (16.3.3) goes to zero as $\lambda \to 0$ if $\omega \neq \omega'$.

Lemma 16.3.1 (Special case of the Riemann–Lebesgue Lemma). *For any $S, T \in \mathbb{R}$, such that $S < T$, for any family of functions $F_\alpha \in \mathbf{L}^1([S,T])$ such that there is a function $F \in \mathbf{L}^1([S,T])$ such that $|F_\alpha| \leq |F|$, for any function of α, say $n(\alpha)$, such that $n(\alpha) \to \infty$ as $\alpha \to \alpha_0$, one has*

$$\lim_{\alpha \to \alpha_0} \int_S^T dt\; e^{it\; n(\alpha)} F_\alpha(t) = 0\,. \tag{16.3.4}$$

Proof 16.3.2. As $\alpha \to \alpha_0$,

$$\left| \int_S^T dt\; e^{it\; n(\alpha)} F_\alpha(t) \right| = \left| \int_S^T dt\; F_\alpha(t) \frac{\int_0^t ds e^{is\; n(\alpha)} - 1}{i\; n(\alpha)} \right|$$

$$\leq \frac{|S| + |T| + 1}{|n(\alpha)|} \int_S^T dt\; |F_\alpha(t)| \leq \frac{|S| + |T| + 1}{|n(\alpha)|} \int_S^T dt\; |F(t)| \longrightarrow 0\,.$$

References

[ABCR70] S.L. Adler, J.N. Bahcall, C.G. Callan, M.N. Rosenbluth: Phys. Rev. Lett. **25**, 1061 (1970)

[Abr61] A. Abragam: *The Principles of Nuclear Magnetism* (Clarendon Press, Oxford 1961)

[Ac78] L. Accardi: Rend. Sem. Math. Fis. Univ. Polit. Milano **48**, 135 (1978)

[AcArVo95] L. Accardi, I.Ya. Aref'eva, I.V. Volovich: 'The Master Field for Half–Planar Diagrams and Free Non-Commutative Random Variables'. *Proceedings of the 9th International Seminar: Quarks '96, Vol II.* (Institute for Nuclear Research of Russian Academy of Sciences, Moscow 1997) pp. 31–38; Volterra Preprint No. 231 (1996). hep-th/9502092

[AcArVo97] L. Accardi, I.Ya. Aref'eva, I.V. Volovich: Photon Splitting Cascades and a New Statistics; Volterra Preprint No. 286. hep-ph/9708289

[AcArVo00] L. Accardi, I.Ya. Aref'eva, I.V. Volovich: 'Nonequilibirum Quantum Field Theory and Interacting Commutation Relations'. Proc. V.A: Steklov Inst. Math., Vol. 228 (Nauka Publ., Moscow 2000) pp. 116–135

[AcArVo01] L. Accardi, I.Ya. Aref'eva, I.V. Volovich: 'Stochastic limit and interacting commutation relations'. In: *Mathematical Physics and Stochastic Analysis*, ed. by S. Albeverio, Ph. Blanchard, L. Ferreira, T. Hida, Y. Kondratev, R. Vilela-Mendes (World Scientific, Singapore 2000) pp. 18–36

[AcFrLu87] L. Accardi, A. Frigerio, Y.G. Lu: 'On the weak coupling limit problem'. In: *Quantum Probability and Applications IV*, SLNM No. 1396 (1987) pp. 20–58

[AcFrLu90] L. Accardi, A. Frigerio, Y.G. Lu: Comm.. Math. Phys. **131**, 537 (1990), submitted (1987)

[AcFrLu91] L. Accardi, A. Frigerio, Y.G. Lu: J. Math. Phys. **32**, 1567 (1991)

[AcGoLu95] L. Accardi, J. Gough, Y.-G. Lu: Rep. Math. Phys. **36**, 155 (1995)

[AcKo99a] L. Accardi, S.V. Kozyrev: 'The Stochastic Limit of Quantum Spin System'. Invited talk at the 3rd Tohwa International Meeting on Statistical Physics, Tohwa University, Fukuoka, Japan, November 8-11, 1999. In: *Statistical Physics*, ed. by M. Tokuyama, H.E. Stanley, American Institute of Physics, AIP Proceedings 519 (2000), pp. 599–610, Volterra Preprint No. 393 (1999)

[AcKo99b] L. Accardi, S.V. Kozyrev: 'On the Structure of Markov Flows'. Chaos, Solitons, Fractals **12**, 2639 (2001). Volterra Preprint No. 394 (1999)

[AcKo00a] L. Accardi, S.V. Kozyrev: 'Quantum Decoherence and the Glauber Dynamics from the Stochastic Limit'. Volterra Preprint No. 419 (2000)

[AcKo00b] L. Accardi, S.V. Kozyrev: 'Quantum Interacting Particle Systems'. Lectures given by the authors at the Volterra–CIRM International School. In: *Quantum Interacting Particle Systems*, Levico Terme, 23–29 September 2000 (World Scientific, Singapore 2002) pp. 1–193. Volterra Preprint No. 431 (2000)

[AcKoVo97] L. Accardi, S.V. Kozyrev, I.V. Volovich: Phys. Rev. A **57**(3), 2557 (1997)

[AcKoVo98] L. Accardi, S.V. Kozyrev, I.V. Volovich: J. Phys. A **32**, 3485 (1999) q-alg/9807137

[AcKoVo98a] L. Accardi, S.V. Kozyrev, I.V. Volovich: Teor. Matemat. Fiz. **116**(3), 401 (1998)

[AcKoVo99] L. Accardi, S.V. Kozyrev, I.V. Volovich: Phys. Lett. A **260**, 31 (1999)

[AcLu90] L. Accardi, Y.G. Lu: 'On the Low Density Limit of Boson Models'. In: *Quantum Probability and Applications V*, Lecture Notes in Mathematics, No. 1442 (Springer, Berlin 1990) pp. 17–53

[AAFL91] L. Accardi, R. Alicki, A. Frigerio, Y.G. Lu: 'An Invitation to the Weak Coupling and the Low Density Limit'. In: *Quantum Probability and Applications VI* (World Scientific, Singapore 1991) pp. 3–62

[AcLu91a] L. Accardi, Y.G. Lu: Commun. Math. Phys. **637**, 1 (1991)

[AcLu91b] L. Accardi, Y.G. Lu: J. Phys. A **24**, 3483 (1991)

[AcLu91c] L. Accardi, Y.G. Lu: 'The Low Density Limit in the Finite Temperature Case (I)' In: *Gaussian Random Fields*, ed. by T. Hida, K. Ito (World Scientific, Singapore 1991) pp. 70–85

[AcLu91d] L. Accardi, Y.G. Lu: 'From the Weak Coupling Limit to a New Type of Quantum Stochastic Calculus'. In: *Quantum Probability and Related Topics, QP–PQ VII* (1993) Volterra Preprint No. 68 (1991)

[AcLu91e] L. Accardi, Y.G. Lu: Ann. Inst. H. Poincaré, Phys. Th. **54** (1991)

[AcLu92] L. Accardi, Y.G. Lu: Nagoya Math. J. **126**, 70 (1992)

[AcLu96] L. Accardi, Y.G. Lu: Comm. Math. Phys. **180**, 605 (1996). Volterra Preprint No. 126 (1992)

[AcLuMa94a] L. Accardi, V. Mastropietro: Ann. Inst. Henri Poincaré **66**, 215 (1997); Volterra Preprint No. 228 (1994)

[AcLuMa94b] L. Accardi, Y.G. Lu, V. Mastropietro: Ann. Inst. Henri Poincaré **66**(2), 185 (1997); Volterra Preprint No. 205 (1994)

[AcLuMa97] L. Accardi, Y.G. Lu, V. Mastropietro: Inf. Dim. Anal. Quantum Prob. Rel. Top. **1**(3), 467 (1998). Volterra Preprint No. 284 (1997)

[AcLuVo93] L. Accardi, Y.G. Lu, I. Volovich: Matemat. Zam. (1994); Volterra Preprint No. 138 (1993)

[AcLuVo95] L. Accardi, Y.G. Lu, I. Volovich: 'Stochastic Bosonization in Arbitrary Dimensions.' In: *Proceedings of the Conference in Honour of Alberto Frigerio*, Università 'di Udine (1996)

[AcLuVo97] L. Accardi, Y.G. Lu, I. Volovich: *Interacting Fock Spaces and Hilbert Module Extensions of the Heisenberg Commutation Relations.* Publications of IIAS (Kyoto) (1997)

[AcLuVo99] L. Accardi, Y.G. Lu, I.V. Volovich: 'A White Noise Approach to Classical and Quantum Stochastic Calculus'. Volterra Preprint No. 375 (Rome July 1999) cf. also: [Hi00] pp. 3–25

[AcSmo93] L. Accardi, O.G. Smolianov: 'On Laplacians and Traces'. Rendiconti del Seminario Matematico dell'Università di Bari, **250**, 1 (1993). Volterra Preprint No. 141 (1993)

[AcVo98] L. Accardi, I.V. Volovich: In: *Stochastic Analysis and Mathematical Physics*, ed. by R. Rebolledo (World Scientific, Singapore 1998) pp. 1–11

[Adl71] S.L. Adler: Ann. Phys. **67**, 599 (1971)

[AdSc96] S.L. Adler, C. Schubert: *Photon Splitting in a Strong Magnetic Field: Recalculation and Comparison with Previous Calculations.* hep-th/9605035

[AFL82] L. Accardi, A. Frigerio, J. Lewis: Publ. Res. Inst. Math. Sci. Kyoto Univ. **18**, 97 (1982)

[Ag73] G.S. Agarwal: In: *Progress in Optics, Vol. XI.* (North Holland, Amsterdam 1973) pp. 1–75

[AGD75] A.A. Abrikosov, L.P. Gorkov, I.E. Dzyaloshinski: *Methods of Quantum Field Theory in Statistical Physics* (Moscow 1962, Dover 1975)

[AG] M. Aizenman, G.M. Graf: J. Phys. A: Math. Gen. 31, **6783** (1998)

[AHK76] S. Albeverio, R. Hoegh-Krohn: *Mathematical Theory of Feynman Path Integrals*, Springer Series: Lecture Notes in Mathematics, No. 523 (Springer, Berlin 1976)

[Aho98a] L. Accardi, Y. Hashimoto, N. Obata: Inf. Dim. Anal. Quantum Prob. Rel. Top. **1**(2), 201 (1998)

[Aho98b] L. Accardi, Y. Hashimoto, N. Obata: 'Singleton Independence'. In: Proceedings of the Quantum Probability Conference, Gdansk, July 1997. Publications of the Banach Center 43 (1998) pp. 9–24

[AlSt75] L. Allens, J.H. Eberly: *Optical Resonance and Two-Level Atoms.* (Wiley, New York 1975)

[AnSu98] I. Antoniou, Z. Suchanecki: Inf. Dim. Anal. Quantum Prob. Rel. Top. **1**(1), 127 (1998)

[And58] P.W. Anderson: Phys. Rev. **109**, 1492 (1958)

[Ar73] I.Ya. Aref'eva: Teor. Mat. Fiz. **14**, 3 (1973)

[Ar94] I.Ya. Aref'eva: Phys. Lett. B **325**, 171 (1994)

[ArKe64] P.N. Argyres, P.L. Kelley: Phys. Rev. **134**, A98 (1964)

[ArVo00] I. Arefeva, I. Volovich: Inf. Dim. Anal. Quantum. Prob. **3**(4), 453 (2000)

[ArVo91] I.Ya. Aref'eva, I.V. Volovich: Phys. Lett. B **268**, 179 (1991)

[ArVo94] I.Ya. Aref'eva: Phys. Lett. B **325**, 171 (1994); I.Ya. Aref'eva, I.V. Volovich: 'Anisotropic Asymptotics and High-Energy Scattering in QCD'. In: *Quarks 94*, ed. by D.Yu. Grigoriev et al. (World Scientific, Singapore 1995) pp. 155–170

[ArVo96] I.Ya. Aref'eva, I.V. Volovich: Nucl. Phys. B **462**, 600 (1996)

[At73] N.M. Atherton: *Electron Spin Resonance* (Wiley, New York 1973)

[Bal75] R. Balescu: *Equilibrium and Nonequilibrium Statistical Mechanics* (Wiley, 1975)

[BaLaPr82] A. Barchielli, L. Lanz, G.M. Prosperi: Nuovo Cim. **72B**, 69 (1982)

[BaMiSh96] V.N. Baier, A.I. Milstein, R.Zh. Shaisultanov: Phys. Rev. Lett. **77**, 1691 (1996)

[Bar86] A. Barchielli: Phys. Rev. A **34**, 1642 (1986)

[Bar94] A. Barchielli: In: *Stochastic Evolution of Quantum States in Open Systems and in Measurement Processes*, ed. by L. Diosi, B. Lukacs (World Scientific, Singapore 1994) pp. 1–14

[Bar95] M.G. Baring: Astrophys. J. Lett. **440**, L 69 (1995)

[BBT61] V.L. Bonch-Bruevich, S.V. Tyablikov: *Method of Green Functions in Statistical Physics* (Nauka, Moscow 1961)

[Bel80] V.P. Belavkin: Radio Eng. Electron Phys. **25**, 1445 (1980)

[Bel85] V.P. Belavkin: 'Nondemolition Measurement and Control in Quantum Dynamical Systems'. In: *Proc. of CISM Seminar on Information Complexity and Control in Quantum Physics, Udine, 1985*, ed. by A. Blaquiere (Springer, New York 1987) pp. 311–329

[BeMe95] V.P. Belavkin, O. Melsheimer: Quantum Semiclass Opt. **8**, 167 (1995)

[BeLi82] V.B. Berestetskii, E.M. Lifshitz, L.P. Pitaevskii: *Quantum Electrodynamics* (Pergamon Press, New York 1982)

[BeLiPa78] A. Bensoussan, J.L. Lions, G. Papanicolau: *Asymptotic Analysis for Periodic Structures* (North Holland, Amsterdam 1978)

[BHG97] M.G. Baring, A.K. Harding, P.L.Gonthier: *The Attenuation of Gamma-Ray Emission in Strongly-Magnetized Pulsars.* Astro-ph/9704210

[Bo45] N.N. Bogoliubov: *On Some Statistical Methods in Mathematical Physics* (1945)

[Bo48] N.N. Bogoliubov: 'Problemi dinamicheskii teorii v statisticheskoi fiziki' (in Russian). (OGIS, Moscow 1948); English translation in: *Studies in Statistical Mechanics, Vol. I.* ed. by J. De Boer, D.E. Uhlenbeck (1962)

[Bo78] A. Bohm: Lecture Notes in Physics **94**. (Springer, Berlin 1978)

[Bo84] N.N. Bogoliubov, N.N. Bogoliubov Jr.: *Introduction to Quantum Statistical Mechanics* (Nauka, Moscow 1984)

[BoDo98] A. Bohm, H.D. Doebner, P. Kielanowski (eds.): *Irreversibility and Causality* (Springer, Berlin 1998)

[Bog46b] N.N. Bogoliubov: 'An Elementary Example of the Statistical Equilibrium in a System Related with Reservoir'. In: *On Certain Statistical Methods in Mathematical Physics.* (Academy of Sciences of Ukraine SSR, 1946)

[Bog46] N.N. Bogoliubov: *Problems of Dynamical Theory in Statistical Physics.* (Gostehizdat, 1946)

[BoLoOkTo87] N.N. Bogoliubov, A.A. Logunov, A.I. Oksak, I.T. Todorov: *General Principles of Quantum Field Theory* (Nauka, Moscow 1987)

[BoSch87] N.N. Bogoliubov, D.V. Schirkov: *Introduction to Theory of Quantized Fields* (Academic Press, New York 1972, Nauka, Moscow 1987)

[BoTy59] N.N. Bogoliubov, S.V. Tyablikov: Dokl. Akad. Nauk USSR **126**, 53 (1959)

[BoKuSp] M. Bozejko, B. Kümmerer, R. Speicher: Commun. Math. Phys. **185**, 129 (1997)

[Bou93] A. Boukas: Russ. J. Math. Phys. **5**, 179 (1993)

[BrRo79] O. Bratteli, D.W. Robinson: *Operator Algebras and Quantum Statistical Mechanics 1* (Springer, Berlin 1979)

[CaLeg83] A. Caldeira, A.J. Leggett: Physica A **121**, 587 (1983)

[ChBr73] Y. Choquet-Bruhat: *Distributions, Theorie et Problemes* (Paris 1973)

[Che66] G.F. Chew: *The Analytic S-Matrix. A Basis for Nuclear Democracy.* (W.A. Benjamin, New York 1966)

[CheFa96] A.M. Chebotarev, F. Fagnola: J. Funct. Anal. **153**, 382 (1998); Preprint No. 308 (Genova, Maggio 1996)

[CTDG92] C. Cohen-Tannoudji, J. Dupont-Roc, G. Grinberg: *Atom-Photon Interactions, Basic Processes and Applications* (Wiley, New York 1992)

[Dav76] E.B. Davies: *Quantum Theory of Open Systems* (Academic Press, 1976)

[Dio88] L. Diosi: Europhys. Lett. **6**(4), 285 (1988)

[Ex85] P. Exner: Fund. Theor. Phys. (1985)

[Fad63] L.D. Faddeev: Dokl. Akad. Nauk USSR **152**, 573 (1963)

[FaReSa94] F. Fagnola, R. Rebolledo, C. Saavedra: J. Math Phys. **35**, 1 (1994)

[FaMa68] C. Favre, Ph.A. Martin: Helv. Phys. Acta **41**, 336 (1968)

[FaPa99] P. Facchi, S. Pascazio: Physica A **271**, 133 (1999)

[Fei72] E.L. Feinberg: 'A Particle with Non-equilibrium Proper Field'. In: *Problems of Theoretical Physics, Memorial Volume to Igor E. Tamm.* (Nauka, Moscow 1972)

[Fey72] R.P. Feynman: *Statistical Mechanics* (W.A. Benjamin, Reading, Massachusetts 1972)

[FoKr47] V. Fock, N. Krylov: J. Phys. **11**, 112 (1947)

[Fri48] K.O. Friedrichs: Commun. Pure Appl. Math. **1**, 361 (1948)

[Frohlich] H. Frohlich: Adv. Phys. **3**, 325 (1954)

[FrVe79] A. Frigerio, M. Verri: Math. Jap. **24**, 53 (1979)

[FrVe82] A. Frigerio, M. Verri: Math. Z. **180**, 257 (1982)

[FS83] Frölich, Spencer: Commun. Math. Phys. **88**, 151 (1983)

[FSMS88] Frölich, Spencer, Martinelli, Scoppola: Commun. Math. Phys. **101**, 21 (1988)

[GaCo85] M.J. Collett, C.W. Gardiner: Phys. Rev. A **31**(6) (1985)

[Gam98] L. Gammaitoni, P. Hanggi, P. Jung, F. Marchesoni: Rev. Mod. Phys. **70**, 223 (1998)

[Gis82] N. Gisin: Physica **111A**, 364 (1982)

[GKS76] V. Gorini, A. Kossakowski, E.C.G. Sudarshan: J. Math. Phys. **17**, 821 (1976)

[Glau63] R.J. Glauber: J. Math. Phys. **4**, 294 (1963)

[GlJa73] J. Glimm, A. Jaffe: Fortschr. Phys. **21**, 327 (1973)

[GoFri76] A. Frigerio, V. Gorini: J. Math. Phys. **17**, 2123 (1976)

[Hag64] R. Hagedorn: *Introduction to Field Theory and Dispersion Relations.* (Pergamon Press, Oxford 1964)

[Gou96a] J. Gough: Infin. Dimen. Anal. Quantum Prob. Rel. Top. **1**(3), 439 (1998)

[Gou96b] J. Gough: *On the Emergence of a Free Noise Limit from Quantum Field Theory.* Volterra Preprint No. 244 (1996)

[GoWa64] M.L. Goldberger, K.M. Watson: *Collision Theory* (Wiley, New York 1964)

[GPR90] G.C. Ghirardi, P. Pearle, A. Rimini: Phys. Rev. A **42**(1) (1990)

[Gree90] O. Greenberg: Phys. Rev. Lett. **64**, 705 (1990)

[Gri98] M. Grifoni, P. Hanggi: Phys. Rep. **304**, 229 (1999)

[Ha72] W. Happer: Rev. Mod. Phys. **44**, 169 (1972)

[HaBa96] A.K. Harding, M.G. Baring: *Photon Splitting in Soft Gamma Repeaters.* Astro-ph/9603095

[Hak70] H. Haken: *Laser Theory* (Springer, Berlin 1970, 1984)

[HaSch98] M. B. Halpern, C. Schwartz: *The Algebras of Large N Matrix Mechanics.* hep-th/9809197

[HaYa86] H.A. Hans, Y. Yamamoto: Rev. Mod. Phys. **58**, 1001 (1986)

[HBG96] A.K. Harding, M.G. Baring, P.L. Gonthier: *Photon Splitting Cascades in Gamma-Ray Pulsars and the Spectrum of PSR1509-58.* Astro-ph/9609167

[Heit54] W. Heitler: *The Quantum Theory of Radiation* (Oxford University Press, London 1954)

[HeLi75] K. Hepp, E.H. Lieb: 'The Laser: a Reversible Quantum Dynamical System with Irreversible Classical Macroscopic Motion'. In: Springer Lecture Notes in Physics **38**, (1975) pp. 179–207

[HeyHe97] J.S. Heyl, L. Hernquist: *Birefringence and Dichroism of the QED Vacuum.* Hep-ph/9705367

[HKPS93] T. Hida, H.H. Kuo, J. Potthoff, L. Streit: *White Noise. An Infinite Dimensional Calculus* (Kluwer, Dordrecht 1993)

[Hi00] Recent Developments in Infinite-Dimensional Analysis and Quantum Probability. Pazers. In Honour of Takeyuki Hida's 70th Birthday, ed. by L. Accardi, H.-H. Kuo, N. Obata, K. Saito, Si Si, L. Streit (Kluwer 2000); cf. also Acta Appl. Math. **63**, 3 (2000)

[Hi01] T. Hida: *Selected papers* (World Scientific, Singapore 2001) pp. 316–330

[HuPa84] R.L. Hudson, K.R. Parthasarathy: Commun. Math. Phys. **93**, 301 (1984)

[ImYuOh99] K. Imafuku, K. Yuasa, I. Ohba: *Quantum Stochastic Resonance in Driven Spin-Boson System with Stochastic Limit Approximation.* e-Print archive http://xxx.lanl.gov/abs/quant-ph/9910025

[Ing92] R.S. Ingarden: Open Systems and Information Dynamics **1**, 75 (1992)

[JaPi94] V. Jakvsic, C.-A. Pillet: 'On a Model for Quantum Friction II'. In: *Fermi's Golden Rule and Dynamics at Positive Temperature.* Université de Genève, UGVA–DPT 1994/12–874

[Kato66] T. Kato: *Perturbation Theory for Linear Operators* (Springer, Berlin 1966)

[Kha57] L.A. Khalfin: Dokl. Akad. Nauk. USSR **115**, 277 (1957)

[KiYuIm00] G. Kimura, K. Yuasa, K. Imafuku: *Stochastic Limit Approximation for Rapidly Decaying Systems.* e-Print archive http://xxx.lanlgov/abs/quant-ph/0007007.

[Ku96] H.-H. Kuo: *White Noise Distribution Theory* (CRC, Boca Raton 1996)

[LCDFGZ87] A.J. Leggett, S. Chakravarty, A.T. Dorsey, M.P.A. Fisher, A. Garg, W. Zwerger: Rev. Mod. Phys. **59**, 1 (1987)

[Li76] G. Lindblad: Commun. Math. Phys. **48**, 119 (1976)

[Lig85] T.M. Liggett: *Interacting Particle Systems* (Springer, Berlin 1985)

[Lind76] G. Lindblad: Commun. Math. Phys. **48**, 119 (1976)

[Lou73] W.H. Louisell: *Quantum Statistical Properties of Radiation* (Wiley, New York 1973)

[Lu92a] Y.G. Lu: 'Quantum stochastic calculus on Hilbert module'. Volterra Preprint No. 106 (1992)

[Lu92] Y.G. Lu: J. Math. Phys. **33**(8), 3060 (1992)

[Lu94] Y.G. Lu: 'A Note on Free Stochastic Calculus on Hilbert Modules and Its Applications'. Volterra Preprint No. 186 (October 1994)

[Lu95] Y.G. Lu: 'Quantum Stochastic Calculus on Hilbert Modules Driven by Poisson Processes'. Volterra Preprint No. 212 (March 1995)

[Ma83] V.P. Maslov: *Complex Markov Chains and Feynman Integrals* (Nauka, Moscow 1983)

[MaBu] P.A. Martin, E. Buffet: J. Stat. Phys **18**, 585 (1978); Erratum: J.Stat. Phys. **20**, 347 (1979)

[MajZeg96a] A.W. Majewski, B. Zegarlisnki: Rev. Math. Phys. **8**(5), 689 (1996)

[Masl87a] V. Maslov, Embarek Djilali: *Methodes Operatorielles* [Operator Methods] (Editions Mir, Moscow 1987)

[Masl87b] V.P. Maslov: *Asymptotic Methods for Solving Pseudodifferential Equations* (in Russian) (Nauka, Moscow 1987)

[Masl88a] V.P. Maslov, V.E. Nazaikinskii: *Asymptotics of Operator and Pseudodifferential Equations. Contemporary Soviet Mathematics* (Consultants Bureau, New York 1988)

[Masl88b] V.P. Maslov: *Asymptotic Methods and Perturbation Theory* (in Russian) (Nauka, Moscow 1988)

[Mat95] V.A. Matveev: 'Origin of the Quark Counting Laws'. In: *Quarks 94*, ed. by D.Yu. Grigoriev et al. (World Scientific, Singapore 1995), pp. 41–51

[Mats93] T. Matsui: J. Funct. Anal. **116**, 179 (1993)

[Mig58] B. Migdal: Sov. Phys. Jetp. **7**, 996 (1958)

[NaNaPa95] H. Nakazato, M. Namiki, S. Pascazio: Int. J. Mod. Phys. B **10**, 247 (1996)

[Oba94] N. Obata: *White noise calculus and Fock space*, Springer LNM No. 1577 (1994)

[Ohy91] M. Ohya: In: *Springer Lecture Notes in Physics* **378**, 81 (1991)

[OhyPet93] M. Ohya, D. Petz: *Quantum Entropy an Its Use* (Springer, Berlin 1993)

[Ol94] S. Olla: 'Homogenization of Diffusion Processes in Random Fields.' Preprint (June 1994)

[Part92] K.R. Parthasarathy: *An introduction to quantum stochastic calculus* (Birkhäuser, Berlin 1993)

[PrBa59a] I. Prigogine, R. Balescu: Physica **24**, 281 (1959)

[PrBa59b] I. Prigogine, R. Balescu: Physica **25**, 302 (1959)

[PrHe57] I. Prigogine, F. Henin: Physica **23**, 585 (1957)

[PrHe60] I. Prigogine, F. Henin: J. Math. Phys. **1**, 349 (1960)

[Pr62] I. Prigogine: *Non-Equilibrium Statistical Mechanics* (Wiley, New York 1962)

[Prig68] I. Prigogine: In: *Fundamental Problems in Elementary Particle Physics, XIV Conseil de Physique Solvay, Brussels, October 1967.* (Interscience, 1968)

[PriHe69] I. Prigogine, F. Henin: 'Kinetic Theory and Subdynamics'. In: *Problems of Theoretical Physics, Essays Dedicated to Nikolai N. Bogolubov on the Occasion of his Sixtieth Birthday* (Nauka, Moscow 1969) pp. 356–364

[Pro96] R.J. Protheroe: *Origin and Propagation of the Highest Energy Cosmic Rays.* Astro-ph/9610100

[PrRo67] Yu.V. Prokhorov, Yu.A. Rozanov: *Probability Theory.* (Nauka, Moscow 1967)

[RuSh96] V.A. Rubakov, M.E. Shaposhnikov: Usp. Fis. Nauk. **166**, 493 (1996)

[Sak71] S. Sakai: *C^*-Algebras and W^*-Algebras.* (Springer, Berlin 1971)

[Schw60] J. Schwinger: Ann. Phys. **9**, 169 (1960)

[Sew61] G.L. Sewell: In: *Lectures in Theoretial Physics (Boulder), Vol. 3.* (1961) pp. 289–327

[Ske96] M. Skeide: Hilbert Modules in Quantum Electro-Dynamics and Quantum Probability, Volterra Preprint No. 257 (1996) Commun. Math. Phys. **192**, 569 (1998)

[SmSh91] O.G. Smolyanov, N. Schavgulidze: *Feynman Path Integrals* (Nauka, Moscow 1991)

[Sp80] H. Spohn: Rev. Mod. Phys. **52**, 569 (1980)

[Spitz71] F. Spitzer: 'Random Fields and Interacting Particle Systems'. In: *Proc. M.A.A. Summer Seminar* (Math. Ass. Amer., Washington DC 1971)

[SpNe94] R. Speicher, P. Neu: Z. Phys. B **95**, 101 (1994)

[SpNe95] R. Speicher, P. Neu: J. Stat. Phys. **80**, 7279 (1995)

[Sull75] W.G. Sullivan: Commun. Math. Phys. **40**, 249 (1975)

[UeUh75] E.A. Uehling, G.E. Uhlenbeck: Phys. Rev. **43**, 552 (1933)

[vHo55] L. van Hove: Physica **21**, 517 (1955)

[vHo91] *Scientific Highlights in Memory of Leon Van Hove*, ed. by F. Nicodemi. World Scientific Series in 20th Century Physics, Vol. 2 (World Scientific, Singapore 1991)

[vKa54] N.G. van Kampen: Physica **20**, 603 (1954)

[Vl83] V.S. Vladimirov: *Equations of Mathematical Physics.* (Nauka, Moscow 1983)

[Voic91] D. Voiculescu: Invent. Math. **104**, 210 (1991)

[Vol99] I.V. Volovich: 'Models of Quantum Computers and Decoherence Problem'. In: *Quantum Information*, ed. by T. Hida, K. Saito. (World Scientific, Singapore 1999) pp. 211–224

[vonN33] J. von Neumann: *Mathematical Foundations of Quantum Mechanics* (Princeton University Press, Princeton 1955)

[vWa86] W. von Waldenfels: 'Spontaneous light emission described by a quantum stochastic differential equation'. In: *Quantum Probability and Applications II*, Springer LNM No. 1136 (1986)

[WaBl53] R.K. Wangsness, F. Bloch: Phys. Rev. **89**, 728 (1953)

[WaMi94] D.F. Walls, G.J. Milburn: *Quantum Optics* (Springer, Berlin 1994)

[Weg79] F. Wegner: Phys. Rev. B **19**, 783 (1979)

[WiBh97] S.R. Wilkinson, C.F. Bharucha et al.: Nature **387**(5), 575 (1997)

[Yosh66] K. Yoshida: *Functional Analysis* (Springer, Berlin 1966)

[Zu71] D.N. Zubarev: *Nonequilibrium Statistical Thermodynamics* (Nauka, Moscow 1971)

[Zur82] W.H. Zurek: Phys. Rev. **D15**, 479 (1982)

[Zw60] R. Zwanzig: J. Chem. Phys. **33**, 1338 (1960)

[Zw61] R. Zwanzig: In: *Lectures in Theoretical Physics, Vol. 3* (Interscience, New York 1961) pp. 106–141

Index

2-point functions, 73

$\mathcal{A}$- and $\mathcal{A}^+$-type blocks, 267
analytical condition, 163
Anderson model, 352
anisotropic renormalization group method, 370
annihilators, 58
anti-Fock representation, 80
anti-Fock state, 61
asymptotically effective interacting Hamiltonian, 261
asymptotically effective interaction Hamiltonian, 276

backward transport coefficient, 33
basic analytical condition, 120
block, 278
block normal ordering theorem, 449
block principle, 278
blue shifts, 144
Bogoliubov, 49
Bogoliubov–Shirkov approach, 370
Bogoliubov transformation, 269
Bohr frequencies, 122
– positive and negative ∼, 159
Boltzmann algebra, 371
Boltzmann field, 58
Born approximation, 52
Bose–Einstein condensation, 79
Bose–Einstein density, 78
boson field, 58
boson Fock fields, 72
boson Fock white noise, 76
boson Gaussianity, 72
boson, finite temperature white noise, 79
bound states, 120
bra vectors, 269
Breit–Wigner complex poles, 370
Brownian motion, 77

canonical commutation relations, 4, 395
canonical decomposition of $H_0 + H_\mathrm{I}$, 43
canonical representation of the thermal Gaussian state, 80
carrier frequency, 103
cascades with real photons, 387
Catalan number, 448
causal commutator rule, 226
causal Fourier transform, 158
causal Ito table, 172
cavity QED, 169
central limit theorem, 30
chemical potential, 78
Cheshire cat effect, 161, 181
chronological (or time-ordered) product, 237
chronological stochastic limit, 108
classical Gaussian random field, 97
classical Hamiltonian systems, 33
classical jump Markov process, 133
classical Markov (jump) process, 186
classical white noise, 76
collective vectors, 28, 48, 453
commutant of $\mathcal{A}$, 172
composite system, 118
conditional expectation, 206
conjugated vertices, 375
conjugation operator, 269
connected components, 396
connected correlators, 355
connected diagrams, 354, 375
converge, 48
convergence in the distribution sense, 6
convergence in the sense of matrix elements, 28
convergence of chronological correlator, 107
convergence of fields in the sense of correlators, 85
correlators, 58
covariance, 60

creators, 58
cutoff, 124
cutoff-free, 184

damped harmonic oscillator, 100
decay rates, 140
decoherence, 137, 182, 194
δ-function of an operator, 8
δ-functions on the standard simplex, 227
detailed balance, 200
dipole approximation, 116
dipole interaction Hamiltonian, 162
dipole-type interactions, 121
discrete spectrum embedded in the continuum, 94
"disentangling" relation, 389
dispersion relations, 141
double integral lemma, 431
drift, 166
drift coefficient, 128
dual space $\overline{\mathcal{H}}_1$, 269
dynamical system, 11
dynamically q-deformed fields, 290

effective 1-particle Hamiltonian, 126
effective fields, 157
Einstein formula, 167
electric field, 117
electromagnetic field, 253
energy exponent, 375
energy fluctuations, 102
energy shell scalar product, 140
energy shift, 142
energy surface, 140
entangled or interacting commutation relations, 370
entangled rescaled fields, 290, 371
entanglement, 373
environment, 24
equilibrium states at inverse temperature β, 68
evolution on any algebra, 61
expectation value, 57
exponential decay, 135
external lines, 226, 375

factorization theorem, 403
fast degrees of freedom, 14
Fermi block theorem
– first ~, 414
– second ~, 415
Fermi–Dirac density, 78
Fermi momentum, 352
Fermi sphere, 356
Fermi surface, 140
filtering problem, 206, 213
first connected component theorem, 401
first vanishing theorem, 417
flip of spin, 200
flow equation, 11, 47
flow with explosion, 40
fluctuation–dissipation relation, 133
Fock representation, 60
Fock state, 61
Fokker–Planck equation, 35
form factor, 123
free 1-particle Hamiltonian, 62
free algebra, 371
free (or Boltzmannian) algebra, 289
free evolution, 62
free Fock (or Boltzmannian) white noise, 306
free Hamiltonian, 62
free independence, 304
free massive scalar field, 248
free massless scalar field, 250
free master Hamiltonian, 145
free state, 64
free Wick theorem, 289
Friedrichs, 50
fugacity, 78
functional central limit theorem, 30, 32
functional integral approach to the stochastic limit, 247

gamma-ray bursts from neutron stars, 386
gamma-ray pulsars, 386
gamma-ray repeaters, 386
gap at the zero eigenvalue, 136
gauge-invariant state, 61
gauge transformations of the first kind, 61
Gaussian field, 59
general noncrossing principle, 448
generalized dipole approximation, 123
generalized random fields, 58
generalized rotating-wave approximation, 125
generalized susceptibility, 192
generalized susceptivity, 167
Generalized Wick theorem, 60
generic quantum system, 173
Gibbs state, 68
GKSL generator, 132

Glauber dynamics, 200
grand canonical state, 78
Green functions, 354, 380
Green–Kubo–Mori theory, 17
ground state, 353

half-bandwidth, 103
half-planar, 377
Hamiltonian
– free ∼, 4
– rescaled interaction, 19
heat bath, 24
Heisenberg equation, 10
Heisenberg equation in the interaction representation, 11
Heisenberg evolution, 10
Heisenberg–Euler effective Lagrangian, 386
Hilbert modules, 265
Hilbert transform, 142
homogeneization problem, 45
hot free algebra, 321
hot free algebra relations, 333

independent, 59
independent 2-level systems, 174
index of refraction, 386
input algebra, 211
input–output channels, 205
input processes, 212
integrable in the sense of distributions, 87, 156
interacting commutation relations, 372
interacting particle systems, 196
interaction representation, 9
internal lines, 226, 379
interpolation problem, 206
intertwiner property, 337
invariance principle, 30
inverse lifetime of the vacuum, 135
ionized states, 121
Ising model, 199

ket vectors, 269
KMS condition, 68

Langevin equation
– backward or outer white noise, 23
– forward or inner white noise, 23
– quantum ∼, 29
large N limit, 368
large N limit for $SU(N)$ invariant gauge theory, 371
left and right multiplication, 335
Liouville equation, 10
Liouvillian, 33
local quantum field theory, 384
locality, 83
localized left unitary evolution, 211
localized left unitary multiplicative functional, 211
long chain, 425
loop variables, 375
low density, 15
low-density limit, 255

macroscopic quantum coherence, 179
magnetic field, 117
Markov semigroup, 172
Markovian approximation, 27
Markovian cocycle, 145
Markovian generator, 172
martingale term, 166
master equation, 27, 132
master field, 93
Maxwell's equations with radiative corrections, 386
mean zero gauge-invariant white noise, 165
measurement, 216
measurement apparatus, 24
minimal coupling, 113
mixed moments, 58
module commutation relations, 289
multiple integral lemma, 438
multiple-simplex theorem, 435
multiplicative interaction, 121

n-particle vectors, 73
n-point functions, 58
$\mathcal{N}$-type blocks, 267
noise, 24
non-time-consecutive pairings, 222
non-time-consecutive theorem, 417
nonconnected diagram, 354
noncrossing pair partition, 293, 446
nondemolition measurements, 217
nondemolition process, 215
nonequilibrium quantum field theory, 162, 369
nonstationary white noises, 201
normal ordering theorem, 448
normally ordered form of the white noise Hamiltonian, 128
normally ordered free equation, 307
normally ordered Langevin white noise equation, 132

number vectors, 73

off-resonance condition, 182
one-parameter automorphism group, 10
one-parameter unitary group, 4
open system, 24
operator density of particles, 371
operator valued transport coefficient, 17
orthogonalization principle, 280
output field, 212

$\mathcal{P}_{\Omega}$-Fock module, 345
pair partition, 446
pair partitions, 60
parametric down conversion, 390
parity operator θ, 269
partial expectation, 26
partition of the identity, 172
photon splitting, 386
photon splitting in a strong magnetic field, 386
photon-splitting cascades in a magnetic field, 372
plane waves, 138
polarization index, 114
polarization vectors, 114
polaron model, 320
polynomial decay, 319
polynomial interactions, 251
polynomial interactions with cutoff, 262
prediction problem, 206
Prigogine, 55
projection on the diagonal algebra, 42
propagator, 9
pseudo-eigenstates, 137
purely oscillatory behaviours, 135

q-commutation relations, 58
q-deformed correlators, 293
q-deformed module relations, 290
q-deformed module Wick Theorem, 296
q-Gaussian Equilibrium States, 70
QED in an external field, 385
quantum (Boltzmannian or free) stochastic differential equation, 309
quantum central limit theorems, 86
quantum coherence, 195
quantum dissipative and irreversible phenomena, 8
quantum transport coefficient, 16
quasi-free evolution, 67
quasi free states, 64
quasi-standard models, 265

rainbow graphs, 285
random field, 57
red shifts, 144
reduced dynamics, 26
regularization, 307
relativistic dispersion laws, 384
relativistic quantum white noise, 106
rescaled fields, 20, 127
reservoir, 24, 120
resonance surface, 7
resonant frequency, 142
rotating-wave approximation, 117

S-matrix, 370
scattering operator, 259
Schrödinger equation, 4
Schwartz space, 5
Schwinger–Dyson equation, 367
second connected component theorem, 402
second multiple-simplex theorem, 439
second quantization of $\omega(k)$, 62
second quantized representation of the nonrelativistic QED Hamiltonian, 381
second vanishing theorem, 419
semiclassical expansion, 99
semigroup, 35
– dynamical, 26
– quantum Markovian, 26
semiquantum approximation, 99
semiscalar product, 139
sense of the $\mathcal{V}_{\lambda}$-matrix elements, 48
short chain, 425
simplex $\Delta^{(n)}$, 226
simultaneous measurement, 217
slow degrees of freedom, 14
smeared master fields, 168
spectral density, 181
spectral function, 58
spectral matrix, 58
spin–boson Hamiltonian, 118, 178, 187
spin variables, 196
squeezed, 61
SQUID, 179
standard domain, 219
standard model, 265
state, 57
stationary, 58
stochastic bosonization principle, 280
stochastic calculus, 235
stochastic differential, 166

stochastic differential equation, 235
stochastic golden rule, 130
stochastic limit for the Green functions, 372
stochastic limit of interacting fields, 371
stochastic limit of quantum theory, 13
stochastic process, 57
stochastic resonance principle, 153, 280
stochastic Schrödinger equation, 166
stochastic symmetry breaking, 160
structure constants, 172
structure maps, 171
superparity, 281
susceptibility, 192

time-consecutive pairings, 222
time-consecutive principle, 227
time scales, 13
time shift S_t, 146
totally connected, 396
translation-invariant interaction, 371
transport coefficient, 139
tri-linear Hamiltonian, 373
triphon master field, 389
triphons, 372
type I, 229
type-I term theorem, 425
type II, 229

u^0_{-t}-1-cocycle, 47
ultralocality, 83
unitarity condition, 233
universality class, 276, 385
universality class principle, 275

vacuum transition amplitudes, 40
vacuum vector, 73
van Hove, 50
vertex, 375

wave operator, 19
Weak coupling, 15
Weisskopf–Wigner resolvent method, 370
white noise, 76
white noise calculus for nonFock noises, 234
white noise Langevin equation, 131
Wick Theorem for the QED module, 301
Wiener process, 77
Wiener Processes, 77
Wightman functions, 93

X^ε, 58

Yukawa interaction, 384

Production: Druckhaus Beltz, Hemsbach